2nd International Conference on Green Civil and Environmental Engineering (GCEE 2019)

Discovering Innovative and Environmentally-sound Construction for Sustainable Living

IOP Conference Series: Materials Science and Engineering
Volume 669

Malang, Indonesia
4 - 6 September 2019

ISBN: 978-1-7138-0922-7
ISSN: 1757-8981

Printed from e-media with permission by:

Curran Associates, Inc.
57 Morehouse Lane
Red Hook, NY 12571

CURRAN ASSOCIATES INC.
proceedings
.com

Some format issues inherent in the e-media version may also appear in this print version.

Institute of Physics
Dirac House, Temple Back
Bristol BS1 6BE UK

Phone: 44 1 17 929 7481
Fax: 44 1 17 920 0979

techtracking@iop.org

Additional copies of this publication are available from:

Curran Associates, Inc.
57 Morehouse Lane
Red Hook, NY 12571 USA
Phone: 845-758-0400
Fax: 845-758-2633
Email: curran@proceedings.com
Web: www.proceedings.com

TABLE OF CONTENTS

ECONOMIC PROSPECTS OF STEEL REINFORCED FUNCTIONALLY GRADED CONCRETE (SRFGC) BEAM STRUCTURES 1
B S Gan, S Kiryu, A Han, M M A Pratama

CYCLIC BEHAVIOR OF MOMENT-RESISTING FRAMES STRENGTHENED WITH STEEL CURVED DAMPERS 10
H L Hsu, H Halim

EMISSION FACTORS OF NOX AND PM OF HEAVY CONSTRUCTION EQUIPMENT USED IN TOLL ROAD PROJECT IN INDONESIA – CASE STUDY: REMBANG-PASURUAN TOLL SECTION II 15
A M Hajji, A Yulistyorini, Huang Yue, D C Patulak

RAINFALL INDUCED RESIDUAL SOIL SLOPE INSTABILITY: BUILDING CRACKED AND SLOPE FAILURE 21
L J Yue, M F Ishak, M S I Zaini, M F Zolkepli

FORENSIC ASSESSMENT ON LANDFILLS LEACHATE THROUGH ELECTRICAL RESISTIVITY IMAGING AT SIMPANG RENGGAM IN JOHOR, MALAYSIA 32
M S I Zaini, M F Ishak, M F Zolkepli

CRACKING BEHAVIOUR OF GREEN LIGHTWEIGHT CONCRETE ONE-WAY SLABS USING MEDIUM K BASALTIC ANDESITIC PUMICE AND SCORIA 42
H Suseno, A Soehardjono, A Rachmansyah

MIXTURE OF CLAY AND TELETHONG AS ENVIRONMENTAL FRIENDLY BINDER 55
Ruswandono, J Paing, Andaryati

GROUND WATER EXPLORATION USING GEO-ELECTRIC METHOD FOR SUMBER CEMUNG POOL DEVELOPMENT 60
G Idfi, N Suryoputro, Y A Laksono, N L Khomsiati

STAKEHOLDERS INTERACTION FRAMEWORK OF UTILITIES DEVELOPMENT IN AND ADJACENT TO THE PUBLIC RIGHT OF WAY IN INDONESIA 64
I N M Bukit, P F Marzuki, R Z Tamin, Meifrinaldi

REPAIR OF REINFORCED CONCRETE BEAM BY ADDING EXTERNAL REINFORCEMENT 71
S M Dewi, L Susanti, N I Triyuliani, A Saputra

EXPLORATORY STUDY ON THE UTILIZATION OF RECYCLED WOOD AS RAW MATERIAL FOR CROSS LAMINATED TIMBER 78
W A Munandar, R H Purba, A Christiyanto

ONE DAY HIGH-PERFORMANCE CONCRETE MECHANISM OF PLATE, BEAM, AND COLUMN 87
C Niken

CONSTRUCTED FLOATING WETLAND WITH MOORING SYSTEM TECHNOLOGY TO IMPROVE WATER QUALITY IN GEDEBAGE RESERVOIR, BANDUNG 95
Y M Hidayat, W Herwindo, S Bachri

THE EFFECT OF COPPER-BASED CATALYTIC CONVERTER WITH CIRCULAR TUBE SHAPE ON EXHAUST EMISSION OF YAMAHA VIXION 1PA 101
I A N Shoffan, S Sumarli, I M Nauri

THE ANALYSIS OF ALPHA PARAMETER ON NAKAYASU SYNTHETIC UNIT HYDROGRAPH IN TIMOR ISLAND WATERSHEDS 110
D S Krisnayanti, M E Bolla, J K Nasjono, M J M Wangge

THE STRENGTH PROPERTIES OF HIGH CALCIUM FLY ASH GEOPOLYMER SPECIMENS INCORPORATING MUD-VOLCANIC AND LIMESTONE VARIATIONS 119
A Wardhono, Y Risdianto, M F Sofianto, M T Nur, A H Pradipta

SHIFTING CROPPING SHIFTS BY EFFICIENT IRRIGATION WATER TO PRODUCE MAXIMUM RICE PRODUCTIVITY 125
S Azis

THE COMPARATIVE STUDY OF FLOOD MODELLING WITH THE UNSTEADY AND THE STEADY FLOW ON NGOTOK RIVER 133
G Idfi, I D Wahyono, A Yulistyorini, N L Khomsiati

THE EFFECT OF REINFORCEMENT BRIDGING ON THE ELASTIC FRACTURE ENERGY OF CONCRETE 140
A H Patty

LIFE CYCLE ASSESSMENT OF LOW COSTS HOUSING IN INDONESIA 149
S Zuraida, Y S Wahyuni, D Larasati

PERFORMANCE OF AEROBIC AND ANAEROBIC LANDFILL BIOREACTOR FOR MUNICIPAL SOLID WASTE TREATMENT IN JATIBARANG LANDFILL, SEMARANG 159
M A Budihardjo, A R Rizaldianto, F I Muhammad

PLANNED INVESTIGATION OF UTES POTENTIAL IN ALGERIA 165
S Amara, T Baghdadli, B Nordell

PERVIOUS LOW-CALCIUM FLY ASH GEOPOLYMER CONCRETE USING GAP-GRADED LIMESTONE COARSE AGGREGATES IN AMBIENT CURING 173
P H Simatupang, M A J Adu, Y M I Sadukh, Jonbi

COOLING TREATMENT OF ALKALI SOLUTION TO PROLONG THE SETTING TIME OF HIGH CALCIUM FLY ASH BASED GEOPOLYMERS 181
O K Wattimena, A Antoni, R T Surja, R Mintura, D Hardjito

SETTING TIME AND COMPRESSION STRENGTH OF LOW-CALCIUM FLY ASH GEOPOLYMER PASTE WITH NON-CALCINED RED SOIL SUBSTITUTION IN AMBIENT CURING 187
P H Simatupang, V Nine, A E Sinaga

THE USE OF NATURAL FIBER FROM OIL PALM EMPTY FRUIT BUNCHES FOR SOFT SOIL STABILIZATION 197
Y F Arifin, Misnawati, E Normelani

MODAL PARAMETERS IDENTIFICATION OF LIGHT RAIL BRIDGES 205
E Zulkifli, D R Widarda

APPLICATION OF STEADY FLOW MODEL TO OVERCOME INUNDATION BASED ON ECO DRAINAGE SYSTEM IN THE PORT AREA OF DOBO CITY, ARU ISLANDS REGENCY 214
B E Prasetyo, G Idfi, Mujiyono, E Suwarno

BARRIERS AND CRITICAL SUCCESS FACTORS TOWARDS SUSTAINABLE HAZARDOUS WASTE MANAGEMENT IN ELECTRONIC INDUSTRIES – A REVIEW 220
Y H Chin, H N A Yong

BASE FLOW IDENTIFICATION USING CONCEPTUAL HYDROLOGY MODEL 229
N Suryoputro, A Yulistyorini, G Idfi, N E B Alias

THE EFFECT OF WHITE SOIL ON GEOPOLYMER MORTAR POROSITY 238
Y A Priastiwi, Muhrozi, R Y Adi, D Daryanto, Z Salamsyah

DETERMINING PREDICTOR VARIABLES OF HC, CO, AND CO2 EMISSIONS USING DECISION TREE MODELS 246
A Larasati, Y W Chen, A M Hajji, A F P Mahardika, V E B Darmawan

FINANCIAL ECONOMIC COST ON GRAVEL ROAD MAINTENANCE: STUDY USING HDM-4 253
F Suwarto, S Fauziyah

EXPERIMENTAL STUDY OF PRECAST CONCRETE BEAM-TO-COLUMN CONNECTION UNDER SUSTAINED LOADING 261
V A Noorhidana

DEVELOPMENT OF SURFACE GROUND MOTION AND SPECTRAL ACCELERATION BASED ON MODIFIED SHEAR WAVE PROPAGATION ANALYSIS 268
W Partono, F Kistiani, U C Sari, Haryadi, E Putro, W R Ramadhan

EFFICIENCY ANALYSIS OF MUSKINGUM-CUNGE METHOD AND KINEMATIC WAVE METHOD ON THE STREAM ROUTING (STUDY CASE: UPPER CILIWUNG WATERSHED, INDONESIA) 274
D Prawira, H Soeryantono, E Anggraheni, D Sutjiningsih

PERFORMANCE OF SCS UNIT HYDROGRAPH AND KINEMATIC WAVE METHODS APPLICATION ON RURAL AND URBANIZED WATERSHED 285
D Sharaswati, H Soeryantono, E Anggraheni, D Sutjiningsih

DESIGN OF DRAINAGE SYSTEM ON INDUSTRIAL AREA KAMPUNG CINA, DOBO CITY, ARU ISLANDS REGENCY, BASED ON ECO DRAINAGE USING WITH HEC-RAS 4.0 293
B Rizaldi, G Idfi, E Suwarno, Mujiyono

ANALYSIS OF FLOOD INUNDATION IN NORTH SUNTER ON THE NORTH SUNTER POLDER SYSTEM PERFORMANCE 300
S Ignatius, H Soeryantono, E Anggraheni, D Sutjiningsih

GREEN CIMAHI WATERSHED FOR BALANCING WATER SUPPLY AND FLOOD CONTROL 311
A B Safarina, I Karnisah, A S Permana, D Sufianto

A COMPARATIVE STUDY OF DETERMINISTIC APPROACH FOR ASSESSING LIQUEFACTION POTENTIAL 322
H Farichah

RISK STUDY ON SUPPLY CHAIN MANAGEMENT IN CONSTRUCTION (CASE STUDY: BUILDING PROJECTS IN INDONESIA) 330
A Nurdiana, M N Sholeh

EXPERIMENTAL ANALYSIS ON THE PERFORMANCE OF CONCRETE BEAM DUE TO THE EFFECT OF KNITTED BAMBOO REINFORCEMENT RATIO 338
D Setyowulan, D Nuralinah, E Arifi, F D Prasetyo

WATER QUALITY INDICES FOR RAINWATER QUALITY ASSESSMENT IN BANDUNG URBAN REGION 345
N Y Hasan, Driejana, A Sulaeman, H D Ariesyady

THE EXPERIMENTAL COMPRESSIVE STRENGTH OF NORMAL AND SELF-COMPACTING CONCRETE WITH RECYCLED COARSE AGGREGATE 351
D Nuralinah, E Arifi, D Setyowulan

THE EFFECT OF SEA WATER SOAK ON ASPHALT POROUS CHARACTERISTICS WITH AND WITHOUT WASTE OF RUBBER TIRE WASTE AS AN ASPHALT ADDICTIVE 357
O S T Mawarni, M Fauziah

THE GAPS OF PASSIVE STRATEGY OPTION IN INDONESIA'S GREEN BUILDING REGULATION 368
Sahid, Y Sumiyati, R Purisari

THE DOMINANT FACTOR CAUSING POTENTIAL EVAPOTRANSPIRATION IN MALANG RAYA USING GEOGRAPHIC INFORMATION SYSTEMS 375
T Rahayuningsih, G Idfi, A Anindyasari, Sukriyadi

TENSILE TEST ANALYSIS OF COLD-FORMED COLUMN IN CHANNEL PROFILE 385
A A Azlz, N Nlndyawatl, K Karyadl

ACHIEVING OPEN DEFECATION FREE IN SURABAYA CITY BY 2019 391
E Soedjono, N Fitriani, F R E Santoso, R Destio, I Fahmi, A Gemardi, D A Ningsih

COMPRESSION TEST ANALYSIS OF COLD-FORMED COLUMN IN CHANNEL PROFILE CONNECTION 398
D F Rachmahwati, N Nindyawati, K Karyadi

THE ADDITION OF LAMELLA IN ANAEROBIC BAFFLED REACTOR USED FOR DECENTRALIZED MUNICIPAL WASTEWATER TREATMENT 405
A Yulistyorini, M C Yuaniar, M Mujiyono, N Suryoputro, S Sukarni

STRUCTURAL PERFORMANCE OF MULTI-STOREY BUILDING USING FLAT SLAB AND CONVENTIONAL SLAB TO SEISMIC LOADS (CASE STUDY: FACULTY BUILDING OF SPORT SCIENCE IN UNIVERSITAS NEGERI MALANG, INDONESIA) 413
R A Pradhana, M M A Pratama, E Santoso, A Karjanto

FLEXURAL BEHAVIOUR OF THE FUNCTIONALLY GRADED CONCRETE BEAMS USING TWO-LAYERS AND THREE-LAYERS CONFIGURATION 420
M M A Pratama, T Vertian, B S Umniati, W H Yoh

NUMERICAL AND EXPERIMENTAL INVESTIGATION ON THE SHEAR RESISTANCE OF FUNCTIONALLY GRADED CONCRETE (FGC) BEAMS 428
M M A Pratama, W Arifanda, K Karyadi, N Nindyawati, R Sulaksitaningrum, H Prayuda

THE OPTIMAL DAMPER PLACEMENT CONFIGURATION FOR THREE-DIMENSIONAL RC BUILDING 437
R Sulaksitaningrum, B S Umniati, I P Ellsa Sarassantika, C B Casita, M M A Pratama, E Santoso, M Sulton

THE EVALUATION OF GREEN PERFORMANCE OF MIFTAHUL HUDA ISLAMIC BOARDING SCHOOL, MALANG 445
A A Sari, A E Winahyo, D Ariestadi, Imam Alfianto

THE ROLE OF THE FIRE SAFETY MANAGEMENT IN PROVIDING A GUARANTEE OF A FIRE PROTECTION: THE CASE OF GRAHA REKTORAT BUILDING OF STATE UNIVERSITY OF MALANG 451
A A Sari, F K Rafrita, T Rahayuningsih, I Alfianto

CONFINEMENTS OF CONCRETE CYLINDERS USING BAMBOO SPIRAL STIRRUPS 457
N Nindyawati, M Sulton, M M A Pratama, T Rahayuningsih

THE APPLICATION OF FUZZY LOGIC FOR PROFIT OPTIMIZATION TO CONTRACTOR PROJECT CASH FLOW 463
B Djatmiko, A Yulistyorini, R M Sugandi, E Setyawan

EFFECTS OF SEGREGATION ON REINFORCED CONCRETE COLUMN, A NUMERICAL APPROACH 473
W Damayanthi, Y Novitasari, M M A Pratama

CONCEPTS OF CONSTRUCTABILITY FOR PROJECT CONSTRUCTION IN INDONESIA 480
M M A Ansyorie

Author Index

IOP Conf. Series: Materials Science and Engineering **669** (2019) 011001 doi:10.1088/1757-899X/669/1/011001

PREFACE

The 2nd International Conference on Green Construction and Environmental Engineering (GCEE 2019) has been scheduled and successfully taken place at Atria Hotel and Convention, Malang, Indonesia on 4-6 September 2019. GCEE 2019 was organised by Department of Civil Engineering, Faculty of Engineering, Universitas Negeri Malang, Indonesia. The theme of this international conference was ***"Discovering Innovative and Environmentally-sound Construction for Sustainable Living"*** with topic areas including green building, green material, green construction and technology, a sustainable technique in building and infrastructure, sanitation system and technology options, water resource management, solid and hazardous waste management.

We would like to acknowledge all of those who supported GCEE 2019:

- The entire steering, scientific, and organising committee members of GCEE 2019 for their help and support.
- Faculty of Engineering, Universitas Negeri Malang for the support for GCEE 2019.
- University partners and institutions
- The distinguished keynote speakers for their acceptance to give speeches and intellectual contribution on their respective fields of expertise.
- The participants of GCEE 2019 for their contribution and sharing knowledge.
- The sponsors for cooperation and their consideration to support GCEE 2019.

Each individual and institutional help were very important for the success of this conference:

IOP Conf. Series: Materials Science and Engineering **669** (2019) 011002 doi:10.1088/1757-899X/669/1/011002

LIST OF COMMITTEE

The 2nd International Conference on Green Civil and Environmental Engineering (GCEE 2019)

Organised by:
Department of Civil Engineering, Faculty of Engineering
Universitas Negeri Malang, Indonesia

Steering Committee
Prof. A.H. Rofiuddin – Universitas Negeri Malang, Indonesia
Prof. Marji – Universitas Negeri Malang, Indonesia
Arif Nur Afandi, PhD. – Universitas Negeri Malang, Indonesia

International Scientific Committee
Prof. Huang Yue – University of Leeds, UK
Assoc. Prof. Dr. Julius Marpaung – Northeastern University, USA
Dr. Dewi Larasati – Institut Teknologi Bandung, Indonesia
Dr. Li Siong Wee – Universiti Teknologi Mara, Malaysia
Prof. Sri Murni Dewi – Universitas Brawijaya, Indonesia
Prof. Tavio – Institut Teknologi Sepuluh November, Indonesia
Prof. Shih-Huang Chen – National Central University, Taiwan
Vera Agustriana Noorhidana, Ph.D – Universitas Lampung, Indonesia
Dr. Davide Pogio – University of Sheffield, UK
Altinay Perendeci, Ph.D – Akdeniz University, Turkey
Prof. Sofiane Amara – University of Tlemcen, Algeria
Prof. Stefanus Adi Kristiawan – Universitas Sebelas Maret, Indonesia
Prof. Wen Hsiang Hsieh – Formosa University, Taiwan
Prof. Han Ay Lie – Universitas Diponegoro, Indonesia
Dr. Antoni – Universitas Kristen Petra, Indonesia
Dr. Philip van den Heede – Ghent University, Belgium
Eddy Setiadi Sudjono, Ph.D – Institut Teknologi Sepuluh Nopember, Indonesia
Dr. Abdullahi Ahmed – Coventry University, UK
Dr. Monita Olivia – Universitas Riau, Indonesia
Rr. Poppy Puspitasari, PhD – Universitas Negeri Malang, Indonesia
Aisyah Larasati, PhD – Universitas Negeri Malang, Indonesia

Chairman
Assoc. Prof. Apif Miptahul Hajji – Universitas Negeri Malang, Indonesia

Vice-Chairwoman
Anie Yulistyorini, Ph.D – Universitas Negeri Malang, Indonesia

Secretary
M. Mirza Abdillah Pratama – Universitas Negeri Malang, Indonesia

Treasury
Dr. Nindyawati – Universitas Negeri Malang, Indonesia

IOP Conf. Series: Materials Science and Engineering **669** (2019) 011003 doi:10.1088/1757-899X/669/1/011003

Peer review statement

All papers published in this volume of *IOP Conference Series: Materials Science and Engineering* have been peer reviewed through processes administered by the proceedings Editors. Reviews were conducted by expert referees to the professional and scientific standards expected of a proceedings journal published by IOP Publishing.

Economic prospects of Steel Reinforced Functionally Graded Concrete (SRFGC) beam structures

B S Gan[1], S Kiryu[2], A Han[3] and M M A Pratama[4]

[1]Department of Architecture, College of Engineering, Nihon University, Koriyama, Japan
[2]Department of Architecture, Graduate School of Engineering, Nihon University, Koriyama, Japan
[3]Civil Engineering Department, Diponegoro University, Semarang, Indonesia
[4]Department of Civil Engineering, Faculty of Engineering, State University of Malang, Malang, Indonesia

buntara@arch.ce.nihon-u.ac.jp

Abstract. Engineering materials, for instance, ceramics and concrete, are manufactured with uniform properties purposefully. Because the uniformity is a convenient measurement to ensure the safety design calculation of a structure to be built. The non-uniform material property of concrete can be resulted from mixing, placing, consolidation, and curing processes during the manufacturing of structural members. Therefore, using concrete as a material could lead to an uneconomical use of natural resources because of non-uniformity problems that occur in the structural members. Ideally, in the Functionally Graded Concrete (FGC) material, the properties (elasticity, density, rigidity) of concrete material have to be smoothly graded in the projected directions (longitudinally, transversally, diagonally) by following a governing equation. Hence, FGC has significant advantages over the existing concrete materials in the prominence of cement optimization. In recent progress, many researchers have been studied the mechanical behaviors and production methodologies of FGC enthusiastically. In the state-of-the-art of the FGC developments, two objectives need to be achieved: 1. The manufacturing procedure of FGC material, and 2. The method of analysis to design FGC material. In current developments, the technologies for manufacturing FGC material are still not available for the practitioners. The objective of this paper is to present a method of analysis to design a Steel Reinforced Functionally Concrete Graded (SRFGC) beam element based on the SNI (Indonesian National Standard) of Structural Concrete Requirements for Buildings code 2847:2013. The effects of the non-uniformity of concrete strengths are incorporated into the analysis. Because of the crack-opening assumption in the tensile region of the concrete, the neutral axis position is no longer at the mid-height of the beam cross-section. Hence, an iterative technique has to be employed in the design. An iterative technique is necessary because the internal forces of the beam have to be calculated at the cross-section being designed. The application of the method of analysis is highlighted by designing various type of beam cross-sections, which show the economic benefits of SRFGC beam compared to both the normal and high strength concrete's material.

1. Functionally graded material

Concrete and ceramic are the industrial materials which are designed and manufactured intentionally to have homogeneous mechanical properties. The material homogeneity is a worthwhile criterion to ensure the safety of members of a structure. The non-homogeneous material property is a result of method of mixing, placing method, consolidation process, and curing process, including the effects of segregation and accumulation of the aggregates during the mixing. Thus, the non-homogeneity could show uneconomical use of natural resources due to high-stress concentrations that occur in the boundary between structural elements. On the opposing to the homogeneity assumed in the design and analyses, the steel Reinforced Concrete (RC) structure element in the built structures are mainly found as a graded concrete material [1,2].

Functionally Graded Material (FGM) is a new phrasing of a composition between two or more different materials where the essential properties are graded over a particular orientation to gain some desired functional properties [3,4]. In the compositions of FGM, the two or more material properties are merged functionally to enhance the individual material property. In nature, bamboo is one among typical materials that show a radial gradient as a result of the evolutionary process to adapt to their growing environment.

Compared to the experimental works, there have been a vast number of researches doing the analyses of FGM numerically [5-14], from where the analyses of FGM can be used to analyze the Functionally Graded Concrete (FGC).

By today, FGC has not been implemented widely in construction projects. Two major problems in implementing FGC are: there is no producing method established yet, and there is no building code available for designing FGC elements in structures. Variation of elastic moduli and strengths of concretes in FGC element needs an accurate method of analysis to predict its strength and behavior [15].

In this paper, the method and analysis of the ACI 318M-11 [16] to design a Steel Reinforced Functionally Concrete Graded (SRFGC) beam with various cross-sections subjected to a constant bending moment are highlighted. By using the iteration method, the SRFGC can be designed similarly with the conventional steel RC beam. In the end, a comparison study on the cost of material is showed to highlight the economic prospects of the SRFGC.

2. Functionally graded concrete (FGC)

Researches on FGC, however, are still inadequate. The efforts [4,17-18] to manufacture an FGC material had confronted one challenging difficulty: to produce a smooth graded between two different properties of concrete strengths. The continuity of graded concrete has not been obtained. Besides, laminated or composite concrete strengths were found. In the laminated or composite concrete strengths, the stress concentration between the concrete layers will degrade the quality of the FGC material.

Experimentally [19-21], these studies have revealed that the ultimate strengths of the FGC were restricted by the lowest concrete strength of the constituents in the FGC, but the stiffness of the FGC is close to the stiffness of the highest concrete strength in the mixture.

2.1. SRFCG beam

The application of FGM concept to the RC beam can be achieved by producing SRFGC. By varying two different types of the compressive strengths in the vertical direction of the cross-section of a beam, a potential approach to reduce the unnecessity of concrete strength at the bottom fiber under the tensile stress and increasing the necessity of concrete strength at the top fiber under compressive stress. Preferably, there will be no reduction of the beam's bending strength, and at the same time can potentially reduce the cost of RC material. Similar ideas could be applied for the improvement of other building members.

The Young's modulus of the FGC beam is expressed as follow:

$$E_C(y) = E_{Cb}\left[1 + \frac{E_{Ct} - E_{Cb}}{E_{Cb}}\left(\frac{y}{h}\right)^p\right] \tag{1}$$

where $E_C(y)$ is the concrete elastic modulus at y-ordinate; E_{Cb} is the concrete elastic modulus at the bottom of the beam cross-section; E_{Ct} is the concrete elastic modulus at the top of the beam cross-section; h is the height of the cross-section and p is the gradient of the modulus variation. y is measured from the bottom fiber of the cross-section.

2.2. Design of SRFGC beam cross section

The kinematic assumptions of SRGFC beam cross-section under a bending moment is illustrated in Fig. 1. The compression forces consist of the un-cracked concrete portion and compression steel bar, while the tensile force is only given by the steel bar in the cracked concrete portion. The design of the beam cross-section is calculated iteratively by balancing the moment strengths resulted by the concrete and steel.

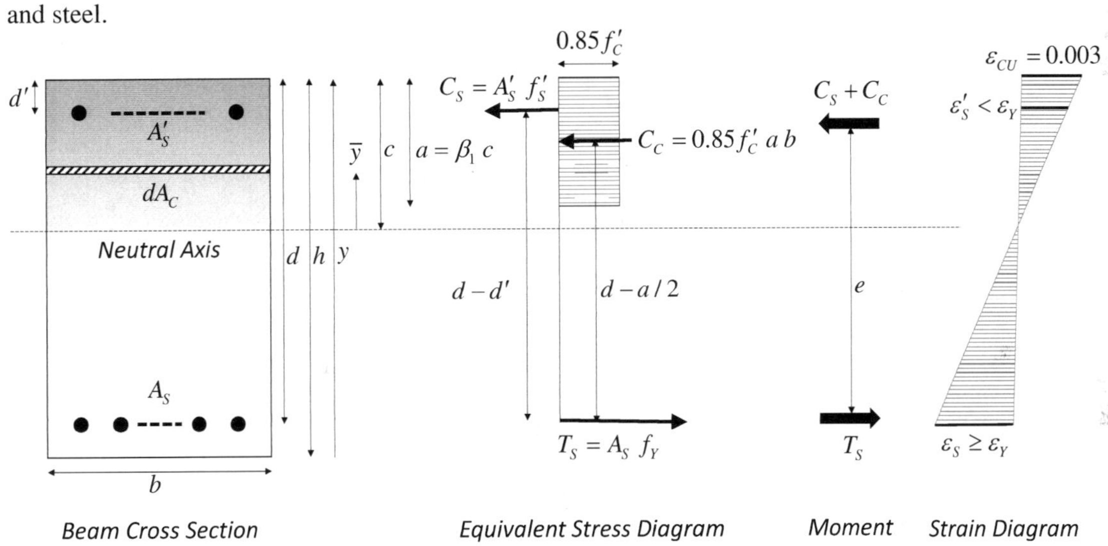

Figure 1. Stress and strain distribution of the cracked SRFGC cross-section

In the case of a beam subjected to a bending moment, the compressive force is resisted by the concrete part above the neutral axis and the compressive rebar, and the tensile force is resisted by the tensile rebar below the neutral axis. Beam cross-section design is iteratively calculated by taking into consideration the balance of moment applied to the concrete and rebar steel. The elastic modulus ratio used in this calculation can be defined as a function of the distance from the bottom fiber as follows.

$$n(y) = \frac{E_S}{E_C(y)} \tag{2}$$

The equilibrium state of stress between concrete and rebar steel is reached by assuming the yielding of compressive concrete and tensile steel bar which can be expressed as follows, except for the tensile concrete part below the neutral axis is not considered.

$$T_S = C_C + C_S \tag{3}$$

where,

$$T_S = f_Y \sum_{i=1}^{ns} A_{Si}$$

$$C_S = E_S\left(1 - \frac{\beta_1 d'}{a}\right) \times 0.003 \times \sum_{i=1}^{nS'} A_{S'i}$$

$$C_C = 0.85 f'_{Ct} \int_{\beta_1 c} b \, d\bar{y}$$

where, T_S is the total tensile force from the steel bars; C_S is the total compressive force from the steel bars; C_C is the total compressive force from the uncracked concrete portion; f'_{Ct} is the stress of the concrete fiber at the top of the beam's cross-section; c is the height of the uncracked steel reinforced FGC section from the neutral axis; $\bar{y}$, $\bar{y}_S$ and $\bar{y}_{S'}$ are measured from the neutral axis of the cracked beam cross section; nS, nS' are the number of tensile and compressive steel bar and b, A_{Si} and $A_{S'i}$ are the width of the uncracked concrete, i-th tensile steel bar and compressive steel bar, respectively.

The force equilibrium of the cross-section is then can be computed by,

$$E_S\left(1-\frac{\beta_1 d'}{a}\right)\times 0.003\times\sum_{i=1}^{nS'} A_{S'i} + 0.85 f'_{Ct}\int_{\beta_1 c} b\, d\bar{y} + f_Y \sum_{i=1}^{ns} A_{Si} = 0 \quad (4)$$

From where the value of a can be obtained by iteratively solving the above equilibrium equation.

By using the equivalent block ultimate stress of steel bar and concrete (compressive side), the bending-resistant moments of the SRFGC beam cross-section can be determined from the lowest value between :

$$M_{Sr} = C_S\left(d-d'\right) = E_S\left(1-\frac{\beta_1 d'}{a}\right)\times 0.003\times\sum_{i=1}^{nS'} A_{S'i}\left(d-d'\right) \quad (5)$$

$$M_{Cr} = C_C\left(d-\frac{a}{2}\right) = 0.85 f'_{Ct}\int_{\beta_1 c} b\, d\bar{y}\left(d-\frac{a}{2}\right) \quad (6)$$

where, M_{Sr} and M_{Cr} are the equivalent resisting moments of steel and concrete, respectively.

3. Material cost comparison study

In this study, a homogenous RC beam strength is at 69MPa and 24 MPa, the SRFGC beams of varying concrete compressive strengths from the bottom to the top of the beam cross section which is designed based on the equivalent block ultimate stress design method (see Fig. 1), to calculate the cost of material for comparison. Two types of cross sections (see Fig. 3) are considered: rectangular and trapezoidal cross sections.

3.1. Parameter of material properties

The graded modulus function in Eq. (1) of the concrete compressive strengths is selected to follow the degree of polynomial order of p = 1 (linear), 2 (quadratic), and 3 (cubic) cases variation. Fig. 2 shows two homogeneous (p = 0) RC beams of 69 MPa and 24 MPa concrete compressive strengths, namely Case A and Case C, respectively. Case B-1, B-2 and B-3 in Fig. 2 show SRFGC cross-sections of functionally graded concrete compressive strengths which vary from 24 MPa at the bottom fiber and 69 MPa at the top fiber of the beam cross-section folling the polynomial order of p.

The following equation determines the Young's modulus of concretes.

$$\begin{aligned} &For\ F_C \le 36\ \text{N/mm}^2 : E_C = 2.10\times 10^4\left(\frac{\gamma_C}{23}\right)^{3/2}\left(\frac{F_C}{20}\right)^{1/2} \\ &For\ F_C > 36\ \text{N/mm}^2 : E_C = 3.35\times 10^4\left(\frac{\gamma_C}{24}\right)^{2}\left(\frac{F_C}{60}\right)^{1/3} \end{aligned} \quad (7)$$

where E_C is the elastic modulus of concrete in N/mm^2; F_C is the compressive design strength of concrete in N/mm^2 and γ_C is the weight density of concrete in kN/m^3. The weight densities of $\gamma_C = 23$ kN/m^3 and $\gamma_C = 25$ kN/m^3 for concrete with compressive strengths of $F_C = 24$ N/mm^2 and $F_C = 69$ N/mm^2, respectively. The weight densities of $\gamma_S = 78$ kN/m^3 is assigned to the steel bar.

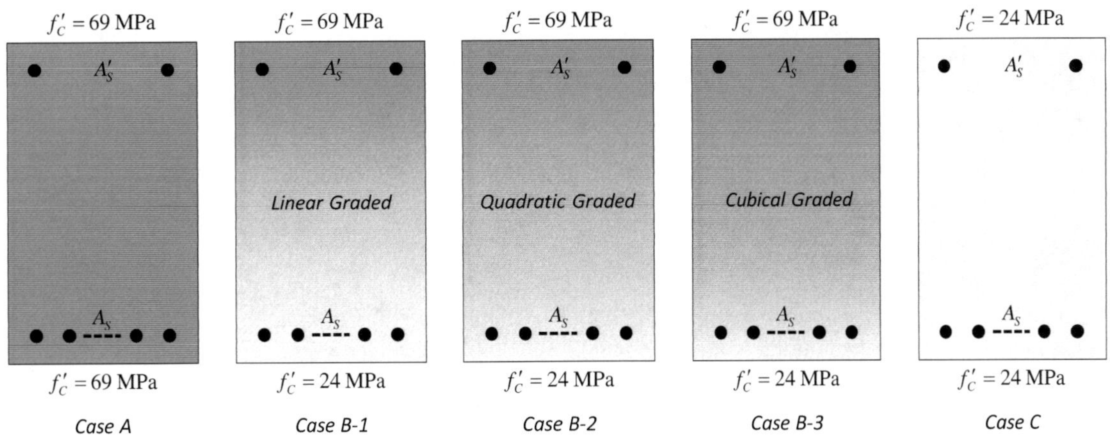

Figure 2. Cases for a comparison study of RC/SRFGC beams

3.2. Volume fraction of FGC beam

To calculate the cost of material for each case, the volume fraction of the respective material constituent in the SRFGC should be computed from the weight density distribution in Eq. (1). The SRFGC weight density distribution of concrete part can then be given as,

$$\gamma_C(y) = \gamma_{Cb}\left[1 + \frac{\gamma_{Ct} - \gamma_{Cb}}{\gamma_{Cb}}\left(\frac{y}{h}\right)^p\right] \tag{8}$$

where $\gamma_C(y)$ is the weight density of concrete at y ordinate; γ_{Cb} is the weight density of concrete at the bottom fiber of the beam; γ_{Ct} is the weight density of concrete at the top fiber of the beam.

By integrating Eq. (8) along to the depth of the SRFGC beam's cross-section, ones can obtain the total volume per unit depth V of the concrete part of the SRFGC from,

$$V = \int_0^h \left\{\gamma_{Cb}\left[1 + \frac{\gamma_{Ct} - \gamma_{Cb}}{\gamma_{Cb}}\left(\frac{y}{h}\right)^p\right] b(y)\right\} dy \tag{9}$$

where, $b(y)$ is the y-variable width of the SRFGC beam cross-section.

3.3. Comparison of material cost

The average unit price of homogeneous concrete material and steel bar available in the standard material construction in Japan at present (2019) are shown in Table 1.

Table 1. Concrete and steel bar unit price (2019 average price market in Japan)

Strength Fc' (MPa)	Price (Yen/m^3)	Steel bar SD-345 (Yen/Ton)
24	17,000	67,000
69	34,700	

As the basis of the comparison study, the high strength concrete of RC beam (Case-A) is designed to determine the reference bending moment for loading, its dimension of the cross-section and required reinforcing steel bars are designed based on the ACI 318M-11 [1]. The results are then used for designing the other cases to find the optimal beam cross-section dimensions and necessary reinforcing steel bars. The equivalent block stress of concrete is determined from $f'_{Ct} = F_c/3$, and the equivalent block stress of steel bar is limited by $f_Y = 215$ N/mm^2 (Steel Grades Carbon Steel SD345). The design is optimized by finding a suitable loading of bending moments that are close to Eqs. (5-6), the dimensions of the cross-section and requirement of reinforcing steel bars within the equivalent block stress limit of concrete and steel bar materials.

Two beam cross-sections, as shown in Fig. 3 are selected for comparison study. A rectangular and a trapezoidal cross-section with the dimension of $b \times h$ and $(b+b') \times h$, respectively, being reinforced by double steel bars.

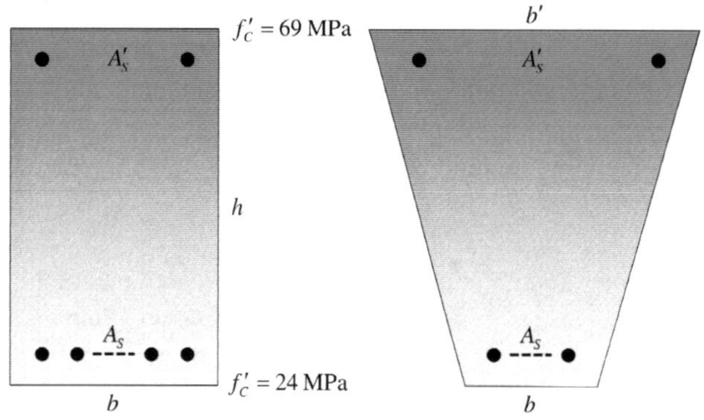

Figure 3. Various SRFGC beams cross-sections

The results of design and analysis of various the RC and SRFGC beam subjected to bending moment are tabulated in Tables 2 and 3. The beam cross-section dimensions, required steel bars, and the total cost of material are also shown in the tables.

Table 2. Material cost comparison between RC and SRFGC rectangular cross-section beam subjected to bending moment (M_{design} = 870 kN.m)

Case	b x h (in mm)	Concrete volume (m³) per unit length		p	Steel bars	Total price (yen/ m³)
		f'c = 28 MPa	f'c = 69 MPa			
A	400 x 400	0.0000	0.1600	0	1 x 2 x Ø16 1 x 4 x Ø29	7,143 (+17.5%)
B-1	400 x 420	0.0840	0.0840	1	1 x 2 x Ø16 1 x 4 x Ø29	5,934 (+2.4%)
B-2	400 x 430	0.1173	0.0587	2	1 x 2 x Ø16 1 x 4 x Ø29	5,621 (-7.5%)
B-3	400 x 440	0.1350	0.0450	3	1 x 2 x Ø16 1 x 4 x Ø29	5,447 (-10.4%)
C	400 x 570	0.2640	0.0000	0	1 x 2 x Ø16 1 x 4 x Ø29	6,079

In the study, the volume of the concrete is calculated from the fraction volume which is given by Eq. (9) multiplied by the corresponding weight density and unit price shown in Table 1. The price of steel bars per unit length, which is calculated from the unit price are also shown in Table 1.

IOP Conf. Series: Materials Science and Engineering **669** (2019) 012001 doi:10.1088/1757-899X/669/1/012001

Table 3. Material cost comparison between RC and SRFGC trapezoidal cross-section beams subjected to bending moment (M_{design} = 870 kN.m)

Case	b x b' x h (in mm)	Concrete volume (m^3) per unit length		p	Steel bars	Total price (yen/ m^3)
		f'c = 28 MPa	f'c = 69 MPa			
A	350 x 400 x 460	0.0000	0.1725	0	2 x Ø16 4 x Ø29	7,577 (+25.2%)
B-1	350 x 400 x 480	0.0843	0.0807	1	2 x Ø16 4 x Ø29	5,824 (-3.8%)
B-2	350 x 400 x 490	0.1169	0.0556	2	2 x Ø16 4 x Ø29	5,507 (-9.0%)
B-3	350 x 400 x 510	0.1368	0.0432	3	2 x Ø16 4 x Ø29	5,416 (-10.5%)
C	350 x 400 x 660	0.2625	0.0000	0	2 x Ø16 4 x Ø29	6,053

4. Summary and discussion

From Tables 2 and 3, in general, Cases B-3, which are the cubical SRFGC cross-sections, shows the lowest cost of material compared to both high-strength RC (Case A) and the normal RC (Case C). The high strength RC (Case A) has highest material price compared to the normal RC (Case C). In Figs. 4 and 5, the lowest cost of material was found in Case B-3 for all type of cross-sections where the gradation function of compressive strength is following the 3rd degree (p = 3) of the polynomial.

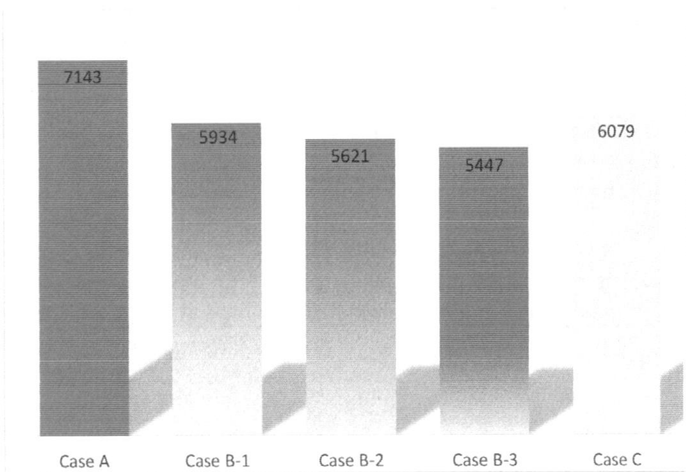

Figure 4. Comparison of concrete material (rectangular cross-section)

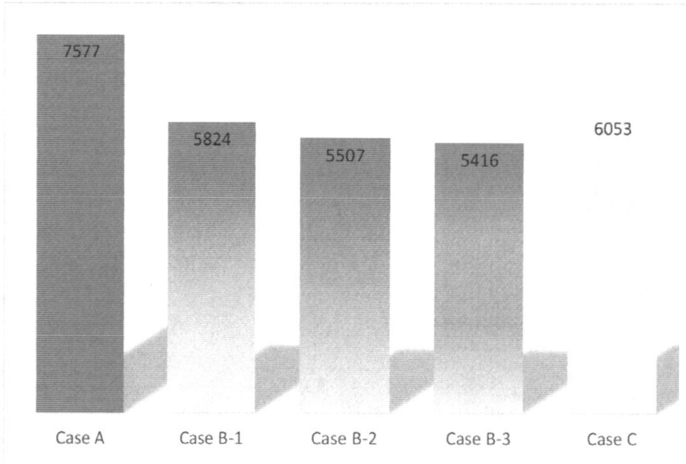

Figure 5. Comparison of concrete material (trapezoidal cross-section)

The following conclusions can be drawn from these findings.

1) When comparing the cost of material, the SRFGC beams (Case B-1, B-2 and B-3) exhibit more economical than both the normal RC beams (Case C) and high-strength RC beams (Case A). Moreover, among the SRFGC beams, the cost of material of gradation p = 3 (Case B-3) shows the lowest and the most economical case.
2) The most effective graded function of the SRFGC beam was found to be p = 3. Therefore, the difficulty to create a linear (p = 1) SRFGC beams in practice is not necessary.
3) One additional benefit of using FGC in the SRFGC was pointed in previous study [10,17,20], where the compressive strength of the FGC was limited by the lowest compressive strength part and the rigidity of the FGC was following the highest compressive strength part.

Finally, the results of comparisons between the weight per unit length of the most economical (Case B-3) SRFGC beams to the normal RC beam (Case C) and high-strength RC beam (Case A) are summarized in Table 4. From the Table, it was shown that the weight of the SRFGC beam (Case B-3) is lighter than that of the normal RC beam and almost the same weight with the high-strength RC beam in any type of cross-sectional shapes under the same bending moment. Especially for the normal concrete, the weight is mostly reduced by 29 % and very close when compared to the high-strength beam (Case A).

Table 4. Rate reduction of weight with the reference weight of Case-B-1C of SRFGC beams subjected to the same bending moment (M_{design} = 870 kN.m)

Comparison between Case B-3 with	Cross-section	
	Rectangle (%)	Trapezoid (%)
Case A (High strength concrete)	+4.0	-3.2
Case C (Normal concrete)	-28.8	-28.5

Therefore, the SFRGC of Case B-3 exhibits the lower weight due to smaller beam cross-section required compared to the normal concrete (see Table 4 for Case C), while cheaper material costs compared to both the normal and high strength RC beams (see Tables 2 and 3 for Cases A and C).

References

[1] Gan B S, Han A, Pratama M M A 2015 The Behavior of Graded Concrete, an Experimental Study *Procedia Eng.* vol 125 p 885–891

[2] Han A, Gan B S, Suryanto B, Priastiwi Y A 2017 Influence of the Stiffness Modulus and Volume Fraction of Inclusions on Compressive Strength of Concrete *Procedia Eng.* vol 171 p 760-767

[3] Koizumi M 1997 FGM Activites in Japan, *Compos. Part B Eng.* vol 28 (1–2) p 1–4
[4] Pratama M M A, Suhud R K, Puspitasari P, Kusuma F I, Putra A B N R 2019 Finite element analysis of the bending moment-curvature of the double-layered graded concrete beam *IOP Conf. Series Mat. Sci. Engrg.* vol 494 p 1-7
[5] Gan B S, Nguyen D K 2014 Large displacement analysis based on the co-rotational approach for functionally graded planar beam structures *Proc. of the 12th Int. Conf. on Steel, Space and Composite Structures*
[6] Gan B S, Nguyen D K 2014 Large Deflection Analysis of Functionally Graded Beams Resting on a Two-Parameter Elastic Foundation *J. Asian Archit. Build Eng.* vol 13:3 p 649-656
[7] Gan B S, Trinh T H, Le T H, Nguyen D K 2015 Dynamic response of non-uniform Timoshenko beams made of axially FGM subjected to multiple moving point loads *Struct. Eng. Mech.* vol 53:5 p 981–995
[8] Gan B S, Trinh T H, Nguyen D K 2017 Post-buckling Behaviour of Axially FGM Planar Beams and Frames *Procedia Eng.* vol 171 p 147-158
[9] Gan B S, Nguyen D K, Le T H 2017 Effect of intermediate elastic support on vibration of functionally graded Euler-Bernoulli beams excited by a moving point load *J. Asian Archit. Build Eng.* vol 16:2 p 363-369
[10] Gan B S, Trinh T H, Hara T, Nguyen D K, Tran T T 2017 Effects of support conditions to the post-buckling behaviors of axially functionally graded material rods *Key Eng. Mater.* vol 730 p 502-509
[11] Nguyen D K, Gan B S, Le T H 2013 Dynamic Response of Non-Uniform Functionally Graded Beams Subjected to a Variable Speed Moving Load *J. Comput. Sci. Technol.* vol 7:1 p 12-27
[12] Nguyen D K, Gan B S 2014 Large deflections of tapered functionally graded beams subjected to end forces *Appl. Math. Model.* vol 38:11-12 p 3054-3066
[13] Nguyen D K, Gan B S, Trinh T H 2014 Geometrically nonlinear analysis of planar beam and frame structures made of functionally graded material *Struct. Eng. Mech.* vol 49:6 p 727–743
[14] Nguyen D K, Gan B S, Trinh T H, Alexandrov S 2015 Post-buckling Behavior of Elastic-Plastic Functionally Graded Beams Subjected to Eccentric Axial Load *J. Multidisciplinary Eng. Sci. and Technol.* vol 2:5 p 1129-1135
[15] Han A, Gan B S, As'ad S, Pratama M M A 2015 Parametric Study of the Load Carrying Capacity of Functionally Graded Concrete of Flexural Members *Int. J. Eng. Technol. Innov.* vol 5:4 p 233-241
[16] Reported by ACI Committee 318 2011 Building Code Requirements for Structural Concrete *American Concrete Institute*
[17] Pratama M M A, Gan B S, Han A, Putra A B N R 2018 A Numerical Analysis of The Modulus of Elasticity of The Graded Concrete *Adv. Soc. Sci. Edu. and Humanities Res.* vol 242 p 118-120
[18] Pratama M M A, Sri Umniati B, Wulandari B A M, Han A, Gan B S, Zhabrinna 2018 Modulus elasticity of the graded concrete, a preliminary research *MATEC Web of Conf.* vol 195:01005 p 1-8
[19] Han A, Gan B S, Pratama M M A 2016 Effects of Graded Concrete on Compressive Strengths *Int. J Technol.* vol 5 p 530-538
[20] Pratama M M A, Han A, Gan B S, Umniati B S, Risdanareni P, Fauziyah S 2017 Effect of concrete strength gradation to the compressive strength of graded concrete, a numerical approach *AIP Conf. Proc.* vol 1887:1:020029
[21] Pratama M M A, Sri Umniati B, Prasiwi G, Han A, Gan B S, Risdanareni P, Zhabrinna 2018 Effect of monotonic lateral load on the performance of reinforced graded concrete column, an experimental study, a preliminary research *MATEC Web of Conf.* vol 195:02022 p 1-8

IOP Conf. Series: Materials Science and Engineering **669** (2019) 012002 doi:10.1088/1757-899X/669/1/012002

Cyclic behavior of moment-resisting frames strengthened with steel curved dampers

H L Hsu[1] and H Halim[1]

[1] Dept. of Civil Engineering, National Central University, Taoyuan City, Taiwan

hhsu@ncu.edu.tw

Abstract. Steel moment frames have been widely used for earthquake-resistant purposes. These systems possess significant ductility, however, excessive deformation due to the lower structural stiffness. In order to improve the behavior of moment-resisting frames when subjected to cyclic loads, a new design that combined steel curved damper and brace member was proposed in this study. The curved damper was cut from steel plate and was placed in the center portion of the brace member. The new device was attached to the frame and was used to dissipate energy throughout the load application. A series of loading tests on steel frames strengthened with the proposed design were carried out. The strength and energy dissipation capabilities of frames with and without the strengthening device were compared. It was found from the comparisons that the behavior of framed structures was significantly improved, thus justified the effectiveness of the proposed design.

1. Introduction

Steel moment frames have been widely used for earthquake-resistant purposes [1-3]. These systems possess significant ductility, however, excessive deformation due to the lower structural stiffness. Application of brace members to the steel moment frames has been considered effective approach to reduce the excessive deformation, as structural stiffness can be significantly enhanced. This approach subsequently raises concerns in loss of structural deformability due to excessive structural stiffness and buckling of brace member at large deformation [4-5]. Therefore, a remedy, i.e. development of a brace with adequate stiffness and deformability simultaneously, to further improve the structural behavior is essential.

It has been investigated in the authors' previous study [6] that when steel plate was fabricated in curved shape and loaded by in-plane bending, the device would exhibit sufficient strength, sustained stability at large deformation and dissipated significant energy through inelastic deformation. This device was named steel curved damper, as shown in Figure 1. To further improve the behavior of moment-resisting frames when subjected to cyclic loads, a new design that combined steel curved damper and brace member, forming A-Brace as shown in Figure 2, was proposed in this study. The curved damper was placed in the center portion of the brace member and mobilized by the lever mechanism of the A-Brace. The new device was attached to the frame and was used to dissipate energy throughout the load application.

A series of loading tests on the steel curved dampers and steel frames strengthened with the proposed design were carried out. The strength and energy dissipation capabilities of frames with and without the strengthening device were compared to evaluate the effectiveness of the proposed method.

IOP Conf. Series: Materials Science and Engineering **669** (2019) 012002 doi:10.1088/1757-899X/669/1/012002

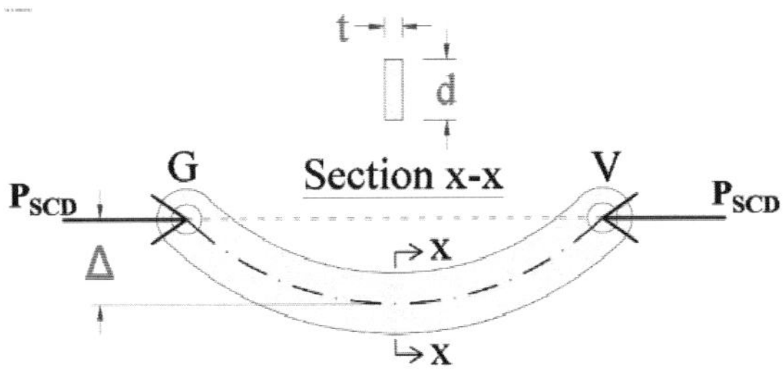

Figure 1. Definition of steel curved damper [6].

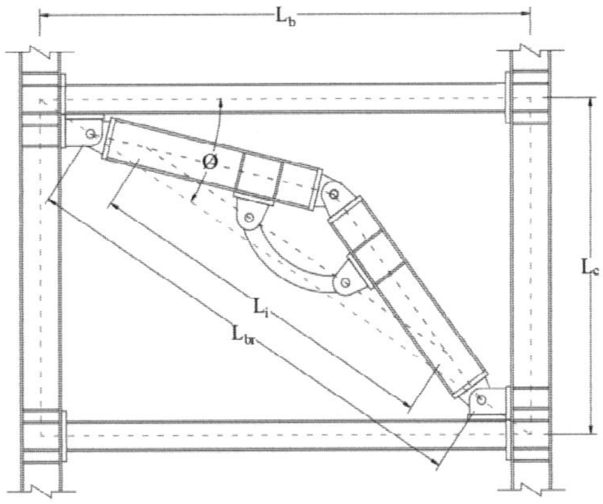

Figure 2. Relationship among frame, A-Brace and steel curved damper

2. Strength of steel curved dampers

Steel curved damper was cut from steel plate with desired geometry. As shown in Figure 1, the maximum eccentricity between the center point of the damper and the axis that connecting the two ends, i.e. Δ , will cause extra moment ($P\Delta$) when subject to axial load P. In this regard, the maximum stress at the center point of the curved damper can be evaluated by the following equation [6]:

$$\sigma_{max} = \frac{M(\frac{d}{2})}{I} + \frac{P}{A} \tag{1}$$

In which, d and t are the damper depth and thickness, respectively.
Since the curved damper was bent about the strong axis of the plate, the above-mentioned stress can be further expressed as follow:

$$\sigma_{max} = \frac{P\Delta(\frac{d}{2})}{\frac{1}{12}d^3t} + \frac{P}{dt} \tag{2}$$

Therefore, the strength of steel curved damper (PSCD) that will cause yielding of curved damper, can be obtained when the maximum stress was equal to the plate yielding stress

$$P_{SCD} = \frac{d^2t}{6\Delta+d}\sigma_y \tag{3}$$

Figure 3 shows the strength of steel curved damper, PSCD, with various eccentricities. It can be found from the figure that damper with larger eccentricity will incur lower strength.

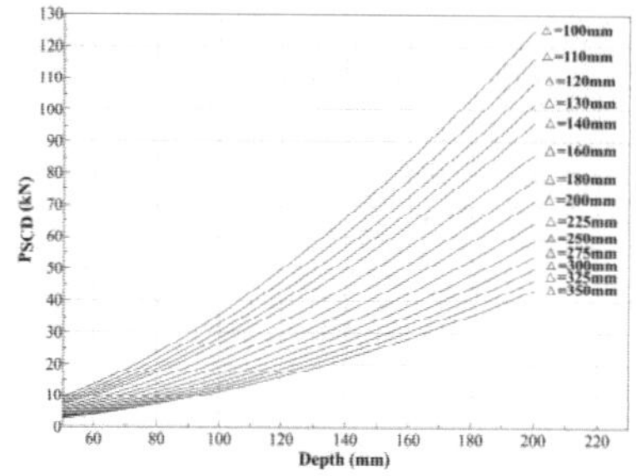

Figure 3. Damper strength with various eccentricities

3. Evaluation of frame performance

As mentioned in earlier section, the structural behavior would be improved by the application of the proposed A-Brace. To evaluate the effectiveness of the proposed method, the design and load-resisting mechanism of the A-Brace must be investigated. The A-Brace is composed of steel curved damper (SCD) and two rigid members. The two members were used to transmit force from brace to damper, therefore should be designed to remain elastic at the ultimate strength level.

As shown in Figure 2, the length of beam, height of column, length of brace and length of intersections between axes of SCD and A-Brace can be represented by Lb, Lc, Lbr, and Li, respectively. The strength relationship between A-Brace (Pbr) and steel curved damper (PSCD) can be described using the following equations:

$$P_{br}\frac{L_i}{L_{br}}P_{SCD} \quad (4)$$

For adequate design, the damper strength should be determined not to damage the brace member. The typical hysteretic behavior of A-Brace, evaluated by ANSYS [7], can be seen in Figure 4.

It can be found from the figure that the strength of A-Brace in compression and tension is slightly different. This is due to the changing of eccentricity of curved damper during various deformation load. Nevertheless, the A-Brace exhibited improved stable behavior over the tradition braceS, under tension and compression.

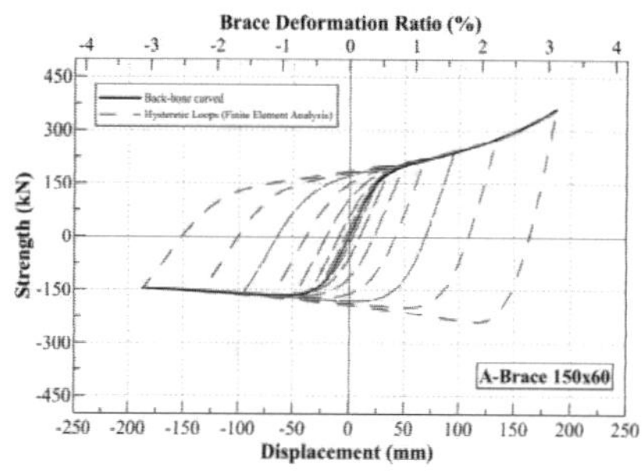

Figure 4. Hysteretic behavior of A-Brace.

4. Experimental verification

4.1. Test Speciments

Four test frames including one moment frame and three strengthened frames, as listed in Table 1, were fabricated to investigate the effectiveness of A-Brace as strengthening scheme. The detail of test frame can be seen in Figure 5. The columns and beams were made of SN400 H250x250x9x14 and H175x175x7.5x11, respectively. A strengthened panel zone in the beam-column joint was applied by adding 10 mm plate to prevent shear failure in this area. 30-mm-thick steel curved damper with length equaling 700mm was attached to the center region of A-Brace. Depths of curved dampers were 120 mm, 100 mm and 90 mm, respectively. The test frames were labeled according to the size of damper geometry. For example, AF 120x30 indicated a steel moment frame equipped with A-Brace and curved damper with depth and thickness equaling 120 mm and 30 mm, respectively. Each test frame was loaded by a servo-controlled hydraulic actuator under a series of cyclic displacement commands according to FEMA 461 [8] protocol until 5% story drift ratio.

Table 1. Details of test specimens.

Specimen	Dimension of the curved damper	
	Depth d (mm)	Thickness t (mm)
MF	N.A	N.A
AF 120x30	120	30
AF100x30	100	30
AF 90x30	90	30

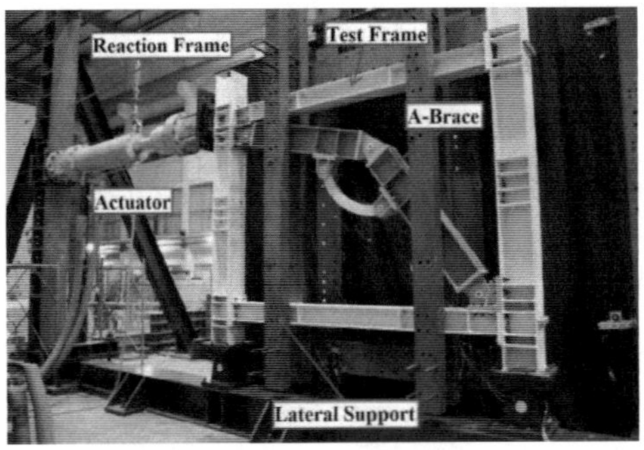

Figure 5. Detail of test frame.

4.2. Results and comparisons

Failure patern of frame with A-Brace is shown in Figure 6. It was observed from the test that the failure began with the yielding in the inner side of the curved damper. Further observations validated that no fracture or damper plate buckling was developed during the loading process thus proving the ductility and stability of the damper. The hysteretic loops for all test frames are shown in Figure 7. It can be found from the figure that all frames exhibited stable behavior when the frame story drift reached 5% with no degradation in performance. Higher strength can be achieved when A-Brace is attached to the steel moment frame. The strength improvement of the frame after adding A-Brace can reach up to 2.31 times. Frame strength and energy dissipation were further correlated and shown in Figure 8. It can be found from the comparison that significant energy dissipation gain was achieved, for example, 1.94 times in frame AF 120x30 than that of the moment frame, when the proposed device was adopted. These phenomena validated the effectiveness of the proposed method to structural strengthening design.

Figure 6. Typical failure pattern of frame with A-Brace.

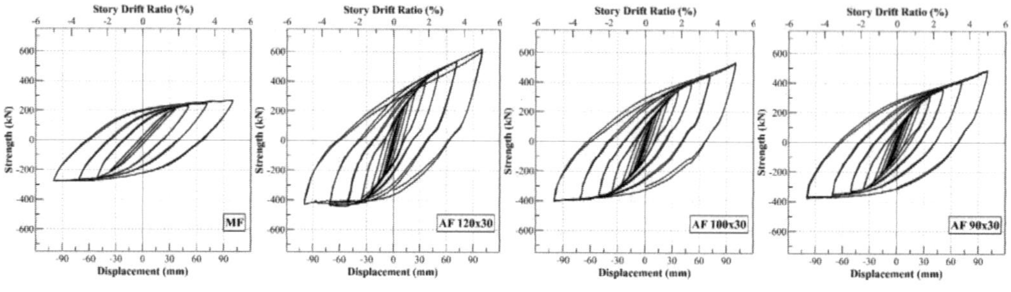

(a) (b) (c) (d)

Figure 7. Hysteretic loops of test frames: (a)MF; (b)AF120x30; (c)AF100x30; (d)AF90x30

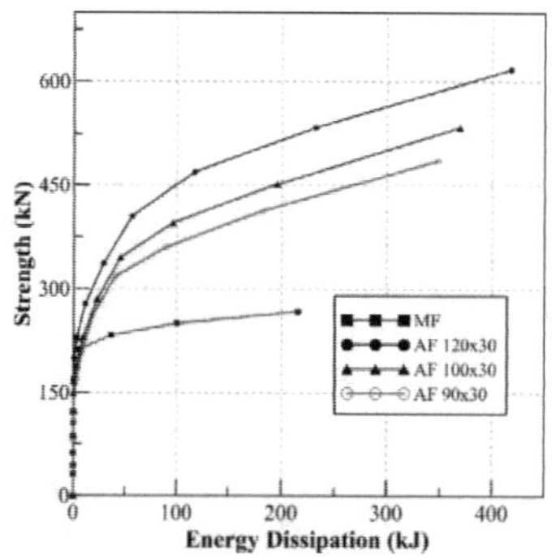

Figure 8. Performance of frames with various A-braces.

5. Conclusions

This study investigated the behavior of steel curved damper and its applicability to structural strengthening. A series of loading tests on steel frames strengthened with the proposed design were carried out. It was found from the tests that all frames exhibited stable behavior when the frame story drift reached 5% with no degradation in performance. Strength improvement of the frame after adding A-Brace can reach up to 2.31 times. Further comparison on the test results showed that significant energy dissipation gain was achieved, for example, 1.94 times in frame AF 120x30 than that of the moment frame, when the proposed device was adopted. Simultaneous enhancement in strength and energy dissipation validated the effectiveness of the proposed method to structural strengthening design.

References

[1] Black EF 2012 *Inelastic parameter estimates for regular steel moment-resisting frames* Eng Struct **34**, pp.33-39.

[2] Kim J and Park J 2008 *Design of steel moment frames considering progressive collapse.* Steel and Compos Struct **8**, pp.85-98.

[3] Hu FX, Shi G and Shi YJ 2017 *Experimental study on seismic behavior of high strength steel frames: Global response* Eng Struct **131**, pp.163-179.

[4] Choi KS, Park JY and Kim HJ 2017. *Numerical investigation on design requirements for steel ordinary braced frames.* Eng Struct **137**, pp.296-309.

[5] Dehghani M and Tremblay R 2018 *Design and full-scale experimental evaluation of a seismically endurant steel buckling-restrained brace system.* Earth Engng Struct Dyn **47**, pp.105-129.

[6] Hsu HL and Halim H 2017. *Improving seismic performance of framed structures with steel curved dampers.* Eng. Struct **130**, pp.99-111.

[7] ANSYS 2016 *Workbench user's guide*, ANSYS Inc.; Canonsburg, PA, USA.

[8] FEMA 2007 *Interim protocols for determining seismic performance characteristics of structural and nonstructural components.* Federal Emergency Management Agency; Washington, D.C., USA.

IOP Conf. Series: Materials Science and Engineering **669** (2019) 012003 doi:10.1088/1757-899X/669/1/012003

Emission factors of NOx and PM of heavy construction equipment used in toll road project in Indonesia – Case study: Rembang-Pasuruan Toll Section II

A M Hajji[1], A Yulistyorini[1], Huang Yue[2] and D C Patulak[1]

1 Civil Engineering Dept., *Universitas Negeri Malang*, Indonesia
2 Intitute of Transport Studies, University of Leeds, UK

apif.miptahul.ft@um.ac.id

Abstract. Heavy duty diesel (HDD) equipment play an important role in constructing Indonesian infrastructure. This paper proposes a methodology for recording an emission inventory for HDD equipment that formulates emission factors as a baseline of current emissions quantities. The sample case of the proposed emissions inventory technique is based on a small part of Indonesia's national toll road project: Rembang-Pasuruan toll road project. Rembang-Pasuruan toll road is a part of Gempol-Pasuruan toll road, a piece of national infrastructure project called Trans-Java Toll road that runs from Merak, Northwestern end of Java, to Banyuwangi, the eastern end of the island. This toll road connects almost all major cities and landmarks of the island with the total length of the road is nearly 1,167 km. The length of Gempol-Pasuruan toll road is about 34.15 km, and consists of three section: Section I, Gempol-Rembang 13.9 km, Section II, Rembang-Pasuruan 6.6 km, and Section III, Pasuruan-Grati 13.65 km. This toll road project utilizes a HDD fleet that includes 9 excavators, 2 bulldozers, 2 roller compactors, 1 motor-grader and 1 paver. This equipment consumes large quantities of diesel fuel and release air pollutants, including nitrogen oxides (NOx) and particulate matters (PM). The proposed inventory technique in this paper addresses the key attributes of the equipment on how long it is used during its operational life and how emission rates of NOx and PM are estimated. By using emission factors calculated from equipment's key attributes, it is revealed that all equipment in the project release approximately 452.12 gr/hr of NOx or 3400.97 gr/day on average, and emit 31.99 gr/hr of PM or equals to 252.92 gr/day.

1. Introduction

Heavy duty diesel (HDD) equipment play an important role in constructing Indonesian infrastructure. Most of this equipment fleets are owned and operated by public entities, such as local governments and government-owned construction companies, that have limited budgets and are wondering how to acquire, maintain, and update their fleets. Not only does this equipment needs comprehensive fleet management but it also has a significant impact on national energy consumption in the form of diesel fuel, and an impact on the environment in the form of hazardous air pollutants.

This paper proposes a methodology for recording an emission inventory for HDD equipment that formulates emission factors as a baseline of current emissions quantities. The emission inventory technique proposed in this paper can help fleet managers or equipment operator to quantify emissions air pollutants for each individual item of equipment and possibly the entire fleet used in an infrastructure project. Furthermore, this technique can also help some public entities or stakeholders determine strategies to make infrastructure project more environmentally-sound. The sample case of the proposed emissions inventory technique is based on a small part of Indonesia's national toll road project: Rembang-Pasuruan toll road project. Rembang-Pasuruan toll road is a part of Gempol-Pasuruan toll road, a piece of national infrastructure project called Trans-Java Toll road that runs from Merak, Northwestern end of Java, to Banyuwangi, the eastern end of the island. This toll road connects almost all major cities and landmarks of the island with the total length of the road is nearly 1,167 km. The length of Gempol-Pasuruan toll road is about 34.15 km, and consists of three section: Section I, Gempol-

Rembang 13.9 km, Section II, Rembang-Pasuruan 6.6 km, and Section III, Pasuruan-Grati 13.65 km. This toll road project utilizes a HDD fleet that includes excavators, bulldozers, roller compactors, and pavers. These equipment consume large quantities of diesel fuel and release air pollutants, including nitrogen oxides (NOx) and particulate matters (PM). The proposed inventory technique in this paper addresses the key attributes of the equipment on how long it is used during its operational life and how emission rates of NOx and PM are estimated.

2. Related studies

HDD equipment consumes large amounts of fuel and releases significant quantities of pollutants. The partial load of the HDD engine, which associates with working modes, had the greatest impact on fuel use and emissions rate estimates [1], [2]. By using information regarding working modes and partial loads of the HDD engine, the equipment managers can estimate the energy and environmental impacts of their equipment fleet [3]. Measuring and monitoring real-time emissions will provide practitioners with information to assess environmental impacts and improve the sustainability of construction [4], [5]. Monitoring of the environmental impacts of the use of HDD equipment will also help country's transportation agencies and boards to plan investigation strategies to reduce emissions from construction and rehabilitation of toll road infrastructures [6-8].

Models and other prediction models have been developed to estimates the emissions from HDD equipment, such as discrete event simulations [9] or regression models from HDD equipment productivity rate [10]. Some case studies are also conducted in observing construction activities associated with the emission potentials [11], [12]. The estimates of emissions released by HDD equipment can also be predicted during planning stage of the project. In this stage, the estimates of emissions can be taken off from construction project scope, engineering schedules, working volumes, or specifications [13-17].

3. Method

To develop an emission inventory of NOx and PM, the key attributes of selected HDD used in the project were collected. According to the EPA *Construction Fleet Inventory Guide* [18], these attributes included equipment type, engine model year, engine horsepower, and annual activity. In addition to these attributes, this inventory techniques also takes into account the engine tier classification; engine tier classifications - a hybrid attribute based on horsepower rating and model year - were identified. Engine tiers are emissions standards adopted by EPA for all new nonroad diesel engines [18].

In order to develop the inventory for Rembang-Pasuruan toll road project, emission factors of HDD fleet were needed. These factors are approximations of the amount of pollutants emitted by a particular type of equipment during a unit of use. The factors used for this inventory were based on calculations and the methodology employed by the EPA NONROAD model. For pollutants, emission factors are reported in grams per horsepower-hour (g/hp-hr). For NO_x, HC, and CO, the emission factor for a specific type of nonroad equipment with a particular model year and age is calculated as follows:

$$EF_{adj\,(NOx)} = EF_{ss}\ x\ TAF\ x\ DF \qquad (1)$$

where:

EF_{adj} = final emission factor used in NONROAD, after adjustments for transient operation and deterioration (g/hp-hr)

EF_{ss} = zero-hour, steady-state emission factor (g/hp-hr)

TAF = transient adjustment factor (unitless)

DF = deterioration factor (unitless)

The zero-hour, steady-state emission factor (EF_{ss}) is a function of the engine's model year and horsepower rating, which defines the engine tier category. Transient adjustment factors are applied to Tier 0, 1, 2, and 3 engines but are not applied to Tier 4 engines because transient emission controls will be a part of all Tier 4 engine design considerations. Transient adjustment factors are calculated as the ratio of the transient emission factor to the corresponding steady-state emission factor and may be greater than or less than 1.0. Deterioration factors are used to account for increases in emissions over time above a new engines base emission level. Emissions may increase over time for numerous reasons including engine wear, poor maintenance, or modification of emission control systems. Emissions performance typically deteriorates at a slow rate for well-maintained engines but rapidly for poorly-maintained engines [18], [19].

Since PM emissions are dependent on the sulfur content of the fuel consumed by the engine, the equation for the PM emission factor is modified from Equation 1 as follows:

$$EF_{adj\,(PM)} = EF_{ss} \times TAF \times DF - S_{PMadj} \quad (2)$$

where:
S_{PMadj} = adjustment to PM emission factor for variations in fuel sulfur content (g/hp-hr)

For BSFC (or fuel use factor), deterioration factors are not applied, thus, the equation is simplified as follows:

$$EF_{adj\,(BSFC)} = EF_{ss} \times TAF \quad (3)$$

The engine rated horsepower is the maximum level of power that an engine is designed to produce at its rated engine speed. Nonroad equipment seldom operates at its rated power for extended periods and frequently operates at a variety of speeds and loads. NONROAD uses a load factor (*LF*) to indicate the average proportion of rated power used to account for the effects of operation at idle and partial load conditions. For example, a 100 hp engine with a load factor of 0.3 (or 30%) will produce an average of 30 hp over the course of normal operation. Depending on equipment usage patterns, load factors may vary widely for nonroad engines and can be difficult to quantify. Since equipment usage patterns for Rembang-Pasuruan tool road project fleet were not observed or measured, the NONROAD model default load factors were used. The NONROAD model also provides default values for average annual activity for many types of equipment; however, the default values were not used here. For the Rembang-Pasuruan project fleet, average annual activity was determined based on the total usage divided by the age of the equipment. This information was obtained from the fleet management database as well as field data collection; thus, the emissions results presented here are specific to the project fleet.

4. Results

The nonroad HDD fleet used in Rembang-Pasuruan Section II toll road project consists of 15 units, including nine excavators, two bulldozers, one motor grader, and two roller compactors, and one paver. The individual equipment attributes of each unit is provided in Table 1.

Of the 15 HDD equipment used in the project, 80% are categorized as new equipment manufactured in 2008-2015, 13% made in 2005-2007, and only 1% (one roller compactor) released by the factory in 2000-2004. As for the engine size, 53% HDD equipment have 100-175 HP, 27% equipped with 175-300 HP of engine, 13% between 50-75 HP, and only one HDD equipment has 76-100 HP (one excavator). Based on the model year and engine size, the majority of HDD equipment deployed in the project or 53% are categorized as Tier 4 engine (8 equipment), 40% at Tier 3, and 7% (one roller compactor) at Tier 1. The newest HDD equipment in the project is 54 HP excavator, which is manufactured in 2015 (Tier 4), while the oldest one is 131 HP roller compactor, which is produced in 2002 (Tier 1).

As displayed in Table 2, following the EPA NONROAD's estimates of cumulative hours, all HDD equipment have an accumulated total activity of 130,370 hours. While observed at the odometer reading of each equipment, the total accumulated total activity of those 15 HDD equipment is 84,203 hours. The oldest equipment, a 131 HP roller compactor does not have the longest hours of activity, which is only 6,420 hours during its operational age, compared to the activity hours estimated by the EPA NONROAD (13,680 hours). The newest equipment, a 54 HP excavator has longer activity hours (7,093 hours) compared to those estimated by the EPA NONROAD (5,460 hours). The other HDD equipment which have longer activity hours than its estimates are 148 HP and 156 HP excavators. Some equipment are also found rarely used in its operational age compared to its estimate hours, they are 147 HP (2010 model), 201 HP, 268 HP, and 91 HP excavators, all two bulldozers and roller compactors, and one motor grader. The total activity hours for paver is not available.

Table 1. HDD fleet in Rembang-Pasuruan Section II

Equipment Type	Make	Type	Horsepower (hp)	Model year	Engine Tier
Excavator 1	Komatsu	PC200	147	2010	3
Excavator 2	Komatsu	PC490LC	201	2013	4
Excavator 3	Kobelco	SK330	268	2014	4
Excavator 4	CAT	312E	91	2014	4
Excavator 5	Hyundai	R220-9SH	148	2014	4
Excavator 6	Komatsu	PC200	147	2010	3
Excavator 7	Komatsu	PC78US	54	2008	3
Excavator 8	Komatsu	PC78UU	54	2015	4
Excavator 9	Sumitomo	SH210	156	2014	4
Bulldozer 1	Komatsu	D65PX	207	2008	3
Bulldozer 2	Komatsu	D68ESS	174	2007	3
Motor grader 1	Komatsu	GD511A	135	2013	4
Roller Compactor 1	Dynapac	CA25	131	2002	1
Roller Compactor 2	Ingersroll Rand	SD1000	124	2008	3
Paver 1	Wirtgen	SP500	176	2015	4

Table 2. Cumulative hours of HDD fleet in Rembang-Pasuruan Section II

Equipment Type	Horsepower (hp)	Model year	Engine Tier	Cumulative hours	Odometer hours
Excavator 1	147	2010	3	10920	1578
Excavator 2	201	2013	4	7644	1398
Excavator 3	268	2014	4	6552	1092
Excavator 4	91	2014	4	6552	1491
Excavator 5	148	2014	4	6552	9661
Excavator 6	147	2010	3	10920	4606
Excavator 7	54	2008	3	13104	14876
Excavator 8	54	2015	4	5460	7093
Excavator 9	156	2014	4	6552	23210
Bulldozer 1	207	2008	3	10788	2805
Bulldozer 2	174	2007	3	11687	3354
Motor grader 1	135	2013	4	6734	1829
Roller Compactor 1	131	2002	1	13680	6420
Roller Compactor 2	124	2008	3	9120	4790
Paver 1	176	2015	4	4105	0

Table 3 presents a summary of emissions inventory of NOx and PM for HDD equipment fleet deployed in Rembang-Pasuruan Section II toll road project. The emission factors (EF) are shown for each item of equipment, along with the total emission rates (per-hour and per-day) of each unit. By using emission factors calculated from equipment's key attributes, it is revealed that all equipment in the project release approximately 6,377 grams per-hour or 51,015 grams of NOx per-day, and 480 grams of PM per-hour, which equals to 3,839 grams per-day.

Table 3. Emission factors and emission rates of NOx and PM of HDD fleet in Rembang-Pasuruan Section II

Equipment Type	HP	Model year	EF NOx (g/hp-hr)	EF PM (g/hp-hr)	NOx (gr/hr)	NOx (gr/day)	PM (gr/hr)	PM (gr/day)
Excavator 1	147	2010	2,629	0,513	386,42	3091,37	75,47	603,78
Excavator 2	201	2013	2,620	0,000	526,64	4213,12	0,00	0,00
Excavator 3	268	2014	2,617	0,000	701,42	5611,34	0,00	0,00
Excavator 4	91	2014	3,141	0,000	285,80	2286,41	0,00	0,00
Excavator 5	148	2014	2,617	0,000	387,35	3098,80	0,00	0,00
Excavator 6	147	2010	2,629	0,513	386,42	3091,37	75,47	603,78
Excavator 7	54	2008	3,161	0,504	170,71	1365,70	27,05	216,37
Excavator 8	54	2015	3,137	0,012	169,41	1355,28	0,67	5,34
Excavator 9	156	2014	3,141	0,000	489,95	3919,56	0,00	0,00
Bulldozer 1	207	2008	2,628	0,342	544,07	4352,58	70,71	565,66
Bulldozer 2	174	2007	2,631	0,528	457,75	3661,98	91,92	735,32
Motor grader 1	135	2013	2,618	0,000	353,39	2827,12	0,00	0,00
Roller Compactor 1	131	2002	5,593	0,605	732,63	5861,00	79,22	633,77
Roller Compactor 2	124	2008	2,624	0,479	325,37	2602,99	59,35	474,78
Paver 1	176	2015	2,611	0,000	459,50	3676,00	0,00	0,00

On an individual basis, all Tier 4 engines (201 HP, 268 HP, 91 HP, 148 HP, 54 HP, and 156 HP excavators) emits very low or zero PM emission. Meanwhile, two oldest equipment (174 HP bulldozer and 131 HP roller compactor) release the highest PM emissions, which are 735 grams and 634 grams per day respectively. It shows that the newer the engine, the Tier level will be higher, and the PM emissions will be lower. The same proof is also shown in NOx emissions. 131 HP roller compactor release the highest NOx emission rate, which is 5,861 grams per-day. High NOx emission rate is also occurred at the biggest engine size (268 HP excavator), which is 5,611 grams per-day. It shows that the emission rate of NOx is not only determined by the age of equipment, but also the size of the engine. The other case is 207 HP bulldozer, which also emits significantly high NOx emission per-day (4,353 grams). The lowest emission rate of NOx is released by the newest equipment with the smallest engine size (54 HP excavator 2015 model), which is 1,355 grams per-day.

5. Conclusions

The toll road project of Rembang-Pasuruan Section II has a substantial fleet of nonroad HDD equipment with respect to the diversity and total number of units, annual activity, and emissions of NOx and PM. The environmental inventory methodology presented is this paper was successful at quantifying the fleet emissions by using the data that was available from the data that was collected by visual inspection from the actual equipment. By examining the results of environmental inventory, it is possible to determine which equipment subgroups, as well as which individual equipment units, are responsible for the most pollutants emitted. For example, roller compactor and bulldozer rank first and second, respectively, in NOx and PM emissions based on equipment type. This type of information is useful in identifying which equipment has the highest environmental impacts and can be used as a new metric for fleet management decisions.

6. References

[1] Lewis P, Fitriani H, Arocho I 2015 *Transportation Research Record*. **2482** (1) 8-15.

[2] Cao T, Durbin, TD, Russell, RL, Cocker III DR, Scora G, Maldonado H, Johnson KC 2016 *Atmospheric environment*. **147** 234-245.
[3] Lewis P, Rasdorf W 2016 *Journal of Management in Engineering*. **33** (2) 04016038.
[4] Heidari B, Marr LC 2015 *Journal of the Air & Waste Management Association*. **65** (2) 115-125.
[5] Hajji AM, Muladi, Larasati A 2016 *AIP Conference Proceedings*. **1778** (1) 030008.
[6] Quiros DC, Thiruvengadam A, Pradhan S, Besch M, Thiruvengadam P, Demirgok B, Hu S 2016 *Emission Control Science and Technology*. **2** (3) 156-172.
[7] Ma F, Sha A, Lin R, Huang Y, Wang C 2016 *International journal of environmental research and public health*. **13** (3) 351.
[8] Kar SS, Behl A, Shukla A, Jain PK 2015 *J. Civil Environ. Eng*. **5** (198).
[9] Limsawasd C, Athigakunagorn N 2017 *Engineering Journal*. **21** (7) 197-211.
[10] Hajji AM, Lewis MP, Larasati A 2017 *AIP Conference Proceedings*. **1887** (1) 020051.
[11] Wang X, Duan Z, Wu L, Yang D 2015 *Journal of Cleaner Production*. **103** 705-714.
[12] Liu Y, Wang Y, Li D 2017 *Journal of cleaner production*. **144** 337-346.
[13] Park JY, Lee DE, Kim BS 2016 *KSCE Journal of Civil Engineering*. **20** (6) 2162-2169.
[14] Lim TK, Gwak HS, Kim BS, Lee DE 2016 *KSCE Journal of Civil Engineering*. **20** (4) 1211-1220.
[15] Rasdorf W, Lewis P, Arocho I, Hummer J 2015 *Smart and Sustainable Built Environment*. **4** (3) 315-328.
[16] Setiawati A, Prasetyo SCA, Hatmoko JUD, Hidayat A 2015 *Jurnal Karya Teknik Sipil*. **4** (1) 83-92.
[17] Wirahadikusumah RD, Sahana HP 2012 *Journal of Civil Engineering*. **19** (1) 25-36.
[18] EPA 2010 Exhaust and Crankcase Emission Factors for Nonroad Engine Modeling - Compression Ignition.
[19] Hajji AM, Lewis P 2013 *Smart and Sustainable Built Environment*. 2 (1) 84-100.

Acknowledgement
The authors express gratitude to the Institute of Research and Community Services, Universitas Negeri Malang for their generousness and cooperation in providing research fund and management in order to complete this study.

IOP Conf. Series: Materials Science and Engineering **669** (2019) 012004 doi:10.1088/1757-899X/669/1/012004

Rainfall induced residual soil slope instability: building cracked and slope failure

L J Yue[1], M F Ishak[1], M S I Zaini[1] and M F Zolkepli[1]

[1] Faculty of Engineering Technology, Universiti Malaysia Pahang, Malaysia

fakhrurrazi@ump.edu.my

Abstract. Rainfall is one of the factors of slope failures due to the intensity of rainfall that will seep and increase the moisture content of geo-materials. One of the important role in slope stability is matric suction. This study present the relationship between rainfall and suction of soil which induced the slope instability and the slope stability analysis involving a building which may cause by instability due to suction variation. The slope had separated into three parts which are top of slope, middle of slope and toe of slope. Every part has been inserted with three tensiometer with 0.3 m, 0.45 m and 0.9 m depth. Suction of soil on slope was carried out by using tensiometer. Intensity of rainfall was collected by using rain gauge which placed at open area which to compare with matric suction of soil. Several site and laboratory tests were carried out to identify the soil properties. Stability of slope was analyzed using Slope/w and Manual Calculation via input parameters collected from laboratory test and field test. Fellenius' method was used to interpret the factor of safety for slope study. This includes the analysis of the factor of safety of slices at the selected area with its matric suction value. It was found that when the rainfall increased, the suction of soil decreased and affected to decrease of factor of safety. Hence, the constructed slope was suggested to be redesign to improve its safety factor.

1. Introduction

Slope instability can cause by three different factor or categories. Some of the main factors has being categorized by geological, climate change and human activities. From the geological perspective, soil layer and profile were crucial in influencing the slope instability. Intensity of rainfall, earthquake, flooding and those related disaster are some of the crucial factors of slope instability from the perspective of climate change. During the rainfall, rain water may distribute into two medium on surface and subsurface via surface runoff and infiltration/seepage phenomenon. Shallow failure of slope will occurred due to the less rainwater infiltration volume and surface runoff. Meanwhile, deep-seated slope failure will occurred due to the influence of the river and high groundwater level supplied by the excessive rainfall water.

Rainfall induced slope instability create a common geotechnical hazard in tropical region such as Malaysia [1]. The intensity of rainfall will affect the measurement of rain gauge and the suction of soil because of the measure amount of the rain that falls over time. Rainfall was one of the major cause of slope failures or slope instability when the area or region experience heavy rainfall. Therefore, the characteristic of the rainfall should be considered in the analysis of the slope instability.

Suction plays an important role in the stability of slope especially in tropical region. Hence, the evaluation on the measurement of suction in soil of slope stability is needed [1]. Rainfall which happening at the area will cause the moisture content increased, the matric suction is reduced which reduces the intergranular of the soil and the effective stress also reduced. Since the effective stress decreased so the shear strength also reduced [4]. Matric suction is forms at soil-air interface because of the surface tension which result in reduced vapor pressure in the water. Vapor pressure decrease or become negative, the matric suction pressure increases [3]. The reduction of matric suction occurred when the rainfall which cause wetting of influencing the reduction of unsaturated shear strength of soil slope. This is called rainfall infiltration [2].

The aim of this study is to determine matric suction of residual soil slope with building structures. Moreover, the study was able to demonstrate the slope instability due to the rainfall events.

2. Methodology

The data collection area was located at a slope with building which show in Figure 1 and 2. Visual Inspection is to contribute for an initial view of the natural and structural at surrounding of the study area.

Figure 1. Location of the study area from google maps view

Figure 2. Location of the study area from front view

In order to determine the soil type and soil sample for further testing, sample of soil was collected by boring machine and Trial Pits according to British Standard Code of Practice BS 5930: 1999 AND BS 1377: 1990.

2.1. Suction and Rainfall
The slope was divided into three parts which are top, middle and toe of slope for suction testing with three different tensiometer depth (0.3 m, 0.45 m and 0.9 m below the ground surface). The location of tensiometer was shown in Figure 3. Rain gauge was placed at the open area of the site or installed it at the centre of research area.

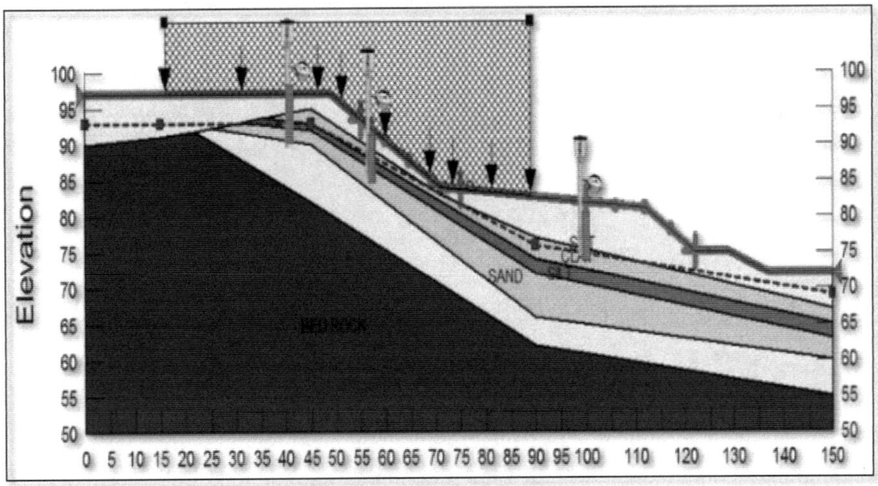

Figure 3. Location of tensiometer

2.1.1. Apparatus used for in situ test. Jet fill tensiometer was used in this study to measure the soil suction. The reading of suction was taken from dial gauge at tensiometer daily from 9 a.m. to 10 a.m. If the soil is too dried, water is added and make sure the water added is enough for to saturate the soil completely. If the soil had completely saturated, the gauge reading on tensiometer will drop to zero. Tensiometer used in the study was shown in Figure 4. Rain gauge is an instrument to collect rainfall. The decision of rain gauge installation location was due to the suitable open area, offset from buildings and free form above obstacles (trees, roof, etc.). This is because the overhead will cause the inaccuracy of the measurement. The reading was recorded daily. The example of the rain gauge is show in Figure 5.

Figure 4. Tensiometer

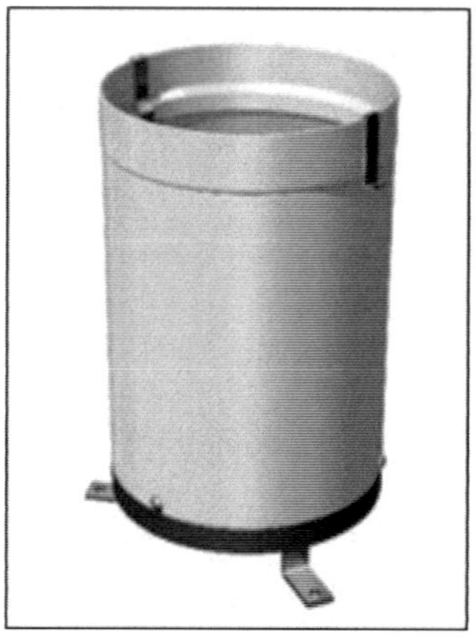

Figure 5. Rain gauge

2.2. Laboratory test

The laboratory test is carry out to measure the soil properties. Some of the soil tests measure direct properties of the soil, while others measure "index properties" which provide useful information about the soil without directly measuring the property desired. Five (5) laboratory soil testing were performed via particle size distribution, Atterberg limit, standard proctor, unconsolidated undrained triaxial and consolidated drained triaxial (see Table 1). Unified Soil Classification (USCS) with American Society for Testing and Materials (ASTM) standards was used to classify the soil type at slope studied.

Table 1. Laboratory test

No.	Type of Laboratory Test	Function	Standard
1	Particle Size Distribution (PSD)	To determine the size range of the soil and the apparatus used for this test is sieve shaker	ASTM
2	Atterberg Limit Test (AL)	To categorised soil into types and provide their engineering properties such as permeability, strength and compressibility	ASTM D41318
3	Standard Proctor Test	To determine the optimum moisture and maximum dry unit weight of the sample	ASTM 698
4	Unconsolidated Undrained Triaxial Test (UU)	To obtain shear strength parameters for a variety of soil types under undrained soil	ASTM D2850
5	Consolidated Drained Triaxial Test (CD)	To obtain shear strength parameters for a variety of soil types under drained soil	ASTM D7181

2.3. Data analysis

2.3.1. Stability analysis method: SLOPE/W. Geo-Studio Software or SLOPE/W (2004) was performed to determine the stability of soil slope with input parameters from Mohr Coulomb in site investigation (SI) report. The soil input parameters used are cohesion c' and effective friction angle Φ'. Particular research, ordinary method is used in analyse the slope with the direction of movement, left to right and slip surface option which is entry and exit. Figure 6 shows the SLOPE/W analysis of the study area.

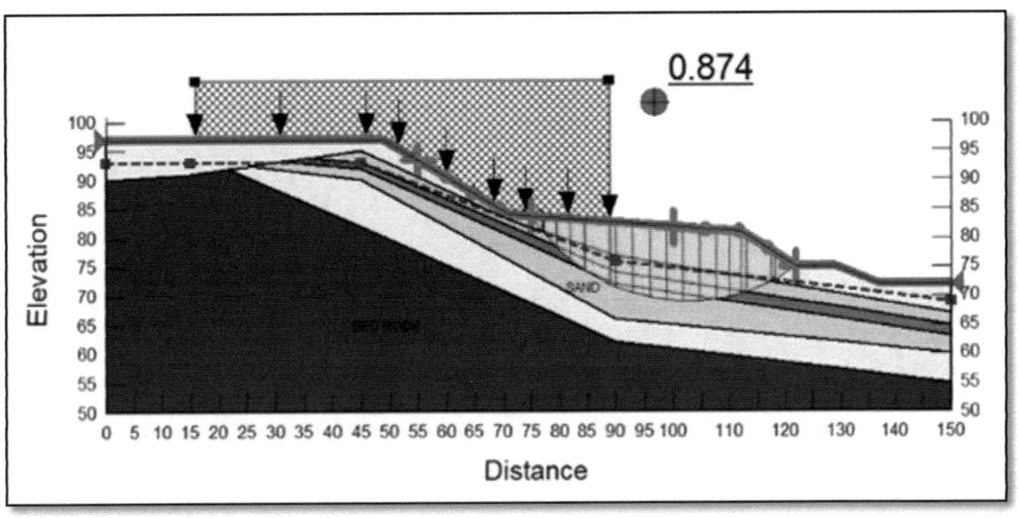

Figure 6. SLOPE/W analysis

2.4. Stability analysis: Fellenius Method

The parameters (cohesion c', effective friction angle Φ', suction and etc.) carried out from laboratory test was calculated using in the formula with the slices selected in each part of the slope studied. The Fellenius' method of Slices was calculated via Microsoft Excel with the formula in table to define the factor of safety (FOS) at each part of slope. The Fellenius' Method formula was shown below:

$$F = \frac{\sum c'l + (W cos\alpha - ul) tan\varphi'}{\sum W sin\alpha} \quad (1)$$

3. Results and Analysis

3.1. Consolidated drained triaxial test (CD)

Three (3) specimens from trial pits sample were used to determine the shear parameters which useful for input parameters for Geo-Slope (SLOPE/W) software analysis.

Table 2. Summary of CD's result for TP-1

Shear Strength	
Cohesion, c_u	Angle Resistance (°)
C' = 9 kPa	ϕ' = 32°

Table 3. Summary of CD's result for TP-2

Shear Strength	
Cohesion, c_u	Angle Resistance (°)
C' = 8 kPa	ϕ' = 35°

Table 2 and 3 shows the result of (CD) at the study area that were used in analysis for slope stability.

3.2. Relationship between rainfall and suction

Rainfall intensity has influence on the potential failure of slope due to the water content and shear stress developed along the surface. The result and relationship between two parameters are shown in Table 4.

According to Table 4, it was found that the suction value decreased due to increase of water content in soil during the high rainfall and number of precipitation.

Table 4. Relationship between rainfall and suction at different location

Slope	Relationship between Rainfall and Suction
Top	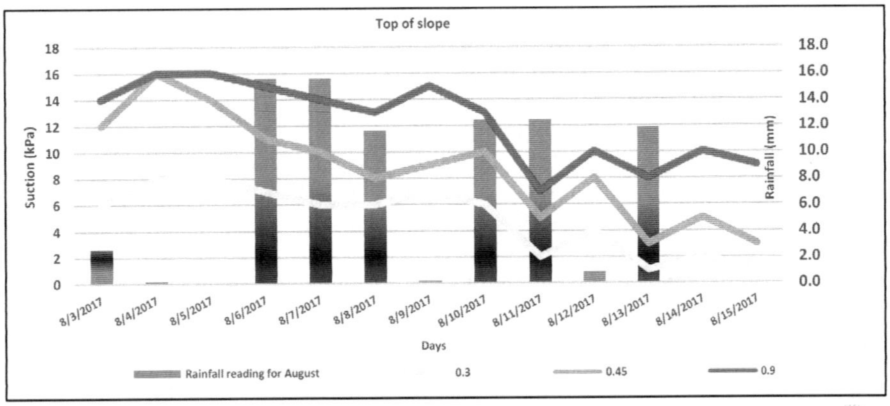
Middle	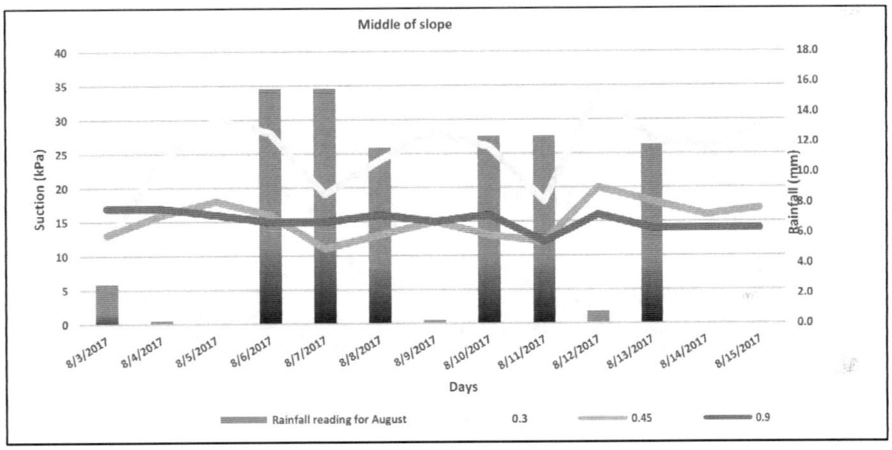
Toe	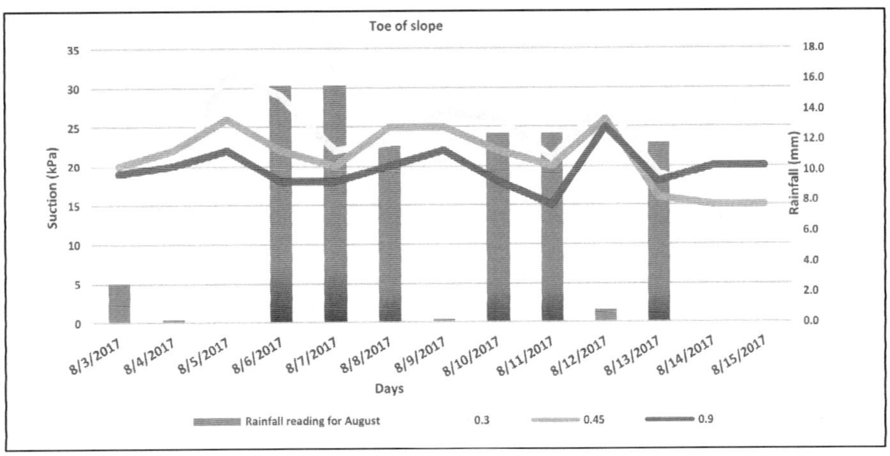

3.3. Slope stability analysis

Table 5 shows the soil properties for slope design which all the parameters stated in Table 5 were used to analyse the slope stability in SLOPE/W software as shown in Figure 7.

Table 5. Soil properties for slope design

Parameters Adopted in Slope Stability Analysis				
Region	Slope Description	Unit Weight (Kn/M^3)	Cohesion,C' (Kpa)	Friction Angle, Φ (°)
1	Sandy CLAY	19.4	9	32
2	Sandy SILT	22.2	9	34
3	Silty SAND	19.9	8	35
4	GRANITE (Bedrock)		-	

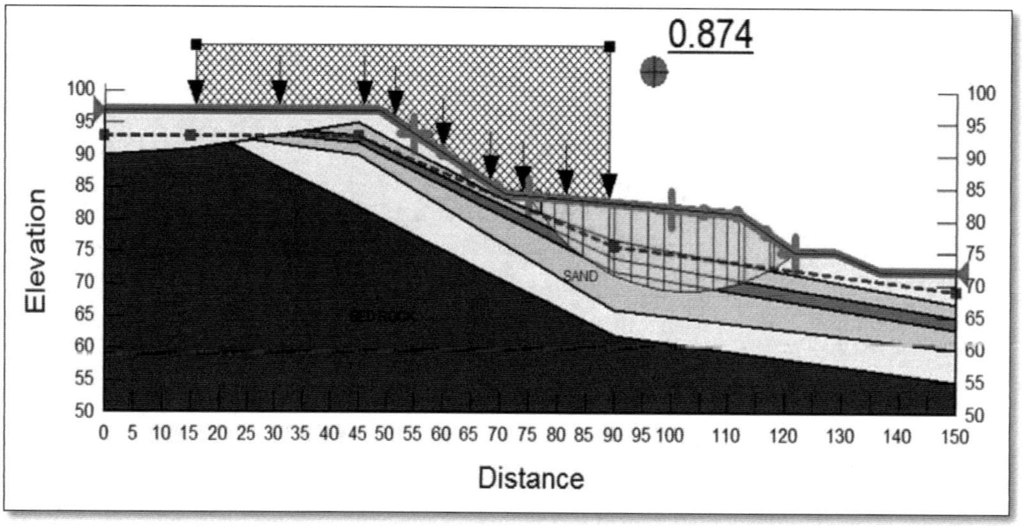

Figure 7. Analysis result of critical FOS with building surcharged on toe of slope

3.4. Comparison value of factor of safety from slope/w and Microsoft Excel

Table 6. Comparison value of factor of safety

Slope	Factor of Safety		Percentage Different (%)
	Slope/W	Excel	
Top	2.04	2.4	15
Middle	1.963	2.65	25.92
Toe	0.874	2.3	49.62

From the analysis by using Slope/w software, on top of slope, the factor of safety on the top of the slope is higher than factor of safety at middle and toe of slope. It was found that slope was not stable due to the factor of safety value of less than 1.3 at toe of the slope studied. This may happen Due to the seepage force in the sloping alignment together with the gravity force thus promoting to the slope instability. The factor of safety at middle of slope was determine by 1.963 via Slope/W analysis. Since at the middle of slope present a building structure the factor of safety may higher than expected factor of safety due to the disturbance of soil structure via earthwork (cut, fill and compaction of soil).

Factor of safety at middle of slope which calculate by Fellenius' Method is higher than top and toe of slope due to the surcharge load. Since the existing of the present buildings, the stability should higher but the soil at middle of slope may influences by the instability of soil at toe of slope. Furthermore, the building had cracked and the road at toe of slope collapsed. The suction value had assumed as 0 to define the factor of safety where suction of soil is zero.

According to Table 6, there is a huge different between these two methods. The result may occurred due to the influence of soil layer. The different layer of soil had substituted into software to analyse the factor of safety. As a result, the analysis has contained different type of soil but Fellenius' method used slices method to analyse the factor of safety. Every slice may contain different type of soil but in Table 6 every slice can only be considered having one type of soil despite of every slice had different type of soil. It was found that the different soil type has shown different shear strength properties which had affected the factor of safety of the slope studied.

3.5. Relationship between factor of safety, suction and rainfall

Factor of safety (FOS) is directly proportional to suction of soil. Table 7 shows the graph of FOS calculated with suction of soil in specific slices against suction at toe of slope that had been plotted to show the relationship between factor of safety, suction and rainfall.

It was found that the trend line for suction and FOS at toe of slope is most critical compared to FOS at middle and top of slope shown in Table 7. The trend line for suction and FOS in these three graphs are having same extent. On 11th August, the volume of the rainfall is same as 10th August for three locations and the FOS and Suction had dropped gradually. While on 12th August, value of FOS and suction at toe of slope had rose rapidly compared to FOS and suction at middle and top of slope. Different location and depth of soil had different suction and FOS due to infiltrate of rainwater with different type of soil and groundwater. In conclusion, the rainfall had influenced the changes of suction and FOS. The suction of soil and FOS are interrelated affect each other.

Table 7. Relationship between factor of safety, suction and rainfall at three different location with 0.9m depth

<table>
<tr><th>Slope</th><th>Relationship Between Three Parameters at 0.9m Depth from Ground</th></tr>
<tr><td>Top</td><td>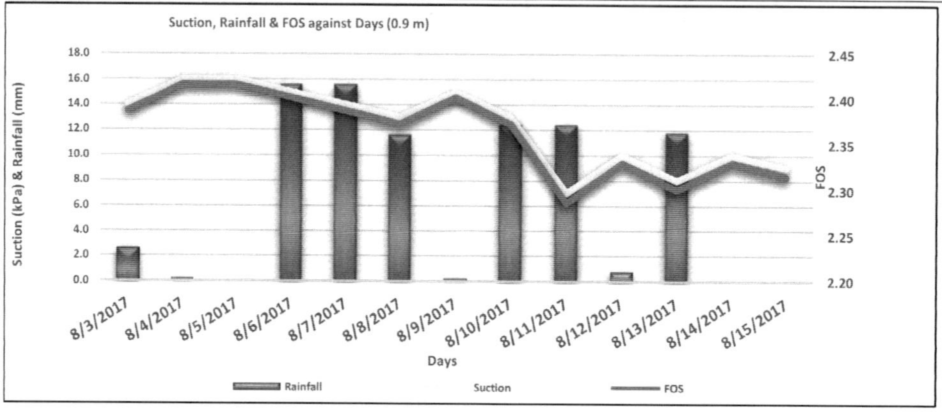
</td></tr>
<tr><td>Middle</td><td>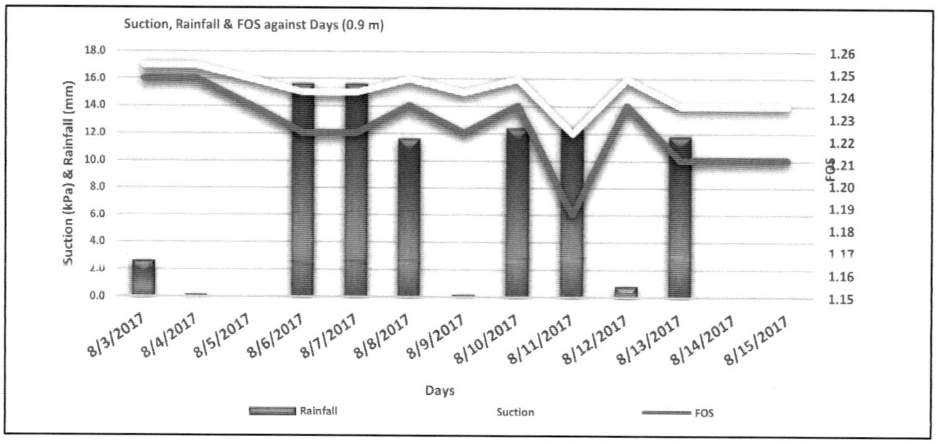
</td></tr>
<tr><td>Toe</td><td>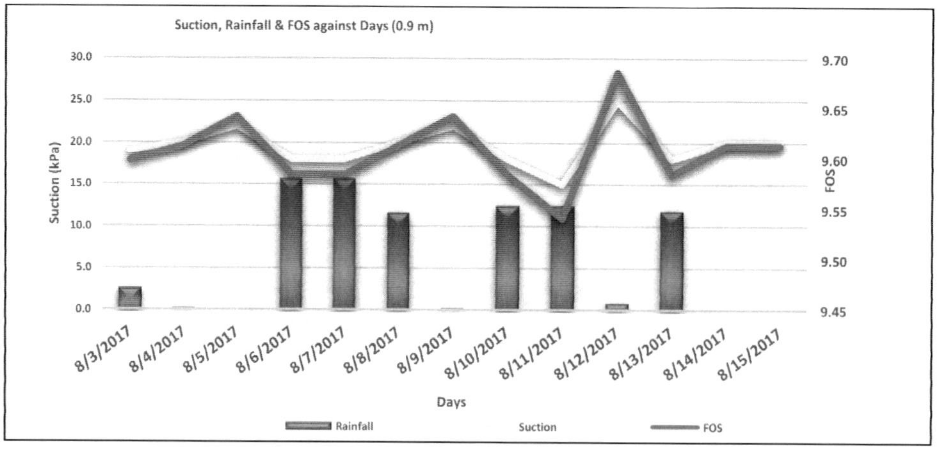
</td></tr>
</table>

IOP Conf. Series: Materials Science and Engineering **669** (2019) 012004 doi:10.1088/1757-899X/669/1/012004

4. Conclusion

The paper had shown the relationship between suction, rainfall and FOS. FOS was determined using SLOPE/W and Fellenius' Method with Microsoft Excel formula. Matric suction was influenced by rainfall because when rainfall intensity increased, matric suction of soil decreased. Factor of safety is interrelated with rainfall and suction of soil. Suction of soil increased and thus influence the increasing of the factor of safety. The factor of safety at specific slice was analysed using the suction which affected by rainfall. Finally, the changes of rainfall intensity caused the factor of safety to reduce, the lower factor of safety had caused the instability of slope. Hence, the constructed slope was suggested to be redesign to improve its safety factor.

References

[1] Gofar N, Lee M and Kassim A 2008 *Response of suction distribution to rainfall infiltration in soil slope* J. Geotech. Eng., 1-13
[2] Mahmood K, Kim J and Ashraf M 2016 *The Effect of Soil Type on Matric Suction and Stability of Unsaturated Slope under Uniform Rainfall* Journal of Civil Engineering, 1-6
[3] Tarantino A and Mongiovi L 2003 *Calibration of tensiometer for direct measurement of matric suction* Geotechnique, 137-141.
[4] Varsei M, Bourasset C and Miller G 2013 *The effects of soil suction on shallow slope stability* (Midwest City: The Oklahoma Transportation Center)
[5] ASTM
[6] ASTM D41318
[7] ASTM 698
[8] ASTM D2850
[9] ASTM D7181
[10] USCS

Acknowledgement

The authors would like to thanks University Malaysia Pahang, Malaysia by providing the funds for this project.

IOP Conf. Series: Materials Science and Engineering **669** (2019) 012005 doi:10.1088/1757-899X/669/1/012005

Forensic assessment on landfills leachate through electrical resistivity imaging at Simpang Renggam in Johor, Malaysia

M S I Zaini[1], M F Ishak[1] and M F Zolkepli[1]

[1] Faculty of Engineering Technology, Universiti Malaysia Pahang, 26300 Kuantan, Pahang, Malaysia

fakhrurrazi@ump.edu.my

Abstract. From years to years, the usage of electrical resistivity imaging (ERI) method dominated by geophysicist has increased tremendously in the application of geotechnical engineering owing to the effectiveness of the method in term of time, cost and also data coverage. The forensic assessment performed with respect to the particular reference to ERI in assessing the landfills leachate at Simpang Renggam, Malaysia. There were two lines of ERI performed at the study area using ABEM Terrameter LS 2 set of equipment based on Schlumberger protocol. Furthermore, the analysis made based on the electrical resistivity anomaly managed to detect the presence of chemical contaminants in the soil with particular reference to the chemicals resistivity values which was associated to low resistivity anomaly. Besides, the heterogeneous of the subsurface material presented in the paper by using integrated ERI analysis enabled forensic assessment of the leachate. The leachate from the landfills was believed to occur owing to the failure of the systems in accommodating and treating the waste which became worst with the present of heavy rainfall. In retrospect, the ERI result was applicable to be adopted in detecting the leachate and thus can assist the authority in taking immediate measure to prevent the frequent water disruptions at the study area.

1. Introduction

Over the past few years, the consequences and frequency related to the human activities and natural event around the whole world has increased rapidly with continuous trend. In engineering, forensic assessment is the investigation of structures, products, materials or breakdown components which may lead to property damage and cause injury [1]. As were already known, forensic engineering in geotechnical field is growing in most of the countries where the soil failures may lead to criminal action and even litigation [1, 2]. Related to the study, in most of the cases, some countries may have different soils but similar patterns of rainfall or some countries may have different rainfall but similar soils [3]. As stated by [4, 5], the main source of environmental problems is from the development of residential areas which resulted to the high production of domestic waste through daily activities. Lack of systematically efficient systems in developing countries to prevent the solid waste disposal problem leading to the open dumping areas which were considered as a cost-effective and speedy solution. The soil, surface water and lastly to the groundwater which are the areas normally surrounding the waste disposal sites will be polluted through the open dumping activities [4, 6].

The next point that need to be highlighted is solid waste are the production of the daily basis as a result of unescapable human activities. The increment volume of solid waste all over the world despite of current level of industrialization and global technological advancement occurred owing to the intensity of human's activities. Dumpsite or landfill acts as the important recipient of municipal solid waste. Theoretically, it is a common waste management employed by most of the nations in any parts of the world [7, 8]. Besides, the waste dump in this process causes a variety of public health and aesthetic problems which will also attracts rodents, insects and various disease vectors [7, 9, 10]. In this dumping process, the solid waste undergoes slow, anaerobic decomposition over a period of time and generate large amount of leachate with decomposing products such as heavy metals, leachate, landfill gas and varieties of hazardous contaminants which may seep into underground aquifers from the landfill site thus polluting the urban water resources [7, 11].

Based on [4] leachate is delineated as any liquid that is contaminated or wastewater which is produced from rain water percolating through accumulating contaminants, solid waste materials and moving into subsurface and surrounding areas. Owing to the presence of dissolved salts and has higher conductivity, leachate is formed which is in the liquid state from the decomposed waste [7]. Moreover, as stated by [12], leachate is an influence of surface run-off, precipitation and intrusion or infiltration of groundwater percolating through the landfill, inherent water content of wastes and biochemical processes in wastes cells themselves which is produced when liquid or water comes into contact with the wastes. The landfill leachate is a serious issue pointed out due to the continuous inflow of the leachate into subsurface materials [13]. Next, [13] also stated that the process of diffusion of the leachate occurs quickly and is extensive due to the increasing value in the groundwater flux following the heavy rain. On the other side of discussion, the landfill leachate possibly results in a groundwater contamination and the supply of surface water which can affects human health when migrating into neighbouring lands and waters occurring from the landfill [12].

Additionally, when dealing with the leachate analysis, certain number of complicated factors should be taken into account such as solid waste composition, operation of the landfill, age of the waste, hydrogeological conditions in vicinity of the site, temperature, moisture content, rate of the water movement through the waste, biological activities, landfill chemicals and seasonal weather variations which varies for four seasons country [12], [14-17]. The method of treatment and characterisation of landfill leachate have been developed for the past 40 years [7, 12], [18-21] and various techniques can be applied to describe the properties of the landfill, regional contamination from offsite seepage or leachate circulation [22].

Compare to groundwater resistivity, the leachate electrical resistivity is much lower and contaminants and pollutants released into the environment barely remains at the point of discharge. Most of them are transported via the porous media by four mechanisms such as molecular diffusion, advection, adsorption and mechanical dispersion [22-23]. Adsorption which determine the chemicals contribution in the soil water environment and affect the fate of chemicals in soils is one of the most influential transport mechanisms [22, 24]. The groundwater was polluted owing to the infiltration of contaminants and percolation of fluvial water via the soil in waste disposal sites [22, 25] where the leachate of the landfill is associated with high ion concentrations which resulted to low value of resistivity of the rock formations containing them [22, 26]. Electromagnetic and electric geophysical methods such as ERI have been broadly used for investigating and exploring various works especially in investigating the groundwater [7, 28], geological [19, 29-31] and environmental features [32, 33-36] Therefore, geoelectrical techniques or mostly known as geophysical methods are the most suitable for mapping the leachate contamination around the landfills [22, 27].

Geophysical methods play an important role in finding the hidden subsurface of a soil accurately and adequately without bore holing or drilling [37]. This method proves the fact that the electrical resistivity (reciprocal of electrical conductivity) of the landfill leachate is much lower than that of a clean groundwater by which small changes are detectable owing to the high data resolution [38-40] and variations in ERI may demonstrate changes in composition contaminant or layer levels [41]. ERI or Electrical resistivity tomography (ERT) is recognized [43] to be one of the best tool for site characterization at disposal site owing to the pollutant of the leachate which are rich in soluble ions that will facilitate the flow of an electrical current hence lowering the resistivity of the medium [44]. Last but not least, geophysical investigation offers a faster, cheaper [42] and has the ability to cover greater areas more thoroughly [45-46] by getting credible and detailed information about the subsurface [42]. Various researchers from all over the world especially in geophysics and geotechnical field have exposed that the integration of geotechnical data and geophysical survey can provide a useful data and interpretation for subsurface profile characterization [47-50]. Hence, a forensic assessment is conducted on landfills by ERI at Simpang Renggam in Johor, Malaysia. Basically, this study was conducted aiming to bestow to those related parties on the good anticipation of ERI as an alternative method in forensic leachate of the landfill investigations and exploration.

2. Methodology

The ERI method is classified into the group of geoelectrical methods commonly used in geophysical engineering which measures the differences in electrical resistivity of the ground [17, 52-54].ERI method was chosen in this research owing to the abilities of the leachate ions to facilitate the current movement in the medium. The study involves three phases which are through desk study, field measurement and data processing. Desk study starts with gathering necessary information and data regarding the study area thru maps and journals in order to obtain early information such as topography, geology of the sites, etc. in global and localize scale as presented in subsection 2.1. Later, the field measurement was conducted by using ERI and finally raw data from the field measurement was analyzed and processed by using RES2DINV software as illustrated in subsection 2.2. In this method, subsurface potentials that was measured owing to the passage of direct current are utilized in calculating the apparent resistivity and later applied as inputs in inverse modeling to construct a subsurface true resistivity model [51]. It is more accurate when the 2D model of the subsurface having the resistivity changes in the vertical direction along the survey line [55] and the 2D measurements via ERI should be taken along profiles in chronological order [48].

2.1. Study area and geologic setting

The study area was located at Simpang Renggam landfill site which was located in Simpang Renggam, Kluang, Johor, with the coordinate of 1°53'41"N, 103°22'35"E as shown in Figure 1. The location of the study area is about two kilometres away from Simpang Renggam town. With a total of six hectares in size, the landfill receives 400-500 tonnes of wet solid wastes per day which is from three different areas such as Simpang Renggam, Batu Pahat and Kluang. Based on the previous research conducted by the othe researchers, it was claimed that the leachate at the landfill is severe and disturbed the nearest river which became the sources of water supply to the residents of Simpang Renggam.

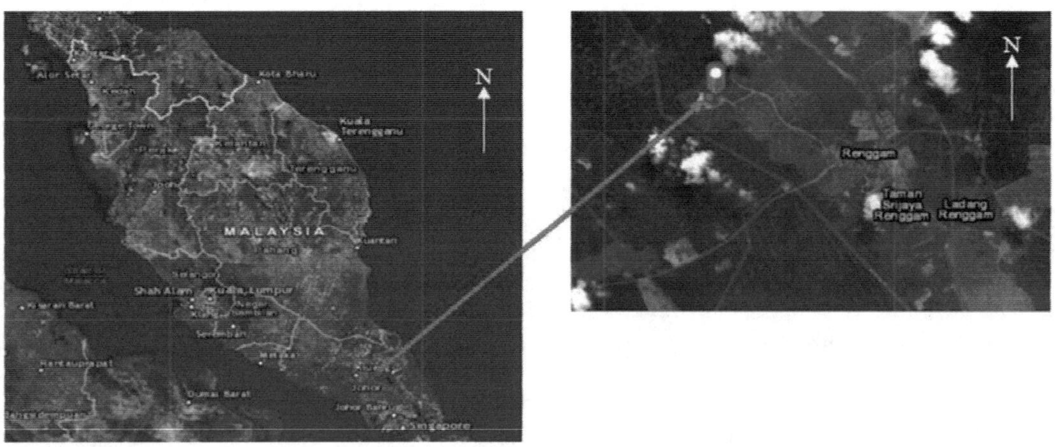

Figure 1. Study area

2.2. Equipment

The ERI equipment consists of three main components such as record, inducer and source [1]. The ERI source was set up by using a DC battery of 12 volt. A 61 of steel electrode was used as a current inducer and ABEM Terrameter LS 2 was used to record and plot the apparent resistivity value and finally the raw data obtained from the field measurement was analysed and clarified by using RES2DINV software.

2.3. Data acquisition and processing

By using ABEM Terrameter LS 2 equipment set, the ERI was conducted with a total number of two (2) spread lines of ERI survey were conducted (Resistivity Line, RL1 and Resistivity Line 2, RL2) at the study area as shown in Figure 2. The test configuration of the equipment was conducted based on the Schlumberger array by using four (4) resistivity cables, sixty one (61) numbers of electrode and sixty four (64) numbers of jumper cable. Electrode spacing and total electrical survey length are illustrated in Table 1. The field arrangement of the ERI was illustrated in Figure 3. Besides, the Schlumberger array was used during the data acquisition owing to the ability of the protocol to provide dense near-surface data of the resistivity.

Table 1. Electrode spacing and total electrical resistivity length

Resistivity Cable	Electrode Spacing (m)	Takeout Number	Total Length (m)
Cable 1 (C1)	5.0	20	100
Cable 2 (C2)	2.5	20	50
Cable 3 (C3)	2.5	20	50
Cable 4 (C4)	5.0	20	100
Total ERI Length (m)			**300**

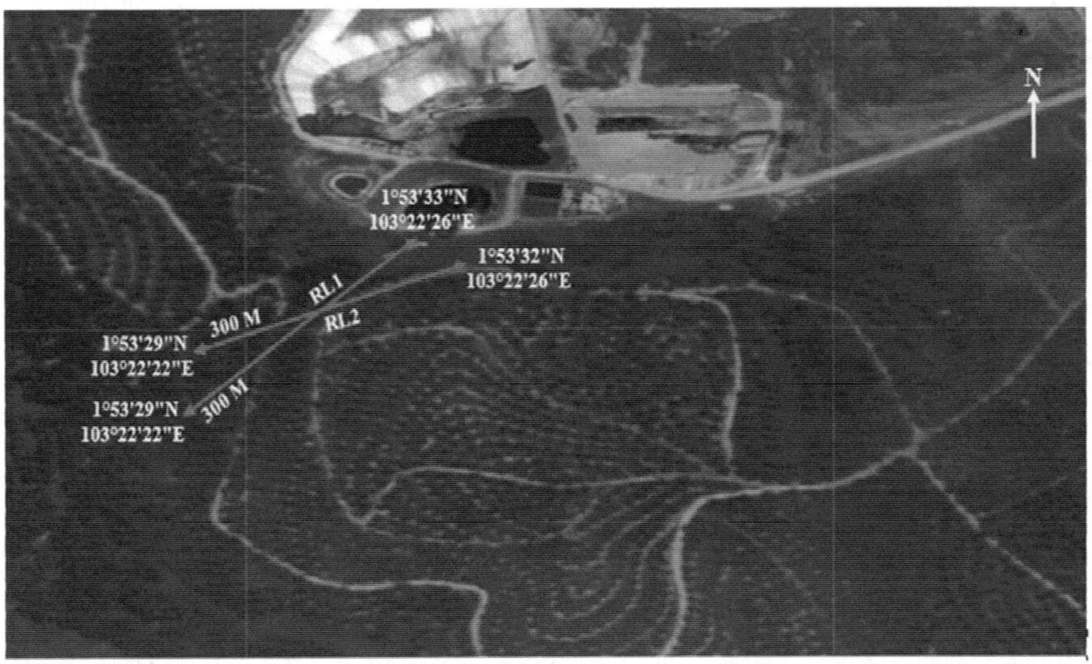

Figure 2. Entire location of resistivity lines (spread lines) conducted at Simpang Renggam Landfill, Johor.

As exposed by [57], the Schlumberger array can give a clear image and can provide a good vertical resolution which is suitable to be used to detect the leachate of the landfill. Additionally, raw data obtained from the data acquisition were being processed by using commercialize RES2DINV software of [51] to present an inverse model that closely shown the actual subsurface structure. From the input resistivity data which is by using the inversion code of RES2DINV, the 2D inverse resistivity models of the subsurface was obtained and the RES2DINV can automatically determines the 2D resistivity models from the resistivity data input by using smoothness constrained least squares method [53-57]. Figure 2 shows the location of the entire resistivity lines (spread lines) conducted at Simpang Renggam landfill, Johor.

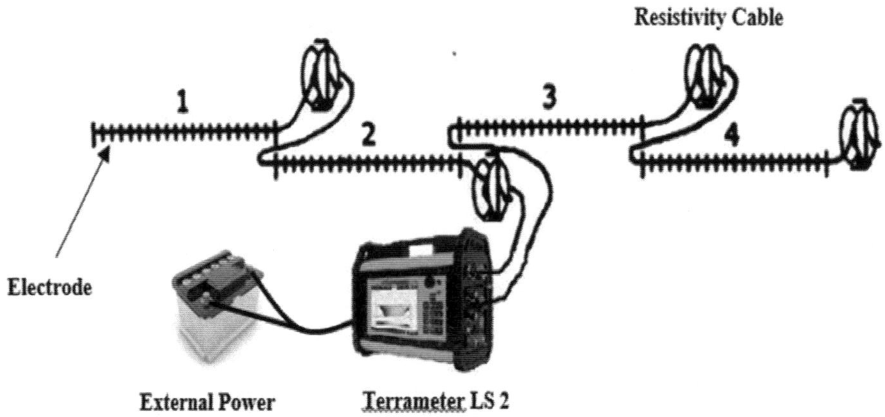

Figure 3. Four cables configuration of field arrangement in electrical resistivity imaging (ERI).

3. Results and Discussions

A total of two (2) resistivity spread lines were modelled by using RES2DINV software as shown in Figure 2. At the study area, the lines are randomly located occupying the side part of the landfill as well as certain part of the nearest oil palm plantation. The interpretation of the inverse resistivity models are basically based on the resistivity values of the mixture between leachate with clay, clayey sand and sandy clay experimentally measured by previous researcher in a small scale laboratory studies. Generally, the range of resistivity values observed and recorded in both of the of the inverse models differ from 1.0 Ωm to about 120 Ωm which describe the contaminated (less than 10 Ωm) or uncontaminated (more than 10 Ωm) soil and groundwater from the surface to a maximum depth approximately about 34 m in the ground.

Basically, the resistivity distributions can be grouped into three (3) different major zones interrelated with high resistivity values which is between 50 Ωm to 100 Ωm that represent the decomposed waste or waste pile covered with soil at the depth of 0-2.5 m. Noted that, in the areas that is away from the landfill which is at resistivity line 2 (RL2) located in the palm oil plantation, the high resistivity top layer can be correlated to the dry soil situated above the water table. Next, a zone with much lower resistivity value was observed with a range of value of 1 Ωm to 10 Ωm at a depth of 0-5 m below the surface coincide to sandy clay and clay soils based on [53].

Figure 4 and Figure 6 shows the inverse model of resistivity of RL 1 and RL 2 respectively which is in sync with the graph of resistivity (Ωm) versus distance (m) at a certain depth of 0.5 m, 4.5 m, 2.5m, 4.0 m and 5.0 m as shown in Figure 5 and Figure 7 respectively. The dotted line in Figure 4 and Figure 6 represents the depth of the study for the leachate at the landfill which is starting from 0.5 m until 5.0 m depth. Besides, high leachate contents is always described by a much lower resistivity values as stated by previous researcher which is lower than 10 Ωm. Therefore, the high concentrated leachate zone could barely detected in the inverse model from both of the survey lines conducted in clay areas where both the leachate and clay have quite similarity in resistivity values. Based on the resistivity value obtained in RES2DINV along RL1, it can be observed that there is a presence of the leachate with resistivity values around 2-9.5 Ωm in resistivity models of profiles RL1 measured across some parts of the active landfill towards the southwest borders of the oil palm plantation. The leachate was detected at a distance of 86.0 m to 91.0 m and at the depth of 1.5 m to 4.0 m. Figure 5 shows the resistivity value plotted against the distance of the ERI survey by taking the depth of exploration into consideration for RL1. The graph (Figure 5) was correlated with the inverse model plotted with RES2DINV software as shown in Figure 4.

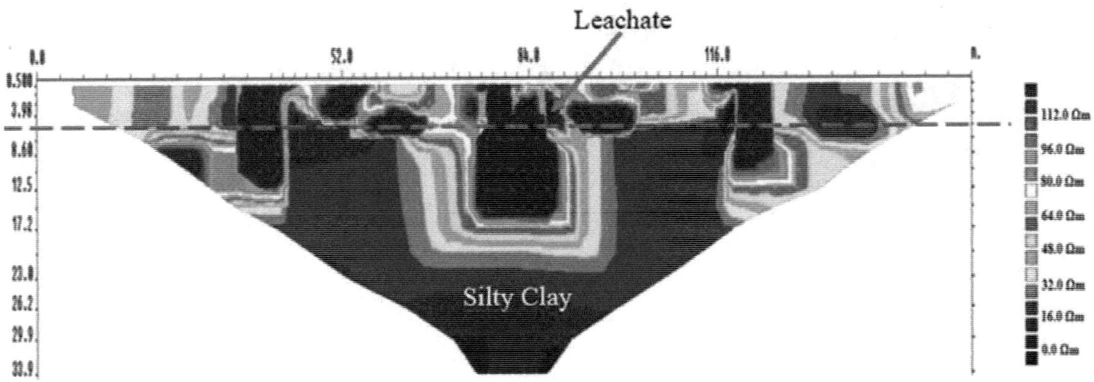

Figure 4. The inverse model of resistivity for Resistivity Line 1 (RL 1).

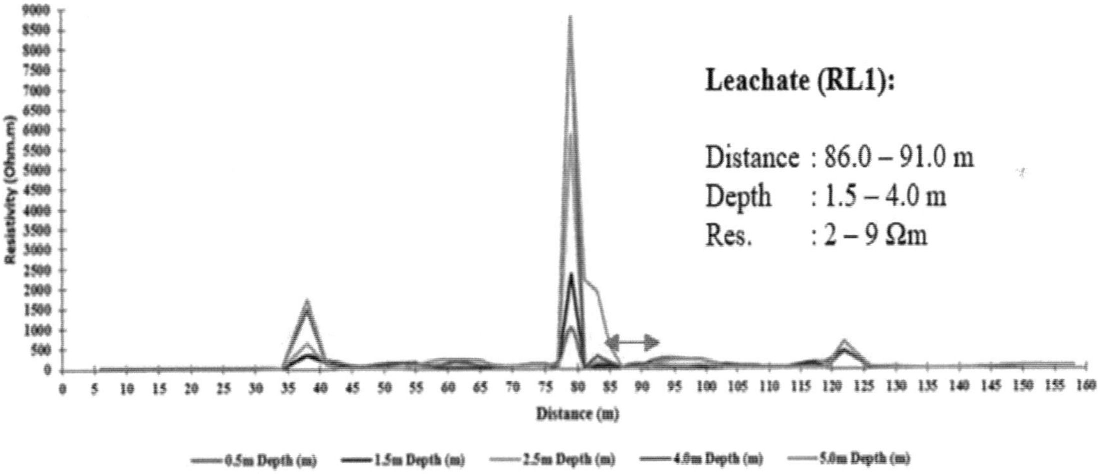

Figure 5. Graph of resistivity (Ωm) versus distance (m) at a certain depth of 0.5 m, 4.5 m, 2.5m, 4.0 m and 5.0 m for Resistivity Line 1 (RL 1).

One line (RL2) of resistivity imaging was carried out outside of the landfill area which is situated in the oil palm plantation. Based on the resistivity value obtained in RES2DINV along RL2, it can be observed that there is a presence of the leachate with resistivity values around 1.5-9.5 Ωm in resistivity models of profiles RL2 measured across outside of the active landfill towards the northeast borders of the landfills. The leachate was detected at a distance of 87.0 m to 89.0 m and at the depth of 1.5 m to 5.0 m. RL2 was carried out near the drainage which the water from the landfill was discharge. Figure 6 shows the leachate that was detected in the inverse model of the resistivity of RL2. Figure 7 shows the resistivity value plotted against the distance of the ERI survey by taking the depth of exploration into consideration for RL2. The graph (Figure 7) was correlated with the inverse model plotted with RES2DINV software as shown in Figure 6.

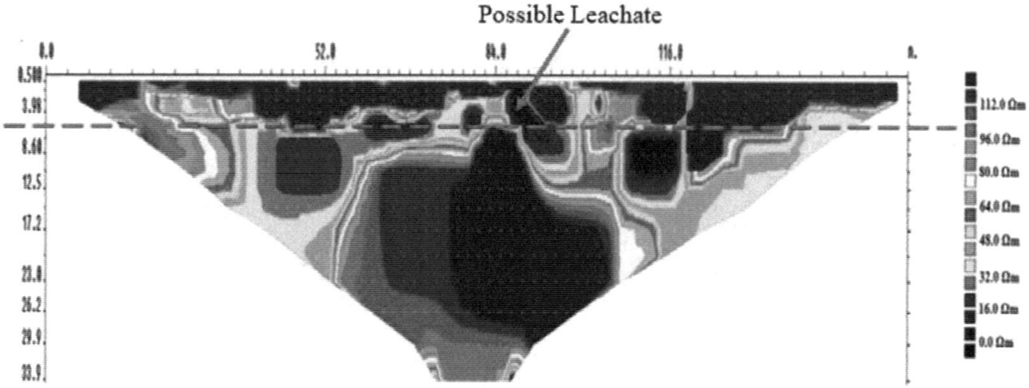

Figure 6. The inverse model of resistivity for Resistivity Line 2 (RL 2).

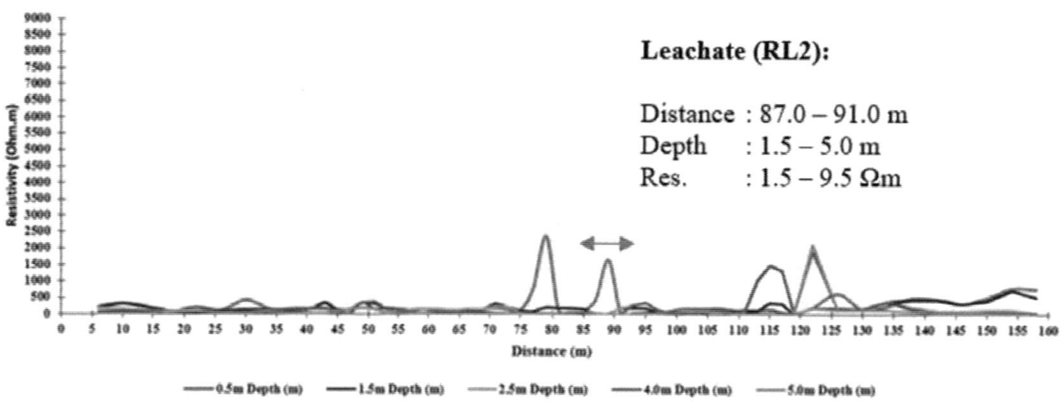

Figure 7. Graph of resistivity (Ωm) versus distance (m) at a certain depth of 0.5 m, 4.5 m, 2.5m, 4.0 m and 5.0 m for Resistivity Line 2 (RL 2).

4. Conclusions

The 2D electrical resistivity imaging conducted at Simpang Renggam, Johor managed to prove the existing and the leachate movement from the landfill into the surrounding of soil and water. Based on the analysis conducted the leachate also affected the nearby oil palm plantation which was situated southwest of the landfills area. Generally, the maximum leachate infiltration was detected at 1.5 to 4.0 m depth vertically below the surface of the soil and 80 m laterally away from it. From 1 to 10 Ωm of the resistivity values observed which described the most contaminated zone in both of the inverse model of resistivity (RL1 and RL2). Finally, by detecting the presence of the leachate under the ground, it can be confirmed that the ERI technique or 2D electrical resistivity imaging is among the effective technique for detecting the presence of the leachate.

References

[1] Abidin M H Z, Madun A, Tajudin S A A and Ishak M F 2017 *Forensic Assessment on Near Surface Landslide Using Electrical Resistivity Imaging (ERI) at Kenyir Lake Area in Terengganu,Malaysia* In Procedia Engineering. https://doi.org/10.1016/j.proeng.2017.01.354

[2] Poulos H G 2008 *A Framework for Forensic Foundation Engineering*, in: *Proc.* International Conference on Forensic Engineering: From Failure to Understanding, London, pp. 247.

[3] Nordiana M M, Azwin I N, Nawawi M N M and Khalil A E 2017 *Slope failures evaluation and*

landslides investigation using 2-D resistivity method. NRIAG Journal of Astronomy and Geophysics. https://doi.org/10.1016/j.nrjag.2017.12.003

[4] Hamzah U, Jeeva M and Ali M N A 2014. *Electrical resistivity techniques and chemical analysis in the study of leachate migration at Sungai SEDU landfill.* Asian Journal of Applied Sciences. https://doi.org/10.3923/ajaps.2014.518.535

[5] Keller E A 2000 *Environmental Geology.* 5th Edn., Merrill Publishing Co., Ohio, USA.

[6] Martinho E and Almeida F 2006 *3D behaviour of contamination in landfill sites using 2D resistivity/IP imaging: Case studies in Portugal.* Environmental Geology. https://doi.org/10.1007/s00254-005-0151-7

[7] Ganiyu S A, Badmus B S, Oladunjoye M A, Aizebeokhai A P, Ozebo V C, Idowu O A, and Olurin O T 2016 *Assessment of groundwater contamination around active dumpsite in Ibadan southwestern Nigeria using integrated electrical resistivity and hydrochemical methods* Environmental Earth Sciences. https://doi.org/10.1007/s12665-016-5463-2

[8] Singh B J and S 2009 *CHARACTERISTICS OF LEACHATE FROM BHALASWA LANDFILL SITE, AND GROUNDWATER SAMPLES FROM NEARBY LOCATIONS* Concentration in groundwater sample at radial distance from landfill facility. World Academy of Science, Engineering and Technology International Journal of Civil and Environmental Engineering.

[9] Olayinka A I and Olayinwola M A 2001 *An integrated use of geoelectric Imaging and hydrogeochemical methods in delineating limits of polluted surface and groundwater at a landfill site in Ibadan area, Southwestern, Nigeria.* J. of Mining and Geology, Vol 37 (1): pp 53-68.

[10] Adeyami O, Oloyede O B and Oladiji A T 2007 *Physicochemical and microbial characteristics of leachate- contaminated groundwater.* Asian. J. Biochem, 2: 343- 348.

[11] Bhuiya M A H, Huq N E and Hossain M M 2002 *Unplanned waste disposal and its possible impacts on subsurface environment of Dhaka City, Bangladesh.* In Bangladesh Environment 2002. Vol 2(ed) pp 7231- 7331.

[12] Giang N V, Kochanek K, Vu N T and Duan N B 2018 *Landfill leachate assessment by hydrological and geophysical data: case study NamSon, Hanoi, Vietnam.* Journal of Material Cycles and Waste Management. https://doi.org/10.1007/s10163-018-0732-7

[13] Park S, Yi M J, Kim J H and Shin S W 2016 *Electrical resistivity imaging (ERI) monitoring for groundwater contamination in an uncontrolled landfill, South Korea.* Journal of Applied Geophysics. https://doi.org/10.1016/j.jappgeo.2016.07.004

[14] Matias M S, Marques da Silva M, Ferreira P and Ramalho E 1994 *A geophysical and hydrogeological study of aquifers contamination by a landfill.* J Appl Geophys 32: 155- 162.

[15] Abudeif A M 2015 *Integrated electrical tomography and hydrochemical analysis for environmental assessment of El-Dair waste disposal site, west of Sohag city.* Egypt Environ Earth Sci 74(7): 5859- 5874. https://doi.org/10.1007/s12665-015-4610-5

[16] Bhalla G, Kumar A and Bansal A 2011 *Assessment of groundwater pollution near municipal solid waste landfill.* Asian J Water Environment Pollut 8(1): 41- 51.

[17] Sharma P V 1997 *Environmental and engineering geophysics Cambridge University Press*, Cambridge.

[18] Hoang V Y, Vasel J L and Le V C 2009 *Application of data analysis and experimental planning method to study partial nitrification in treatment of ammonium in landfill leachate in Vietnam.* 20th international waste management and landfill symposium, Sardinia, p 8.

[19] Chambers J C, Kuras O, Meldrum P I, Ogilvy R D and Hollands J 2006 *Electrical resistivity tomography applied to geologic, hydrogeologic, and engineering investigations at a former waste disposal site.* Geophysics 71(6): B231- B239.

[20] Khanh N H 2007 *Research and comparison of landfll leachate treatment system Vietnam and the world to choose suitable one to meet level of Vietnamese standard for landfll site NamSon in Hanoi*, report of the project, Institute of Environmental Technology, VAST.

[21] Paul T I, Debra R R, Marja E, Roger G, Nitin G B H, Sreeram J, Timothy G T and Ramin Y 2007

Review of state of the art methods for measuring water in landflls. Waste Manag 27:729–745.

[22] Giang N V, Duan N B, Thanh L N and Hida N 2013 *Geophysical techniques to aquifer locating and monitoring for industrial zones in North Hanoi*. Vietnam Acta Geophys 61(6):1573–1597. https://doi.org/10.2478/s11600-013-0147-8.

[23] Marylyn V Y 1990 *Modeling microbial transport in soil and groundwater.* Vol 56. No. 6.

[24] Kah M and Brown C D 2007 *Change in pesticides adsorption with time at high soil to solution ratio.* Chemisphere 68 1335-1343.

[25] Abdullahi N K, Osazuwa I B and Sule P O 2011 *Application of integrated Geophysical techniques in the investigation of groundwater contamination.* A case study of municipal solid waste leachate. Ozean Journal of Applied Science 4 (1), 7-25.

[26] Cristina P, Cristina D, Alicia F and Pamela B 2012 *Application of Geophysical methods to waste disposal studies, municipal and Industrial waste disposal.* Dr Xian-Ying Yu (Ed) ISBN 978-953-51-0501-5. In Tech.

[27] Bernstone C and Dahlin T 1999 *Assessment of two automated DC Resistivity Data acquisition systems to landfill location surveys.* Two case studies "Journal of Environmental and Engineering Geophysics. 4(2): 113-121.

[28] Gupta G, Patil J, Maiti S, Erram V C, Pawar N, Mahajan S and Suryawanshi R 2015 *Electrical resistivity imaging for aquifer mapping over Chikotra basin*, Kolhapur district, Maharashtra. Environ Earth Sci 73:8125–8143.

[29] Bernstone C, Dahlin T, Ohlsson T and Hogland W 2000 *DC-resistivity mapping of internal landfll structures: two pre-excavation surveys.* Environ Geol 39(3–4):360–368.

[30] Cardarelli E and Fischanger F 2006 *2D data modelling by electrical resistivity tomography for complex subsurface geology.* Geophys Prospect 54:121–133

[31] Kumar D 2012 *Efcacy of ERT technique in mapping shallow subsurface anomaly.* J Geol Soc India 80:304–307.

[32] Pantelis S, Nikos P, Ilias P, Maria K, Filippos V, Apostolos S and Thrassyvoulos M 2010 *Application of integrated methods in mapping waste disposal areas.* Environ Geol 53:661–675. https://doi.org/10.1007/s00254-007-0681-2.

[33] Guerin R, Begassat Ph, Benderitter Y, David J, Tabbagh A and Thiry M 2004 *Geophysical study of the industrial waste land in Mortagne-du-Nord (France) using electrical resistivity.* Near Surf Geophys 3:137–143.

[34] Porsani J L, Filhob W M, Vagner R E, Shimelesa F, Douradob J C and Moura H P 2004 *The use of GPR and VES in delineating a contamination plume in a landfll site: a case study in SE Brazil.* J Appl Geophys 55:199–209.

[35] Maillet G M, Rizzo E, Revil A and Vella C 2005 *High resolution electrical resistivity tomography (ERT) in a transition zone environment: application for detailed internal architecture and inflling processes study of a Rhône River paleo-channel.* Mar Geophys Res 6:317–328. https://doi.org/10.1007/s11001-005-3726-5.

[36] Martinho E and Almeida F 2006 *3D behaviour of contamination in landfll sites using 2D resistivity/IP imaging: case studies in Portugal.* Environ Geol 49:1071–1078.

[37] Sk M, Ramanujam N, Champoil V, Biswas S K, Rasool Q A and Ojha C 2018 *Identification of Groundwater in Hard Rock Terrain Using 2D Electrical Resistivity Tomography Imaging Technique: Securing Water Scarcity at the Time of Seasonal Rainfall Failure* South Andaman. International Journal of Geosciences, 9, 59-70. https://doi.org/10.4236/ijg.2018.91004.

[38] Buselli G and Lu K 2001 *Groundwater contamination monitoring with multichannel electrical and electromagnetic methods.* J. Appl. Geophys. 48, 11–23.

[39] Hagrey S A a and Petersen T 2011 *Numerical and experimental mapping of small root zones using optimized surface and borehole resistivity tomography*. Geophysics 76, G25–G35.

[40] Slater L and Sandberg S 2000 *Resistivity and induced polarization monitoring of salt transport under natural hydraulic gradients.* Geophysics 65, 408–420.

[41] Telford W M and Sheriff R F 1984 *Applied Geophysics. Cambridge University Press.*

[42] Osinowo O O and Falufosi M O 2018 *3D Electrical Resistivity Imaging (ERI) for subsurface evaluation in pre-engineering construction site investigation.* NRIAG Journal of Astronomy and Geophysics. https://doi.org/10.1016/j.nrjag.2018.07.001.

[43] Batayneh A T and Barjous M O 2005 *Resistivity surveys near a waste-disposal site in the Qasr Tuba area of central Jordan.* Bull. Eng. Geol. Environ., 64: 287-294.

[44] Aristodemou E and Betts A T 2000 *DC resistivity and induced polarisation investigations at a waste disposal site and its environments.* J. Applied Geophys., 44: 275-302.

[45] Hazreek Z A M, Rosli S, Chitral D C, Fauziah A, Azhar A T S, Aziman M and Ismail B 2015 *Soil Identification using Field Electrical Resistivity Method*, Journal of Physics: Conference Series. 622 1-7.

[46] Khatri R, Shrivastava V K and Chandak R 2011 *Correlation between vertical electric sounding and conventional methods of geotechnical site investigation*, Int. Journal of Advanced Engineering Sciences and Technologies. 4 042-053.

[47] Almalki H, El-Werr A K and Abdel-Rahman K 2011 *Estimation of near-surafce geotechnical parameters using seismic measurements at the proposed KACST expansion site*, Riyadh, KSA, Arabian Journal of Geosciences. 4 1131-1150.

[48] Sudha K, Israil M, Mittal S and Rai J 2009 *Soil characterization using electrical resistivity tomography and geotechnical investigations*, Journal of Applied Geophysics. 67 74-79.

[49] Oh S and Sun C G 2008 *Combined analysis of electrical resistivity and geotechnical SPT blow counts for the safety assessment of fill dam*, Environmental Geology. 54 (2008) 31-42.

[50] Hack R 2000 *Geophysics for Slope Stability*, Surveys in Geophysics. 21 423-448.

[51] Loke M H, Acworth I and Dahlin T 2003 *A comparison of smooth and blocky inversion methods in 2D electrical imaging surveys.* Explorat. Geophys., 34: 182-187.

[52] Keller G V and Frischknecht F C 1966 *Electrical methods in geophysical prospecting.* Pergamon Press Inc., Oxford.

[53] Loke M H and Barker R D 1996 *Rapid least squares inversion of apparent resistivity pseudosections by a quasi-Newton method.* Geophys Prospect 44:131–152.

[54] Smith R C and Sjogren D B 2006 *An evaluation of electrical resistivity imaging (ERI) in Quaternary sediments*, southern Alberta, Canada. Geosphere 2(6):287–298.

[55] Loke M H 2000 *RES2DINV version 3.44 for windows 95/98 and NT: rapid 2D resistivity and IP inversion using the least squares method.* Advanced Geosciences Inc., Austin.

[56] Sasaki Y 1992 *Resolution of resistivity tomography inferred from numerical simulation.* Geophys. Prospect, 40: 453-464.

[57] Hamzah U, Yaacup R, Samsudin A R and Ayub M S 2006 *Electrical imaging of the Groundwater Aquifer at Banting, Selangor, Malaysia*, Environmental Geology. 49 1156-1162.

Acknowledgements
The authors would like to thanks University Malaysia Pahang owing to the provided funding (RDU1703205) throughout this project.

Cracking behaviour of green lightweight concrete one-way slabs using medium K Basaltic Andesitic pumice and Scoria

H Suseno[1], A Soehardjono[1] and A Rachmansyah[1]

[1] Department of Civil Engineering, Faculty of Engineering University of Brawijaya Jl. M T Haryono 167 Malang 65145 Indonesia

hendros@ub.ac.id

Abstract. This paper investigates experimentally the cracking behaviour of green lightweight concrete one-way slabs where coarse aggregates comprised medium K basaltic andesitic pumice and scoria. Three points bending test was performed on 24 one-way slabs of both lightweight concrete with three compressive strengths and four reinforcement ratios, and 1 one-way slab of normal concrete as control. Responses consisted of crack width, mid-span reinforcement strain and mid-span deflection that were measured for observation purposes. To show the cracking behavior, their low tensile strengths were evaluated by representing their compressive strengths. Load versus crack width and reinforcement strain diagrams are simultaneously used to evaluate their cracking behaviors. Effects of reinforcement ratio, reinforcement stress, type of lightweight concrete and compressive strength to the maximum crack width are also investigated. Similarly, cracking patterns as well as failure modes are observed to show their appropriate performances. The results show that the reinforcement ratio and reinforcement stress are the main factors that affect significantly the cracking behavior and maximum crack width. However, the type of lightweight concrete and compressive strength affect them but less significantly. All one-way slabs exhibited a typical flexural crack, at the first time the reinforcement yields and then the concrete compressive zone crushes.

1. Introduction

One of the problems in a structural lightweight concrete is the use of artificial lightweight coarse aggregates that require high thermal energy to produce them [1] so that it does not support energy saving programs. In addition, during manufacturing it releases air pollution so that these products are less environmentally friendly. The utilization of volcanic lightweight coarse aggregates such as pumice and scoria are considered as one of the solutions to overcome the disadvantages of this lightweight concrete [2]. These natural aggregates are abundant and spread evenly in the existing check dams so it is easy to explore them. Several studies on pumice and scoria lightweight concretes were conducted previously [3-7] and the results fulfilled the requirements of structural lightweight concrete with the reduction of density is approximately 20 %. Thus, this typical lightweight concrete can be considered as green lightweight concretes that save energy, are more environmentally friendly and can be used for structural elements of the building.

Kelud Volcano is located in East Java Indonesia, it ejects simultaneously medium K basaltic andesitic pumice and scoria in certain period as part of the explosive eruption products. These vesicular rocks differ only in color however their chemical composition, mineralogy and texture are practically similar. Furthermore, their specific gravities are greater than one [8] so that they are directly submerged in water.

The study of these volcanic rocks as coarse aggregates on structural lightweight concrete was conducted by [9, 10] and the results showed that these aggregates were categorized as the lightweight coarse aggregates and had unique properties that differed from existing pumice and scoria. The production of these pumice and scoria structural lightweight concrete used PPC, without admixture and coarse aggregates were presoaked for 18 hours. They yielded a maximum compressive strength of 30 MPa with approximately 20 % reduction of density. Their chord modulus of elasticity were approximately (73-76) % of normal concrete as control. However, their splitting tensile strength and modulus of rupture were relatively low where their magnitudes ranged (81-84) % of control.

The use of dolomite limestone as coarse lightweight aggregates on reinforced concrete beams indicated a similar behavior in flexure compared with reinforced normal concrete beams [11]. Meanwhile, the utilization of scoria lightweight aggregate on reinforced concrete beams also showed that similarity as structural elements [12]. The application of medium K basaltic andesitic pumice and scoria as coarse aggregates on reinforced concrete one-way slabs was carried out by [10, 13]. The results showed that these lightweight coarse aggregates can be utilized practically on reinforced concrete one-way slabs with sufficient performance. These kind one-way slabs also showed the typical flexural behavior on the reinforced concrete structures with initial cracking loads, ultimate loads and stiffness were lower than those of normal concrete one-way slab as control. By considering their reduction of density mentioned previously, the tall building self-weight, for example, may be significantly reduced. The building dimensions can be reduced and overall costs may also be significantly reduced, then the building mass decreases, the horizontal and inertial forces decrease so that the risk of damage due to the earthquake also decreases [2]. Furthermore, these typical one-way slabs may support a green building program and it can be produced as a precast slab element that can reduce their self-weight because the slab is a structural element that dominates the total volume of building.

One of the characteristics of the serviceability of reinforced concrete structures is the maximum crack width that must be evaluated due to the low tensile strength [14, 15]. The cracking behavior can be investigated from load versus crack width diagrams obtained by bending test of reinforced concrete structures while the maximum crack width determined at the specified service load. Steel reinforcement stress and reinforcement ratio are the main factors that affect significantly the cracking behavior and maximum crack width [16]. However, the mechanical characteristics of concrete, i.e. the compressive strength, tensile strength as well as modulus of elasticity also influence them but less significantly. The maximum crack width depends on the maximum crack spacing which is influenced by the bond strength between the steel reinforcement and the concrete surrounding it [17]. To simplify the calculation, usually the bond strength is expressed proportionally by the concrete tensile strength [18-20], when the tensile strength decreases the bond strength also decreases. Meanwhile, for increasing bond strength, the maximum crack spacing decreases and the maximum crack width also decreases.

As mentioned previously, the tensile strength of these typical pumice and scoria lightweight concrete are relatively low, this was due to their high porosities and amorphous glass microstructures [9, 10]. The Splitting tensile strength and modulus of rupture were approximately 9 % and 12 % of their compressive strength and lower than normal concrete as control. Furthermore, both tensile strengths correlated with their compressive strength and were expressed precisely in power function of the compressive strength [21]. The failure mode of lightweight concrete also differs from normal concrete, initial cracks will occur in coarse aggregates because they are weaker than cement paste and transition zone [10, 22]. The proportion of these coarse aggregates occupied a large part of their total volume, then if they are used as the material of structural elements, these disadvantages should be considered. The cracking behavior as well as the maximum crack width of one-way slabs using both lightweight concretes, may be affected similarly by the weaknesses of the pumice and scoria coarse aggregates.

The objective of this study is to investigated experimentally the effect of pumice and scoria concrete disadvantages on cracking behavior and maximum crack width of one-way slabs. These disadvantages are only emphasized on the tensile strengths which will be represented by their compressive strengths. For comparison, the cracking behavior and maximum crack width are also evaluated by varied reinforcement ratio of one-way slabs and measured tensile strain on the steel reinforcement level. In this study, other factors such as steel bar diameter, concrete cover and reinforcement spacing are not considered and specified for all testing specimens.

2. Experimental investigation

2.1. Construction materials

The lightweight coarse aggregates were produced from medium K basaltic andesitic pumice and scoria that obtained from lava catchment area or check dams of Kelud volcano. They consisted of 4 fractions with maximum particle size was 19 mm, their grading was designed according to the requirements of ASTM C330-04 [23] and the fineness modulus were 6.69 [9]. The photograph of these typical pumice and scoria is presented in Figure 1, while the grading of both lightweight coarse aggregates is presented in Figure 2. The local crushed stones were used as control of normal aggregate with similar previous grading. The light river sand was used as fine aggregates with its grading was according to ASTM C330-04 [23] and the fineness modulus was 2.61 [9]. The binders consisted of Portland Pozzolana Cement (PPC) and clean water for drinking, while admixtures were not used in all concrete mixtures. The steel deformed bars D13 were utilized as the longitudinal reinforcement while the transversal reinforcements were plain bars ϕ6. The mixed proportion of structural lightweight concrete using these typical pumice and scoria were designed in accordance with ACI 211.2-98 (R2004) [24]. Pumice and scoria lightweight concrete comprised three mixed proportions with determined compressive strengths of 20 MPa, 25 MPa and 30 MPa, respectively. Meanwhile, the control of normal concrete with local crushed stone as coarse aggregate, and its mixed proportion was designed according to ACI 211.1-91 (R2002) [25] with specified compressive strength of 25 MPa. The slump value for all fresh concrete was determined approximately (60-70) mm with 3 % estimated air content

Figure 1. Medium K basaltic andesitic pumice and scoria.

Figure 2. The grading of both lightweight coarse aggregates.

2.2. Fabrication of one-way slab specimens

Specimens comprised three groups, group A (PLCS) consisted of 12 one-way slabs made with pumice lightweight concrete, group B (SLCS) consisted of 12 one-way slabs made with scoria lightweight concrete and group C (NCS) was 1 one-way slab made with normal concrete as control. Groups A and B included three variations of previous compressive strength and each consisted of four variations of reinforcement ratios. The reinforcement ratios were designed between rho-min and rho-max, such that they comprised 3, 4, 5 and 6 steel deformed bars. The detail of the experimental design is presented in Table 1, whereas the detail of one-way slab dimensions is presented in Figure 3 [13]. Five transversal reinforcements bounded the longitudinal reinforcements that assembled according to the previous amount of deformed steel bar. The tensile strain in steel bar level were measured by a strain gauge pasted at mid-span of the one of longitudinal reinforcement.

Table 1. Detail of experimental design.

Group	Lightweight coarse aggregates	Concrete compressive strength (MPa)	Label of one-way slab	Number of reinforcement	Reinforcement ratio ρ
A	Pumice	20	PLCS11	3	0.00671
			PLCS12	4	0.00895
			PLCS13	5	0.01119
			PLCS14	6	0.01342
		25	PLCS21	3	0.00671
			PLCS22	4	0.00895
			PLCS23	5	0.01119
			PLCS24	6	0.01342
		30	PLCS31	3	0.00671
			PLCS32	4	0.00895
			PLCS33	5	0.01119
			PLCS34	6	0.01342
B	Scoria	20	SLCS11	3	0.00671
			SLCS12	4	0.00895
			SLCS13	5	0.01119
			SLCS14	6	0.01342
		25	SLCS21	3	0.00671
			SLCS22	4	0.00895
			SLCS23	5	0.01119
			SLCS24	6	0.01342
		30	SLCS31	3	0.00671
			SLCS32	4	0.00895
			SLCS33	5	0.01119
			SLCS34	6	0.01342
C	Local crushed stone	25	NCS22	4	0.00895

Before concrete mixing, a presoaking treatment on pumice and scoria coarse aggregates were performed during 18 hours in order to reduce high absorption as well as high rate absorption. All one-way slabs were casted in two layers on plywood moulds and compacted carefully by a vibrator. Four 150x300 mm cylinders were casted simultaneously for each casting of half one-way slab in order to measure their compressive strength and equilibrium density. Demolding of concrete cylinder were performed after 24 hours casting. All specimens were cured by covering them within wet burlaps during 7 days and then placed in a dry room until 28 days, whereas demolding of one-way slabs were carried out after 21 days. The curing of equilibrium density as well as its testing were conducted in according to ASTM C567-00 [23].

2.3. Instrumentation and testing procedure

Three points bending test diagram is presented in Figure 4, where all one-way slabs were assumed as a simple beam with 2000 mm span. The testing was carried out at 28 days by a point load at mid-span. The vertical loading was generated by a manual handy pump to load cell through hydraulic jack, then a rigid lateral loading spreader transferred it to slab surface. The strain gauges pasted previously measured tensile reinforcement strains, whereas LVDT measured the mid-span deflections that was installed at the central bottom side. To remove anything slack on the supports, all one-way slabs were subjected to an initial loaded approximately 0.5 kN and then all instruments were initialized. A USB digital microscope photographed the maximum crack widths and then a complementary software incorporated on the equipment was used to measure them precisely. During testing, the load indicator recorded loading data outputs, the strain-meter recorded the mid-span reinforcement strains whereas the data logger measured the mid-span deflections. All responses were simultaneously recorded at 2 kN interval of the monotonic loading until failure. After failure, the total number of cracks were appropriately counted on the two side slabs. Compressive tests were also carried out on concrete cylinders at the similar day to determine their compressive strengths as well as their chord modulus of elasticity. The testing instrumentation set up is presented in Figure 5.

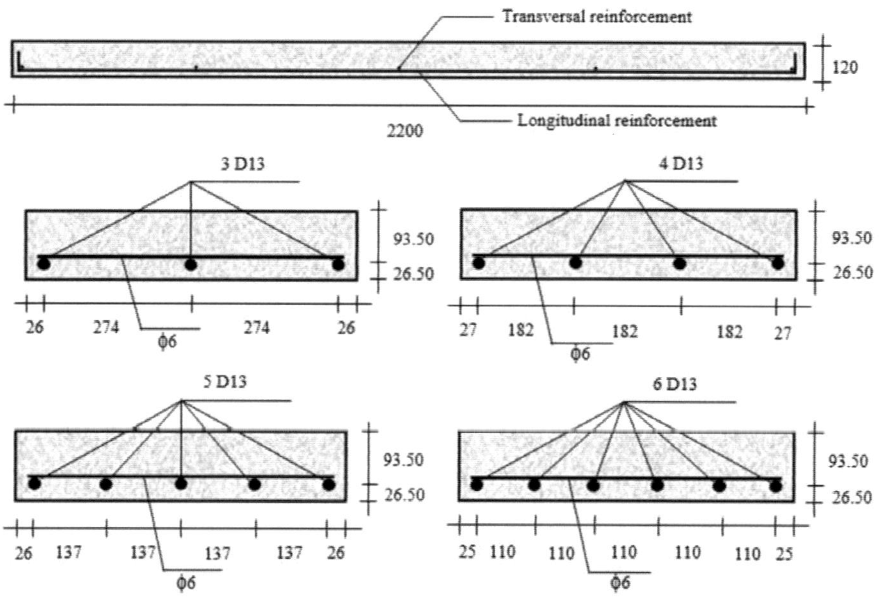

Figure 3. Reinforcement of one-way slab sections.

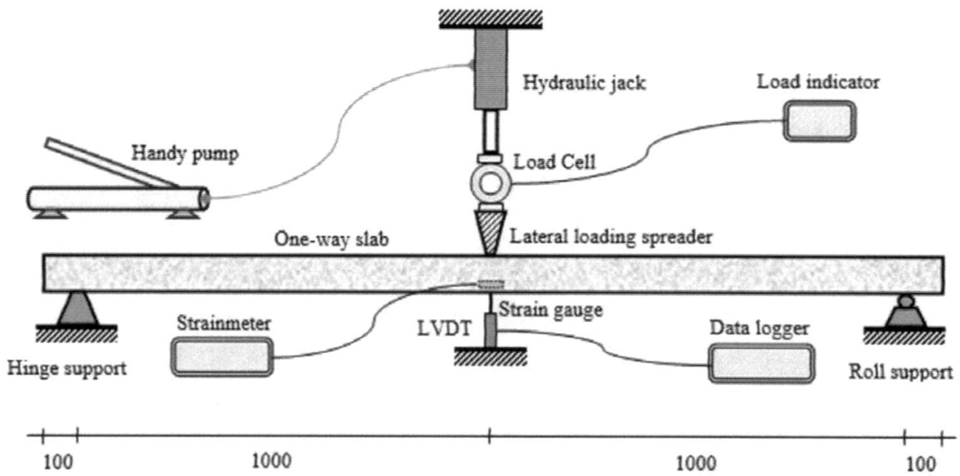

Figure 4. Three points bending test diagram.

Figure 5. Testing instrumentation set up

3. Results and discussion

3.1. Construction materials

The equilibrium density of pumice and scoria lightweight concretes ranged (1867-1878) kg/m3 and (1891-1919) kg/m3, respectively. These are lower than 1920 kg/m3 and then satisfy to requirements of structural lightweight concrete [2]. The density measured at 28 days of normal concrete as control is 2385 kg/m3, this means that the reduction of density of both lightweight concretes is approximately 20 %. The compressive strength of these lightweight concrete ranged (22-30.50) MPa and (23-31) MPa, respectively, these mean that the measured compressive strengths satisfy those determined in their mix designs. The chord modulus of elasticity of this lightweight concrete ranged (12293-14942) MPa and (12367-15159) MPa, these are approximately 73 % of normal concrete as control, i.e. 18630 MPa. The yield tensile strength of steel reinforcement was 458 MPa whereas ultimate tensile strength was 648 MPa, modulus of elasticity was 205 GPa and yield tensile strain was 0.00224.

3.2. Load versus crack width diagrams

Figure 6 shows an example of load versus crack width diagrams for the second compressive strength of both lightweight concretes. Increasing the crack width due to the given load yield a different slope of the curve referenced from the initial cracking point. When the reinforcement ratios increase, the maximum crack widths decrease significantly and the curve becomes nonlinear. The type of lightweight concrete, i.e. pumice and scoria lightweight concretes may be said to affect the crack widths less significantly. Furthermore, the magnitude of initial cracking loads differ less significantly. Similarly, for the first and third compressive strengths also indicate a similar trend.

3.3. Load versus mid-span reinforcement strain diagrams

Figure 7 constitutes an example of load versus mid-span reinforcement strain diagrams for the second compressive strength of both lightweight concretes. The diagrams are characterized by three different curve segments, namely the linear curve from the origin to the initial crack point, linear curve with a lower slope to yield reinforcement point and third curve until its failure. When the reinforcement ratios increase, the mid-span reinforcement strains decrease significantly, the slope of the curve also increases so that the one-way slab stiffness increases. The magnitude of the reinforcement strain at the initial cracking load differ less significantly, while the type of lightweight concrete also does not affect them. The tensile strain correlates proportionally with the tensile stress on this reinforcement so that their diagrams may also similar. Similarly, for the first and third compressive strengths also indicate the similar trend.

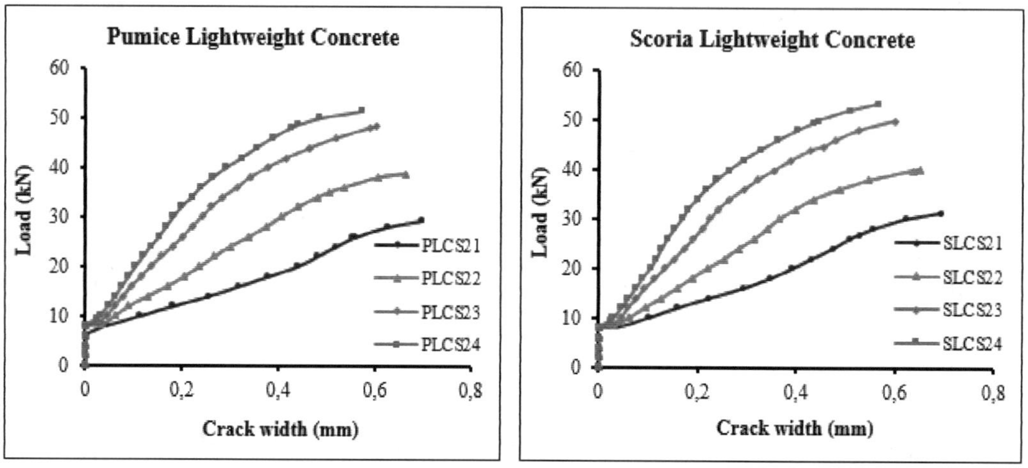

Figure 6. Load versus crack width diagrams.

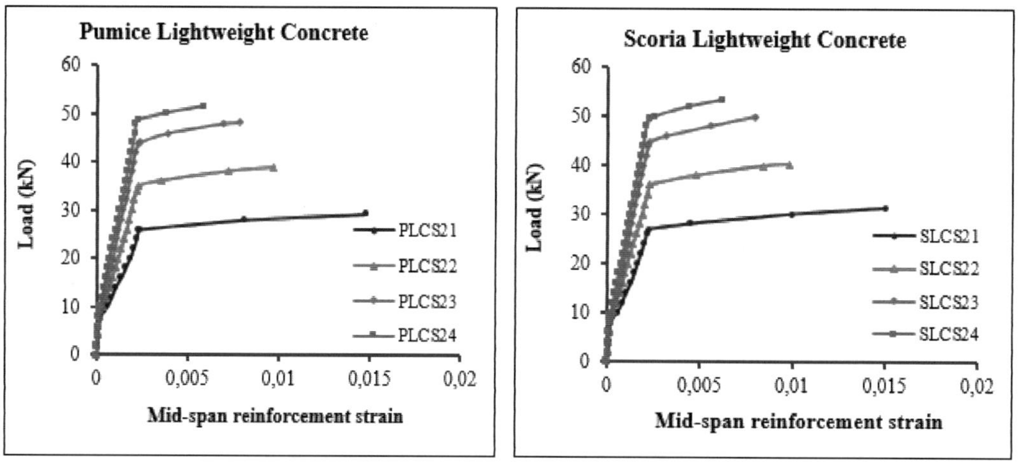

Figure 7. Load versus mid-span reinforcement strain diagrams.

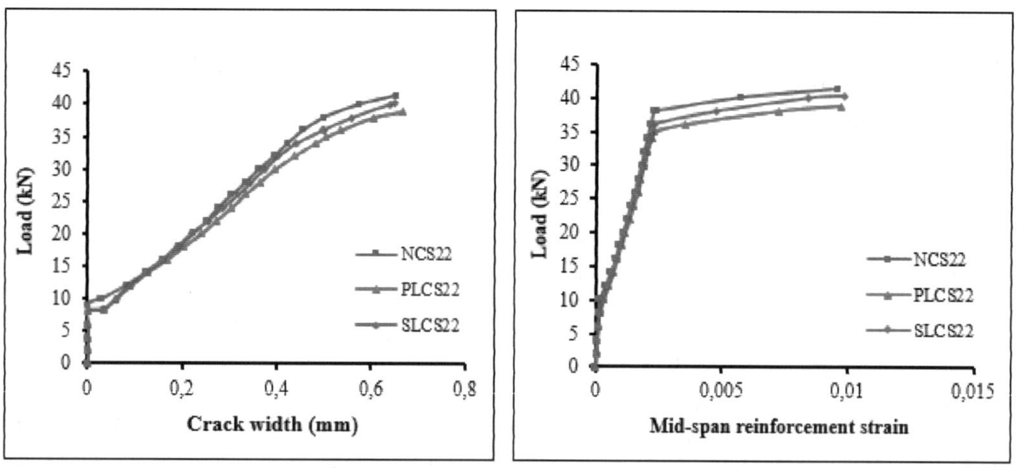

Figure 8. Comparison with control.

3.4. Comparison with control
Figure 8 shows a comparison with normal concrete one-way slab for two sample diagrams presented above. The trend of increasing crack width may not be significantly different between three types of one-way slabs, however the initial cracks of one-way slabs made with pumice and scoria lightweight concretes occur earlier than control. Similarly, the trend of increasing of mid-span reinforcement strain also indicates the similar trend.

3.5. Results of three points bending test
Table 2 shows the results of three points bending test for one-way slabs utilizing pumice and scoria lightweight concretes and control. The cracking loads were measured when the initial crack occurs, the ultimate loads were measured when the failure of one-way slab occurs and the numbers of crack were calculated on both sides. The service loads are determined based on the assumption that its magnitude is the ultimate load divided by 1.6 [24], while the maximum crack widths, mid-span reinforcement stresses and mid-span deflections are related to these service loads. The mid-span reinforcement stresses are calculated based on the elastic property before their yield points with modulus of elasticity presented previously. The mid-span deflections are determined from the load versus mid-span deflection diagram which were also measured simultaneously in the three points bending test. The comparison of load versus mid-span deflection diagram of both lightweight concretes with control is presented in Figure 13. It can be seen that the mid-span deflections of one-way slabs using both lightweight concretes are lower than the control so that their stiffness are also lower. This may be due to their modulus of elasticity which are lower than the control as mentioned previously.

3.6. Effect of reinforcement ratio and compressive strength
Figure 9 shows the effect of reinforcement ratio on the maximum crack width for three variations compressive strength of both lightweight concretes. The maximum crack widths decrease significantly to the increase in the reinforcement ratios with the average decrease is approximately 18 %. When the reinforcement ratio increases, the number of reinforcement bars also increases and the distance of each longitudinal reinforcement decreases. The tensile force due to flexure may spread according to the amount of reinforcement so the tensile force on each reinforcement bar decreases. The capability of each reinforcement bar to bond with lightweight concrete increases and the tendency to slip decreases so that the maximum crack width is also significantly reduced. Meanwhile, when the compressive strength of both lightweight concretes increases, the maximum crack width decreases, but these effects are less significant.

Figure 10 shows an example of the effect of compressive strengths on the maximum crack width for the second reinforcement ratio. The maximum crack widths decrease less significantly to the increase in the compressive strengths with the average decrease is approximately 5.64 %. When the compressive strength increases, the tensile strength increases so that its bond strength also increases. The capability of each reinforcement to bond with lightweight concrete also increases and the tendency to slip decreases so that the maximum crack width decreases less significantly. Thus, the reinforcement ratio affects the maximum crack width more significant than the compressive strength. Similarly, the other reinforcement ratio also indicates similar trends.

Table 2. Results of three points bending test.

Label	Cracking load P_c (kN)	Ultimate load P_u (kN)	Number of crack	Service load P_s (kN)	Maximum crack width w_s (mm)	Mid-span reinforcement stress f_s (MPa)	Mid-span deflection Δ_s (mm)
PLCS11	7.85	28.43	13/13	17.77	0.380	337.46	10.48
PLCS12	7.98	37.32	15/17	23.33	0.335	316.37	10.87
PLCS13	8.34	46.56	18/18	29.10	0.275	301.83	11.72
PLCS14	8.96	50.63	16/19	31.64	0.225	284.43	11.84

Table 2. Results of three points bending test (cont.)

Label	Cracking load P_c (kN)	Ultimate load P_u (kN)	Number of crack	Service load P_s (kN)	Maximum crack width w_s (mm)	Mid-span reinforcement stress f_s (MPa)	Mid-span deflection Δ_s (mm)
PLCS21	7.98	29.32	13/11	18.33	0.343	332.96	9.91
PLCS22	8.25	38.84	17/18	24.28	0.308	312.89	10.61
PLCS23	8.83	48.22	17/18	30.14	0.252	297.94	11.64
PLCS24	9.38	51.76	19/19	32.35	0.213	283.20	11.79
PLCS31	8.12	30.72	15/17	19.20	0.328	332.14	9.48
PLCS32	8.51	39.78	16/17	24.86	0.287	311.46	10.53
PLCS33	9.15	50.13	18/20	31.33	0.238	293.03	11.20
PLCS34	9.78	52.72	18/21	32.95	0.205	281.56	11.29
SLCS11	8.16	29.14	12/13	18.21	0.373	337.05	10.37
SLCS12	8.47	39.82	12/12	24.89	0.332	315.76	10.66
SLCS13	8.72	47.61	14/15	29.76	0.268	303.88	10.99
SLCS14	9.31	51.33	16/16	32.08	0.223	284.22	11.35
SLCS21	8.14	31.26	14/17	19.54	0.341	331.73	9.90
SLCS22	8.34	40.34	16/16	25.21	0.304	309.20	10.58
SLCS23	8.77	50.00	18/18	31.25	0.245	296.92	11.24
SLCS24	9.43	53.37	17/18	33.36	0.208	282.79	11.37
SLCS31	8.24	32.68	15/17	20.43	0.316	329.68	9.46
SLCS32	8.50	41.88	17/19	26.18	0.281	305.11	10.49
SLCS33	9.26	51.41	17/20	32.13	0.237	290.77	11.18
SLCS34	9.80	54.44	19/18	34.03	0.204	278.49	11.26
NCS22	9.98	41.34	16/16	25.84	0.304	307.97	8.96

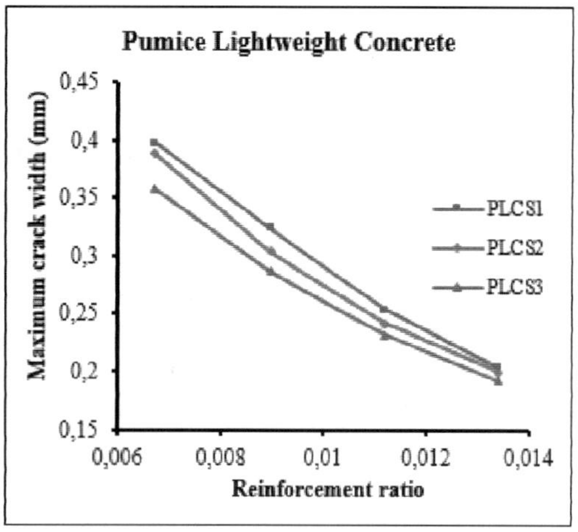

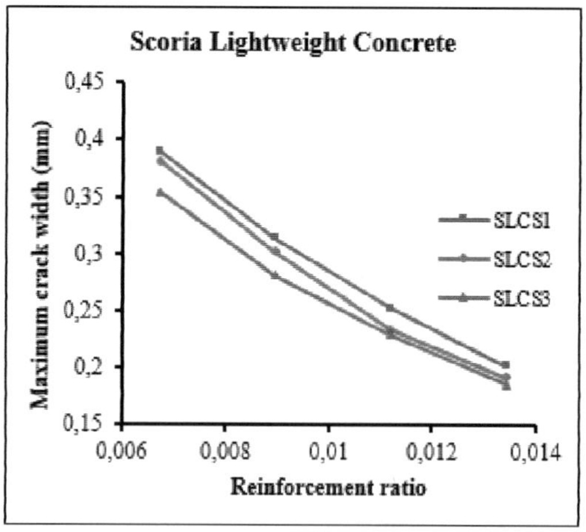

Figure 9. Maximum crack width versus reinforcement ratio diagrams.

3.7. Effect of mid-span reinforcement stress

Figure 11 shows an example of the effect of mid-span reinforcement stress on the maximum crack width for the second compressive strength. The maximum crack widths increase significantly to the increase in the mid-span reinforcement stress with the average increase is approximately 20 %. When the mid-span reinforcement stress increases, the given load increases and the maximum crack width also increases significantly. Meanwhile, the reinforcement ratio increases, the mid-span reinforcement stress decreases and the maximum crack width also decreases. It can be also seen that the type of lightweight concrete does not affect significantly the maximum crack width. Similarly, the first and third compressive strengths also indicate the similar trend.

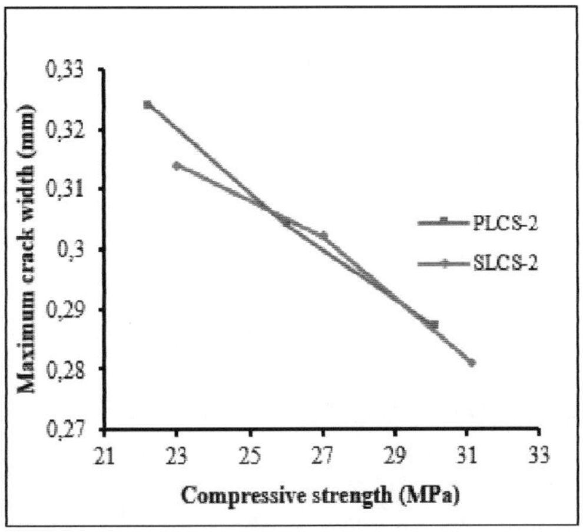

Figure 10. Maximum crack width versus ccompressive strength diagram.

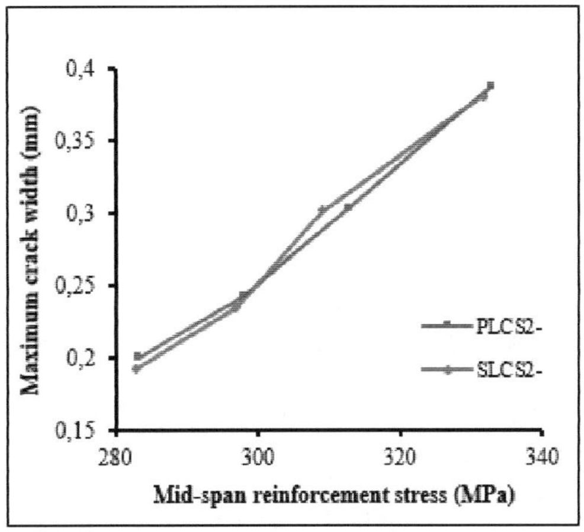

Figure 11. Maximum crack width versus mid-span reinforcement stress diagram.

3.8. Effect of typical lightweight concrete

Figure 12 shows the effect of reinforcement ratio on the maximum crack width for the second compressive strength in considering the typical lightweight concrete. The maximum crack width for scoria lightweight concrete is lower than that for pumice lightweight concrete, but its difference is less significant, i.e. approximately 2 %. This may be due to the difference of tensile strength that does not differ significantly so that the maximum crack width also does not vary significantly. Similarly, the first and third compressive strength also indicate same trends. The results of this evaluation are precisely similar of those presented by previous researchers and literatures reviewed, that the reinforcement ratio and reinforcement stress are the main factor that affect the maximum crack width. However, the compressive strength or tensile strength of lightweight concrete and the type of lightweight concrete affect it less significantly.

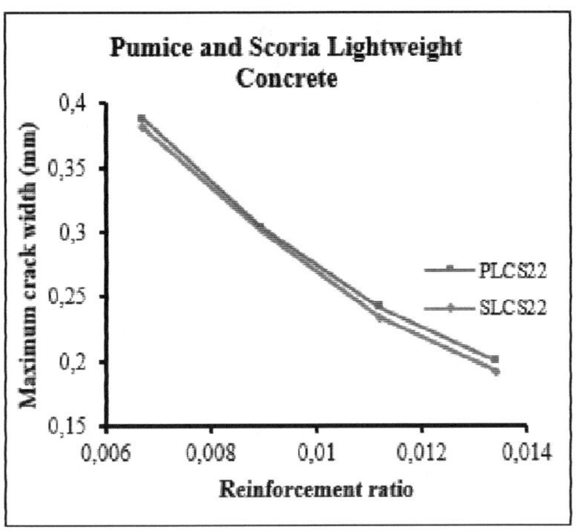

Figure 12. Maximum crack width versus typical coarse aggregates.

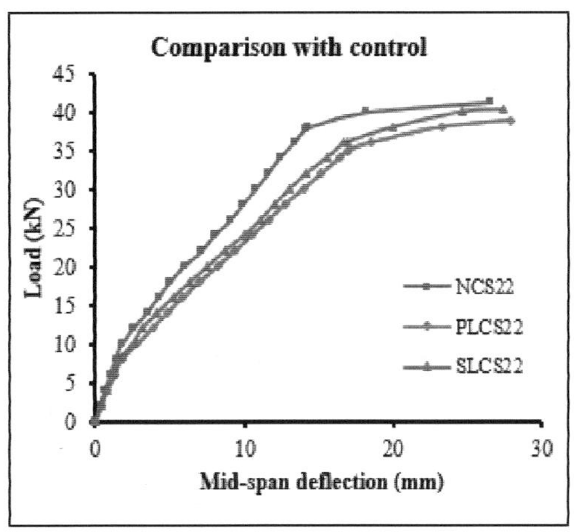

Figure 13. Maximum crack width versus mid-span deflection diagram.

3.9. Cracking patterns and failure modes
Figure 14 shows the cracking patterns for one-way slabs made with pumice and scoria lightweight concretes, and control. All cracking patterns exhibited a typical flexural crack while the number of cracks on both sides of one-way slabs varies as presented in Table 2. It can be seen that for the second compressive strength as well as the second reinforcement ratio, the number of cracks for pumice lightweight concrete is greater than the control whereas for scoria lightweight concrete is similar. The failure modes of all one-way slabs were similar with previous design, i.e. at the first time the reinforcement yields and then the concrete compressive zone crushes.

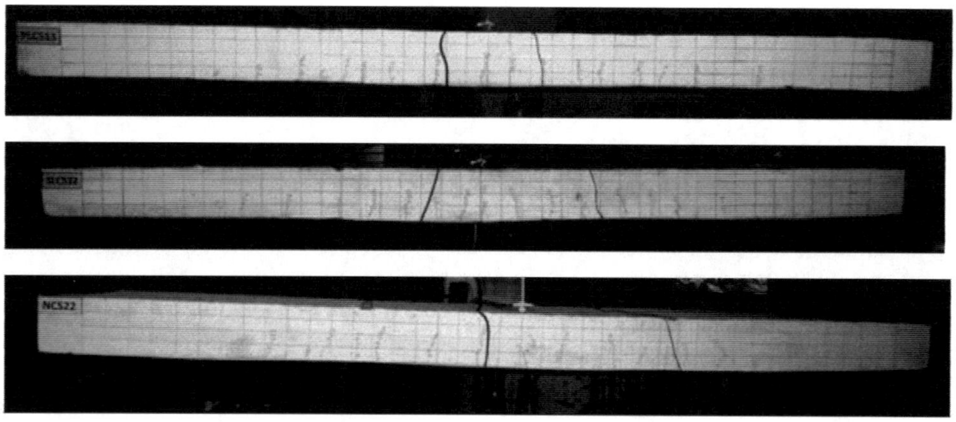

Figure 14. Three typical one-way slabs cracking patterns.

4. Conclusions
This paper shows that the disadvantages of lightweight concrete made with medium K basaltic andesitic pumice and scoria, affect less significantly on the cracking behaviour and maximum crack width of one-way slabs. This is confirmed by the investigation of the compressive strength of lightweight concrete which is related analytically to their tensile and bond strength. In addition, the reinforcement ratio, steel reinforcement stress and the type of lightweight concrete were also investigated for observing its effect on cracking behavior and maximum crack width. From this evaluation reported on this study, five following conclusions are drawn:

- The compressive strength of both lightweight concretes may have a low effect on the cracking behavior and maximum crack width measured on one-way slabs. Therefore, the tensile and bond strengths may also have a less significant effect.
- The reinforcement ratio and steel reinforcement stress constitute the main factors that contribute more significant on the cracking behavior and maximum crack width measured on one-way slabs.
- The type of lightweight concrete may be said not affects the cracking behavior and maximum crack width measured on one-way slabs.
- The one-way slab cracking patterns are typical flexural cracks whereas the failure modes are similar with previous design, i.e. at the first time the reinforcement yields and then the concrete compressive zone crushes.
- As natural resources that are more environmentally friendly, both typical pumice and scoria can use practically for producing lightweight concrete of one-way slabs as precast elements of green building.

References
[1] Chandra S and Berntsson L 2003 *Lightweight Aggregate Concrete: Science, Technology and Applications* (Norwich-New York: Noyes Publications/William Publishing)

[2] ACI 213 R-03 2003 *Guide for Structural Lightweight Aggregate Concrete* (Detroit: American Concrete Institute)
[3] Hossain K M A 2004 *Properties of Volcanic Pumice Based Cement and Lightweight Concrete Cement and Concrete Res.* 34 283-91
[4] Hossain K M A 2004 *Properties of Volcanic Scoria Based Lightweight Concrete Magazine of Concrete Res.* 56 2 111-20
[5] Kilic A, Atis C D, Teymen A, Karahan O and Ari K 2009 *The Effect of Scoria and Pumice Aggregates on The Strengths and Unit Weights of Lightweight Concrete Scientific Res. and Essay* 4 10 961-65
[6] Shannag M J 2011 *Characteristics of Lightweight Concrete Containing Admixtures Construction and Building Mat.* 25 658-62
[7] Al Naaymi T A 2015 *Assessment of Pumice and Scoria Deposits in Dhamar-Rada' Volcanic Field SW-Yemen, as a Pozzolanic Materials and Lightweight Aggregates Inter.* J. Innovative Sci., Eng. & Tech. 2 9 386-02
[8] Bourdier J L, Pratomo I, Thouret J C, Boudon G and Vincent P M 1997 *Observation, Stratigraphy and Eruptive Processes of The 1990 Eruption of Kelut Volcano, Indonesia J. Volcanology and Geothermal Res.* 79 181-03
[9] Suseno H, Soehardjono A, Wardana I G N and Rachmansyah A 2017 *Suitability of Medium-K Basaltic Andesite Pumice and Scoria as Coarse Aggregates on Structural Lightweight Concrete Inter.* J. Engineering and Technology 9 4 3318-29
[10] Suseno H 2018 *Perilaku Retak dan Prediksi Lebar Retak Pelat Satu Arah dari Beton Ringan Batu Apung dan Skoria Bertulang* (Indonesian version) Dissertation Unpublished (Malang: Jurusan Teknik Sipil Fakultas Teknik Universitas Brawijaya)
[11] Jomaa'h M M, Daham H A and Rao'of S M 2012 *Flexural Behavior of Lightweight Concrete Beams.* Euro. J. Scientific Res. 58 4 582-92
[12] Al Nasser S, Shannag M J and Charif A 2014 *Proc. The Second Inter. Conf. on Advanced in Civil, Structural and Environmental Engineering-ACSEE* (Zurich: Institute of Research Eng. and Doc., USA) p 176
[13] Suseno H, Soehardjono A, Wardana I N G and Rachmansyah A 2018 *Performance of Lightweight Concrete One-way Slabs Using Medium-K Basaltic Andesite Pumice and Scoria.* Asian J. Civil Engineering-Building and Housing 19 4 473-85
[14] McGregor J G 1997 *Reinforced Concrete, Mechanics and Design* (New Jersey: Prentice-Hall International Inc)
[15] Setareh M and Darvas R 2017 *Concrete Structures* (Switzerland: Springer International Publishing)
[16] Park R and Paulay T 1986 *Reinforced Concrete Structures* (New York: John Wiley & Sons Inc)
[17] Borosnyoi A and Balazs G L 2005 *Models for Flexural Cracking in Concrete: The State of The Art Structural Concrete* 6 2 53-62
[18] Marti P, Alvares M, Kaufmann W and Sigrist V 1998 *Tension Chord Model for Structural Concrete Structural Eng. Inter.* 8 4 287-298
[19] Gilbert R I 2008 *Control of Flexural Cracking in Reinforced Concrete ACI* Structural J. 105 3 301-07
[20] Marzouk H, Hossin M and Hussein A 2010 *Crack Width Estimation for Concrete Plates ACI* Structural J. 107 3 282-290
[21] Suseno H, Soehardjono A, Wardana I N G and Rachmansyah A 2019 *Relationships Between Mechanical Properties of Lightweight Concrete Made with Medium-K Basaltic Andesitic Pumice and Scoria* J. Engineering Sci. and Tech. 14 2 698-11
[22] Mehta P K and Monteiro P J M 1993 *Concrete Structure Properties and Materials* (New Jersey: Prentice Hall)
[23] ASTM C 567-00 2000 *Standard Test Method for Determining Density of Structural Lightweight Concrete* (West Conshohocken: ASTM International)

[24] Lim H S, Wee T H, Islam M R and Mansur M A 2011 *Reinforced Lightweight Concrete Beams in Flexure ACI* Structural J. 108 1 3-12

Acknowledgement
The authors thank especially to technical staffs of Laboratory of Structures and Construction Materials, which is part of Department of Civil Engineering Faculty of Engineering University of Brawijaya Indonesia.

IOP Conf. Series: Materials Science and Engineering **669** (2019) 012007 doi:10.1088/1757-899X/669/1/012007

Mixture of clay and *telethong* as environmental friendly binder

Ruswandono[1], J Paing[2], and Andaryati[2]

[1] Veterinary Faculty, Wijaya Kusuma University
[2] Engineering Faculty, Wijaya Kusuma University

atmajaros@yahoo.com

Abstract. This research was inspired by the phenomenon of local wisdom that has existed in the archipelago. In Bali using cow manure as stucco for finishing their walls. In the village of Teraseng, Menganti - Gresik Regency uses the waste as a cover for the walls of their houses from woven bamboo. In the village of Sade - Mataram, Lombok Island, a mixture of cow dung waste (*telethong*) has been used as stucco on the floor of the house. This paper examines to the extent that scientific evidence can be made of existing local wisdom. The combination of silicate-rich clay and *telethong* has been tested on mortar specimen. Sample mortar was physically and mechanically tested at 3, 7, 14 and 28 days. The results of the trials showed that the binding material produced by a combination of clay and *telethong* was still below the PC cement sample strength which was used as a control variable. In combination of clay and *telethong*, compressive test results show that the greater the *telethong* level, the lower the compressive strength produced. The test sample with a percentage of 25% gives the maximum results with ratio of 6% of the PC cement sample value as the control variable.

1. Background

The rapid pace of development has resulted in the destruction of the karst ecosystem as raw material for cement production. Therefore, it is necessary to look for alternative types of cement that do not damage the ecosystem. This type of cement by some researchers is called eco-cement or environmentally friendly cement.

In Indonesia the use of industrial and agricultural waste materials such as rice husks, wood scraps, sawmills, shells and coconut fibers, peanut shells, bagasse and others, to be processed into building materials, most of which have been tried. In order to develop local building materials that are non-conventional and non-traditional, Puslitbang Pemukiman has conducted research on the waste mentioned above. Some of the studies that have been carried out include the use of rice husks for making stone and cement boards, using adhesive materials consisting of a mixture of trash, lime, with or without Portland cement.

On the other hand, local wisdom turned out to be quite advanced. With the facts - the use of organic waste for building material needs. In Bali using cow manure as stucco for finishing their walls. In the village of Teraseng, Menganti Sub-district, Gresik Regency uses the waste as a cover for the walls of their houses from woven bamboo. In the village of Sade - Mataram, Lombok Island, a mixture of cow dung waste has been used as stucco on the floor of the house. This paper examines to the extent that scientific evidence can be made of existing local wisdom.

2. Previous research

Research on eco-cement began in 1992 in Japan. The results of the study stated that the ash from the combustion of waste can be used as the main ingredient in making cement. Comparison of composition of eco-cement making whose chemical compounds are almost similar to normal Portland cement, which is 58.2% rubbish ash, 40% limestone, 0.3% iron sand and 1.3% clay with the final burning of echoes at 1350°C. [1-5]

Eco-cement research has also been carried out [6-9] showed that the best composition of eco-cement was 55% organic ash and 45% limestone, with final combustion at 900°C. this study produced concrete with compressive strength 72.6 Kg/cm^2 at 7 day.

According to Neli Susanti (2009), the best composition for cement is 30% organic ash, 40% limestone and 20% clay. The rest is the addition of gypsum, iron sand and magnesium oxide (MgO). The results of the compressive test are 81.57 Kg/cm^2 at 28 days.

3. Material and methods

The main ingredient of this research is *telethong* which is the result of a waste from the PLTT (*telethong* Power Plant). Physically, the *telethong* used is no gas element. The dried *telethong* material is then finely ground until it passes the no. 200 or 212 μm diameter. For mixing soil clay is used with the chemical formula $Al_2O_3.nSiO_2.KH_2O$. This type of clay has a dominant element of alumina and silica. The variations in percentages or proportions are 25%, 50%, 75% and 100% *telethong*. This mixture is added with sugar cane waste as an additional ingredient as much as 10%. Variable control, which consists of water-cement-sand with a ratio of 0.5:1:3. Test specimens were printed on cylindrical PVC pipe molds (50/100 mm). The process of making this mortar takes around 15 to 30 minutes. Meanwhile, the treatment process (curing) is carried out since the mortar is one day to 28 days soaked in water at a temperature of 20 ± 2°C as water curing. Press is done on days 3, 7, 14 and 28 days after producing. All tests were carried out with 3 specimens.

4. Material examination result

Material examination results include examination of fine aggregates (sand), cement, *telethong* and clay. For fine aggregates, the water content value is 1.76%, for the specific gravity of the saturated surface dry condition, the value is 2.578. While the weight of the sand is 2,603 Kg/m^3 with 1.8% sludge content and fineness modulus value of 2.641. From the results of this examination it can be stated that the fine aggregate material (sand) that will be used meets the requirements to be used as a concrete constituent. Physical and chemical components of cement, *telethong* and clay are listed in table 1.

Table 1. Chemical and physical components

Chemical composition (%)	Cement	*telethong*	Clay
SiO_2	26.14	41.70	V
Al_2O_3	6.34	6.23	V
Fe_2O_3	4.08	4.33	V
CaCl	49.13	18.20	X
MgO	2.99	7.33	V
K_2O	0.55	2.58	V
SO_3	0.67	8.62	X

If the above table is observed, it will appear that the components of $SiO_2 + Al_2O_3 + Fe_2O_3$ for cement amounted to 36.56%, while the components of $SiO_2 + Al_2O_3 + Fe_2O_3$ for eco-cement were 52.26%. This does not conflict with previous research that has been done, either by Shimoda, Neli Susanti or Siswanti. In accordance with ASTM C-618 standards, if the amount of silicate, aluminate and ferro content exceeds 50%, then eco-cement is categorized as class C. The amount of eco-cement mixture has more silicate and aluminate elements because of each component (clay and *telethong*) has a fairly high silicate and aluminate element.

With a total calcium content of 18.2% in *telethong*, the clay and *telethong* mixture as eco-cement does not require the addition of more lime, as the previous researchers took precedence. For activator functions, if the previous researchers used sodium hydroxide and sodium silicate chemicals, then in this study used molasses waste which has a high sucrose content, which is 48-55%. In the industrial world, this waste is widely used as a raw material for cooking/MSG (Mono Sodium Glutamate).

Tabel 2. *Physical components*

	Cement	Eco-Cement
Specific Gravity	2.86	2.75
Initial setting (min)	90	325
Final setting (min)	165	510

Physically, cement and eco-cement do not have a significant difference in the specific gravity properties. The difference is only 0.11. However, in the aspect of initial setting time there are very striking differences. Portland cement only takes 90 minutes to start the hydration reaction and form a bond of cement paste. While eco-cement requires a much longer time, which is 325 minutes to start the hydration reaction and form eco-cement paste bonds. Likewise, for the final setting, Portland cement takes 165 minutes. While eco-cement takes 510 minutes.

Even though it is longer, but in the aspect of implementation, this is still acceptable because the producing process is given a final binding time of up to 1 day (24 hours) before the mold is released and water curing is carried out. The compressive test results on samples with different *telethong* levels can be seen in table 3. The data shows the effect of *telethong* percentage and the age of the sample on strength to pressure.

Table 3. Compressive strength at various ages

%	Age (days)			
	3	7	14	28
25	0	1,5	3,6	4,2
50	0	1,4	3,3	4
75	0	1,3	3,2	3,8
100	0	1	3	3,5
PC	24	52	65	75

From the table above, it appears that the maximum strength achieved by cement is 4.2 MPa at the age of 28 days. This strength was achieved by eco-cement mixture with a percentage of 25%. While the use of pure thelethong (without clay) produces a compressive strength of 3.5 MPa at 28 days. There is a difference of 17%. It appears that there is a similar pattern, where the addition of *thelethong* percentage is inversely proportional to eco-cement strength. The more *telethong* percentage the lower the compressive strength produced.

For an analysis of *telethong* percentage compared with strength, it appears that the addition of *telethong* by 2x fold (50%) results in a decrease in strength of (4.2-4) / 4.2 = 4.7%. For *telethong* addition as much as 3x (75%) resulted in a decrease in strength of (4.2-3.8) / 4.2 = 9.5%. Whereas if using theletong as a whole (100%) results in a decrease in strength of (4.2-3.5) / 4.2 = 17%. That is, the use of eco-cement mixture (thelethong + clay) will reduce the strength of the bond of pure clay (without mixture) by 17%. This means that at the centers of cattle farms with very abundant *telethong* waste there is no need to add more clay, because the difference in compressive strength produced is not too significant. For the percentage of strength at 28 days can be seen in the table below.

Table 4. *Percentage of strength against at 28*

%	age (days)			
	3	7	14	28
25	0	35	85	100
50	0	35	82,5	100
75	0	34	84	100
100	0	28	85	100
PC	30	70	90	100

In general, the rate of addition of eco-cement strength is indeed slower than the rate of addition of strength in Portland cement. At 3 days, eco-cement did not show any strength at all (0%) whereas in Portland cement the theory had given a rate of around 30%. The strength of eco-cement only appears at 7 days, with a strength percentage of about 35% and will be around 85% at 14 days. For Portland cement, the strength at 7 days has reached 70% and 90% at 14 days. The value of the rate of increase in strength per day is calculated by the formula = difference in percentage / difference in days. So, the power rate for a period of 0 - 3 days is (30% - 0%) / 3 days = 10 MPa/day. For complete information, see the table below.

Table 5. *The rate of increase in strength per day*

age	Power rate (MPa/day)	
	Eco-cement	PC
0 – 3	0	10
3 – 7	8,6	10
7 - 14	7,1	2,9
14 - 28	1,1	0,6

If the data in the table above is depicted on a graph it will look like in the picture below.

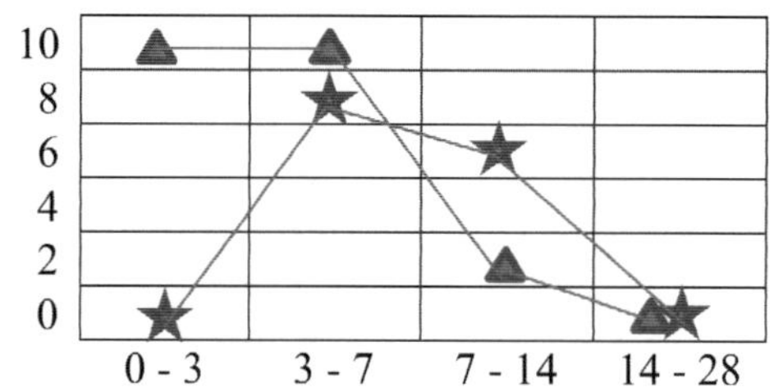

Figure 1. Graph power rate/day

Legend:
★ is Eco-cement
▲ is PC

From the graph above, it appears that PC cement has a high rate of addition of strength at the start of the hydration reaction and is constant (10 MPa/day) until the specimen is 7 days. After 7 days the value of this force starts to decrease to 2.9 MPa/day and even 0.6 MPa/day at 28 days. For the eco-cement, the old initial setting has caused the rate of increase to indicate a value of 0 MPa or no rate at all. Only after 3 days and exceeding the initial binding time did a hydration reaction occur with a rate of increase in strength reaching 8.6 MPa/day. This value is still below the reaction rate of PC cement hydration of 10 MPa/day. After 7 days, the rate of increase in strength is still quite high even though there is a slight decrease, which is 7.1 MPa/day. This rate decreases to 1.1 MPa/day at 14-28 days.

5. Conclusion

Based on the results of the above analysis, the use of this eco-cement mixture is recommended to be used in mortar work which does not emphasize compressive strength and in an area that has a fairly large cattle breeding center with adequate clay type support. In addition, there is a possibility that eco-cement will still show an adequate increase in strength at more than 28 days.

References

[1] Anonim 2003 *American Standart for Testing Material*, Cement; Lime; Gypsum , vol. 04.02., Annual Book of ASTM standard, America merican Standard For Testing Materials (ASTM)
[2] Anonim 2006 *Eco-cement : New Recycling Resources Reborn for an Affluent Future*
[3] Bellizia R P R, Thomas I and Patnaikuni 2012 *Cattle feedlot waste ash blended cement in concrete and mortar,* Journal Solid waste Tech. Mange, 28:182-188.
[4] Luga E and Atis C. D 2016 *Strength Properties of Slag / Fly Ash Blends Activated with Sodium Metasilicate and Sodium Hydroxide+Silica Fume,* Periodica Poytechnica Civil Engineering, 2016; 60 (2): 223-228. http://dx.doi.org/10.331/PPci.8270
[5] Neli Susanti 2009 *Pembuatan Ekosemen dari Abu Sampah dan Uji Aplikasinya untuk Panel beton*, Thesis, Sekolah Pasca Sarjana Universitas Sumatera Utara, Medan.
[6] Nurchasanah Y 2011 *Addition of Lime to Improve the Characteristic of Tulakan Soil as Natural Pozzolan to Substitute Portland Cement as Contruction Material*, Journal Civil Engineering Dynamics, Universitas Muhammadiyah Surakarta, Vol.11, No.2-Mei 2011.
[7] Nurchasanah Y 2013 *Characteristic of 'Tulakan' Soil as Natural Pozzolan to Substitute Portland Cement as Contruction Material*, Procedia Engineering at Elsevier Ltd. Procedia Engineering – Vol 54, Pgs 1 - 884. (2013) - ScienceDirect.com.
[8] Siswanti N D, Nanda R and Anggarini 2009 *Pembuatan Ekosemen dari Sampah Organik,* Jurnal Teknik Kimia, Vol.3 No.2 Jurusan Teknik Kimia, UPN Veteran, Surabaya, Jawa Timur.
[9] Shimoda S and Yokoyama 2009 *Ecocement : A new Portland Cement to Solve Municipal and Industrial Waste Problem,* Proc. Of International Conggress with Concrete, Dundee.

Ground water exploration using geo-electric method for Sumber Cemung pool development

G Idfi [1], N Suryoputro [1], Y A Laksono [2], N L Khomsiati [1]

[1] Civil Engineering Department, State University of Malang, Indonesia
[2] Physics Department, State University of Malang, Indonesia

Gilang.idfi.ft@um.ac.id

Abstract. The existence of springs cannot be obtained in each region; therefore it will be very detrimental if the presence of springs is not optimally utilized. This encourages Sumbercemung Villager, Wlingi Regency to take advantage of the potential of the existing spring. It is planned that in this area a tourist pool will be built using water sources. So there needs a research to support the developement of the pool. This study aims to determine the potential of ground water that will be planned for the location of pool. With the existence of this pool, it is expected to be a source of income for the Sumbercemung villager. This study was conducted by measuring soil resistivity using the Geo-electric method to determine the condition of groundwater around a spring. Analysis of geo-electric data uses Zondres software in 2D which is then analized in 3D using Voxler software. Based on the results of the analysis of groundwater from geo-electric data, no artesian aquifer were found in the location reviewed and only a water basin with a volume of 25291 m^3 was found. Thus, the needs of pool water is filled with dependable flow.

1. Introduction

Water is a vital element for human life. Humans cannot survive without water, because every sector of human life need water for its survival. The existence of a spring cannot be obtained in every region, therefore it will be very disadvantageous if the existence of a spring is not utilized optimally. In today's society, awareness of the sport of water is increasing and becoming a human lifestyle today.

The catchment area at spring location in Wlingi covering 34 hectares, with the highest elevation at +298 m and the lowest elevation at +283 m. The direction of water flow from the North to the South. The average land slope is 1.61% in the flat category.

It's encourages residents of Sumbercemung, Beru Village, Wlingi Sub-District to take advantage of the potential of spring water in the village. So far, potential of water in the area is only used for irrigating fields and fulfilling amount of villager water required only, so that a lot of water is wasted and not utilized. Amount of water that is wasted and not utilized is encouraging the villagers to build a tourist pool by utilizing the existence of a spring in the area.

So there needs a research to support the developement of the pool. This study aims to determine the potential of ground water that will be planned for the location of pool. With the existence of this pool, it is expected to be a source of income for the Sumbercemung villager.

2. Metodology

Geoelectric resistivity is one of the geophysical methods that used to determine the subsurface condition based on the resistivity value of the rocks. Resistivity, the inverse of electrical conductivity, is the resistance of the geologic medium to current flow when a potential difference is applied [4]. The purposes of the geoelectric survey is to quantify the resistivity value of the rock based on the measurement at the surface. Resistivity values between about 500 and 1000 ohm-meters are interpreted as weathered zones, fractures zones, or faults within the hard carbonates [6]. The actual resistivity value can be obtained from the measurement value [2]. This is the easiest and cheapest method that not required too high specification and software. The reading that can be obtained from this method are aquifer parameter, groundwater table depth, salinity rock permeability and bedrock [1].

Some of the most common electrode arrays are Wenner, Schlumberger, pole-pole, pole-dipole and dipole-dipole array [3]. However, these arrays are most sensible to noise, whereas the often used Wenner array and the gamma array are less sensible to noise. This method was introduced by Wenner (1915). Wenner configuration is one configuration that is often used in geoelectric exploration with a spacing arrangement of equal length (r1 = r4 = a and r2 = r3 = 2a). The distance between the current electrodes is three times the distance of the potential electrode, the potential distance with the sounding point is 2 / a, then the distance of each current electrode with the sounding point is 2/3a. The depth target that can be achieved in this method is 2 / a. In field data acquisition the current and potential electrode arrangements are placed symmetrically with sounding points [5].

In the Wenner configuration the distance between the current electrode and the potential electrode is the same. As shown in figure 1.

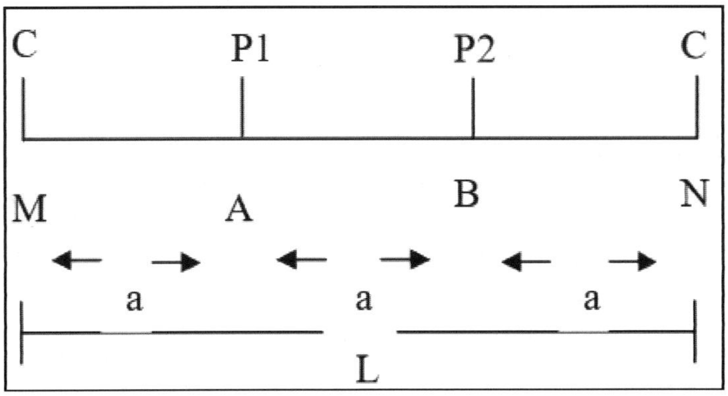

Figure 1. Current and potential electrodes in Wenner configuration.

From the picture above it can be seen that the distance AM = NB = a and distance AN = MB = 2a, using equation (1.3) is obtained:

$$K = \frac{2\pi}{\left[\left(\frac{1}{a}-\frac{1}{2a}\right)-\left(\frac{1}{2a}-\frac{1}{a}\right)\right]} \quad (1)$$

$$K = 2\pi a \quad (2)$$

So, the geometry factor for Wenner configuration is:

$$K_w = 2\pi a \text{ and } \rho_w = K_w . R \quad (3)$$

3. Results and Discussions

The survey location is in Sumbercemung village located at 8 ° 05'04.1 "LS, 112 ° 19'08.2" BT as shown in Figure 2. Sumbercemung has four water source locations. To obtain a water deposit, a geoelectric survey was conducted using the Wenner method with 4 (four) tracks. Given the terrain is the planting season rice field, electrode installation cannot be done in a straight path. the survey line can be seen in Figure 1 which is marked with a red line with the names L1, L2, l3, and L4. L1 track to the west and L4 to the east. The geoelectric results data are processed with Zondres, obtained contours as shown in Figure 3.

Figure 2. Survey Location

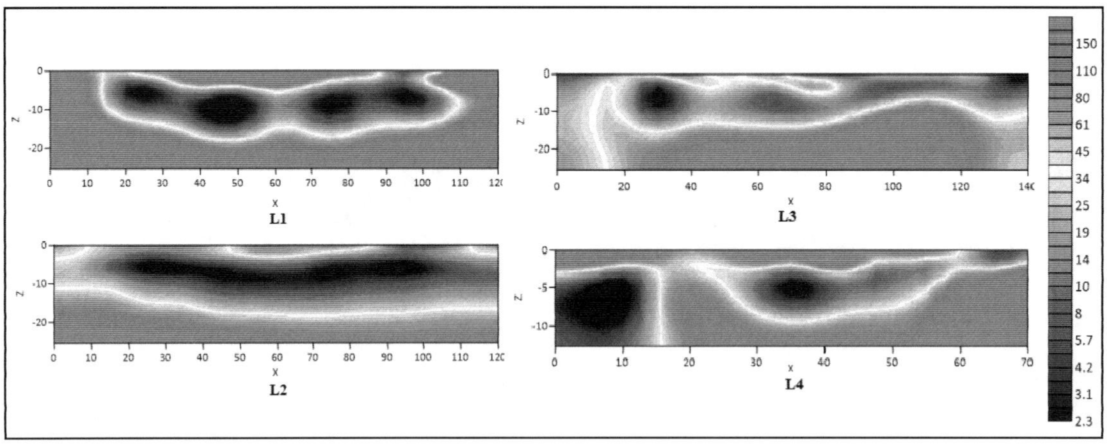

Figure 3. 2D contours of each survey line based on resistivity values

The contour in Figure 3 shows the underground structure L1 to L4. Groundwater resistivity values are between 10-100 Ωm in color scales ranging from green to orange. Pink is a rock region. The contour shows that the Sumbercemung spring is only a rock basin that holds water. There is no artesian and it is water trapped in a rock basin. Possible artesian are in other locations outside the survey area. The results of this contour are then made into a grid and processed by Voxler to determine the volume of groundwater and the results can be seen in Figure 4.

The volume of groundwater is calculated using Voxler. Voxler calculation shows that the volume for resistivity 0 - 100 is 42221 m^3, and the volume for resistivity below 10 is 16930 m^3 so the volume of groundwater is around 25291 m^3.

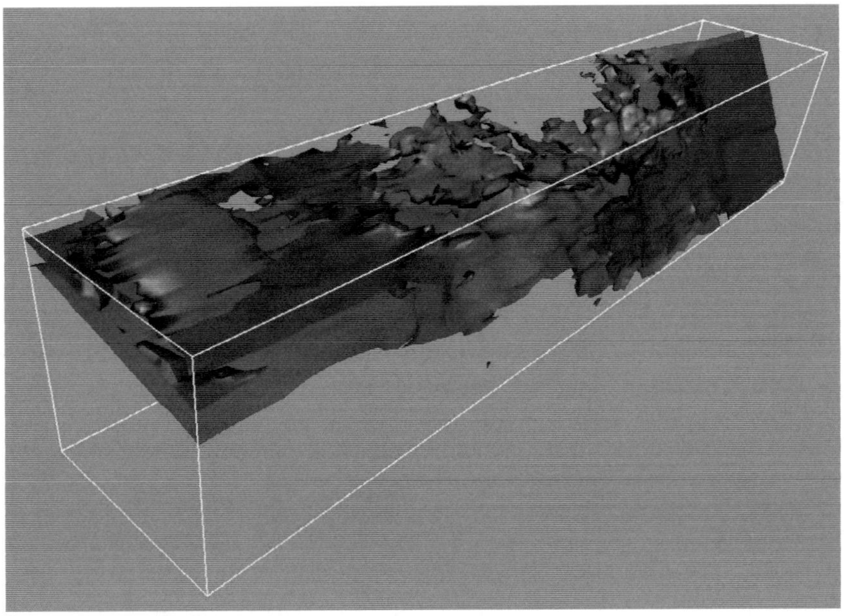

Figure 4. The isosurface results from the Source 3D contour are displayed by displaying a resistivity limit of 100 for dark blue, and 10 for black

4. Conclusion

Based on the results of the ground water analysis from geoelectric data, there is no artesian found and there was only a water basin. Thus, the needs of pool water is filled with dependable flow. Catchment area must be maintained to preserve water resources. It is not permissible to use large-powered water pumps, because it will disturb and even damage ground water or water sources.

References

[1] Siddalingamurthy S and Balasuraramanian A 2014 *Studies of Groundwater exploration In Chamaranjagar Taluk, Using Electrical Resistivity Tehniquies* International journal of Geology, Earth & Environmental Sciences, pp 123-137

[2] Lubis G 2017 *Exploration of Groundwater Potential with Geoelectric Resistivity Method in Tanjung Kuba, Batu Bara Regency* Vol 5 (IOSR Journal of Applied Geology and Geophysics (IOSR-JAGG)) pp 71-73

[3] Gusfan H and Joko W S 2008 *Pendugaan Potensi Air Tanah dengan Metode Geolistrik Konfigurasi Schlumberger di Kampus Tegal Boto Universitas Jember* (Media Teknik Sipil, 109)

[4] Shankar K R 1994 *Groundwater exploration* (20th WEDC Conference. Colombo Srinlanka)

[5] Todd D K 1980 *Groundwater Hydrology* (John Willey & Sons. Inc. New Work), 2d.ed

[6] Tony M and Jim H 2018 M Deep Groundwater Exploration Using Geophysics (Hasbrouck Geophysic, Layne Christensen Company

IOP Conf. Series: Materials Science and Engineering **669** (2019) 012009 doi:10.1088/1757-899X/669/1/012009

Stakeholders interaction framework of utilities development in and adjacent to the public right of way in Indonesia

I N M Bukit[1,2], P F Marzuki[2], R Z Tamin[2] and Meifrinaldi[2]

[1] Civil Engineering Doctoral Program, Faculty of Civil & Environmental Engineering, Institut Teknologi Bandung, Indonesia
[2] Construction Engineering and Management Research Group, Faculty of Civil & Environmental Engineering, Institut Teknologi Bandung, Indonesia

ipbukit@gmail.com

Abstract. Public infrastructure development in Indonesia is an integral part of the national development planning system, and the government attempts to ensure the process is done rigorously to fulfil the system's objectives. Part of public infrastructure is roads, and the utilities that are installed nearby. The development of utilities in and adjacent to the public right of way in Indonesia involves of various stakeholders that should be managed properly to avoid unnecessary interface problems. Those stakeholders include government agencies, community groups, private enterprises, and state or regional owned enterprises. The interaction between stakeholders of utilities in and adjacent to the public right of way should be identified clearly, and for sustainability objective, an identification of the process and interaction first should be done in the low level of the government i.e. district area. This paper is a preliminary study of a research about stakeholders' interaction structure in Indonesia's district public infrastructure development process. This paper focuses on utilities that installed in and adjacent to the district's road in Indonesia. The purpose of this paper is to explain the interaction framework of stakeholders involved, and the methods that are used to describe the relationship.

1. Introduction

Infrastructure is defined as physical facility that is intentionally built to help people fulfill their needs in daily basic. Public infrastructures are infrastructures that are provided by government as part of government's services for its people. Goodman and Hastak [1] define public infrastructure as follow: *"Public works is applied to facilities that usually require substantial capital investment; provide public services or solve problems perceived to be the public responsibility; and are planned, designed, constructed, and operated by or under the supervision of government agencies"*. Moreover, the American Public Work Association defines public work as physical structures and facilities that are provided by government bodies in order to administer the government functions by providing clean water, electrical/energy, waste disposal facilities, transportation infrastructure and other services for social and economic purposes [2].

Based on the Indonesian presidential decree No. 38/2015, public infrastructure is defined as engineering facilities, physical or system form, hard or soft type, that is required for the community services and support structural system in order to ensure the social and economic community objectives will be achieved [3]. Government has an important responsibility in public infrastructure development, and becomes a main key in its process. In order to do its functions, the government assigns various governmental bodies which each has specific role in development process. Indonesian legislation rules that every region has their own autonomous government called Province that is divided into districts [4]. Districts are provincial coverage with smaller areas and have their own autonomous government as well. In these areas, the development process begins as part of the people's wills to fulfill their necessities. The event of this start point of development is called District's development plan deliberation (Musrenbang Kabupaten).

The Musrenbang process involves stakeholders from various origins and interests, such as governmental bodies (government agencies), private enterprises, community groups, professionals, state owned enterprises, regional owned enterprises, environmental groups, media, and so forth. They are all interrelated and interdependent during planning phases, and have even more complicated relationships during execution phase of the development projects.

This raises our awareness that there are interactions between the stakeholders in district infrastructure development process, and these interactions could be overlapping in the field. This paper will discuss the interaction framework of stakeholders for public infrastructure development in district's right of way network involving utilities facilities (water, electrical and telecommunication) which implicate their organizer such as Local Government Public Work Unit, state owned enterprises for telecommunication and electrical (PT Telkom and PT PLN Persero), regional owned enterprises for water supply (PDAM), and other community groups.

This paper discusses how the interaction framework of stakeholders should be created as the initial activity of making the coordination system and to support the research on coordination that currently being carried out. This paper is a literature research which is used as preliminary study of a current research for academic purposes in expertise group of Construction Engineering Management, Faculty of Civil & Environmental Engineering, Institut Teknologi Bandung.

2. Background and problems

This research is started with idea that the number of stakeholders affects a project and it is a great deal in achieving project success. Shokri [5] states that the large number of stakeholders could drive project into problems. Public infrastructure development involves many activities that are done by different specific stakeholders. Those stakeholders are interdependence and interacted in a platform called interface. It is defined as the contact point between two or more self-directed organizations which having interdependent relationship and interacting each other in order to reach the system objectives as a whole [6]. In a development system, the interface will occur between different sectors of organizations whose task to organize different kind of infrastructures.

This contact point between such organizations/stakeholders is vulnerable to friction because each stakeholder has their own interest according to their orientation towards their products. Frictions between stakeholders can be emerged from several causes, especially potential causes happen in the project's site. Some different infrastructures are shared in a same location. For example, utilities are, for some reasons, placed along and under the road sides. District road is described as a public road that serves as a local road in primary road network system [7]. Some utilities shared location in the road side for efficiency reasons; however the efficiency is not in line with how it was done. The way those utilities installed overlapping with others existing infrastructures. The typical problem in site is lack of communication and lack of information sharing that lead to risks.

The road authority for district's road is the public work services i.e. Bidang Bina Marga. This is a local government agency whose responsible in organizing the road under and responsible to the Mayor/Bupati as the District's head of government. The district government is also dealing with province and national road authority, resulting in overlap in distribution of responsibilities. Other governmental bodies who have interest are Bidang Cipta Karya, who is affiliated with district's government secretary in order to provide clean water supply along with a regional owned enterprises i.e. local water company (PDAM). Moreover, there are two other types of utilities installed upper and under the road side such as electrical company (PT PLN Persero) and telecommunication infrastructures company (PT Telkom). All of them work and place their products at the same location and during their project life cycle they could meet and overlap. The frictions can be caused by lack of coordination between stakeholders whose work on the road, including the road authority.

Risks that exist in this location will be the damages that occur to other infrastructures during a project execution. Those damages could be one or combination of damages as follows: 1) road shoulder damage due to excavation of water pipe or cable optic or electric poles installation; 2) water pipe leak or cable optic is disconnected when road improvement work is conducted; 3) electric poles that have to be removed; etc. The implication of these damages is citizens' complain that lead to customers dissatisfaction, traffic problems, and dispute between stakeholders.

In order to reduce the damages and to improve the quality of inter-public infrastructure development in the district area, there should be a coordination system among those stakeholders; therefore each stakeholder would be able to share information about their current active infrastructure and future work plan. Afterward, the system will be used to organize the project life cycle of future public infrastructure development planning in a particular range of time.

3. The role of interface management

Interface management is a branch of management discipline that is focused on communication management and responsibility amongst parties in a project organization, thus they can be coordinated to achieve the desired results [8]. Shokri et al [6] consider interface management as an organizing process of communication, responsibility, and coordination amongst parties involved in a project, phases, or other physical entities which is interdependent. In a nutshell, interface management is a process to organize communication, responsibility and coordination between interdependence organizations in construction project and/or inter-construction projects that is related one another.

Interface management will be valuable to obtain a more structured picture of interface and its relationship among organizations. We would like to understand the interaction of the stakeholders from the interface side. Therefore, it is considered as the basis theory for this research.

Interaction between stakeholders can be identified using interface management theory. According Chan [8] there are 4 steps on the interface management procedure: 1) interface identification; 2) simplify the interface; 3) prioritize the interface; and 4) interface matching. To understand the interaction framework of stakeholders, we have to recognize all steps above respectively.

The common issues between road's authority and utility companies are as follow [9]:

- Utility relocations cause the road projects delay and require additional costs.
- The quality data infrastructures are often poor.
- Cost is incurred due to the decrease in road service life as a result of utilities cutting newly constructed pavement.
- Efforts are needed to obtain equitable cost sharing in terms of the right of way management, pavement degradation and utility relocation.

4. Methodology

The study methodology will include literature review and survey on the district's road authority and utility companies in Indonesia. After literature review, a survey will be conducted to gather interface problems and recent communication structure between stakeholders. Questions will be developed to examine [9]:

- The relationship and issues between road's authority and utility companies.
- Legislation and policy
- Project planning
- Design and construction
- Standard utility alignment
- Working in the right of way

The survey will be scheduled in 5 districts and 5 cities in Indonesia. The research methodology is described in figure 1.

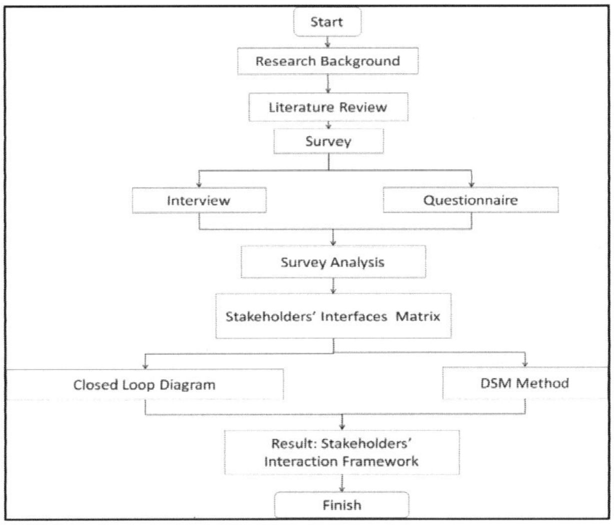

Figure 1. Research methodology

5. Discussion on interaction framework

In order to identify the interface, we should determine all the stakeholders involved in the public right of way. As aforementioned above that the stakeholders are including public work services (Dinas PU Bidang Bina Marga) as the road authority, Bappeda Kabupaten, PT Telkom, PT PLN Persero, PDAM and others community organization (NGO, environmental group and so forth). Each stakeholder has specific characteristics, orientation on project life cycle, orientation and interest towards project results, and possesses specific technical standards. Those characteristic should be listed and used as valuable information to determine interface for each other.

What is the meaning of interaction framework?

Interaction (noun form of interactive) is described as influencing each other by two way communication [10]. According to William–Webster online dictionary, framework is a basic conceptual structure or ideas. Therefore; interaction framework is the conceptual structure of communication between two or more organizations that have specific relationship. Interaction can be illustrated as a closed loop graph, a static call graph or a dependency structural matrix. This study intends to arrange those interactions in a dependency structural matrix (DSM) model. The process to make a DSM is explained as follows:

First, we list all stakeholders involve in a project development. In this activity, we use government's business process for road construction project life cycle based on district's development planning stipulated in the national development planning system, which is called *Rencana Kerja Pembangunan Daerah Kabupaten/Kota 1 tahun*, one year district regional work plan (RKPD) as a new timeframe. It is important to make the later timeframe because each stakeholder has its own project life cycle which is completely different from other stakeholder's. By doing that, we could limit the number of stakeholders involve in one year, and the timeframe will help the researchers to identify relationship between stakeholders based on time. The result of this activity is coordination level based on project life cycle and activities that are related to the coordination level (WBS).

Second, we configure relationship between stakeholders. This is a delicate action; we have to do a deep interview with all stakeholders to obtain accurate answer about this matter. The relationship should not be ambiguous nor are they unclear. Some relationships are sometimes quasi-relationship, meaning that two organizations probably related, but perhaps not actually. Even though, it is important to know such relationship to understand the stakeholder characteristics. At the end we want to make a simple closed loop diagram that provides enough accurate information in an effort to obtain the structure of the interaction among stakeholders.

Third step is to develop a dependency structure matrix (DSM). The DSM model shows interaction between organizations, phases/schedule, and tasks of activity which is involved in a project. Eppinger mentions that the dependency relationship among the design activity consists of three – i.e. independent, dependent and interdependent [11]. Relation between stakeholders in right of way is interdependent. The old fashioned project management tools such as CPM, Gantt chart and PERT only work for sequential and parallel process, but not for interdependent relation [12], so it cannot be modeled [11]. DSM is an appropriate method to describe interaction framework of interdependency. It is a method of sensible shared information involving a complex task or team relationship and makes a sequence activity that can be transformed into a simple model.

5.1. Previous studies

In the United States, each transportation agency has the responsibility to maintain highway right of way under its jurisdiction and to preserve the operational safety, integrity and function of the highway facilities [13]. Therefore, for utilities that are located or intersected on the highway right of way must be regulated and authorized by the road authority. Transportation agencies authorize and regulate the use of utilities within highway right of way, and they are obliged to supervise and control the use made of right of way acquired for public highway purposes [13]. Despite of the different law and regulation in each state, the regulation on right of way engineering requirement has reasonable uniformity that should be employed by transportation agencies to regulate highway right of way by utility companies. The AASHTO guide general condition explains the utility in and adjacent right of way should consider following aspects: safety, design, location, preservation and reservation, and visual quality. Utility accommodation should not unfavorably affect highway constructability, operations, maintenance, and safety [13].

In Canada, the road authority has responsibility to ensure the safety and convenience of road users, including utilities within right of way. One must consider the long term needs of utility companies to install, operate and repair their facilities in a coherent, safe and efficient manner while minimizing the disruption to road users and other utility infrastructure [9]. Disputes occur as result of the lack of information sharing and understanding due to no join agreement on road of way, and in terms of cost sharing. Many disputes are resolved outside the courtroom with the help of justice department, but some must be decided in the federal courts. The study found that the important aspect concerning the efficient of the system is relationship developed between utility companies and road authority [9].

5.2. Developing the interaction framework

Interaction framework is illustrated by graph representation. As aforementioned, relationship in a system has three configurations i.e. independent, dependent and interdependent. Inter-public infrastructure is an interdependent relationship, and it can be described as closed loop diagram (figure1). That means each stakeholder could have more than one relation, including relation which is overlapped.

The steps of dependency relationship structure for this study are as represent in figure 3.

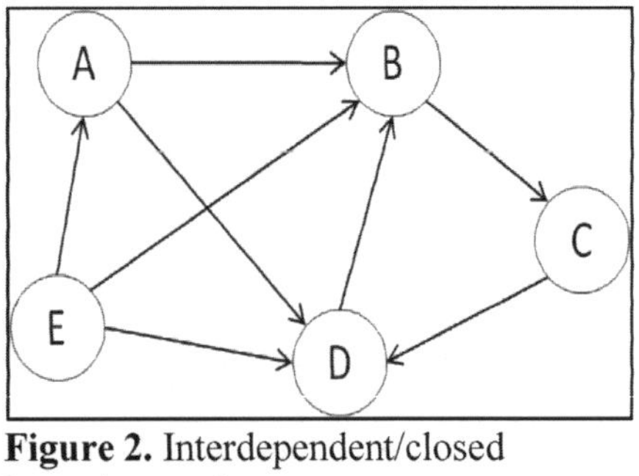

Figure 2. Interdependent/closed loops interaction

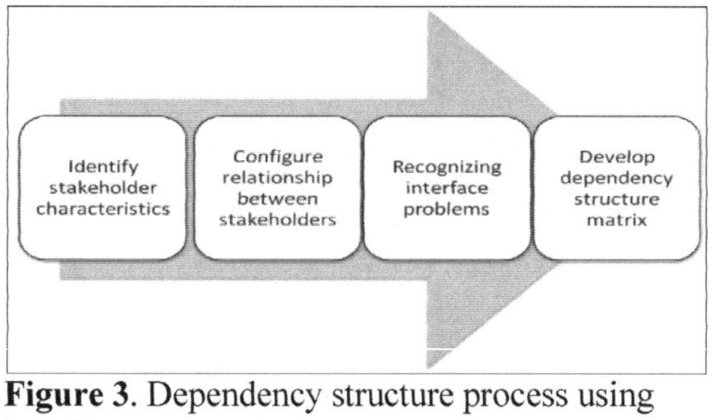

Figure 3. Dependency structure process using interface management

From figure 3, interaction framework is carried out in the second step; i.e. relation configuration between stakeholders. This step is a brainstorming to develop an interface sequence table as shown by the example in table 1. The principle of this activity is the entire sequence of dependencies is carried out within the time frame according to RKPD, and conducted by the entire project team. It is needed to comprehend that all the information in the example (table 1) merely the sample data for this paper only, NOT the actual conducted data survey.

Table 1. Example: stakeholders interfaces on the district road

No	Code	Stakeholders	Sequence/interdependent	
			Need information from	Give information to
1	A	Dinas PU Kab	C, D, E	B
2	B	Bappeda Kab	A, C, D, E	F
3	C	PT PLN Persero	A, D, E	F, H, A
4	D	PT Telkom	A, C, E	F, H, A
5	E	PDAM	A, C, D	F, H, A
6	F	Community group	-	A, B
7	G	Environment group	F	A, B
8	H	Costumer C, D, E	C, D, E	C, D, E

Activity of making interface sequence table and identifying interface problems is done carefully by involving the entire project team. Furthermore, the matrix is transform to the closed loop graph that will be looked like the following graph (figure 4):

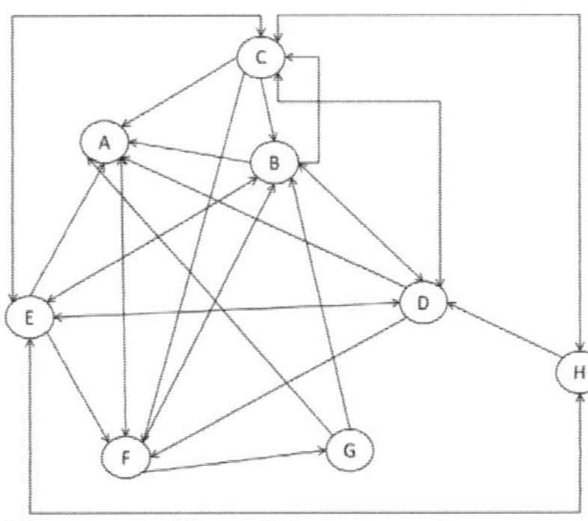

Figure 4. Closed loop graph

	A	B	C	D	E	F	G	H
A			X	X	X	X	X	
B	X		X	X	X	X	X	
C	X			X	X			X
D	X		X		X			X
E	X		X	X				X
F		X	X	X	X			
G						X		
H			X	X	X			

Figure 5. DSM model

This graphic in figure 4 looks complicated, especially if there are more stakeholders involved. Since a closed loop graph cannot be modelled, then DSM will answer how the dependency model should be made (figure 5). Dependency matrix simplify the closed loop graph, thus the relationships are recognized easily. The rows and columns axis shows the stakeholders in sequential same order based on how much interests they have on the project. The horizontal position of stakeholders explains stakeholders who provide information to certain stakeholders, while the vertical position explains stakeholders who obtain information from certain stakeholders. It can be seen that the DSM model is much simpler compared to the closed loop graph in figure 4. Figure 5 is just the initial of the DSM modelling; and the model is not completed yet. The completion of the model is not further discussed in this paper, but it will be an integral part of more in-depth research.

6. Conclusion

In conclusion, the interaction between stakeholders in public right of way can be identified using interface management knowledge; because it helps to understand the relation, information exchange, and problems more clearly. Subsequently, the interaction framework can be compiled using DSM model, which avoid the complexity of closed loop graph, especially when involving numerous potential stakeholders in the field. DSM model shows a simple representation of the relationships, with better understanding on how information exchanged amongst stakeholders. Eventually, the interaction framework will be helpful in designing a coordination system that is more beneficial for local government in development planning.

References

[1] Goodman A S and Hastak M 2015 *Infrastructure planning, engineering and economics* (McGraw-Hill Education, USA)

[2] Uddin W, Hudson W R and Haas R 2013 *Public infrastructure asset management* (McGraw Hill Education, USA)

[3] Indonesian Presidential decree 2015 *Peraturan presiden republik indonesia no. 38 tahun 2015 tentang kerjasama pemerintah dengan badan usaha dalam penyediaan infrastruktur* (Indonesian: Jakarta)

[4] Indonesian lagislation 2014 *Undang-undang Republik Indonesia No. 23 tahun 2014 about pemerintahan daerah* (Jakarta)

[5] Shokri S, Ahn S, Lee S, Haas C T and Haas R C G 2016 *Interface management process for managing risks in complex capital project* (J. of Constr. Eng. Mgt) 142 2 (@ASCE)

[6] Shokri S, Safa M, Hass C T, Hass R C G, Maloney K and MacGilivray S 2012 *Interface management model for mega capital project* (Construction Research Congress 2012) p 447-456.

[7] Undang-undang 2004 *Legislation of Republik of Indonesia No. 38/ 2004 about road* (Indonesian: Jakarta)

[8] Chan W T, Chen C, Messner J I, and Chua D K H 2005 *Interface management for China's Build-Operate-transfer Projects* (J. of Constr. Eng. Mgt) 131 No. 6. p 645-655 (@ASCE)

[9] Barril A and Project Steering Commitee 2008 *Management utilities in and adjacent to the public right of way; survey of practices* (Canada: TAC)

[10] Blockley D 2005 *New dictionary of civil engineering* (London: Penguin Books)

[11] Maheswari J U and Varghese K 2005 *A Structured approach to form dependency structure matrix for construction project* (Italy: 22nd Int. Symp. on Automation on Robotics in Construction 2005)

[12] Yasine A A 2004 *An introduction to modelling and analyzing complex product development process using the design structure matrix (DSM) method* (Urbana-Champaign: Researchgate)

[13] ASSHTO Technical Commitee on Geometric design 2005 *A guide for Accomodating utilities within highway right of way* (USA: American Association of Staet Highway and Transportation Official)

Acknowledgement

This paper is part of Construction Engineering and Management Research Group, Faculty of Civil & Environmental Engineering, Institut Teknologi Bandung (ITB) from P3MI research scheme; therefore we would like to show our gratitude and appreciation to ITB for providing the opportunity to conduct the research.

IOP Conf. Series: Materials Science and Engineering **669** (2019) 012010 doi:10.1088/1757-899X/669/1/012010

Repair of reinforced concrete beam by adding external reinforcement

S M Dewi[1], L Susanti[1], N I Triyuliani[1] and A Saputra[1]

[1] Civil Engineering Department, Brawijaya University,Malang,Indonesia

srimurnid@ub.ac.id

Abstract. The negligence in construction are usually due to error in replace or calculation of required reinforcement. The mistake known after construction and must be corrected as soon as possible. Some actions should be taken namely, add some pretension, augmented the beam size, use the FRP, and add some external reinforcement. Studies on rehabilitation and retrofitting are gaining importance due to the need for restoration of partially damaged structures due to wind load and earthquake. To prevent any disaster during the future earthquakes, the existing deficient buildings need to be retrofitted. The external tension reinforcement added when the flexural beam capacity inadequate, and the stirrups added when the shear beam capacity inadequate. The adding with external reinforcement also done after some tensile failure or cracks in the beam. In this case some damage of concrete must being consider in reducing stiffness and strength. The objective of this research is studied about the adding of external tensile reinforcement and stirrups after tensile failure and explore the influence of those action to the beam and then compared the strength, stiffness, ductility and crack pattern of beams before and after retrofit. The specimens were 12 original beam, and 12 retrofitted beam. The original beams have dimensions of 15 cm x 15 cm x 100cm, the tensile reinforcement were 4ϕ6 and compression reinforcement 2ϕ6. The stirrups were ϕ6-200 along the beam. Retrofit strategies were adding bottom tensile reinforcement 2ϕ6 and 3ϕ6, and adding stirrups ϕ6-60 in the shear span. The experiment shown the increase of beam ultimate load after retrofitting, but decrease the ductility and stiffness of the beam. The pattern of cracks also changed between original and retrofitted beam.

1. Introduction

In recent years, repair and retrofit of existing structures have been among the most important challenges in civil engineering. Strengthening or retrofitting purpose enhancing the structural performance of an existing structure. By strengthening of some selected elements, the life of the whole structure may be extended. Studies on rehabilitation and retrofitting are gaining importance due to the need for restoration of partially damaged structures due to wind load and earthquake. Many existing buildings lack collapsed or were damaged during the earthquake. To prevent any disaster during the future earthquakes, the existing deficient buildings need to be retrofitted. Low steel ratio also causes of cracks and needs some repair.

Many researchers have worked on rehabilitation and retrofitting of beams. Many methods have been developed in the recent past. Kothandarman et al [1] used an external bar and showed the reduced in crack width and increase the beam capacity. Alanwar et al [2] used U-shape jacketing mesh and steel bars, showed the increase in both flexural and shear capacity of reinforced concrete beam after retrofit. Narayanan et al [3] introducing additional continuous bottom bars. For an effective concrete jacket, the use of closed stirrups as dowels involves closely spaced drilling of the existing beam. The other options of retrofitting a beam are the fiber-reinforced polymer (FRP) wrapping. Many researchers used this method i.e. Krishna [4], Soliman [5], and Ramana [6]. Basha et al [7] used Cementitious composite material to retrofitted the beam after preloading about 58%of failure load. Zhang [8], used the near surface mounted (NSM) FRP strengthening technique.

Ground Granulated Blast Furnace Slag(GGBFS) and Fly Ash (FA) has been widely used in some countries to partially replace the Ordinary Portland Cement (OPC) as Pozzolanic material. Many research studied the behavior of GGBFS as replacing part of OPC. Saresh [9], Kapoor [10], and Arivalagan [11] reported the benefit of using GGBFS in concrete. Some advantages in using cementitious material like GGBFS and Fly Ash are increasing concrete durability and reduce the crack width.

In Indonesia GGBFS produce by PT Krakatau Semen Indonesia. For these product, Dewi et al [12], showed that 40%GGBFS from PT Krakatau Semen Indonesia results an optimum workability and strength of concrete.

This study presents investigation about the repair on 12 beams with external reinforcement. The main objective was to explore the effectiveness of adding external reinforcement to repair and increase the beam capacity. The main parameter investigated were concrete quality, failure mechanism, ultimate load, ductility, and stiffness. The other parameter was looks is the difference between normal concrete and concrete had adding of cementitious aggregates.

2. Experimental Program

2.1. Material and specimen fabrication
The normal concrete and concrete added with GGBFS and Fly Ash were used to make beam specimen. There were 4 types of concrete (1) normal concrete, (2) used 20% FA, (3) used 40% GGBFS, and (4) used 40% GGBFS + 20% FA. All concrete mixed to had same concrete quality fc 22.5 MPa.

2.2. Original beam specimen
Three cylinder samples of each concrete type were test in compression, the results shown in Table 1.

Table 1. Concrete strength

Concrete Type	Concrete strength MPa	Average
A1, Type-1	21.8	
	23.2	22.46
	22.4	
A1FA, Type-2	22.1	
	23.4	22.7
	22.6	
A3, Type-3	23.1	
	22.8	22.67
	22.1	
A3FA, Type-4	23.1	
	22.7	22.8
	22.6	

2.3. Retrofitted Beam Specimen
After tested in four points load and had bending failure, all the beams were retrofitted by throw away the broken part of concrete, restore deflection to previous. For ensuring the composite action then it was infilled with Sika-Grout mortar. On the beam that has been repaired the additional tensile reinforcement then glued to with Sika-bond. After that the additional external U-stirrups envelop the

beam, glued with Sika-bond. There were two type of additional tensile reinforcement 2ф6 for type-1 and type-3 concrete, and 3ф6 for type-2 and type-4 concrete. The arrangement of retrofit beam shown in Figure 1. The sectional details of the specimen are shown in Figure 2, and the image of retrofitted bar showed in Figure 3.

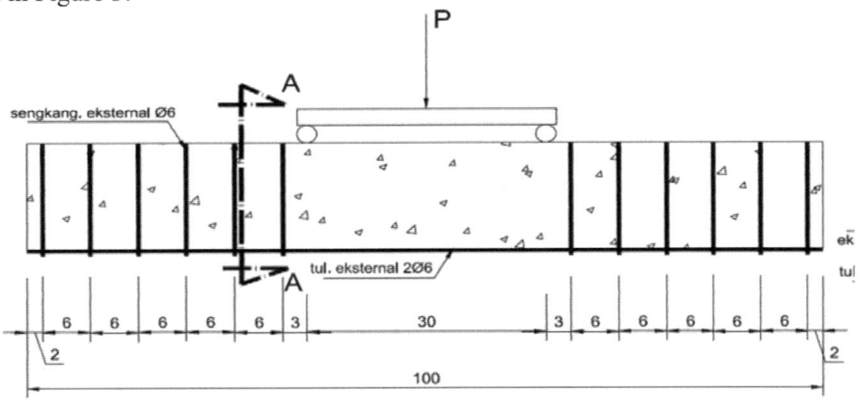

Figure 1. Retrofitted Stirrups and Bottom Tensile Steel

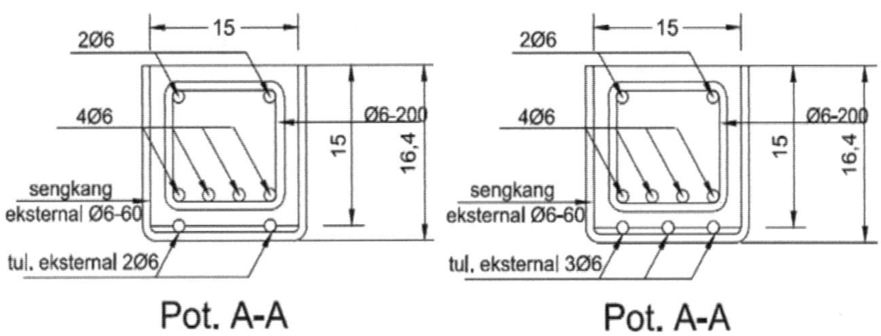

Figure 2. B1 and B2 Retrofitted Cross Section

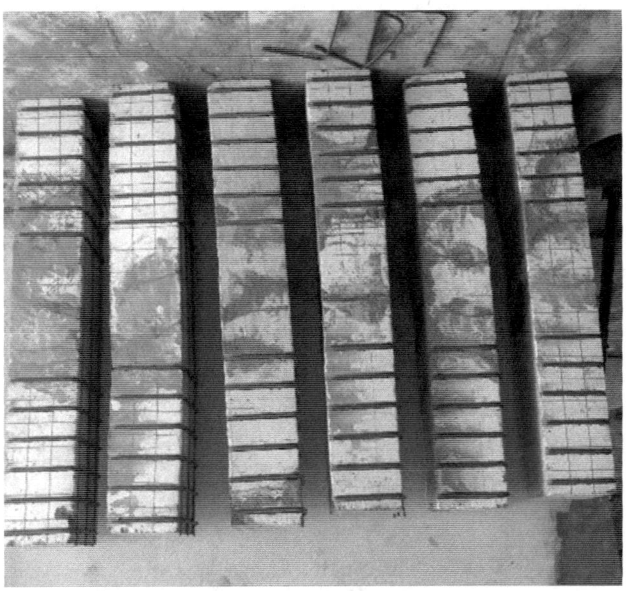

Figure 3. Retrofitted stirrups and bottom tensile steel

3. Test setup

Simply supported beam specimens are tested under a two-point loading to study the mid-span positive flexural capacity. An actuator was used to apply displacement controlled loading. The applied load was measured by a load cell. Linear variable differential transducers (LVDTs) were used to measure the deformations. The arrangement of loading setup showed in Figure 4.

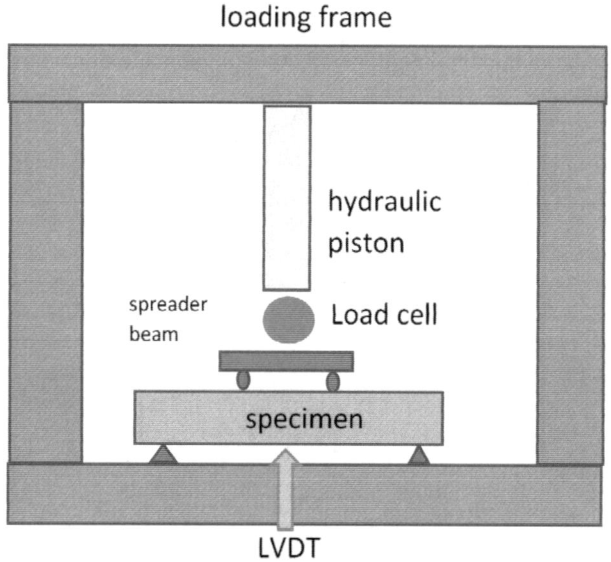

Figure 4. Test setup of beam specimen

4. Test results

4.1. Loading of original specimen

Ten of twelve specimen had flexural failure mode, one had shear failure mode, and one had transition from flexural to shear failure mode. The beam with flexural failure mode had average ultimate load 40 kN, beam with shear failure mode had ultimate load 35 kN. The Load deformation curve shown in Figure 5.

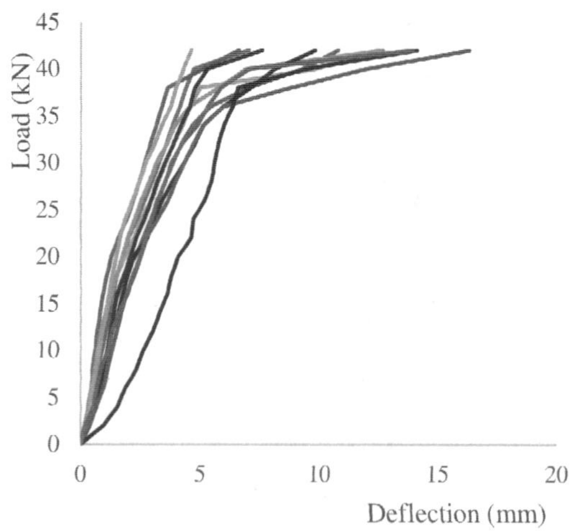

Figure 5. Load deflection curve of original beam

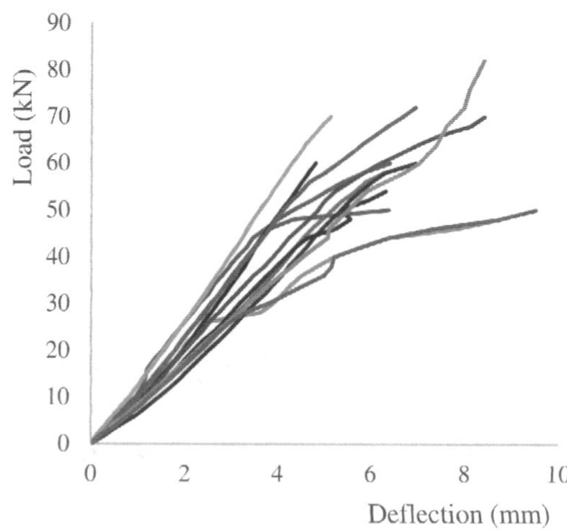

Figure 6. Load deformation curve of retrofitted beams

4.2. Loading of retrofitted specimen

The retrofitted specimen loaded with the same load position with original specimen. The load deformation curve of retrofitted beams showed the less ductility than the original ones except two beams, as shown in Figure 6.

4.3. Comparison between original and retrofitted beam

The ultimate loads and crack loads of original and retrofit specimen shown in Table 2. The average increase of ultimate loads and crack loads were 123% until 163%.

Table 2. Flexural loads original and retrofitted beams

Specimen	Original beam load kN		Retrofitted beam load kN		Ratio	
	failed	crack	failed	crack	failed	Crack
A1	42	21	60	31	1.25	1.27
B1	40	21	44	23		
Type-1	48	24	59	30		
A1FA	47	23	50	25	1,3	1.31
B2	46	23	50	25		
Type-2	49	24	85	42		
A3	50	24	63	31	1,2	1.23
B1	52	25	53	27		
Type-3	48	23	64	31		
A3FA	50	25	78	38	1.57	1.63
B2	48	23	75	37		
Type-4	46	21	77	38		

To compare the results, four sample of load deflection curve original beam and retrofitted beam shown in Figure 7.

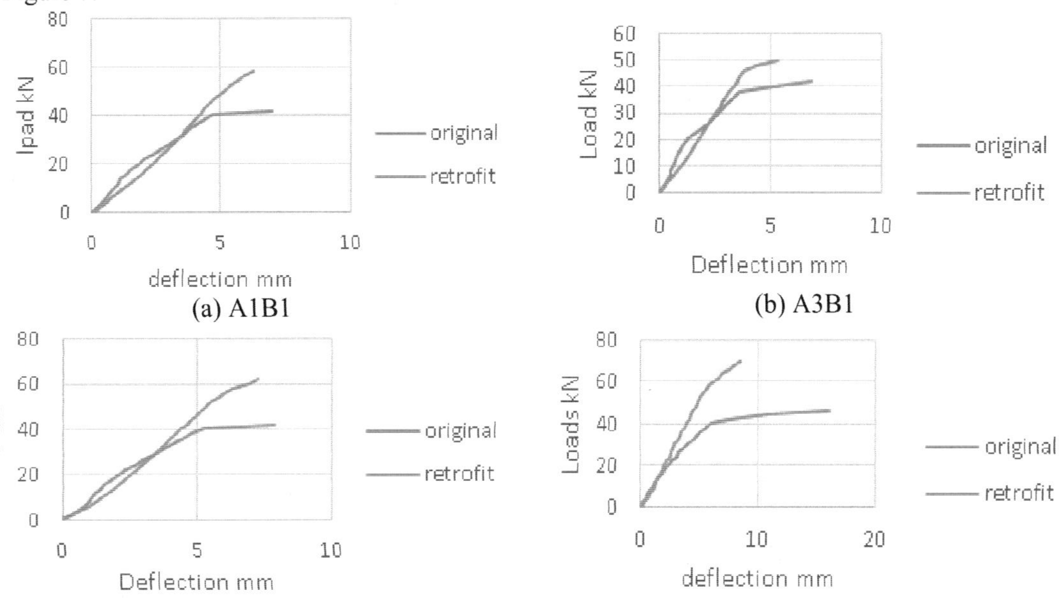

(a) A1B1 (b) A3B1

(c) A1B2FA (d) A3B2FA

Figure 7. Sample comparison of load deflection original and retrofitted beam

It is shown in Figure-7 that there were a slightly reduce of stiffness before cracks, then it was increased until the yield point. Use of BBGFS and Fly Ash increase the capacity of original beams but there was no influence in capacity increasing after retrofitted. Number of steel bar adding influence the increasing capacity after retrofitted.

Failure mode of all specimens were flexural failure with crack number less than the original specimen. Positions of cracks also different than the original position. It was indicating that grouting in retrofitting process were satisfied. The crack width of retrofitted beam also less than the crack width of the original beam, because adding of reinforcement increase the steel ratio then reduce the ductility. Two sample of beam cracks mode before and after retrofitting shown in Figure 8.

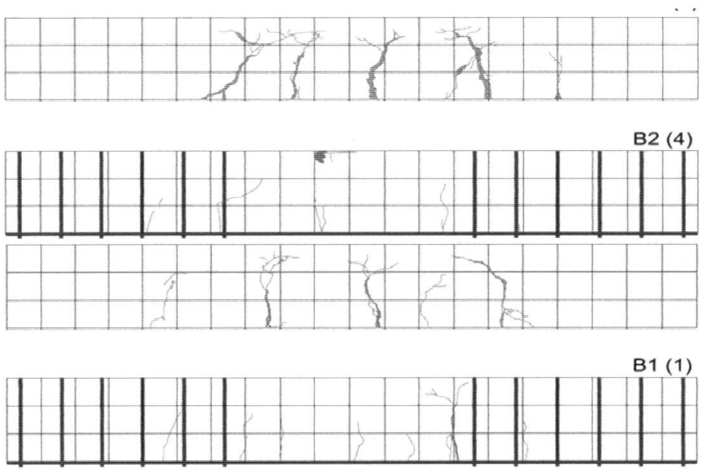

Figure 8. The cracks pattern of original and retrofitted beam

5. Conclusion

The present research has investigated the effect of retrofitting by repair cracks and adding stirrups and bottom tensile bars on the beam. Performance of beams strength, ductility, stiffness, and cracks pattern has been illustrated. Following conclusions have been obtained from the present research.

- The repair of cracks by grouting reduce small stiffness before crack.
- Crack repair with Sika-Grout increase the cracks load.
- Adding of bottom tensile steel should increase the strength but reduce the ductility.
- Adding of stirrups prevent the flexural failure of the beams.
- Crack width were reduced after retrofitting because it had been increase of steel ratio.
- Changed position of cracks before and after retrofitting indicate satisfied performance of grouting adhesive.
- Cementitious material has no effect on crack width.

References

[1] Kothandarman S and Vasudevan G 2010 *Flexural Retrofitting of RC Beams Using External Bars at Soffit Level* Construction and Building Materials **Vol 24** pp 2208-2216

[2] Abo-Alanwar M M and Elbatal S A 2015 *A Smart Reinforced Steel Wire Mesh U-Shape Jacketing Technique in Strengthening and Retrofitting RC Beams,International Journal of Engineering Research & Technology (IJERT)* ISSN: 2278-0181 **Vol 4** Issue 12 December 2015

[3] Badari N V , Amlan K S and Satish K S 2012 *Seismic Retrofit of Beams in Buildings for Flexure Using Concrete Jacket, ISET Journal of Earthquake Technology* Paper No 518 **Vol 49** No 1-2 March-June 2012 pp 1–22

[4] Krishna K J, Reddemma S and Murthy N P 2015 *Retrofitting of Reinforced Concrete Beam with Externally Bonded CFRP, IJIRST* International Journal for Innovative Research in Science & Technology **Vol 2** Issue 07 December 2015 ISSN (Online): 2349-6010

[5] Judy M I, Tarek K H, Abdelrahman A A, Osama H and Sami H R 2017 *Behavior and Flexural Prediction of Special Cementitious Bonding Material for Fiber Reinforced Strengthening Systems* Journal of Civil Engineering and Architecture 11 doi: 10.17265/19347359/2017.03.000

[6] Venkata R V, Meher L P, Markandeya R P and Pavan K M 2017 *Shear Retrofitting of Reinforced Concrete Beams* International Journal of Academic Research, ISSN: 2348-7666 **Vol 4** Issue-2 (1) February 2017

[7] Ali B, Sabry F and Galal E 2019 *Flexural Behavior of Cracked RC Beams Retrofitted with Strain Hardening Cementitious Composites* KSCE Journal of Civil Engineering Copyright ⓒ 2019 Korean Society of Civil Engineers DOI 10.1007/s12205-019-1874-4

[8] Zhang S S, Yu T and Chen G M 2017 *Reinforced concrete beams strenghtened in flexure with near-surface mounted (NSM)* CFRP strips: Current status and research needs Composites Part B 131 (2017) 30e42 journal homepage: www.elsevier.com/locate/compositesb

[9] Suresh and Nagaraju K 2015 *Ground granulated blast slag (GGBFS) in concrete – a review* IOSR Journal of Mechanical and Civil Engineering (IOSR-JMCE) **Vol 12(4)** Ver VI pp. 76–82

[10] Garg E K and Kapoor E K 2016 *A review on ground granulated blast-furnace slag as a cement replacing material* International Journal of Engineering Research and Management (IJERM) **Vol 3(07)** 2016 pp. 214–217

[11] Arivalagan S 2014 *Sustainable studies on concrete with GGBFS as a replacement material in cement* Jordan Journal of Civil Engineering **Vol 8(3)** 2014. pp. 263–270

[12] Dewi S M, Susanti L and Suseno H 2018 *Effects of GGFBS to The Compressive Strength, Workability and Time Span Between Mixing and Compacting of Concrete Paste* International Journal of Civil Engineering and Technology (IJCIET) **Vol 9** Issue 7 July 2018 pp. 1632–1640

Acknowledgments
The research was held and supported by Ministry of Research Technology and Higher Education Republic of Indonesia. Researchers wish to acknowledge PT. Sika Indonesia and PT Krakatau Semen Indonesia for providing material support during the research.

IOP Conf. Series: Materials Science and Engineering **669** (2019) 012011 doi:10.1088/1757-899X/669/1/012011

Exploratory study on the utilization of recycled wood as raw material for cross laminated timber

W A Munandar[1], R H Purba[2] and A Christiyanto[3]

[1] PT. Graha Karunia Pratama, Jl. Kalideres Permai, Citra Garden 2 Blok B-20, Jakarta, 11830, Indonesia. Formerly, undergraduate research assistant, Department of Civil Engineering, Universitas Bandar Lampung (UBL)
[2] Department of Civil Engineering, Universitas Bandar Lampung (UBL), Jl. Z. A. Pagar Alam No. 26, Bandar Lampung, Lampung, 35142, Indonesia
[3] PT. Bank Central Asia, Jl. Tiang Bendera V, Batavia Blok L-M No. 42, Jakarta, 11230, Indonesia. Formerly, undergraduate research assistant, Department of Civil Engineering, Universitas Bandar Lampung (UBL)

ronnypurba@ubl.ac.id

Abstract. Cross-Laminated Timber (CLT) has been gaining interest in recent years as an attractive and environmental-friendly building material. As such, CLT has been used for structural components in building constructions such as apartments, commercial buildings, or offices. CLT predominantly uses newly harvested lumber as a raw material. In this study, however, CLT fabricated using recycled wood was explored for possible utilization in wood structures. The recycled wood was obtained from wooden boxes, wooden pallets, and scrap wood from lumber suppliers, and their mechanical properties were tested. The average tension-parallel-to-grain strength of these recycled woods was 52.2 MPa, while their average compression- perpendicular-to-grain and compression- parallel-to-grain strength were 20.6 and 4.0 MPa, respectively. Four CLT beams of 2 m-length, 100 mm-width, and 130 mm-depth glued with epoxy adhesive were tested to measure their flexural strengths. A set of two CLT beams was loaded on upright and lying positions, and their average flexural strengths were 8.1 and 7.6 MPa, respectively. The effects of lamina lap splices and adhesives on CLT beam flexural strength were studied. Apparently, adding several variations of lap-splices and adhesives has marginal impact on CLT beam flexural strength.

1. Introduction

Indonesia currently faces several problems relating to environment and forests. One of these problems is deforestation, of which Indonesia has reached an alarming condition (figure 1). It was reported that Indonesia's forest cover is decreasing around 6,840 sq. km annually [1]. Furthermore, the Global Forest Resources Assessments (FRA) of the United Nations Food and Agriculture Organization (FAO) stated that Indonesia ranked the second highest of forest loss (deforestation) among countries in the world, after Brazil took the first place [1].

Among many factors, the high deforestation rate in Indonesia was triggered by the increasing demand of timber every year. In the last 14 years, the timber demand has far exceeded the available quantity of certified timber [2]. The Ministry of Environment and Forestry of Indonesia reported that volume of certified timber from managed forests is 16.17 million cubic meters as of May 2018 [3], 2018), while the amount of wood consumption is 219 million cubic meters [2].

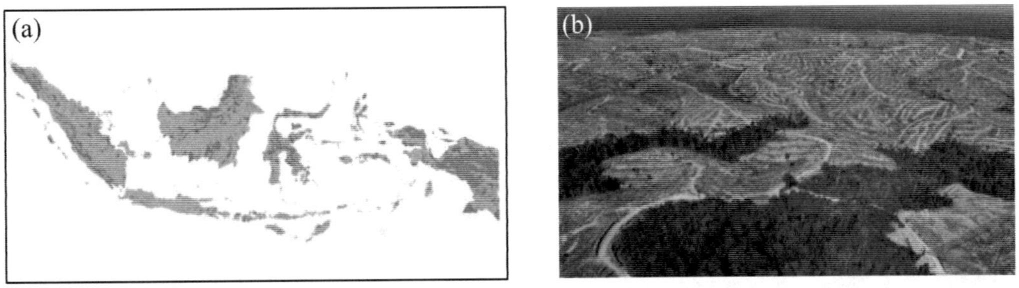

Figure 1. Deforestation condition in Indonesia: (a) map of protected forest cover [4]; (b) aerial view showing deforestation [5]

To alleviate this scarcity of lumber, engineered wood process such as that of cross laminated timber (CLT) products can be applied to manufacture sustainable construction material from reclaimed or recycled wood. The widely known CLT technology has been reported able to improve quality of soft or fast-growing wood to be equivalent to high-quality hardwood products [6, 7]; and CLT products have been applied in many constructions. This research was conducted to examine application of CLT technology on recycled wood, obtained from wooden fruit boxes, wooden pallets, and scrap wood from lumber suppliers, to produce high-quality construction wood.

2. Cross laminated timber and utilization of recycled wood

CLT is an innovative wood product composed of several layers of laminas or thin boards of which each layer is arranged perpendicular to adjacent layers and glued to each other on its wide faces [8]. A typical CLT layer configuration is shown in figure 2. Several advantages of CLT include significant load-bearing capacity and structure stability it can provide to buildings [6, 8]. Also known as X-LAM, CLT generally is arranged in an odd number of layers such as in 3-, 5-, or 7-layer configurations. The dimension of single CLT panel can reach up to 18 m long, 3 m wide, and 300 mm thick, while its lamina varies between 12 to 45 mm thick [9], depending on design requirements and construction needs of certain projects.

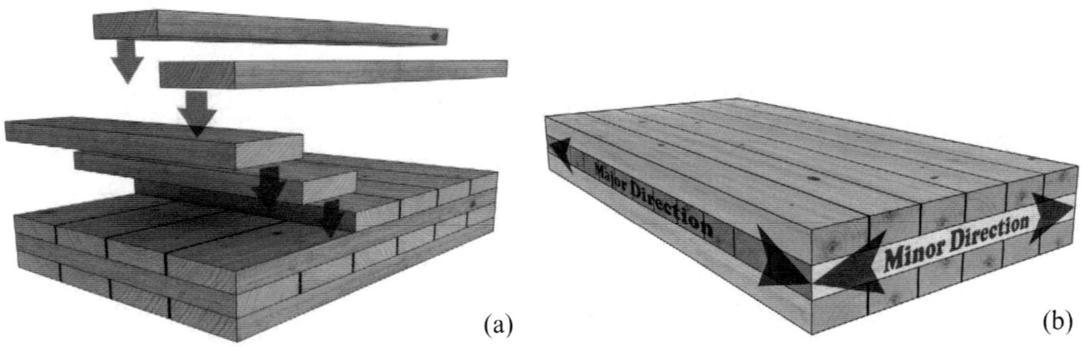

Figure 2. Cross laminated timber: (a) typical layer configuration [10]; (b) layers in the minor and major directions [11]

CLT could provide alternative solutions to needs for large-diameter solid sown timber and high-quality timber, which has been increasingly become limited [12]. As a building material, CLT could facilitate design and shape flexibility of buildings [9, 13, 14], is a good insulator to provide energy efficient in buildings [15], has adequate fire resistance [16, 17] and earthquake resistance due to its flexibility and relatively light weight in comparison to concrete or steel materials [18], and is a renewable and sustainable construction material that would help in carbon sequestration and reducing the effects of global warming [13, 19]. CLT products have been implemented to residential buildings,

buildings, offices and industrial buildings in many countries [13, 14] and has been widely applied in structures as bearing walls, floors, roofs, and stairs (figure 3). In addition, CLT also had been utilized in constructions of bridges [20]. Further discussion on the development of CLT research and its implementations are available elsewhere (e.g., [9, 18, 21]).

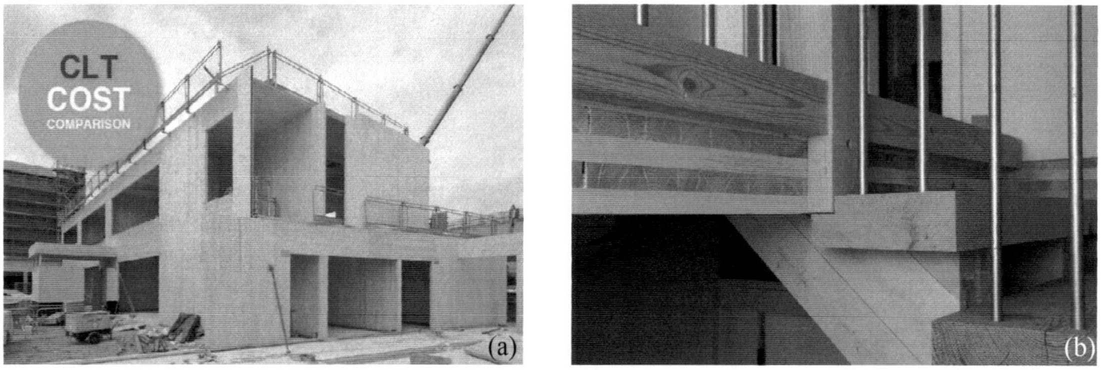

Figure 3. Application of CLT products: (a) applied as bearing walls and floors [22]; (b) applied as stairs [23]

In addition to laminas as raw materials, adhesive is a key factor to achieve a good quality of CLT products. Adhesive application between layers would create a composite cross section of which its mechanical properties such as stiffness and strength would increase significantly in comparison to that of an individual layer of lamina. Several commonly used adhesives in the manufacturing of wood products include aliphatic based adhesives (samples of marketed brand are Crona 234 and Presto), epoxy adhesives (e.g., Avian Epoxy), PVAc or polyvinyl acetate adhesives (e.g., Crossbond X4 and Titebond III), polyurethane adhesives (e.g., Foam Glue), and cyanoacrylate adhesives (e.g., Super Glue). Initially in this study, epoxy adhesive was used because of its high bond-strength and durability, quick initial tack, tolerable to a wide range of wood moisture content, and low-pressure requirements during curing process [24].

CLT predominantly uses newly harvested lumber as a raw material. In this study, however, CLT fabricated using recycled wood was explored for possible utilization in wood structures. The recycled wood was obtained from wooden boxes for packing fresh vegetables or fruits, wooden pallets to secure fragile goods during shipments, and scrap wood from local lumber suppliers. In the current practice, wooden boxes and pallets, after used several times, are typically disposed to a landfill as waste materials or burn as firewood, which in large quantity both could pollute the environment. Alternatively, these woods could be repurposed by making simple household furniture, shelves or drawers. Based on the idea of seeking an environmental-friendly option for this wood waste through reducing its environmental waste footprint and increasing its valuable usages, this research is aimed to apply CLT technology on recycled wood. Obtained CLT products are expected to be used as floors, beams, columns and load-bearing walls of wood structures.

3. Preparation of CLT beam specimens and experimental setting

The process of preparing specimens started from collecting recycled wood as raw materials obtained from local markets, lumber suppliers, post offices, and fruit sellers around the city of Bandar Lampung (figure 4a). Selected wood, which was relatively in a good condition and undeformed or without cracks, was cleaned from any sticking dirt and sawed afterward to obtain a uniform size (figure 4b). After cutting and sanding all surfaces (figure 4c), one obtained lamina was 40 cm long, 6 cm wide, and 1 cm thick.

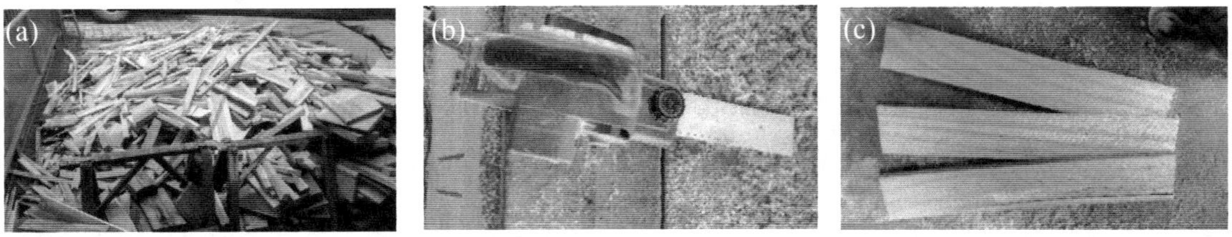

Figure 4. Preparation of laminas: (a) piles of recycled wood; (b) processing; (c) processed lamina

Laminas were arranged in 3 layers, as shown in figure 5. At the top and bottom layers, each lamina was arranged in an upright position and resembled brick arrangement along its length. The resulting panel size for each top and bottom layer was 2 m long, 400 mm width, and 60 mm thick. For the middle layer, each lamina was arranged with wide faces perpendicular to the lamina on the upper and lower layers. Thus, a CLT panel of 2 m long, 400 mm wide and 130 mm thick was obtained. After each lamina was adequately glued with epoxy adhesive applied on one side (single spread) of wide face (figure 6a-b), the bottom layer was cold pressed afterward (figure 6c) and left for approximately 3 hours. The same process was repeated for the top layer. The middle layer was installed after the top and bottom layers was completed (figure 6d-f), and the entire layer was pressed for another 3 hours (figure 6g). The resulting CLT panel (figure 6h) was sanded and cut into 4 CLT beams with an approximate size of 2000×100×130 mm (figure 6i) prepared for bending test.

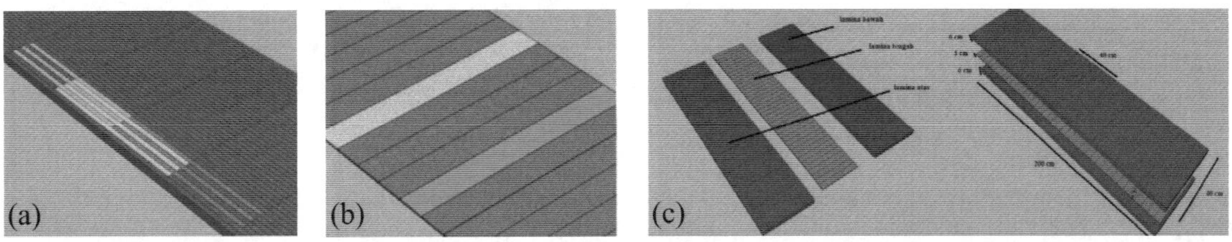

Figure 5. Schematic layout of CLT: (a) top and bottom layers; (b) middle layer; (c) arrangement

Figure 6. Manufacturing of CLT panel: (a)-(b) adhesive application; (c) cold pressing of top and bottom layers; (d)-(e) middle layer instalment; (f) CLT panel; (g) cold pressing of CLT panel; (h) finished CLT product; (i) CLT beams

Meanwhile, tension and compression tests were conducted on recycled wood to obtain its properties. Specimens were prepared from wooden fruit boxes, wooden pallets, and scrap wood. They were tested according to the ASTM standard [25] to obtain tension-parallel-to-grain, compression-perpendicular-to-grain, and compression-parallel-to-grain strengths of recycled wood. The obtained strengths on average were 52.2, 20.6, and 4.0 MPa, respectively. Details about this property tests were available in [26].

Figure 7 shows schematic and test setup for CLT beam 3-point-bending test. The center load at the mid-span of CLT beam was applied by a Universal Testing Machine (UTM) with a displacement control scheme. Two of CLT beams were tested on an upright position (its depth was parallel to the direction of loading) and on a lying position (its width was parallel to the direction of loading). The test was carried out at the Civil Engineering Laboratory of the Universitas Bandar Lampung (UBL).

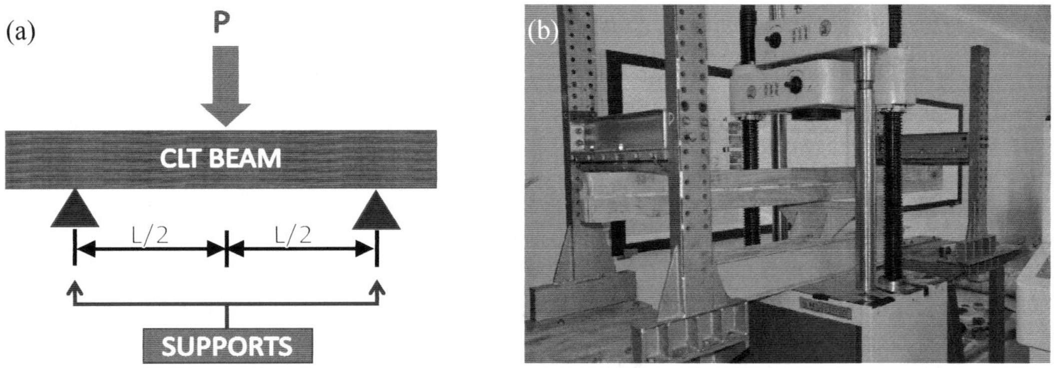

Figure 7. Three-point-bending test: (a) schematic of loading; (b) test setup on UTM

4. Flexural strength of CLT beam

Flexural strength of CLT beams can be calculated using the following equation:

$$\sigma = \frac{My}{I} \tag{1}$$

where M is the flexural moment at the CLT beam mid-span; y is the distance from the center of gravity to the outermost side of the beam, which in this case is half of the beam height; I is the cross section moment of inertia. The maximum moment for the 3-point-bending test can be calculated using the following equation:

$$M = \frac{1}{4}PL \tag{2}$$

where P is the applied load and L is the CLT beam span length. Note that the self-weight of CLT beam was ignored in this case. Substituting equation (2) into equation (1) and using respective moment of inertia for a rectangular cross section, flexural strength (a.k.a. modulus of rapture, MOR) of CLT beam when tested on the upright and lying positions, respectively, can be obtained using the following equations:

$$MOR_{upright} = \frac{3PL}{2bh^2} \tag{3}$$

and

$$MOR_{lying} = \frac{3PL}{2hb^2} \tag{4}$$

where b and h are width and depth of CLT beam, respectively, while other parameters have been previously defined.

IOP Conf. Series: Materials Science and Engineering **669** (2019) 012011 doi:10.1088/1757-899X/669/1/012011

5. Discussion of experimental results

5.1. Bending Test of CLT Beams

Figure 8 (a) and (b) show results of the conducted CLT bending test on the upright and lying positions, respectively. A significant drop of load at about 12 mm deflection (figure 8a) occurred when several laminas started to fail, either due to fracture or delamination of laminas. Specimen B1 dan B3 can maintain stability and the remaining laminas able to sustain additional loads prior to a complete failure at more than 40 mm deflection. However, that was not the case for specimen B2 and B4, first failure of laminas occurred at about 15 mm deflection (figure 8b) and progressively lost their capacity to sustain loads before a complete failure occurred at about 30 mm deflection. The maximum point load when CLT beams tested on the upright and lying positions reached 5.6 and 4.6 kN, respectively. When failure occurred, CLT beams have deformed 60 and 30 mm for the respective positions, which approximately are 1/30 and 1/60 of CLT beam span.

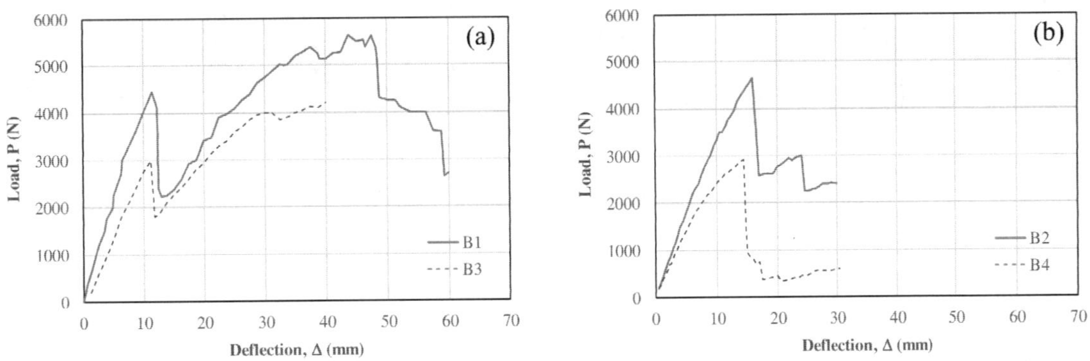

Figure 8. Results of CLT beam bending test: (a) upright position; (b) lying position

Table 1 shows a summary of bending test results for all 4 CLT beams tested. Using equation (3) and (4), the average $MOR_{upright}$ and MOR_{lying} are 8.1 and 7.6 MPa, respectively. These results with larger MOR for the upright position compared to that for the lying position are attributed due to the fact that CLT beam's moment of inertia for the upright position is larger than that of the lying position.

Table 1. Results of CLT Beam Bending Test

Speci men	Dimension				L_{span} (mm)	Test Position	P_{max} (N)	*MOR* (MPa)
	L (mm)	*b* (mm	*h* (mm)	*w* (kg)				
B1	1998	100	130	12.4	1739	Upright	5557	8.58
B3	1998	85	130	9.8	1744	Upright	4205	7.66
B2	1996	100	130	11.3	1740	Lying	4660	9.36
B4	1999	100	130	11.5	1744	Lying	2913	5.86

L = beam length; *b* = width; *h* = depth; *w* = weight; L_{span} = beam span; P_{max} = maximum load

Specimen condition during the test is shown in figures 9 and 10 for B3 and B4 specimens, respectively, while that for specimen B1 and B2 are available in [26]. In general, damages were occurred at the interface/connection between adjacent laminas either that along the length or the depth of CLT beam. The damages occurred on CLT beams tested on the lying position (i.e., B2 dan B4) were relatively more severe than that on CLT beams tested on the upright position (i.e., B1 and B3). This observed behavior leads to a further investigation on the effect of lamina lap splices and type of adhesive material used, as discussed in the next sub-section.

Figure 9. Specimen B3: (a) failure condition at the maximum load; (b) slip between layers; (c) failure of laminas (delamination)

Figure 10. Specimen B4: (a) tested on lying position; (b) delamination at the maximum load (facing south); (c) delamination at the maximum load (facing north)

5.2. Effects of lamina lap splices and adhesives

To determine the effects of lamina lap splices and type of adhesive on CLT beam strength, tension-parallel-to-grain tests were conducted on specimens made of three different types of lamina lap splices (i.e., half , bevel-, and fish-mouth-lap splices) and glued with two different types of adhesive materials (i.e., cyanoacrylate and polyurethane adhesives), as shown in figure 11a. Incidentally, the types of lamina lap splices considered in this research were simple alternatives to a more labor-intensive finger-joint connection. The highest tensile strength of 5.2 MPa was obtained from specimens made of bevel-lap splices and glued with cyanoacrylate adhesive. The strength was approximately twice above that made of the other two lap splices and adhesives. Details about these tests are available in [27].

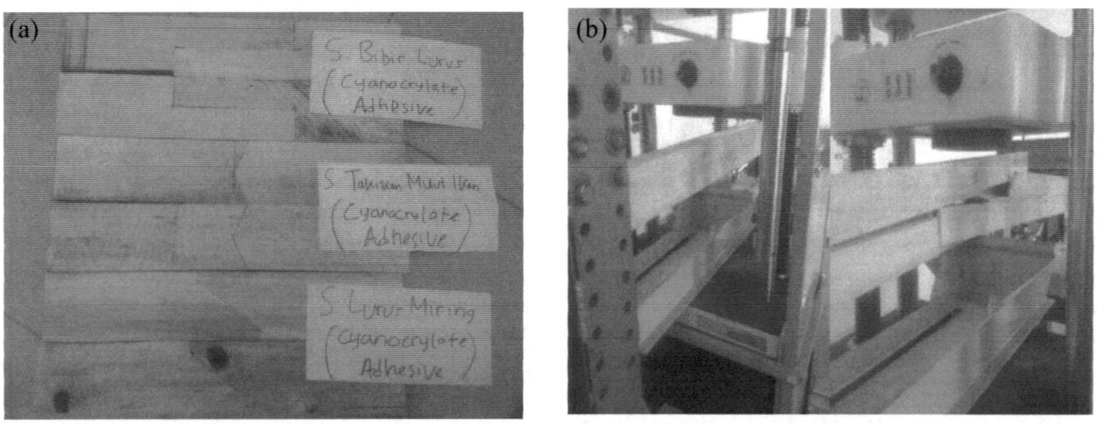

Figure 11. CLT beam with slap splices: (a) lap splices; (b) bending test

Furthermore, based on the aforementioned tensile strength test, another 4 CLT beams were prepared for bending test (figure 11b). The average $MOR_{upright}$ and MOR_{lying} are 2.7 and 3.4 MPa, respectively. Apparently, adding lap-splices and changing adhesives has marginal impact on CLT beam bending strength. It was probably affected by low pressured applied during the manufacturing process of CLT beams or insufficient glue applied to fill gaps and uneven surfaces between laminas, especially at the location of splices. Hence, premature delamination was occurred, as observed from the specimen final condition (figure 11b), that consequently lower the CLT beam strength. However, further research would be needed to confirm this initial finding.

6. Conclusions

This paper establishes an initial investigation on the utilization of recycled wood as raw material for cross laminated timber. Flexural behavior of CLT beams was examined and the effect of lamina lap splices and adhesives on CLT beam flexural strength was studied. While the obtained flexural strength was below the ANSI required strength for CLT, yet the result is an essential starting point to guide future investigation on ways to improve its performance. Apparently, applying several variations of lamina lap-splices and adhesives has marginal impact on CLT beam flexural strength.

In this exploratory research, recycled wood obtained from wooden boxes, wooden pallets, and scrap wood, for expediency, were mixed during the manufacturing process of CLT beams. Further investigation is needed to assess how recycled-wood grouping would improve CLT flexural strength. Non-destructive tests could be implemented on piles of recycled wood to select a more uniform strength quality of recycled wood. Impact of assorting wood strength per layer could also be examined. A comprehensive study on lamina splices would be desirable to find simpler alternatives yet could maintain similar connection strength to a finger joint.

References

[1] Cipto, H. (2016). "Each year, Indonesia loses 684,000 hectare forest." (in Indonesian) <https://regional.kompas.com/read/2016/08/30/15362721/setiap.tahun.hutan.indonesia.hilang.684.000.hektar>, accessed on 9 Oktober 2018.
[2] Medialingkungan (2015). "Timber consumption in Indonesia exceeds available timber stock." (in Indonesian) <http://medialingkungan.com/index.php/news/nasional/penggunaan-kayu-indonesia-melibihi-batas-pasokan-yang-ditetapkan>, accessed on 6 Oktober 2018.
[3] Hadiyantono, T. (2018). "Production of Indonesian timber limited by land usage." (in Indonesian) <http://industri.kontan.co.id/news/produksi-kayu-olahan-indonesia-terkendala-lahan>, accessed on 9 Oktober 2018.
[4] Alisahbana, A. (2014). "Two aspects regarding Indonesian forest moratorium." (in Indonesian) <http://wri-indonesia.org/id/blog/2-things-you-need-know-about-indonesias-forest-moratorium>, accessed on 9 March 2019.
[5] Widodo, A. (2014). "An overview of deforestation condition in Indonesia." (in Indonesian) <https://rifdoisme.wordpress.com/2014/10/06/kerusakan-hutan-deforestasi-di-berbagai-wilayah-indonesia-suatu-tinjauan-sederhana>, accessed on 19 Mach 2019.
[6] Supartini (2012). "Characteristics of cross laminated timber made from fast growing timber with different number of layers." Thesis, Graduate school, Institut Pertanian Bogor (IPB), Bogor, (in Indonesian).
[7] Mutmainnah, Sadiyo, S., and Karlinasari, L. (2014). "Experimental investigation of cross laminated timber shear walls made from three Indonesian timber species." *Jurnal of Forest Sains*, Vol. 11, No. 2, pp. 99-109 (in Indonesian).
[8] Karacabeyli, E. and Douglas, B. (2013). *"CLT Handbook: Cross-laminated timber."* U.S. edition, FPInnovations, Pointe-Claire, Canada
[9] Bradner, R., Flatscher, G., Ringhofer, A., Schickhofer, G., and Thiel, A. (2016). "Cross Laminated Timber (CLT): overview and development." *European Journal of Wood and Wood Products*, Vol. 74, No. 3, pp. 331-351.
[10] Buck, D., Wang, X., A., Hagman, O., and Gustafsson, A. (2016). "Further development of cross-laminated timber (CLT): mechanical tests on 45° alternating layers." *Proceeding World Conference on Timber Engineering*, Vienna, Austria.
[11] Buck, D., Wang, X., A., Hagman, O., and Gustafsson, A. (2016). "Bending properties of cross laminated timber (CLTt) with a 45° alternating layer configuration." *BioResources*, Vol. 11,

No. 2, pp. 4633-4644.

[12] Lestari, R., Y., (2017). "Cross laminated timber: production, characteristics, and development." *Journal of Riset Industri Hasil Hutan*, Vol. 9, No. 1, pp. 41-55 (in Indonesian).

[13] Lattke, F. and Lehmann, S., (2007). "Multi-storey residential timber construction: current developments in Europe." *Journal of Green Building*, Vol. 2, No. 1, pp. 119-129.

[14] Van De Kuilen, J., W., G., Ceccotti, A., Xia, Z., and He., M. (2016). "Very tall wooden buildings with cross laminated timber." *Procedia Engineering, the Twelfth East Asia-Pacific Conference on Structural Engineering and Construction*, Vol. 14, pp. 1621-1628.

[15] Weissensteiner, J. and Barbu, M., C. (2013). "Cross Laminated Timber – European Experinces." *Pro Ligno*, Vol. 9, No. 4, pp. 69-72.

[16] Frangi, A., Bochicchio, G., Ceccotti, A., and Lauriola, M. P. (2008). "Natural full-scale fire test on a 3 storey XLam timber building." *Proceeding of the 10th World Conference on Timber Engineering*, Miyazaki, Japan.

[17] Fragiacomo, M., Menis, A., Clemente, I., Bochicchio, G., and Ceccotti, A. (2013). "Fire resistance of cross-laminated timber panels loaded out of plane." *Journal of Structural Engineering*, American Society of Civil Engineers (ASCE), Vol. 139, No. 12.

[18] Pei, S., van de Lindt, J. W., Popovski, M., Berman, J. W., Dolan, J. D., Ricles, J., Sause, R., Blomgren, H., Rammer, D. R. (2016a). "Cross-laminated timber for seismic regions: progress and challenges for research and implementation." *Journal of Structural Engineering*, American Society of Civil Engineers (ASCE), Vol. 142, No. 4.

[19] Ramage, M., H. Burridge, H., Busse-Wicher, M., Fereday, G., Reynolds, T., Shah, D., U., Wu, G., Yu, L., and Fleming, P. (2017). "The wood from the trees: the use of timber in construction". *Renewable and Sustainable Energy Reviews*, Vol. 68, pp. 333-359.

[20] Fu, M., Liu, Y., Li, N., Zhang, Z., and Siviero, E., (2014). "Application of modern timber structure in short and medium span bridges in China." *Journal of Traffic and Transportation Engineering*, Vol. 1, No. 1, pp. 72-80.

[21] Pei, S., Rammer, D., Popovski, M., Williamson, T., Line, P., van de Lindt, J. W. (2016b). "An overview of CLT research and implementation in North America." *Proceeding World Conference on Timber Engineering*, Vienna, Austria.

[22] Hymas, A., Watts, S., and Harvatt, C. (2018). "Cost gap between CLT and concrete is narrowing." <https://www.trada.co.uk/news/cost-gap-between-clt-and-concrete-is-narrowing>, accessed on 1 April 2019.

[23] Green, J. (2012). "Passivhaus Rainbow Duplex in Whistler". <http://www.jetsongreen.com/2012/04/panelized-passivhaus-rainbow-duplex-british-columbia.html>, accessed on 9 March 2019.

[24] Gurit (2002). "Bonding with epoxy in wood construction <https://www.amtcomposites.co.za/sites/default/files/media/howto/SP_Bonding%20with%20Epoxy%20in%20Wood%20Construction.pdf>, accessed 4 May 2019.

[25] American Standard of Testing Materials (ASTM) D143-14 (2014). "Standard test methods for small clear specimens of timber.", ASTM International, West Conshohocken, PA, www.astm.org.

[26] Munandar, W., A. (2017). "Exploratory Study on the utilization of recycled wood as raw material for cross laminated timber." Senior thesis, Universitas Bandar Lampung (UBL), Bandar Lampung, (in Indonesian).

[27] Christiyanto, A. (2017). "Experimental investigation of cross laminated timber made of recycled wood considering various lamina lap splices and adhesives.", Seninor thesis, Universitas Bandar Lampung (UBL), Bandar Lampung, (in Indonesian).

Acknowledgments

Ruri Damayanti, a research assistant at the Department of Civil Engineering of UBL, assisted at the manuscript preparation and layout editing of this paper. Her contribution is greatly appreciated.

IOP Conf. Series: Materials Science and Engineering **669** (2019) 012012 doi:10.1088/1757-899X/669/1/012012

One day high-performance concrete mechanism of plate, beam, and column

C Niken[1]

[1] Civil Engineering Department, Faculty of Technology, Lampung University, Bandar Lampung, Indonesia

chatarinaniken@yahoo.com

Abstract. Plate, beam, and column have a different mechanical characteristic. Understanding the one-day mechanism of the three concrete structural elements is the objective of this study. The quality of the concrete used is fc' of 60 MPa. Observation of concrete strain was done during 24 hours. Five small specimens measuring of 150 mm × 150 mm × 600 mm with one strain gauges (EVWSG) for each specimen, three specimens in horizontal position as a beam and two specimens in vertical positions as a column; one full-scale beam specimen measuring of 200 mm × 600 mm × 3000 mm (four EVWSG), and one plate specimen measuring of 3000 mm × 1600 mm × 150 mm (four EVWSG) were used. All specimens were covered with styrofoam. One-day behavior was found by computing the average strain change from a small beam and column specimens, and average strain change of the fourth EVWSG in the plate/full-scale beam. As a result, there is a similar type between concrete temperature and the behavior of plate, beam, and column. The behavior of the beam is identically to the plate behavior. Column behavior is more fluctuating than beam and plate. High rate shrinkage is significant in beam and plate, but in the column, expansion was dominated.

1. Introduction

The column is a significant element structure in a building. Design of reinforced concrete for construction should be based on the strong column and weak beam. Lampropoulos and Dritsos [1] has stated that biaxial stress will reduce the strength of the column. Axial load and shrinkage reduce stiffness, and column strength. Dimensions and presence of reinforcement were affected the shrinkage. Nevertheless, the standard specimen for shrinkage test is beam (ASTM C157/C157M-17) [2]. The difference between beam and column behaviour was determined by the both early age behavior. Because at the early age expansion dominates column behavior, in long-term period shrinkage in column become smaller than in beam [3]. Based on beam specimen, Yahia [3] mentioned that early age cracking does not depend on compressive strength of concrete (fc'). These cracks will reduce long-term performance.

Plate and rigid pavement have a largest surface than another element structure. Plate behaviors is the most influenced by surrounding. The factors that influence early-age shrinkage are hydration, relative humidity, self-desiccation, environment factor, water to cement ratio, curing methods, and admixture [5].

In the first one day, the behavior of specimens with covering plastic or styrofoam does not influenced by environment. Water in the air can't penetrate to the concrete. Water is the main factor in hydration process and shrinkage mechanism. The amount of water was calculated carefully in concrete mix design. Some countries have concrete mix design procedures and calculations such as Bureau of

Indian Standards-I.S.10262-2009, American (ACI) method, and Indonesian standard SNI 7656-2012 [6-10]. To maintain the amount of water for the hydration process in accordance with the mix design, it is appropriate to cover the concrete immediately after pouring. Consequence of it, autogenous deformation occurs in the specimen, because autogenous deformation is a shrinkage or expansion without external influences. Right after cement meet with water, chemical reaction occurs. Temperature which appear by this reaction creates physical mechanism. Therefore, in one day age chemical, physical and autogenous shrinkage occur. It is challenging to separate chemical, physical and autogenous shrinkage because they are happened simultaneously.

The first behaviour is expansion and shrinks in macro volume. If the hydration process occurs under lack of water, many pores become empty and change concrete microstructure, such as decreasing space for hydration. The influence of the empty capillary pore is the reduction of internal relative humidity (70% for w/c less than 0.3). The reducing relative humidity process will increase capillary pore water pressure (σ capillary). The pores solution and its radius meniscus influence the magnitude of the pressure. The quality of matrix to be decrease. The consequence is aggregate restrains autogenous shrinkage which occurred in cement paste [11]. High-performance concrete (HPC) is concrete with high strength and durability. HPC makes the element structure becomes smaller than normal concrete, while column need great stiffness. Therefore, the use of HPC to be a crucial of tall building. HPC can only be made with a limited amount of water generally equal or smaller than 0.36; so, not all cement is hydrated; and the matrix does not contribute to the modulus of elasticity. Modulus of elasticity is more influenced by aggregates. Based on this reason, autogenous shrinkage is the major behavior in HPC, while in normal concrete drying shrinkage is a major deformation [3]. In high-strength concrete autogenous shrinkage almost similar to drying shrinkage [12]. In the fact, beside shrinkage, there is expansion in the concrete. The young concrete should be able to handle the two types of deformation which happen alternately to avoid early cracking.

The limited water for hydration and cementitious material influences inner humidity, micro prestresses, degree of hydration and interface structure; therefore, fluctuation of expansion and shrinkage was also affected. Wu et al. and Yahia [4,13] were also studied the influences of limited water and cementitious materials. Chemical and autogenous shrinkage causes self-desiccation because delayed ettringite formation [4]. Due to the limited water-to- cement ratio, HPC develops higher early-age shrinkage which lead initial crack. The initial cracks are easily growth because of self-desiccation. To reduce the risk of random cracking and joint raveling, saw-cutting should be done in rigid pavement in rigid pavement at early age [14].

Aly and Sanjayan [15] was observed the mechanism of early age shrinkage of concrete based on the reason phenomenon that shrinkage at early ages are found to increase with increased curing. Mechanism behavior in one-day age of plate, beam, and column which was has varies size need to study. Size determines of the drying rate. Characteristic of concrete is a strong non-statistical size effect on both the structural strength and on the rate of creep and shrinkage. The size effect is now widely acknowledged to be a serious issue even in the ACI code-making committees, ACI 318 and 445 [16]. Safety factors for quasi-brittle structures should significantly increase with structure size [17].

Understanding of one-day mechanism in column, beam, and plate with laboratory and full-scale specimens give a useful contribution in treating and designing concrete correctly to meet excellent performance and structure safety. Treatment delay time can make raveling, spalling, and random cracking in the concrete surface. The treatment determines the quality and resistance of structural elements to weather and loads throughout service life. Proper treatment can only be done if the behavior of each type of structural element is well understood.

2. Experimental program

The research was conducted in Jakarta, Indonesia, with humid tropical weather.

2.1. Speciments

This research was performed experimentally using one specimen of 3000 mm × 1600 mm × 150 mm as a plate (Fig. 1a), three specimens of 150 mm × 150 mm × 600 mm as a laboratory-scale specimen with one embedded vibrating wire strain gauge (EVWSG) per specimen (Fig 1b); two specimens of 150 mm × 150 mm × 600 mm in vertical position as a column with one EVWSG per specimen (Fig. 1c), and one full-scale specimen of 200 mm × 600 mm × 3000 mm as plate (Fig. 1d). The shrinkage in the beam center was affected by the maximum deflection. Based on this study, the position of the EVWSG was at the end of the beam, 5 cm for laboratory-size of beam and column (Figure 1a) and 70 cm for the full-scale specimen (Figure 1b). HPC with a target compressive strength of 60 MPa and slump flow diameter of 35 ± 2 cm were used.

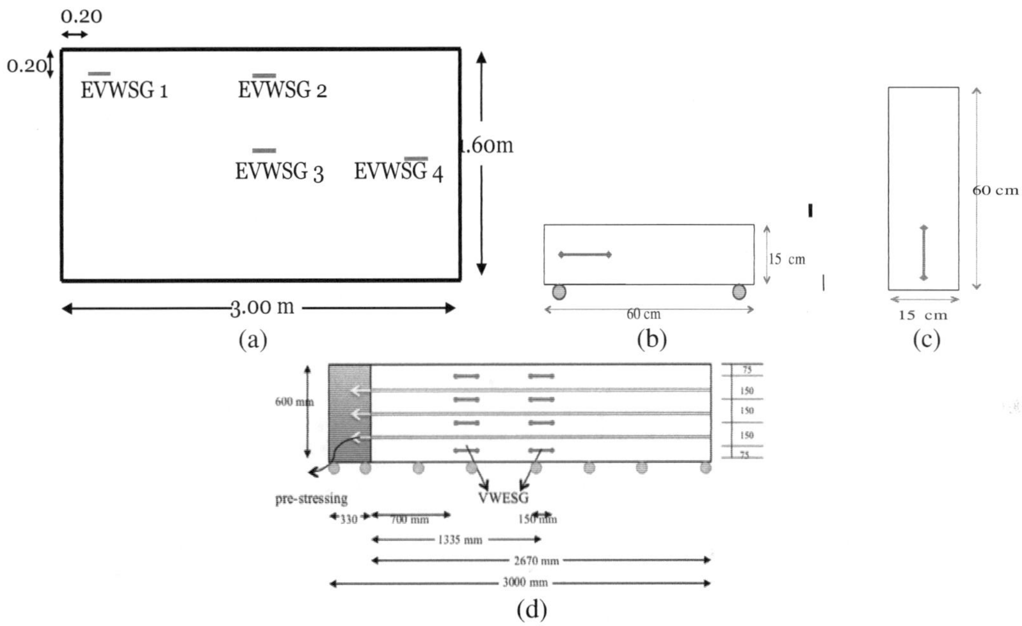

Figure 1. Specimens and placement of EVWSG: (a) Plate, (b) Laboratory-scale beam specimen, (c) Laboratory-scale column specimen, (d) Full-scale beam specimen

2.2. Materials

The mix design was conducted in compliance with ACI 211.4R (1993) with a limit of 500 kg/m3 cement content to meet the shrinkage factor closest to 1 (ACI 209R, 1992). Ordinary Portland Cement (OPC) produced by Indocement Ltd was used. The condition of the aggregate was saturated surface dry (SSD). Fine aggregate in the form of river sand was brought from Sungai Liat (Bangka, Sumatra, Indonesia), specific gravity (SSD) was 2.605, and absorption was 0.4%. The sand had been filtered and cleaned using a mixture of standard graphs obtained from the mid-gradation. Fine aggregate shall be free of the injurious amount of organic impurities (American Society for Testing and Materials International, 2002). Coarse aggregate in the form of volcanic rock fragments was obtained from Banten, West Java, Indonesia. The composition of the coarse aggregate used was 70% sized 13 - 19mm, specific gravity (SSD) of 2.563, absorption of 1.543%, and 30% sized 6 - 12mm, a specific gravity of 2.636, and absorption of 2.26%.

The added material used was silica fume of 8% cement weight, produced by Sika Indonesia Ltd. To achieve the inclanation high strength with a low ratio of water to cementitious material and excellent workability, polycarboxylic super plasticizer under the commercial name Visco Crete 10 from Sika

Indonesia Ltd was added to the concrete mix at the high range water reducer (HRWR). A dose of HRWR of 1.4% cement weight was added according to that generally used in Indonesia. The local water was supplied by the Structure and Material Laboratory of Universitas Indonesia. An electrical scale was used for cementitious material and water to obtain the accurate ratio of water to cementitious material.

The mix composition was 500 kg/m^3 of OPC, 40 kg/m^3 of silica fume, 142.6 kg/m^3 of water, 800 kg/m^3 of sand, 935 kg/m^3 of coarse aggregate, and 7.6 kg/m^3 of HRWR. During the concrete mix design stage, all of the aggregates were assumed to be an under SSD condition. A tilting drum mixture with a 0.3 m^3 capacity was used. The mixing started with all cementitious material in a dry state, followed with 50% fine aggregate. Subsequently, 50% of water was added to the revolving mixture. Then, these materials were mixed for approximately 1½ minutes. Next, 50% of water was slowly poured in, which was homogenously mixed with HRWR. After that, 100% coarse and 50% fine aggregate were added. With all the materials placed according to their order into the mixer, the concrete was mixed for approximately 3 minutes. The slump flow of the mixture was measured before pouring by using an Abrams cone upside down.

2.3. Methods

In this research, shrinkage was measured as strain change against time by installing one EVWSG in each beam and column specimen, and four EVWSGs in a full-scale specimen and plate (Figures 1a and 1b). The EVWSG was able to detect the strain up to 3000 με with an accuracy of about .025% and concrete temperature between -80oC and 60oC with about .5% accuracy. Right after casting, specimens were covered with styrofoam to eliminate water evaporation (Figure 2). The observation was performed right after pouring every 15 minutes until 24 hours using a readout.

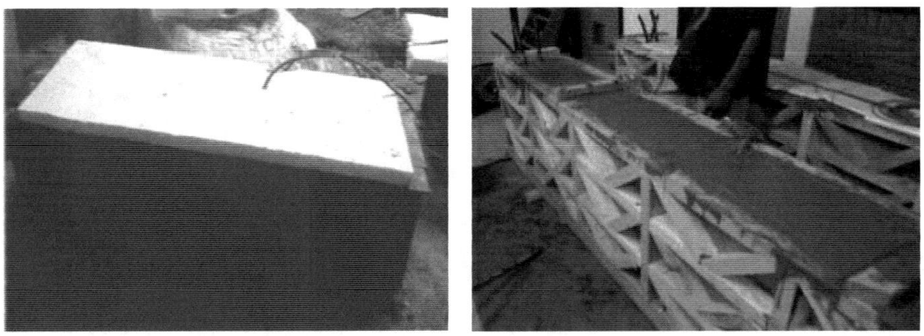

Figure 2. Styrofoam covers the specimens for one day

Data of the three beam laboratory size specimens were analyzed using Dixon's criteria as the standard practice for dealing with outlying observation. ASTM E 178-02 [18] has mentioned that Dixon criteria, based entirely on ratios of differences between the observations may be used in cases where it is desirable to avoid calculation of standard deviation or where quick judgment is called for. For a Dixon test, the sample criterion or statistic changes with sample size. The equations of the Dixon criteria for 3 to 7 samples with $x1 \leq x2 \leq \ldots \leq xn$ are as follows:

- If smallest value is suspected:

$$r_{10}=(x_2-x_1)/(x_n-x_1) \quad (1)$$

- If the largest value is suspected:

$$r_{10}=(x_n-x_{n-1})/(x_n-x_1) \quad (2)$$

Using Equations 1 and 2 for data at ages with a significant difference of shrinkage from the three specimens were computed. Shrinkage of laboratory-size specimens was found by using an average of accepted data. Shrinkage of the full-scale specimen was observed by using the average value of all

EVWSG. The largest of all beam specimens was preferred to reach high safety although Dixon's criteria were used. Column behavior was taken from the average strain changes that occurred in the two specimens. Plate and full-scale beam behavior were made from the average strain changes that happened in all EVWSG in the specimen.

Beam temperature was preferred from the full-scale beam. The temperature in the full-scale beam and plate were taken from the average temperature that occurred in all EVWSG in the specimen. The column temperature was taken from the average temperature that occurred in the two specimens.

By using a literature study, the one-day behavior of plate, beam, and column mechanism can be obtained.

3. Results and discussion

3.1. Result

Data of the three beams laboratory size specimens were analyzed using Dixon's criteria as the standard practice for dealing with outlying observation can be seen in Niken [3]. One day behaviour of plate, beam, and column can be seen in Figure 3a, the inner temperature of the three elements structure and surrounding temperature shown in Figure 3b.

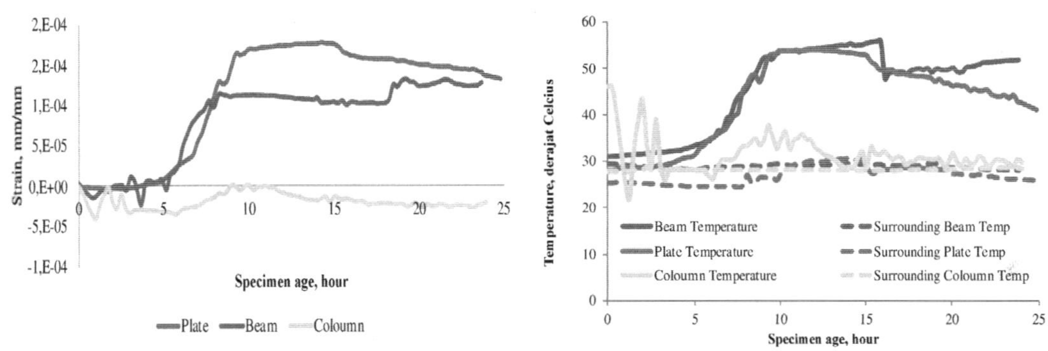

(a) (b)

Figure 3. (a) One day of plate, beam and column, (b) Inner temperature in plate, beam and column and surrounding temperature

One day behavior of the plate, beam and column has similar type to its inner temperature. Inner temperature type of beam is identic to the plate, while internal column temperature shows fluctuation type (Figure 3a and 3b).

3.2. Discussion

Lura and Wyrzykowski [13] have studied the use of low water-to-cement ratio and silica fume in HPC. The both makes internal relative humidity decrease. If no extra water is available for cement hydration, all pore water is progressively consumed. This condition occurs in concrete element with is sealed. Decreasing internal relative humidity leads self-desiccation because the growth of CSH need high internal humidity. Emptying pores leads the increase of capillary pores tension and because of its, external volume contraction become increase. Autogenous deformation to be a major factor with w/c less than 0.40. The change of a part of the water to be a chemically bound in product hydration makes deformation in concrete. Inner temperature, energy, molecules velocity, density and momentum can't be separated from the chemical reaction. The chemical and physical are interrelated. The both bounds water has smaller specific volumes than free water. Contrary to drying shrinkage, autogenous shrinkage occurs without any loss of moisture from the concrete [13].

Internal temperature appears from the hydration process; all concrete behavior follows temperature pattern (Figure 3a, 3b). Inner temperature in plate, beam, and column are higher than its surrounding temperature, with the except at 15-24 hours, the inner temperature of column is equal to surrounding temperature (Figure 3b). Klemczak and Wroble [19] were studied concrete deformation using macroscopic and microscopic phenomena. The macroscopic behavior such as temperature, volume, pressure and mass some time can't be explained deeply concrete deformation mechanism. The good explanation of the mechanism depends on well-known microscopic phenomena. Although the average interatomic spacing is constant, the atoms are in continual vibration; thus, a spring model is useful in some instances. Spring model does permit us to depict the effect of atomic vibration, thermal expansion, activation energies, and other properties. On the basic of the spring model of interatomic forces, it is easy to see how elastic waves travel through solids. By elastic waves, heat is conducted through opaque solids. It will be noted that the spring is non-linear because of the "spring constant" under compression increase rapidly [20]. Particles in the column were under pressure by its weight. The nonlinearity of spring in the column and hydration process caused high fluctuation of temperature and shrinkage (Figure 3a and 3b).

Figure 3a shown there are two types of behaviors: shrinkage (uphill curve) and expansion (decreases curve). Expansion in autogenous deformation is not due to retarder or the increase of bellite in cement, but it was led by re-absorption of water at the start of the cement setting process [21]. At the range time pf 0-5 hour, significant expansion occurs in the column (Figure 3a). At that time solution and dispersion, process occur [22], CH and ettringite formed [23], optimum Ca^{2+} from alite reaction occur [24] and the number of pores decreased slightly [23]. Solution and dispersion process, ettringite, and Ca^{2+} lead a volume expansion. The axial stress alters the interstitial sites. This movement permits additional strain because the lattice distortion is reduced. Axial stress occurs in the column by its weight. Consequently, additional strain occur and the expansion in the column is more significant than in the beam and plate. Inner temperature in column more fluctuating than in plate and beam (Figure 3b). C-S-H begins to form at the age of 3 hours with the rate increase until the period of 28 days. During 3 – 10 hour, the shrinkage rate is high and reaches the highest value for one day of observation. Shrinkage in the column shows the lowest value because of abnormally expansion phenomenon due to chemical reactions that occurs under pressure which inhibits shrinkage due to the formation of C-S-H. Inner temperature in column much lower and more fluctuating than in plate and beam, but with similar pattern. After 15 hours, inner temperature in column similar to surrounding temperature.

At the age of 10 hours, C-S-H begins to thick [25]. When the layer of C-S-H becomes sufficiently thick, the diffusion process occurs, and the porous layer of hydrates covers the surface of the anhydrous cement. The volume of the formed C-S-H is larger than the initial volume of C3S. The fact that the topochemical reaction of the anhydrous product may lead to swelling. According to these data, the volume of C-S-H represents 155% of the volume of the cement anhydrous [26][27]. Capillary stress by emptying pores caused expansion. As a result of this mechanism, autogenous behavior in beams and plates shows a balance between shrinkage and expansion which is shown as a flat line, while in the expansion column it dominates (Figure 3a).

Marušić and Štirmer was mentioned that autogenous shrinkage is a function of compressive strength as follow

$$\varepsilon_{ska\,1} = 0.0046\, f_{cm} - 0.2821 \quad (3)$$

$\varepsilon_{ska\,1}$ is autogenous shrinkage of concrete at one day of age (10^{-3} or mm/m), f_{cm} is mean compressive strength of concrete at 28 days tested on a 15 cm cube (MPa).

By using Equation 3, autogenous shrinkage is equal to 0.0002476. This value has not been reached by all three specimens. The maximum autogenous shrinkage in the plate, beam and column is 70%, 50% and 0,8% of Equation 3 respectively.

4. Conclusion

One day behavior of plate, beam, and column showed a similar type to its inner temperature. There are two types of the one-day behavior: shrinkage and expansion. Shrinkage caused by product hydration growth, and expansion caused by dissolution and dispersion, ettringite, Ca^{2+}, the topochemical reaction of un-hydrous product and special in column plus reducing lattice distortion. One-day deformation in beam and plate almost has similar values, while in the column the autogenous is smaller and more fluctuating than in beam and plate. The minor deformation in the column due to shrinkage by-product hydration growth held by swelling abnormally because the chemical reaction occurs under pressure reducing lattice distortion. Column fluctuation caused by the change of interatomic spacing modeled by the change of "spring constant" due to domination of shrinkage and expansion take turns quickly by the own column weight and product hydration growth.

References

[1] Lampropoulos A P and Dritsos S E 2011 Concrete Shrinkage Effect on the Behavior of RC Columns under Monotonic and Cyclic Loading *Constr Build Mater* Volume 25 (4) pp 1596-1602
[2] ASTM C157/C157M-17 *Standard Test Method for Length Change of Hardened Hydraulic Cement Mortar and Concrete* American Society for Testing and Materials

[3] Niken C, Elly T and Supartono F X 2017 Long-term Deformation of Beam and Column of High - Performance Concrete *IJ Tech* Volume 8 (5) pp 811-819
[4] Yahia A 2009 *Shrinkage development in High-Performance Concrete* ACI – UAE Chapter International 28 slides Available online at https://ecitydoc.com/download/autogenous-shrinkage_pdf
[5] Allena S 2011 *State-of-the-art review on early-age shrinkage of concrete* Indian Concrete Journal 85(7) pp 14-20
[6] Bureau of Indian Standards 2009 *Concrete Mix Proportioning* Gudelines First Revision New Dehli 16 pages
[7] ACI 209R-92 1992 *Prediction of creep, shrinkage, and temperature effects in concrete structure* American Concrete Institute pp 1-47
[8] ACI 211. 4R 1993 *Guide for Selecting Proportion for High Strength Concrete with Portland Cement and Fly Ash* International Concrete Abstracts Portal Volume 90 pp 272–283
[9] ACI 211.1-91 Concrete Mix Design American Concrete Institute
[10] Standar Nasional Indonesia 2847 2013 *Persyaratan Beton Struktural untuk Bangunan Gedung* Badan Standardisasi Nasional 255 pages
[11] Lura P and Wyrzykowski M 2018 *Autogenous Shrinkage* Available online at https://www.empa.ch/web/S38/autogenous-shrinkage
[12] Tazawa E and Miyazawa S 1995 Experimental Study on Mechanism of Autogenous Shrinkage of Concrete *Cem Con Res* Volume 25 (8) pp 1633-1638
[13] Wu L, Farzadnia N, Shi C and Zhang Z 2017 Autogenous Shrinkage of High Performance Concrete: a Review *Const Build Mater* Volume 149 pp 62-75
[14] Raoufi K, Their T, Weiss W J, Olek J and Nantung T E 2009 *Saw-cutting Guidelines for Concrete Pavements for Time and Depth of Saw Cutting* JTRP Technical Report Purdue-e Pub Purdue University pp 1-777
[15] Aly T and Sanjayan J G 2008 Mechanism of early age shrinkage of concrete *Mater Struct* 42 (4) pp 461-468
[16] Hubler M H, Wendner R and Bažant Z P 2015 Statistical Justification of Model B4 for Drying and Autogenous Shrinkage of Concrete and Comparison to Other Models *Mater Struct* Volume 48 (4) pp 797-814
[17] Bažant Z P 2012 *Rational and Safe Design of Concrete Transportation Structures for Size Effect and Multi Decade Sustainability* Technical Report Safetea - LU Project Infrastructure Technology Institute Northwestern University Illinois 26 pages
[18] ASTM E178-02 2002 *Standard practice for dealing with outlying observation* American Society for Testing and Materials pp 1-18
[19] Klemczak B and Knoppik-Wrobel A 2011 Early-age Thermal and Shrinkage Cracks in Concrete Structures Description of the Problem *Architec Civ Eng Environ* No 2 pp 35-47
[20] Van V L 1973 *Materials Science for Engineers* Addison Wesley Publishing Company Canada Philipines pp 1-545

[21] Marušić E and Štirmer N 2016 Autogenous Shrinkage and Expansion Related to Compressive Strength and Concrete Composition *Adv Concr Technol* Volume 14 pp 489-501
[22] Bullard J W, Jenning H M, Livingstone R A, Nonat A, Scherer G W, Schweitzer J S, Scrivener K L and Thomas J J 201 Mechanism of cement hydration *Cem Concr Res* Volume 41 pp 1208-1223
[23] Kurtis K 2015 *Portland cement hydration* School of civil engineering Georgia Institute of Technology Atlanta Georgia, Available online at fliphtml5.com/xtto/ogts/basic 69 slides
[24] Paulini P 1990 Reaction mechanisms of concrete admixture *Cem Concr Res* Volume 20 pp 910-918
[25] Taylor H F 1997 *Cement Chemistry* Thomas Telford 2nd Edition London p 459
[26] ASTM C78/C78M-1 2002 *Standard Test Method for Flexural Strength of Concrete* American Society for Testing and Materials
[27] Barcelo L, Boivin S, Acker P, Toupin, J and Clavaud B 2001 Early age Shrinkage of Concrete: Back to Physical Mechanisms *Concr Sci Eng* Volume 3 pp 85-91

Acknowledgments
We would like to thank the University of Lampung for supporting our research. I express my deepest gratitude to the University of Indonesia, for all the support of laboratory facilities, places, and equipment. I also convey my appreciation to Dr. Yosia Rastandi, who has always lent me a measuring device. Also, our thanks to Sika Indonesia Ltd and Indocement Ltd for their full support materials. Besides, thank you for the technical staff of Structures and Materials Laboratory, Faculty of Civil Engineering, the University of Indonesia for their technical support in conducting the experiments.

IOP Conf. Series: Materials Science and Engineering **669** (2019) 012013 doi:10.1088/1757-899X/669/1/012013

Constructed floating wetland with mooring system technology to improve water quality in Gedebage Reservoir, Bandung

Y M Hidayat[1], W Herwindo[1] and S Bachri[1]

[1] Research Center For Water Resources, Ministry of Public Work and Housing, Jl. Ir H. Juanda No. 193, Bandung 40135

yuliya96119@gmail.com

Abstract. Gedebage Reservoir is built as retention ponds, in the future it will be developed into an integrated religious tourism area, so that the quality of water reservoir must suitable for tourism. Constructed floating wetland with mooring system, besides being used as water quality improvement technology, it will also add aesthetic value. The purpose of this study was to identify the possibility of applying the constructed floating wetland in the Gedebage Reservoir area. This study uses secondary data to determine the design parameters and characteristics of water sources, design parameters are used for location recommendations, while the characteristics of water sources are to determine the accuracy of the technology used. The result shows that the placement of a floating wetland technology with mooring system in Gedebage reservoir has meets the requirements of design parameters than in a biological filter channel. The characteristics of the Gedebage Reservoir water source that have not met requirements consist of several parameters, namely: Ammonia, Chlorine, Mn, DO, Sulfide, Detergent, Oil and Fat, Cd, Pb, DO, BOD, COD, Fecal Coliform and Total Coliform. Whereas, based on the ratio of BOD5 / COD in the dry season that ranged from 0.25 indicating that the water had been toxin, so that it would disrupt the development of biofilms, so in addition to wetland, a pre-treatment was needed before entering the reservoir.

1. Introduction

Gedebage reservoir is located in Gedebage Sub-district, Bandung, basically used to cope the flood of Cinambo River in Citarum Watershed. With the reservoir, flood-prone locations will have clearly boundary, so that the water level, discharge, and volume of water that must be removed can be controlled. The land used as a reservoir is limited by the embankment so that water from outside cannot supply. The reservoir is expected to reduce the impact of flooding that often occurs in Gedebage.

Besides reducing the impact of flooding, the Gedebage Reservoir area will be developed into integrated Religious Tourism with the construction of the Al-Jabbar Mosque and the landscaping surrounding area. The development of the reservoir as a tourism area cause the water quality must support to these activities, so that the water quality in reservoir should refer to water criteria for tourism.

The Gedebage reservoir is supplied by Cinambo River which is contaminated as a result of activities in upstream, both from domestic waste, agricultural, and industry. Contaminated water such as nutrients eq. excessive nitrogen and phosphorus, and other organic materials cause the low value of dissolved oxygen (DO), combined with high levels of BOD, COD, and fecal coliforms bacteria also the disruption of the lives of aquatic ecosystems. In order to prevent the degradation of water quality in the reservoir, it is necessary to do an integrated watershed management. It is also a special concern of the Government to overcome the pollution and damage of the Citarum Watershed which is confirmed through Presidential Regulation 15 of 2018 concerning the Acceleration of Citarum Watershed Pollution and Damage Control [1]. Referring to the regulation Citarum as the longest river in West Java and the dirtiest in the world should be solve whole problem in 7 years.

In supporting for the Presidential Regulation above and to improve water quality, it is necessary to improve water quality in Gedebage Reservoir because it is located in the Citarum Watershed through a technological approach.

Constructed floating wetland with mooring system is a technological innovation produced by the Research Center of Water Resources which will be applied to control water pollution that supply into Gedebage Reservoir. This innovation technology is developed into two technologies namely floating wetland technology and biocord technology.

Floating Wetland is the development of constructed wetlands which is placed flood and quiet stream [2], and as a buffer on the waterfront [3], consisting of plant vegetation grown above media with floating structures above the water surface [4]. In the floating wetland system, the plant stem is above the water surface, the roots grow downward in the water column through growing media, so the plants take nutrients directly from the water column [5].

Biocord is a technological innovation from Bishop Water to increase surface area so that biofilms are formed faster. The results of previous studies, biocords can reduce Suspended Solid by 87.2%, COD, NH3-N, TN, and TP respectively at 19.4-34.4%, 55.2-74.0%, 46.2-55.9%, and 13.1-18.5% [6]. Biocords are developed as mooring system as part of floating wetlands.

Based on the description above, the purpose of this study was to identify the possibility of applying a floating wetland in the Gedebage Reservoir area.

2. Methodology

The study was conducted from August 2018 - April 2019. The research location was located in Gedebage Reservoir, Gedebage Sub-District, Bandung. This study attempts to analyze the possibility of a constructed floating wetland technology with mooring system applied. The data in this research were obtained from secondary data in Citarum River Basin Organization (RBO) and Environmental Services Agency in Bandung.

2.1. Research data

Data related to the location study consisted of: site plan for the Gedebage Reservoir construction, cross section, detailed design which contains topographic analysis, geology, hydrology, storage capacity and simulation pattern. Data for water resources characteristics consist of water quality measurement in Cinambo River.

2.2. Location

Location assessment criteria are based on analysis of the wetland design parameters. These parameters consist of water depth, Hydraulic Retention Time (HRT), and Surface Coverage and Hydraulic Loading Rate. Some references that can be used are as follows:

- Water depth [7-11]
- Hydraulic retention time [8,12,13]
- Surface Coverage [7-10, 14]
- Hydraulic Loading Rate [7, 8, 10, 15]

2.3. Analysis of water quality characteristics

The water quality criteria is refer to Governmental Regulation 81 of 2001 for Class II. Assumptions are made because the future of Gedebage Reservoir will be used as an integrated religious attraction along with the presence of the West Java Grand Mosque.

2.4. Technology selection analysis

The choice of technology is based on the location that has been obtained with conformity to the design parameters and water quality characteristics to be achieved. This analysis determines the treatment of constructed floating wetlands with mooring system in a single treatment or combination with pre-treatment.

3. Results and discussion

3.1. Study

3.1.1. Constructed floating wetland with mooring system placed on channels for biological filters. Study have been carried out on the detailed design of Reservoir Construction (Figure 1), according to the request of Citarum RBO carried out on the channel originally intended for biological filters. The biological filter is a winding channel of 200 - 250 m, width 5 m, which is made with soil excavation where the left and right sides of the channel are for plants commonly used for constructed wetland. The construction of this biological filter is expected that the water before reaching the reservoir has a longer contact time, so that there will be an improvement in water quality when it reaches the reservoir, adding that the emergent plant in both channels will improve the quality of the channel. This floating wetland technology will replace the biological filter originally planned.

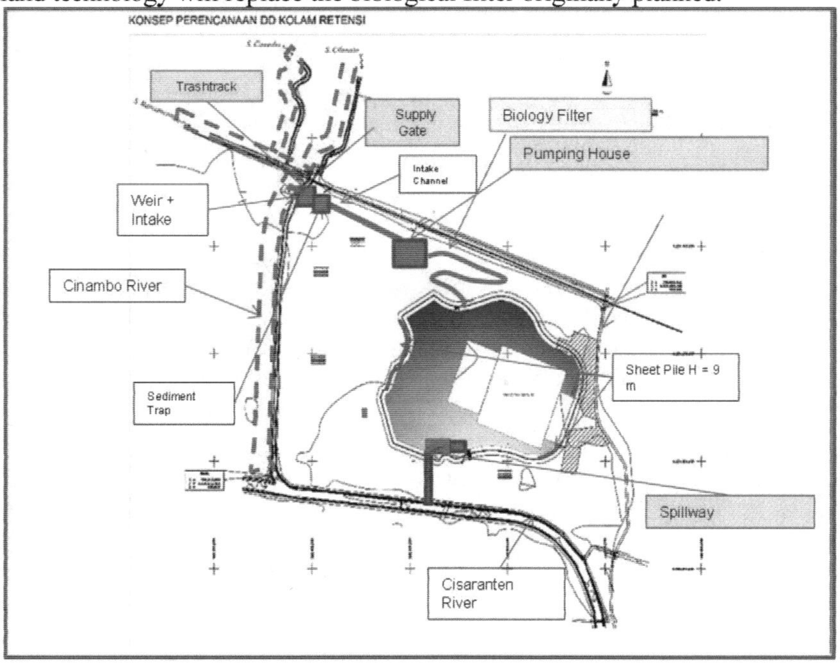

Figure 1. Gedebage Reservoir Planning Concept

Based on the position of the channel level contour in Gedebage Reservoir, the channel depth is still possible, based on water depth on the channel about 1 m. The water depth in the constructed floating wetland is determined based on the maximum length of the root of the plant used [7-10]. The water depth referenced is a minimum of 0.8 m [11], with the aim that the plant roots do not reach the lower zone. If the roots reach the bottom of the root zone will stick and began to take root in the soil or sediment that settles so feared when there are fluctuations in the water then floating wetland will be drawn down and sink [15].

The contact time between pollutant and floating wetland needs to be maximized for optimal processing results [12]. The calculation results show Hydraulic Retention time (HLR) on the channel around 12,500 seconds or 0.14 days. This time is too short, so to increase HRT can be by deepening the channel. However, to deepen the channel is not possible with conditions in the field. The results of previous studies showed that the higher the value of HRT, the greater the efficiency of TP, but not for TN [13]. Other studies recommend maximum HRT for floating wetland is 15 days [8].

Floating Wetland efficiency processing decrease in surface coverage of more than 50% [7-10], but according to other studies 5% surface coverage has been able to process significant nutrients in time (HRT) 15 days with phosphate pollutant concentration of 1 mg/l and Nitrate 3 mg/l [14]. Surface coverage greater than 5% is required when HRT is less than 15 days so surface coverage is recommended between 5-50%, because if it is too high it is feared to cause anoxic conditions in the water body [15]. The surface coverage of 50% on the channel is still possible, because the HRT results are less than 15 days.

Floating wetlands processing efficiency is inside Hydraulic Loading rate (HLR) 0.1 - 0.3 m^3/m^2 [15], based on charts that are processed from various types of research [7, 8, 10]. Calculations show the HLR in a biological filter channel 1 m^3/m^2, a larger HLR indicates that processing with a floating wetland is less effective.

The placement of a floating wetland in a biological filter channel does not meet the design parameter requirements, because of the 4 parameters, the HLR does not meet the requirements, HRT is too short and does not allow it to expand or deepen the channel.

3.1.2. The constructed floating wetland in reservoir Gedebage. Construction of the Gedebage Reservoir as of December 2018 has been completed, at the end of January 2019 impounding has been carried out. After the implementation of the development it turned out that something had changed in the concept of planning, namely the channel for biological filters did not become built. The reservoir that has been built has an area of 7.2 hectares, based on the detailed design data, Gedebage Reservoir has a reservoir volume of 263,524.00 m^3. The calculation results show Hydraulic Retention time (HLR) on the channel around 1, 5, 25 days if rounded 15 days. The time is good enough for a maximum HRT for floating wetland.

Surface coverage of 5% is able to process nutrients that are significant in time (HRT) of 15 days with pollutant concentrations of phosphate is 1 mg/l and nitrate 3 mg /l [14], so as to s urface coverage of wetland floating alone, simply by 5% because the HRT has been 15 days. The addition of mooring system, which can expand the surface so that the biofilm attachment occurs faster, is assumed to be able to reduce the area so that surface coverage can be reduced from 5%.

The calculation shows that the HLR in the Gedebage Reservoir is 36.6 m^3/m^2, the HLR is greater than the Hydraulic Loading rate (HLR) of 0.1 - 0.3 m^3/m^2 reference [15], based on the graphs processed from various types of research [7, 8, 10]. Although this value is greater than the reference HLR, it is still possible, because there is no further research on the effect of HLR on FTW processing efficiency [15]. Reference 0, 1 - 0.3 m^3/m^2 is only based on a regression of the other HLR values that have been done [7, 8, 10].

Placement of a floating wetland at Gedebage Reservoir still meets design parameter requirements, so the application of floating wetlands is more precisely placed in Gedebage Reservoir.

3.2. Water quality characteristics

The results of the quality analysis obtained from secondary data from several stake holders can be seen in Table 1. Specific quality of the Cinambo River originating from the Environmental Services Agency was taken from one data from the highest data series (dry season conditions), and data taken from the downstream near the Gedebage Resevoir, from the upstream-middle-downstream data series.

Data from the Citarum RBO taken in the rainy season. Analysis of the planner consultant data shows that the water quality of the Cinambo River still meets the Class II in Governmental Regulation 81 of 2001 water criteria, but for DO, BOD, COD, Fecal Coliform and Total Coliform parameters still do not meet. The results of the analysis of the data on the detailed design show that several parameters that have met the requirements include: Cd, Pb, DO, BOD and COD. The results of the analysis of data from the Environmental Services Agency show that parameters that did not meet the requirements included: Ammonia, Chlorine, Mn, DO, sulfide, detergent, Oil and Fat, BOD, and COD.

Table 1. Water quality of the Cinambo River

	Parameter	Unit	Criteria Clas II (Gov. Reg. 81 of 2001)	Data of Enviromental Service Agency	Data of Citarum RBO	Data of DED			
						Sample Number			
						I	II	III	IV
Physics	Temperature	^{0}C				27.5	28.1	28	27.5
	TDS	mg/L	1,000	510	345	387	525	488	447
	TSS	mg/L				43	27	17	28
	DHL	mmhos/cm	2,250	1,018		<0.05797	<0.05797	<0.05797	<0.05797
Chemistry	Hg	mg/L	0.001	<0.0004	-				
	Ammonia	mg/L	0.02	0.0136					
	As	mg/L	0.05	<0.0021	-				
	Ba	mg/L	1	<0.01741	-				
	Fe	mg/L	5	0.4652	< 0.021				
	Boron	mg/L	1	<0.00726	0.392	< 0.07232	0.14681	0.11408	0.09661
	Cd	mg/L	0.01	<0.00928	< 0.032	0.05118	0.07066	0.07075	0.04317
	Cloride	mg/L	600	86.43	52.2				
	Chlorin	mg/L	0.003	0.05	-				
	Cobalt	mg/L	0.2	<0.0033	-	0.00747	0.0114	0.01059	0.00623
	Chromium Heksavalen	mg/L	0.05	0.0136	<0.012	< 0.0154	< 0.0154	< 0.0154	< 0.0154
	Mn	mg/L	0.5	0.57199	0.21				
	Ni	mg/L	0.5	<0.01485	-				
	Nitrate	mg/L	10	129,530	0.05	4.5911	5.7765	6.1042	7.4485
	Nitrite	mg/L	0.06	0.0501	0.003				
	DO	mg/L	>3	2.65	1.2	2.6	1.5	3.2	2.5
	Total Phosphate	mg/L	0.2			0.0402	0.198	0.1442	0.074
	pH		6,0-9,0	7.445	7.6	7.3	7.33	7.28	7.02
	Selenium	mg/L	0.01	<0.0013	-	< 0.01797	< 0.01797	< 0.01797	< 0.01797
	Zn	mg/L	0.02	<0.01894	0.006	<0.01852	<0.01852	<0.01852	<0.01852
	Cyanide	mg/L	0.02	<0.005	< 0.003				
	Sulfate	mg/L	400	280.62	12.4				
	H_2S	mg/L	0.002	0.0181	-				
	Cu	mg/L	0.02	<0.00819	< 0.013	< 0.00527	0.0062	0.00957	0.00618
	Pb	mg/L	0.03	<0.01039	< 0.025	0.06717	0.08193	0.08039	0.06294
	Mercury	mg/L	0.001			< 0.01320	< 0.01320	< 0.01320	< 0.01320
	Deterjent	mg/L	0.2	21,195	0.793				
	Phenol	mg/L	0.001	0.0014	0.024				
	Oil and Fat	mg/L	nihil	4.00	<0,1				
	BOD_5	mg/L	6	179.14	23	12.88	62.59	9.63	18.58
	COD	mg/L	10	476.43	93	22.2115	115.91	18.1731	37.1688
Microbiology	Fecal Coliform	Total/100	2,000	3	600,000	210	290	460	240
	Coliform	Total/100	10,000	7,000	6,000,000	2400	2400	2400	1100

Fluctuating water quality at any time depends on the input of wastewater entering at that time, from the analysis can be concluded that the water of the Cinambo River as a source of water for the Gedebage Reservoir has not met the requirements of several parameters including: Ammonia, Chlorine, Mn, DO, Sulfide, Detergent, Oil and Fat, Cd, Pb, DO, BOD, COD, Fecal Coliform and Total Coliform. The high value of detergents and oils and fats shows high concentration of pollutant. The presence of elements Mn, Cd and P b indicates that the Cinambo River water has been contaminated with industrial waste. Levels of Mn are slightly above requirements, levels of Cd ranging from 5 to 7 x exceed the standardized criteria, while Pb levels range from 2.1 - 2.7 above the required criteria.

3.3. Technology selection

Analysis of data from the report on the collection of detailed design shows that BOD5 / COD radios ranged from 0.50 - 0.58, indicating that it is still possible to process using wetlands. Analysis of the data showed that BOD5 / COD radios ranged from 0, 38 , indicating that it is still possible to process wetlands, because in biofilms it can still live under these conditions, even though the BOD5 / COD ratio has dropped. The BOD5 / COD ratio can still be accommodated by biofilms because the ratio is still in the rainy season range. But the results of the analysis of the data from Environmental Services Agency carried out during the dry season shows that ratio BOD5 / COD ranged between 0.25 indicates that the water has toxins, it will disrupt the development of biofilms, so in addition to the required pre-wetland treatment before entering reservoir.

4. Conclusion

Based on the analysis of secondary data that has been carried out, it can be concluded that the location of the Constructed Floating Wetland which is placed in Gedebage Reservoir is more suitable compared to the placement in the biological filter channel. Cinambo River as water supply for Gedebage Reservoir has not meet the requirements of several parameters including: Ammonia, Chlorine, Mn, DO, Sulfide, Detergent, Oil and Fat, Cd, Pb, DO, BOD, COD, Fecal Coliform and Total Coliform. Ratio of BOD5 / COD in the dry season ranges from 0,25 indicates that the water has toxins, so it will disrupt the development of biofilms, so in addition to wetland will require pre-treatment in advance before entering the reservoir.

References

[1] Republic of Indonesia 2018 *Presidential Regulation Number 15 of 2018 concerning Accelerationof Citarum Watershed Pollution and Damage Control* .(Indonesia)
[2] Bernie M 2010 *Water Quality Improvement from Vegetated Floating Islands in Natural and Constructed Wetlands* Enviro 2010 Conference. FIA Technology Pty Ltd; 2010. p. 2010
[3] Sample D, Wang C and Fox L 2016 *Innovative Best Management Fact Sheet No.1: Floating Treatment Wetland* [Internet] 2016 p. 2016. Available from: http://www.ext.vt.edu
[4] Headley T R and Tanner C C 2008 *Floating Treatment Wetlands: an Innovative Option for Stormwater Quality Applications* 11th International Conference on Wetland Systems for Water Pollution Control November 1-7, Indore, India; 2008. p. 2008
[5] Kresh E, Xuan Z, Islam K and Marimon Z 2012 *Floating Wetland Systems for Nutrient Removal in Stormwater Pond* Wanielista MP, Chang NB, Chopra M, editors. Summary of Final Report, BDK78 985-01, Stormwater Management Academy at the University of Central Florida. September 2012
[6] Yuan X, Qian X, Zhang R, Ye R and Wei H 2012 Performance and microbial community analysis of a novel bio-cord carrier during treatment of a polluted river *Bioresources Technology* 117: 33–9
[7] Boonsong K and Chansisri M 2008 Domestic Wastewater Treatment Using Vetiver Gras Cultivated with Floating Platform Technique *J Technoogy* 12: 73–80
[8] Li X N, Song H L, Li W, Lu X W and Nishimura O 2010 An Integrated Ecological Floating Bed Bed Plant, Freswater Clam and Biofilm Searcher for Purification of Eutrofic Water *Ecol Eng* 36: 382–90
[9] Lu H ling, Ku C R and Chang Y H 2015 Water Quality Improvement with Artificial Floating Islands. Ecological Engineering 74: 371–375 *Ecol Eng* 74: 371–5
[10] Sukias J P, Park J, Hadley R and Tanner C C 2011 Nutrient Removal from Eutrophic Waters by Floating Treatment Wetlands *Total Environ Sci*
[11] Chua L H , Tan S B and Goyal M K 2012 Treatment of Baseflow From the Urban Catchment by a Floating Wetland System. *Ecol Eng* 49 (170-180)
[12] Headley T R and Tanner C C 2006 Application of Floating Wetlands for Enhanced Stormwater Treatment: A Review *ARC Technical Publication* No 324 Aukland
[13] Chang N B, Xuan Z, Marimon Z, Islam K and Wanielista M P 2013 Exploring Hydrobiogeo chemical Processes of Floating Treatment Wetlands in A Subtropical Stormwater Wet Detention Pond Ecological Engineering 54: 66-76 *Ecol Eng* 54: 56–66
[14] White S A and Cousins M M 2013 Floating Treatment Wetland Aided Remediation of Nitrogen And Phosphorus from Simulated Stormwater Runoff *Ecol Eng* 61: 207–2015
[15] Pusparinda L and Santoso R I 2016 Floating Treatment Wetland Planning Literature Study in Indonesia *J Tek ITS*. 5 (2016): A471–5

The effect of copper-based catalytic converter with circular tube shape on exhaust emission of Yamaha Vixion 1PA

I A N Shoffan[1,2], S Sumarli[1] and I M Nauri[1]

[1] Department of Mechanical Engineering, Faculty of Engineering, Universitas Negeri Malang
[2] Bachelor Program, Department of Mechanical Engineering, Faculty of Engineering, Universitas Negeri Malang

shoffanilham@gmail.com

Abstract. Indonesia is one of the countries with the third highest air pollution in the world. The biggest contributor air pollution is produced by motorbikes by 60-70%. One the causes of increased exhaust emission is widespread use absorptive exhaust that designed without catalytic converter. The aims of this research is to determine the effect of copper-based catalytic converter with circular tube shape on exhaust emission of Yamaha Vixion 1PA. This study used an experimental research design. The objective of this research is copper catalyst with circular tube shape model. The research instrument used the gas analyzer HG-510 as a test tool for data collection of exhaust emission. Data analysis using inferential statistical techniques paired sample T test through the SPSS 24 statistical analysis application. The results showed that the addition of copper catalytic converter with a circular tube shape model can reduce CO exhaust emission by 16.67%, whereas for exhaust emission HC can be reduced by 32.54%.

1. Introduction

As technology develops, productivity in various fields is increasing especially in the field of transportation which facilitates mobility of people from one place to another. Motorbike is one of the transportation driving motor fuel by utilizing fossil energy. The combustion engine is a machine that converts fuel (fossil energy) into heat energy and produces mechanical energy [1]. Every work cycle, the combustion engine produces exhaust gas from the combustion process in the combustion chamber which contains chemical compounds such as a carbon monoxides (CO), hydrocarbons (HC), and nitrogen dioxides (NO_x). These compounds are very dangerous for the environment and surrounding life, so motorbikes can be said as transportation which contributes to air pollution. Based on data input conducted by the Central Statistics Agency 2017 [2], the number of motorbikes used in Indonesia reached 113,030,793 units, most of which still use fossil fuels. The use of motorbikes with this amount makes Indonesia being one of the third highest levels of air pollution in the world [3]. The biggest contributor air pollution is produced by motorized vehicles with a percentage around 60-70%, while the remainder by industrial. One the causes of increased exhaust emission is widespread use absorptive muffler due to trend of modification in the automotive sector.

Absorptive muffler is a type of exhaust that uses absorption material to muffle sound [4]. Along with modification trend in the automotive sector, absorptive muffler much demand by the public. The use of absorptive exhaust is believable to increase the motorbikes power that affects to motorbikes speed when driving. Based on the fact, absorptives muffler didn't design with catalytic converter as manufacturer's mufflers, so it can increase the exhaust emission. Catalytic converter is a device in the exhaust gas system that use to convert harmful pollutants such as CO, HC, and NO_x into harmless gases such as carbon dioxide (CO_2), water (H_2O) and nitrogen (N_2) through chemical reactions [5]. Based on the construction, the catalytic converter is composed of 3 elements namely, platinum, rhodium and palladium. Platinum is an element that used to reducing and oxidizing compounds in exhaust gases. Rhodium is an element that used to reducing compounds in the exhaust gases, whereas palladium as an oxidizer for compounds in the exhaust gases. Palladium, platinum and rhodium materials are precious metal that have a special role in efforts to accelerate reduction and oxidation reactions. However, the material's abundance on earth is quite low and its manufacture requires a high cost. Therefore, the use of transition metals that have high abundance and relatively inexpensive prices can be an alternative. The transition metals that can be used as a catalyst is copper which has the ability to assist the oxidation and reduction process of exhaust gas emission compounds [6]. Catalytic converters have a various catalyst model. According to Baharudin & Watson [7], various models of catalysts include, (1) hexagonal-shaped, (2) trigonal shaped, (3) square-shaped, (4) circular shaped, triangular arrangement, (5) circular- shaped, square arrangement, (6) circular shape, circular arrangement.

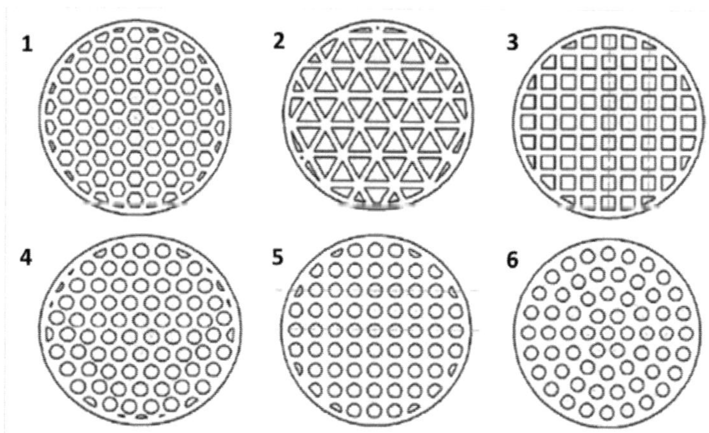

Figure 1. Various models of catalyst

Each model has its own characteristics, one that can affect the performance of each catalyst model is wall thickness and number of catalyst cells. According to Ozhan [8], the more numbers of catalyst cells, had the better the level of conversion efficiency and durability. However, the thicker the surface wall of the catalyst, had a higher the exhaust back pressure which can decreasing engine combustion efficiency.

Research on the copper materials as a catalyst has been carried out as an effort to reduce exhaust emission. Amin et al. [9], copper based catalytic converter. The application of copper-zinc metal as a catalytic converter in motorcycle mufflers to reduce the exhaust emission by Chafidz et al. [10]. Optimum design of manganese-coated copper catalytic converter to carbon monoxide emission on gasoline motors by Irawan et al. [11]. An innovative approach for emission control using copper plate catalytic converters by Choudhury & Deo [12]. I.C. engine emission reduction by copper oxide catalytic converter by Venkatesan et al. [13]. Based on these studies, the addition of copper catalytic converters with a circular tube shape model on the current trend mufllers (absorptive muffler) never been revealed by the researchers. Remembering absorptive muffler much demand by the public, it is hoped that copper-based catalyst can be used as an inexpensive and economical alternative material for absorptive muffler manufacturers to reduce exhaust emission. The aims of this research are to determine the effect of copper-based catalytic converter with circular tube shape on exhaust emission of Yamaha Vixion 1PA. The exhaust emission that measure are CO and HC.

2. Methods

The design of the research used the static comparison design. The subjective of this research is Yamaha Vixion 1PA with Kawahara absorptive muffler, whereas the objective of this research is copper catalyst with circular tube shape model. The research instrument used gas analyzer HG-510 as a test tool for data collection of exhaust gas emission (CO and HC). Data retrieval is done at 2500 rpm to 7500 rpm with a multiple of 500 rpm. Data analysis using inferential statistical techniques paired sample T test through the SPSS 24 statistical analysis application.

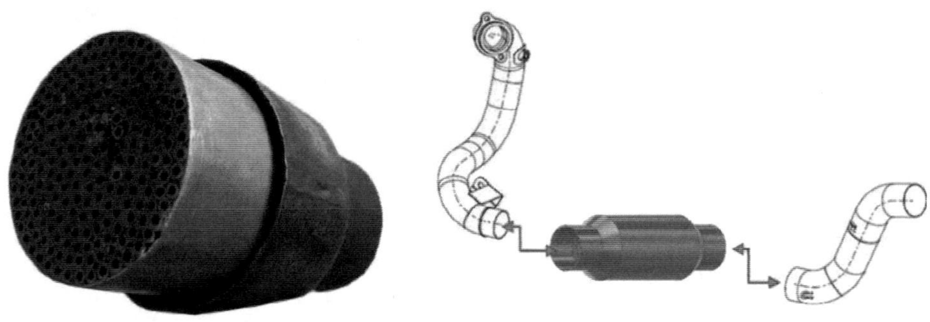

(a) (b)

Figure 2. (a) The circular tube shape catalyst (b) Installation catalytic converter on motorcycle mufller

3. Results and Discussion

Based on research test "The Effect of Copper-Based Catalytic Converter With Circular Tube Shape on Exhaust Emission of Yamaha Vixion 1PA" which has been carried out, obtained data on CO exhaust emission in percent (%) units and HC gas emission in parts per million (ppm) units. The environmental conditions at the time of testing are 30°C with a machine working temperature around 80°C — 90°C. Each testing, machine performance temperature is maintained with the help of fan, so the engine temperature remains stable, the engine temperature is controlled at 80°C in the same point, namely the outside of the combustion chamber near the cylinder head. Temperature controlled/measured by thermometer. Data obtained through twice testing in two treatments (without catalytic converter and using catalytic converter) which will be taken on average. Each data is taken through engine speed from 2,500 rpm to 7,500 with multiples of 500 rpm. The following are research data obtained after testing.

Table 1. CO exhaust gas emission data which obtained by adding copper-based catalytic converter with circular tube shape on Yamaha Vixion 1 PA

Engine Speed (rpm)	CO Emission (%)	
	Without *Catalytic Converter*	Using *Catalytic Converter*
2500	1,32	1,04
3000	1,66	0,91
3500	2,29	1,01
4000	2,01	1,32

Table 1. CO exhaust gas emission data which obtained by adding copper-based catalytic converter with circular tube shape on Yamaha Vixion 1 PA (cont.)

Engine Speed (rpm)	CO Emission (%)	
	Without *Catalytic Converter*	Using *Catalytic Converter*
4500	1,26	0,48
5000	0,62	0,31
5500	0,56	0,28
6000	5,85	5,74
6500	6,18	6,08
7000	6,99	6,72
7500	7,29	6,24
Average	3,24	2,7
Percentage of CO Emission Reduction	$\frac{(3{,}24 - 2{,}7)}{3{,}24} \times 100\% = 16{,}67\%$	

Table 2. HC exhaust gas emission data which obtained by adding copper-based catalytic converter with circular tube shape on Yamaha Vixion 1 PA

Engine Speed (rpm)	HC Emission (ppm)	
	Without *Catalytic Converter*	Using *Catalytic Converter*
2500	488,5	349,5
3000	1851	1114,5
3500	1399,5	1147
4000	767,5	606
4500	701	336
5000	495,5	249
5500	521,5	199,5
6000	378,5	332
6500	286	226
7000	393,5	358,5
7500	356	235
Average	694,41	468,45
Percentage of HC Emission Reduction	$\frac{(694{,}41 - 468{,}45)}{694{,}41} \times 100\% = 32{,}54\%$	

3.1. The effect of copper-based catalytic converter with circular tube shape on CO exhaust emission

Based on Table 3, maximum CO obtained by 7.29% (without catalytic converter) at 7500 rpm and 6.72% (using catalytic converter) at 7000 rpm. While minimum CO obtained by 0.56% (without catalytic converters) and 0.28% (using catalytic converters) in the same engine speed at 5500 rpm. Average of CO obtained in the muffler without catalytic converter is 3.2755%, while average CO that obtained in the muffler with catalytic converter is 2.7391%, so it can be concluded that average CO obtained in the muffler with catalytic converter has decreased compared to average of muffler without catalytic converter. The description is clarified in the following graph.

Table 3. Description of CO exhaust emission data

Descriptives				
			Statistic	Std. Error
Without Catalytic Converter	Mean		3.2755	.81169
	95% Confidence	Lower Bound	1.4669	
	Interval for Mean	Upper Bound	5.0840	
	5% Trimmed Mean		3.2033	
	Median		2.0100	
	Variance		7.247	
	Std. Deviation		2.69207	
	Minimum		.56	
	Maximum		7.29	
	Range		6.73	
Using Catalytic Converter	Mean		2.7391	.83426
	95% Confidence	Lower Bound	.8803	
	Interval for Mean	Upper Bound	4.5979	
	5% Trimmed Mean		2.6545	
	Median		1.0400	
	Variance		7.656	
	Std. Deviation		2.76692	
	Minimum		.28	
	Maximum		6.72	
	Range		6.44	

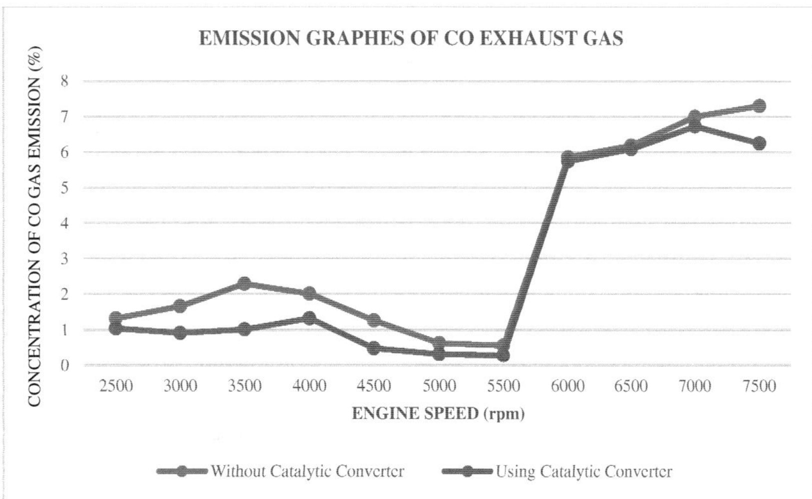

Figure 3. Graph of CO exhaust emission

Based on Figure 3, exhaust gas emission of CO that obtained either through the addition of copper catalytic converters with circular tube shape model or without a catalytic converter in fluctuation. The fluctuating of CO emission caused by incomplete combustion. When CO exhaust emission increase, it means the combustion process is incomplete due to lack of oxygen (O2) during the combustion process, while when exhaust emission CO decrease, it means O2 almost meets the combustion process / combustion process is almost perfect [5]. One of the causes incomplete combustion process is due to the appearance of exhaust back pressure. This is caused by the presence of an area on the catalyst wall that cannot be crossed by the exhaust gas flow, so the exhaust gas flow becomes clogged. Silvaram et al. [14], suggested that the higher exhaust back pressure, can decreasing engine combustion efficiency. This is because the exhaust back pressure allows the exhaust gas to get back into the combustion chamber. When CO exhaust emission increase, it means exhaust back pressure causes a lack of O2 during the combustion process, whereas when CO decrease, it means the exhaust back pressure causes sufficient O2 when the combustion process is fulfilled, it's because the exhaust gas also contains O2. According to Roy et al. [15], a certain level of exhaust back pressure is still needed to improve engine performance and reduce exhaust emission, so it can be concluded that exhaust back pressure can also participate in increasing AFR homogeneity during the process burning.

When compared, the exhaust emission of CO that produced through the addition of copper-based catalytic converter with a circular tube shape model tend to be lower than without a catalytic converter. Total decline of CO exhaust emission through the addition of copper-based catalytic converter with a circular tube shape model by 16.67%. Decreasing CO exhaust emission is due to the oxidation process of CO compounds which are aided by copper catalyst activity. Oxidation reaction is an oxygen binding reaction, when exhaust emission of CO pass through the copper catalyst, the process of reaction between CO and O2 will be more effective due to the characteristics of copper as an oxidation catalyst, so CO exhaust emission can be converted into harmless gas, namely CO2 [10]. The CO oxidation reaction aided by copper catalyst activity is shown in the Figure 4.

Oxidation of CO compounds:

$$2CO + O_2 \xrightarrow{Cu} 2CO_2$$

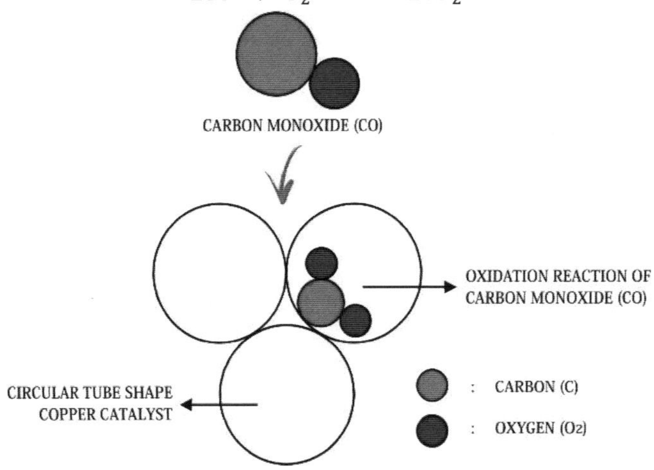

Figure 4. Oxidation of CO compounds

3.2. The effect of copper-based catalytic converter with circular tube shape on HC exhaust emission

Based on Table 4, maximum HC obtained by 1851 ppm (without catalytic converter) at 3000 rpm and 1147 ppm (using catalytic converter) at 3500 rpm. While minimum HC obtained by 286 ppm (without catalytic converter) at 6500 rpm and 199.5 ppm (using catalytic converter) at 5500 rpm. Average of HC emission that obtained in the muffler without catalytic converter is 694,4091 ppm, while average of HC emission that obtained in the muffler with catalytic converter is 468.4545 ppm, so it can be concluded that average of HC emission that obtained on the muffler with catalytic converter has decreased compared to average muffler without catalytic converter. The description is clarified in the following graph.

Table 4. Description of CO exhaust emission data

Descriptives				
			Statistic	Std. Error
Without Catalytic Converter	Mean		694.4091	148.43933
	95% Confidence	LowerBound	363.6656	
	Interval for Mean	Upper Bound	1025.152	
	5% Trimmed Mean		652.8434	
	Median		495.5000	
	Variance		242376.5	
	Std. Deviation		492.3175	
	Minimum		286.00	
	Maximum		1851.00	
	Range		1565.00	
Using Catalytic Converter	Mean		468.4545	104.13731
	95% Confidence	Lower Bound	236.4222	
	Interval for Mean	Upper Bound	700.4869	
	5% Trimmed Mean		445.6995	
	Median		336.0000	
	Variance		119290.3	
	Std. Deviation		345.3843	
	Minimum		199.50	
	Maximum		1147.00	
	Range		947.50	

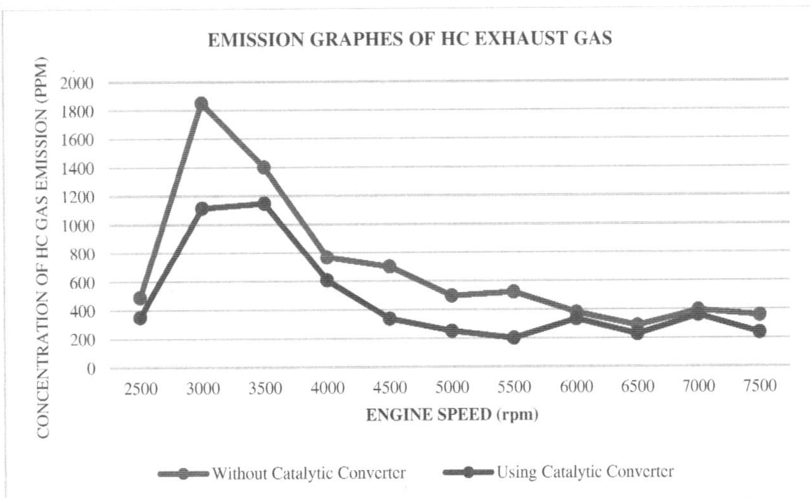

Figure 5. Graph of HC exhaust emission

Based on Figure 5. HC emission that obtained either through the addition of copper-based catalytic converter with circular tube shape model or without catalytic converter in fluctuation. The fluctuating of HC exhaust emission caused by incomplete combustion. When HC exhaust emission increase, it means the combustion process didn't perfectly because there's fuel didn't burn during the combustion process. When HC exhaust emission decrease, it means the fuel can almost completely burn / the combustion process gradually leads to almost perfect [5]. One of the causes incomplete combustion process is due to the appearance of exhaust back pressure. This is due to the presence of areas on the catalyst wall which cannot be crossed by exhaust gas flow, so the exhaust gas flow becomes clogged.

Silvaram et al. [14], suggested that the higher exhaust back pressure, can decreasing engine combustion efficiency. This is because the exhaust back pressure allows the exhaust gas to get back into the combustion chamber. When HC flue gas emission increase, it means exhaust back pressure causes a lack of O2 during the combustion process which causes the fuel not burn, while HC decrease, it means exhaust back pressure causes the O2 to be adequately met, given that the exhaust gas still contains elements O2, so when O2 sufficient is fulfilled during the combustion process, the fuel can almost completely burn. According to Roy et al. [15], a certain level of exhaust back pressure is still needed to improve engine performance and reduce exhaust emission, so it can be concluded that exhaust back pressure can also participate in increasing AFR homogeneity during the process burning.

When compared, the exhaust emission of HC that produced through the addition of copper-based catalytic converter with a circular tube shape model tend to be lower than without a catalytic converter. Total decline of HC exhaust emission through the addition of copper-based catalytic converter with a circular tube shape model is 32.54%. Decreasing HC exhaust emission is due to the oxidation process of HC compounds which are aided by copper catalyst activity. Oxidation reaction is an oxygen binding reaction, when exhaust emission of HC pass through the copper catalyst, the process of reaction between HC and O2 will be more effective due to the characteristics of copper as an oxidation catalyst, so HC exhaust emission can be converted into harmless gas, namely CO2 and H2O [13]. The HC oxidation reaction aided by copper catalyst activity is shown in the Figure 6.

Oxidation of CO compounds:

$$2C_xH_y + (2x + \frac{y}{2})O_2 \xrightarrow{Cu} 2xCO_2 + yH_2O$$

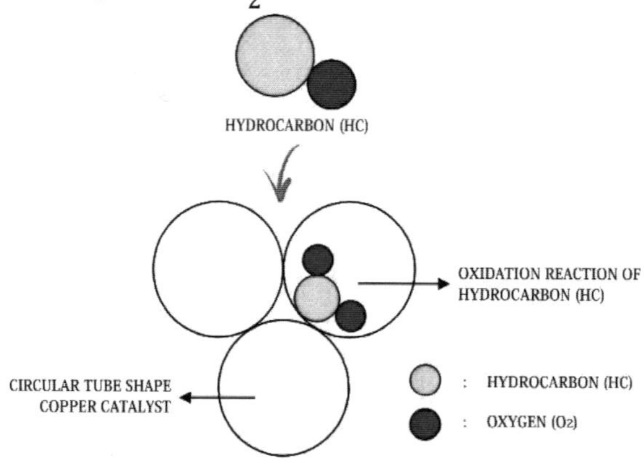

Figure 6. Oxidation of CO compounds

4. Conclusion

Copper-based catalytic converter with a circular tube shape can reduce exhaust emission. Decreasing of CO exhaust emission through the addition of copper-based catalytic converter with circular tube shape by 16.67%, whereas reducing HC exhaust emission through copper-based catalytic converter with circular tube shape by 32, 54%.

References

[1] Seta S and Ilyas M 2014 Analisis Pengaruh Bentuk Permukaan Piston Model Kontur Radius Gelombang Sinus Terhadap Kinerja Motor Bensin Jurnal Ilmiah Teknik Mesin 2(1) pp. 43–49

[2] Badan Pusat Statistik 2017 Perkembangan Jumlah Kendaraan Bermotor Menurut Jenis 1949 Available at: www.bps.go.id.

[3] Irawan B 2012 Rancang Bangun Catalytic Converter Material Substrat Tembaga Berlapis Mangan Untuk Mereduksi Emisi Gas Karbon Monoksida Motor Bensin in Seminar Hasil-Hasil Penelitian – LPPM UNIMUS, pp. 409–422

[4] Daniel P 2005 General Design Principles for an Automotive Muffler in Proceedings of ACOUSTICS Busselton-Western Australia pp. 153–158 Available at: www.daydesign.com.au.

[5] Arianto R and Nauri I M 2018 The Effect of Bed Catalyst Design Variation on Catalytic Converter against Exhaust Emission Level of Carbon Monoxide (CO) and Hydrocarbon (HC) in Gasoline Engine Journal of Mechanical Engineering Science and Technology 2(2), pp. 64–73. doi: 10.17977/um016v2i12018p064

[6] Irawan B and Nurcholis L 2016 Pemanfaatan Logam Transisi Tembaga dan Nikel Sebagai Bahan Katalis Untuk Mereduksi Emisi Gas Karbon Monoksida Kendaraan Bermotor Pemanfaatan Logam Transisi Tembaga dan Nikel Sebagai bahan Katalis in Simposium Nasional Teknologi Terapan (SNTT) 4 2016, pp. 458–468

[7] Baharudin L and Watson M 2017 Monolithic substrate support catalyst design considerations for steam methane reforming operation Reviews in Chemical Engineering doi: 10.1515/revce-2016-0048.

[8] Ozhan C 2015 Multi-scale simulation of automotive catalytic converters UPMC - Paris: HAL. Available at: https://tel.archives-ouvertes.fr/tel-01232031

[9] Amin C M, Rathod P P and Goswami J J 2012 Copper based catalytic converter International Journal of Engineering Research & Technology (IJERT) 1(3)

[10] Chafidz A et al 2018 Application of copper-zinc metal as a catalytic converter in the motorcycle muffler to reduce the exhaust emissions in 8th International Conference on Environment Science dan Engineering (ICESE 2018) doi: 10.1088/1755-1315/167/1/012014

[11] Bagus I R, Purwanto P and Hadiyanto H 2015 Optimum Design of Manganese coated Copper Catalytic Converter to Reduce Carbon Monoxide Emissions on Gasoline Motor in Procedia Environmental Sciences Elsevier B V pp 86–92 doi: 10.1016/j.proenv.2015.01.013

[12] Choudhury P and Deo S 2014 An Innovative Approach For Emission Control Using Copper Plate Catalytic Converter International Journal of Advanced Science, Engineering and Technology 3(2) pp. 19–23 Available at: http://www.bipublication.com

[13] Venkatesan S P et al. 2017 I C Engine emission reduction by copper oxide catalytic Converter in Frontiers in Automobile and Mechanical Engineering doi: 10.1088/1757-899X/197/1/012026

[14] Silvaram A R et al. 2017 Exhaust back pressure effect on the performance features of a diesel engine ARPN Journal of Engineering and Applied Sciences 12(19) pp. 5353–5356

[15] Roy M M, Uzzal M and Uddin S 2010 Effect of Engine Backpressure on the Performance and Emissions of a CI Engine in The 7 th Jordanian International Mechanical Engineering Conference (JIMEC'7) Amman-Jordan

IOP Conf. Series: Materials Science and Engineering **669** (2019) 012015 doi:10.1088/1757-899X/669/1/012015

The analysis of alpha parameter on Nakayasu Synthetic Unit Hydrograph in Timor Island watersheds

D S Krisnayanti[1], M E Bolla[1], J K Nasjono[1] and M J M Wangge[2]

[1] Civil Engineering Department, Nusa Cendana University, Kupang, NTT, Indonesia
[2] Undergraduate of Civil Engineering Department, Nusa Cendana University, Kupang, NTT, Indonesia

denik.krisnayanti@staf.undana.ac.id

Abstract. Benenain and Noelmina river are rivers on Timor Island which have large watershed area. With a large watershed area, it is necessary to observe and calculate the peak flood discharge deeply. One of the methods used for calculate the peak flood discharge Nakayasu Synthetic Unit. In calculating the peak flood discharge with Nakayasu Synthetic Unit Hydrographic Parameters difficulties often encountered in determining the accuracy of the α (alpha) value coefficient. To calculate the α (alpha) value coefficient required data in the form of area and length of river. The results of analysis using software IBM SPSS version 23 the (alpha) value coefficient on Timor Island can be obtained using equation $\alpha = 2{,}461.\ (L)^{-0{,}417}.\ (A)^{0{,}214}$ so that the α (alpha) value for each watershed varies depending on the size of the watershed and the length of the river. Thus, with the confidence level of the results of the analysis is 95% and the coefficient of determination (R2) is 0.826 then the watershed area and river length give an influence of 82.60% on the value of α (alpha) obtained.

1. Introduction

In East Nusa Tenggara Province, especially in Timor Island, there are various methods that have been used in calculating peak flood discharge. Calculation of peak flood discharge can be done empirically, statistically, or by using a Hydrograph. Hydrograph is a graphical representation of a flow element relationship such as discharge (Q) to time (t). The next term called hydrograph is briefly the relationship between discharge and time. The hydrograph forming component comes from runoff or surface flow / direct flow and base flow (formed by intermediate flow and underground flow). The hydrograph consists of three parts, namely the concentration curve / upward curvature, the top part, and the recession arch [1-3].

One of the most commonly used hydrograph methods is the Nakayasu Synthetic Unit Hydrograph. The method formulated by a Japanese national named Nakayasu is a popular method used in many plans in the field of water resources especially in the analysis of immeasurable watershed floods (DAS). Calculation with the Nakayasu Synthetic Unit Hydrograph method requires data on watershed characteristics such as watershed area, river's length and in some cases includes land characteristics. Based on these data, river forms can be described to find out the shape of the river channel, so that the tendency of fluctuations in flood flows that occur in the watershed can be described. Therefore, this method is also known as one of the methods used to simulate the flow of an unmeasured watershed which ranges from 30 km^2 to 30,000 km^2 [2-3]. But in its use are often encountered various difficulties, especially in determining the value of α (alpha) used. The value of α (alpha) itself ranges from 1.5 to 3. But often the value of α (alpha) used does not represent the actual area to be studied or calculated using the Nakayasu Synthesis Unit Hydrograph, so the selection of α (alpha) in a certain area or location becomes inaccurate. The value of α (alpha) is one of the important parameters used in the calculation using the Nakayasu Synthesis Unit Hydrograph method, where the value of α (alpha) actually shows the characteristics of a watershed. To get the value of α (alpha) requires data such as river's length, and watershed area. The accuracy of α (alpha) value will greatly influence the results obtained in calculating peak flood discharge [4-11].

The existence of large rivers in Timor Island, such as the Benenain River Region, and the Noelmina River Region, is often used as a source of life that can be used by humans, especially those living around the river. Within the territory of Timor Island, there are two River Areas, namely the Noelmina River Region which straddles six regencies / cities which are divided into 186 River Subdistricts with a total River Area of 9,360.52 km^2 and the Benenain River Area which spans four districts divided into 46 Sub The River Region with the total area of the River Region within Timor Island is 6,460.12 km^2 while the rest of 3,158.16 km^2 is within the territory of the State of Timor Leste (River Area Station of East Nusa Tenggara II, 2013). So, in an effort to optimize river functions, especially rivers on Timor Island, studies are often carried out regarding major rivers on Timor Island so that they can be used as references that can be used in hydrological planning [12-15].

2. Literature review

2.1. Nakayasu synthetic unit hydrograph

Nakayasu has investigated unit hydrograph in Japan and provided a set of equations to calculate the peak discharge of a unit hydrograph which is formulated as follows:

$$Qp = \frac{C.A.R_0}{3,6(0,3.Tp+T_{0,3})} \quad (1)$$

$$Tp = Tg + 0,8\ Tr \quad (2)$$

$$Tr = (0,5.Tg)\ s/d\ (1.Tg) \quad (3)$$

$$T_{0,3} = \alpha.Tg \quad (4)$$

with:

Qp : peak flood discharge (m^3/s)
C : runoff coefficient
A : catchment area (km^2)
R_0 : rainfall unit = 1 mm
Tp : peak hour (hour)
Tg : the time taken from the center of rain to get to the peak / delay time (hour)
Tr : time of rainfall
$T_{0,3}$: time when debit equals 0.3 times peak discharge (hour)
α : alpha coefficient, ranging from 1.5 - 3.0

Nakayasu also gives several equations to calculate the ordinates forming a hydrograph. The equation can be written as follows [16]:

- For rising limb (Qa)
  $0 < t < Tp$

$$Q_a = \frac{t^{2,4}}{Tp} \quad (5)$$

- For decreasing limb part I (Qd_1)
  $Tp < t < (Tp + T0,3)$

$$Qd_1 = Qp.\,0,3^{\frac{t-Tp}{T0,3}} \quad (6)$$

- For decreasing limb part II (Qd_2)

$(Tp + T_{0.3}) \leq t \leq (Tp + T_{0.3} + 1.5\ T_{0.3})$

$$Qd_2 = Qp.\,0{,}3^{\frac{((t-Tp)+(0{,}5T_{0,3}))}{1{,}5T_{0,3}}} \tag{7}$$

- Decreasing limb part III (Qd_3)
  $t > Tp + T_{0,3} + 1{,}5.\ T_{0,3}$

$$Qd_3 = Qp.\,0{,}3^{\frac{t-Tp+1.5T_{0,3}}{2T_{0,3}}} \tag{8}$$

2.2. The alpha parameter on Nakayasu synthetic unit hydrograph method
One important parameter in calculating the peak flood discharge with the Nakayasu Synthetic Unit Hydrograph method is to find in and find out the magnitude of the value coefficient α (alpha) which is a description of the characteristics of a watershed [1-4].

$$\alpha = (0.47\ (A\ L)^{0.25})\ /\ Tg \tag{9}$$

$$Tg = 0.21.L\ 0.7 \text{ for } L \leq 15 \text{ km} \tag{10}$$

$$Tg = 0.4 + 0.058.L \text{ for } L \geq 15 \text{ km} \tag{11}$$

With:
L : length of the main river (km)
Tg : The time taken from the weight of rainfall to peak discharge / time lag (hour)
With notes:
α : 2.0= The rising part of the hydrograph shows time which tends to be the same as the decreasing part of the hydrograph
α : 1.5= The portion of the hydrograph rises slowly and the part decreases rapidly
α : 3.0= The part of the hydrograph rises fast and the part decreases slowly

3. Methodology of research

3.1. Research location
This research was conducted on the watersheds in Timor Island, East Nusa Tenggara with the distribution of sub-watersheds (Watersheds) taken in the districts/ cities in Timor island. These sub-watersheds are sub-watersheds that are included in the Noelmina and Benenain Watersheds.

3.2. Data analysis
Data analysis carried out included the following:
- Determine and calculate the watershed area.
- Measuring the length of the river.
- Calculate the value of temporary α (alpha).
- Calculate peak flood discharge using the Nakayasu HSS method.
- Analyse the value of α (alpha).
- Calculating the flood peak discharge based on the α (alpha) value that has been obtained.

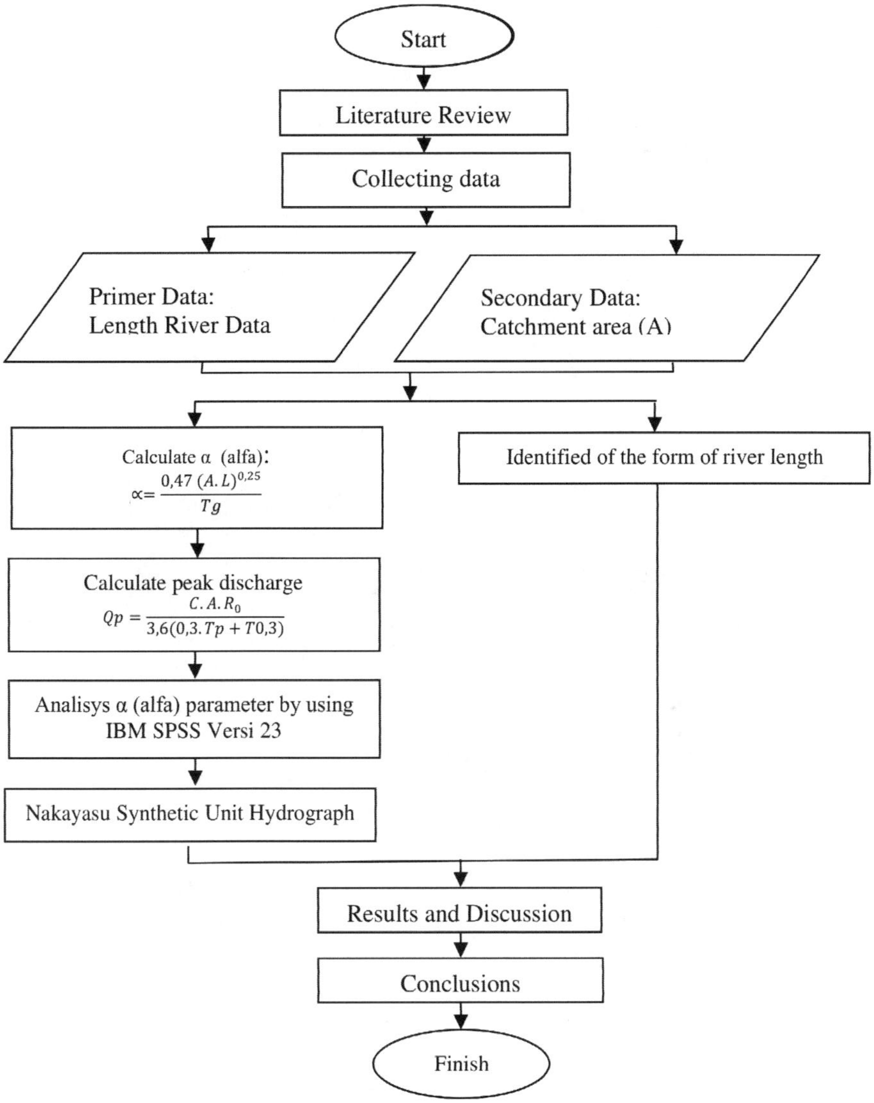

Figure 1. Flow Chart of the research

4. Results and discussion

4.1. Watersheds area and river length

To analyse the value of α (alpha) the parameters used are the watershed area and the length of the river. From the grouping data on the distribution of watershed areas that have been obtained from the Nusa Tenggara River Basin II Territory, 30 sub-basins and river lengths were taken randomly to represent the watershed and the length of the river in Timor Island territory. After that the calculation to get the size of the watershed and the length of the river is obtained using Google Earth software. The reason for choosing used the Google Earth software is because it refers to the initial data obtained from the Nusa Tenggara River Basin II, before the data obtained has been grouped into broad division of watershed using Google Earth software, so that the data is then continued to calculate watershed area and river length. The following is shown a picture of grouping sub-watersheds obtained from the Nusa Tenggara River Basin II, image of the extensive sub-watershed calculation using Google Earth software, and a table of recapitulation of the area of sub-watersheds and the length of the river.

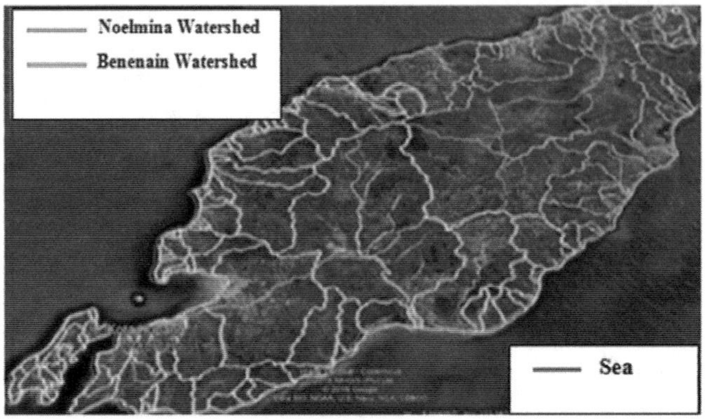

Figure 2. Noelmina and Benenain Sub-Watersheds Grouping

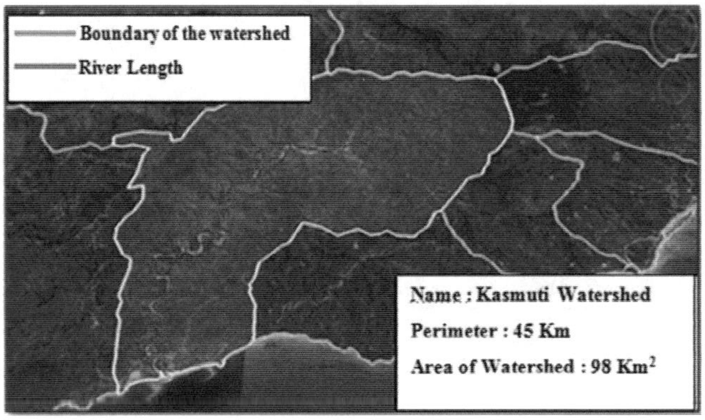

Figure 3. Watersheds Calculations Using Google Earth Software

Table 1. Recapitulation of Catchment Area and River Length in Noelmina and Benenain

No	Name of Sub Watersheds	Name of The River Basin	City / District Administrative Region	Catchment Area (km^2)	River length (km)
1	Biboko	Noelmina	Kupang District	91.7	20.1
2	Kapsali	Noelmina	Kupang District	154	47.9
3	Kenino	Noelmina	Kupang District	170	42.4
4	Kuu Oepaha	Noelmina	Kupang District	78	9.36
5	Manikin Baki	Noelmina	Kupang district and Kupang city	195	25.2
6	Meto batulesa	Noelmina	Kupang district and Kupang city	89.9	17.5
7	Meto	Noelmina	Kupang District	18	9.92
8	Muke	Noelmina	TTS District	531	54.8
9	Naitlopen	Noelmina	Kupang District	177	22.9
10	Nita	Noelmina	Kupang District	78	21.1
11	Noelmina	Noelmina	Kupang district and TTS district	1981	107
12	Otnui Okuuh	Noelmina	Kupang District	23.8	7.71
13	Patukolo	Noelmina	Kupang District	5.37	4.29
14	Sahak Orman	Noelmina	Kupang District	18.9	4.12
15	Sitoto	Noelmina	Kupang District	268	37.4
16	Siumate	Noelmina	Kupang District	98.7	15.5
17	Termanu	Noelmina	Kupang District	428	50.3
18	Kasmuti	Noelmina	Kupang District	90.8	25.9
19	Bisnain	Noelmina	Kupang District	43.8	13.8

Table 1. Recapitulation of Catchment Area and River Length in Noelmina and Benenain (cont.)

No	Name of Sub Watersheds	Name of The River Basin	City / District Administrative Region	Catchment Area (km^2)	River length (km)
20	Metan	Noelmina	Kupang District	224	49.6
21	Benenain	Benenain	Malaka, Belu, TTU and TTS district	3310	107
22	Boking	Benenain	TTS District	44.8	9.96
23	Bone	Benenain	TTS District	135	25
24	Halilanmutu	Benenain	Belu and TTS district	216	33.2
25	Nenoat	Benenain	TTS District	38.3	7.41
26	Oemanu	Benenain	TTU District	399	33
27	Punu	Benenain	TTS District	228	25
28	Rainawe	Benenain	Malaka and Belu District	171	28.1
29	Hasfuik Maubesi	Benenain	Belu and TTU District	193	33.2
30	Umaklaran	Benenain	Malaka and TTU District	341	39.4

4.2. Form of the watersheds

The water flows from the upstream which is located at the top of the hill to the downstream which is located close to the sea. The upstream part of this watershed has a denser density of river flow compared to the downstream. The form of rounded sub-watersheds is caused by having a long river and branching tributaries which are quite large cause this sub-watershed has sufficient fast concentration time, so that flood fluctuations that occur are also high. The flood trend that occurs is located at downstream. Thus, it can be concluded that the shape of the Noelmina sub-watershed is in the form of a radial where the shape of the water flow seems to centre on a point with a fan-like shape or circle with the coming flow time from all corners tends to be almost the same.

Figure 5 shows the shape of the Benenain Watershed located in the Benenain River Region. In the watershed it is seen that water flows from upstream at the top of the hill to the downstream near the sea. In the upstream part there are two quite wide river grooves where the two river channels are bordered by ridges. The two river channels gradually flow at one point to form a main river channel. The Benenain sub-watershed has high flood fluctuations where the tendency for flooding to occur at the meeting point of two river channels. In line with the high flood fluctuations, the shape of the Benenain sub-watershed is far from oval, so the tendency for the watershed to form is like a fan. Based on the characteristics described, it can be concluded that the Benenain sub-watershed with an area of ± 3,310 km2 sub-watersheds has a parallel form in which sub-watersheds are formed by two united paths downstream, causing flooding to occur at the meeting point.

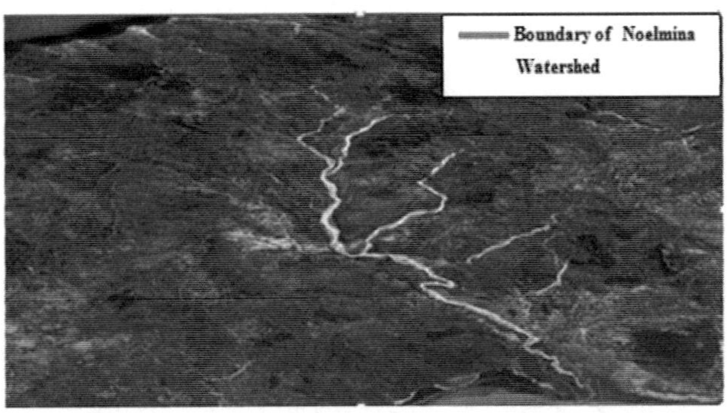

Figure 4. Form of Noelmina sub-watersheds

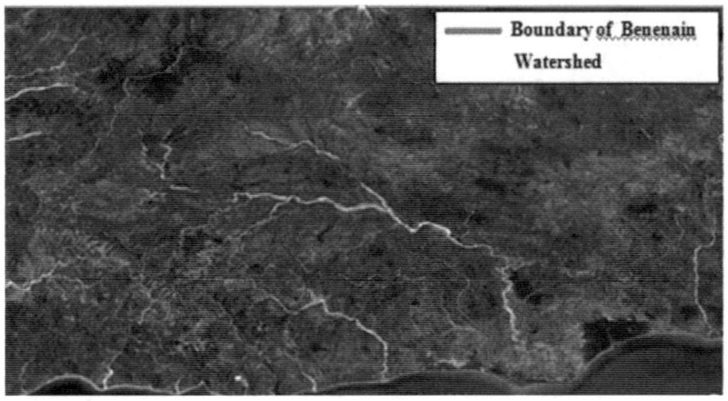

Figure 5. Form of Benenain sub-watersheds

4.3. Calculation of temporary α (Alpha) parameter

- Calculate the amount of time taken from the Heavy Rain Point to the Peak of Debit. In Table 1, the results of the calculation of the river length of the Biboko watershed are shown, the Noelmina River Area is 20.1 km, so Tg can be calculated as follows:

$$\mathrm{Tg} = 0{,}4 + 0{,}058\,(20{,}1)$$

$$\mathrm{Tg} = 1{,}566 \text{ hours}$$

- Calculating temporary α (alpha) value

$$\propto = \frac{(91{,}7.20{,}1)^{0{,}25}}{1{,}566}$$

$$\propto = 1{,}967$$

4.4. Peak flood calculation

- Calculating the Time Required to Reach 30% of Flood Peak (T0,3)

$$\mathrm{T}_{0,3} = \propto .\,\mathrm{Tg}$$

$$\mathrm{T}_{0,3} = 1{,}967\,.\,1{,}566$$

$$\mathrm{T}_{0,3} = 3{,}080$$

- Calculating Rainfall Time (Tr)

$$\mathrm{Tr} = 0{,}8.\,\mathrm{Tg}$$

$$\mathrm{Tr} = 0{,}8.\,(1{,}566)$$

$$\mathrm{Tr} = 1{,}253 \text{ hours}$$

- Calculating the time needed from the start of the rain to the peak of the flood (Tp)

$$\mathrm{Tp} = \mathrm{Tg} + (0{,}8.\,\mathrm{Tr})$$

$$\mathrm{Tp} = 1{,}566 + (0{,}8.1{,}253)$$

$$\mathrm{Tp} = 2{,}568 \text{ hours}$$

- Calculating Peak Flood Discharge (Qp)

$$\mathrm{Qp} = \frac{\mathrm{C.A.R0}}{3{,}6(0{,}3\mathrm{Tp}+\mathrm{T0{,}3})}$$

$$\mathrm{Qp} = \frac{1.91{,}7.1}{3{,}6(0{,}3.2{,}568+3{,}080)}$$

$$\mathrm{Qp} = 6{,}616 \text{ m}^3/\text{second}$$

4.5. Analysis of Alpha Parameters Using the IBM SPSS Version 23 Program

In analyzing this program the temporary α (alpha) value that has been obtained is used as the dependent variable (bound) while the river length and watershed area are used as independent variables (explanatory) so that the assumption that can be built is to find out how much influence given / contribution given river length and watershed area to predict variations in the value of α (alpha) to be searched. Thus input the temporary α (alpha) value, river length and watershed area that has been obtained into the IBM SPSS version 23.

From the results of the iteration, a constant value of A = 2,461, X_1 = -0,417 and X_2 = 0,214 where the level of confidence in the results of the analysis obtained is 95%. Based on the results of parameter estimation, the error predicts the data for A constant value of 0,105, for the X_1 variable of 0,039 and for the X_2 variable of 0,025. It can also be seen that the value of R^2 obtained is 0.826. The value of 0.826 shows a high level of satisfaction because it is above 0.5 and almost close to 1, so that the contribution given parameters given in this case the independent variables along with the equation model entered has reached 82.60%. From the value of R^2 it is also known that the value obtained does not reach 1 and the remaining value is 0.174, so that it can be interpreted to have other factors that influence the value of α (alpha) with a percentage of 17.40%. Thus it can be concluded that the river length and watershed area influence the variation of α (alpha) value of 82.60% with a constant value = 2,461, rank value for river length variable = -0,417 and rank value for watershed area variable = 0,214 analyzed by using equation $\alpha = A.\ \text{River Length}^{-X1}\ .\ \text{Area of watershed}^{X2}$.

4.6. Nakayasu synthetic unit hydrograph

Based on the calculation of the peak flood discharge and hydrograph image in the Biboko Sub-watershed, the Noelmina River Region obtained an α (alpha) value of 1.852 different from the previously calculated α (alpha) value of 1.967. This α (alpha) value affects the results of the calculation of peak flood discharge where the results of the calculation of peak flood discharge are 6.940 m^3/sec different from the results of previous calculations obtained by the value of smaller peak flood discharge which is 6.616 m^3 / second.

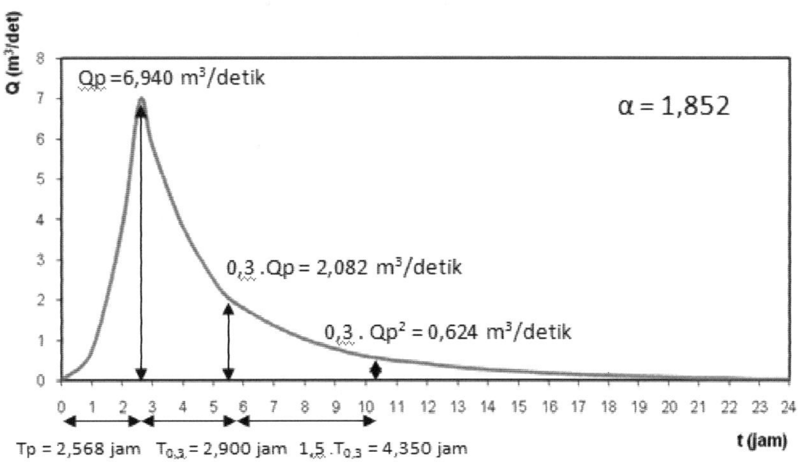

Figure 6. Nakayasu Synthetic Unit Hydrograph

It can be compared also, based on calculations to obtain peak flood discharge and hydrograph produced, in the previous calculation for the same watershed, namely in the Biboko watershed of the Noelmina River Region, the coefficient α (alfa) value was 1.967 and the second coefficient α (alpha) was obtained to 1,852. As for the decreasing part of the hydrograph shown by the amount of time needed to reach 30% of peak flooding ($T_{0,3}$), where the coefficient value of α (alpha) of 1,852 gets a time that tends to be faster that is 2,9 hours so the hydrograph down time faster than the coefficient value of α (alpha) of 1,967 parts decreases to slower that is equal to 3,080 hours.

5. Conclusion

- Curve of the Watershed Sub-Region on Timor Island which is a representation of the Noelmina River Region and the Benenain River Region in the form of radial and parallel.
- The value of α (alfa for the River Region on the Island of Timor can be obtained based on equation $\alpha = 2{,}461 \,.\, (L)^{-0{,}417} .\, (A)^{0{,}214}$ so that the value of α (alfa) varies depending on the size of the watershed and the length of the river. Where the largest α (alpha) value is found in the Sahak Orman River Sub-Region, the Noelmina River Region with a value of α (alpha) of 2.558. While the smallest α (alpha) value is found in the Kapsali River Sub-Region, the Noelmina River Region with a value of α (alpha) of 1,441.
- The confidence level of the results of analysis using IBM SPSS Version 23 software is 95%, with the R^2 value (Determination Coefficient) obtained at 0.826 so that it can be concluded that the watershed area and river length have an effect of 82.60% on the value of α (alpha) obtained

References

[1] Asdak C 2010 *Hidrologi dan Pengelolaan Daerah Aliran Sungai* Edisi Pertama (Yogyakarta: Gadjah Mada University Press)
[2] Chow V T 1964 *Handbook of Applied Hydrology* (New York: McGraw-Hill Book Company)
[3] Chow V T 1988 *Hidrolika Saluran Terbuka* (Jakarta: Erlangga).
[4] Dewi, Rosmalia, Limantara, Lili M, Soetopo and Widandi 2016 *Analisis Parameter Alfa Hidrograf Satuan Sintetis Nakayasu di Sub DAS Lesti* Jurusan Teknik Pengairan Universitas Brawijaya Malang
[5] Ghozali and Imam 2013 *Aplikasi Analisis Multivariate dengan Program IBM SPSS 23* (Semarang: Badan Penerbit Universitas Diponegoro)
[6] Kamiana and I Made 2012 *Teknik Perhitungan Debit Rencana Bangunan Air* (Yogyakarta: Graha Ilmu)
[7] Kodoatie R J and Sugiyanto, 2002 *Banjir, Beberapa Penyebab dan Metode Pengendalian Dalam Perspektif Lingkungan* (Yogyakarta: Pustaka Pelajar)
[8] Kodoatie R J and Sjarief R, 2005 *Pengelolaan Sumber Daya Air Terpadu* (Yogyakarta: Andi)
[9] Lee R 1980 *Forest Hydrology* (Subagio S translation by; Prawirohatmodjo S 1986 Yogyakarta: Gadjah Mada University Press)
[10] Linsley R K 1982 *Hidrologi untuk Insinyur, Edisi Ketiga* (Jakarta: Erlangga)
[11] Ramadani M, Fauzi, Manyuk. Handayani and Yohanna L 2015 *Pemodelan Parameter Alfa Pada Hidrograf Satuan Sintetis Nakayasu (Studi Banding Dengan Hidrograf Satuan Sintetis Gama I)* Jurusan Teknik Sipil Fakultas Teknik Universitas Riau.
[12] Sinukaban N 2007 *Karakteristik Hidrologi dan Daur Limpasan Permukaan Daerah Aliran Sungai Ciliwung. Konservasi Tanah dan Air Kunci Pembangunan Berkelanjutan*. Direktorat Jendral RLPS Departemen Kehutanan.
[13] Sri H 1993 *Analisis Hidrologi* (Jakarta: Gramedia)
[14] Sosrodarsono and Suyono 1999. *Hidrologi Untuk Pengairan* (Jakarta: PT Pradnya Paramita)
[15] Subarkah I 1980 *Hidrologi Untuk Perencanaan Bangunan Air* (Bandung: Ide Dharma).
[16] Triatmodjo B 2009 *Hidrologi Terapan* (Yogyakarta: Beta Offset)

IOP Conf. Series: Materials Science and Engineering **669** (2019) 012016 doi:10.1088/1757-899X/669/1/012016

The strength properties of high calcium fly ash geopolymer specimens incorporating mud-volcanic and limestone variations

A Wardhono [1], Y Risdianto [1], M F Sofianto [1], M T Nur [1] and A H Pradipta [1]

[1] Department of Civil Engineering, Universitas Negeri Surabaya, Indonesia

ariewardhono@unesa.ac.id

Abstract. The use of mud-volcanic and fly ash as cement replacement materials is one alternative solution to overcome environmental issues due to the mudflow disaster and polution from cement production process. This paper objective is to identify the use of mud-volcanic and class C fly ash with high calcium content as waste materials in the development of fly ash geopolymer incorporating limestone. The variations of limestone to fly ash are 0.10:0.60, 0.2:0.5, 0.30:0.40, 0.40:0.30, and 0.50:0.20, respectively. The mud-volcanic ratio was kept at 0.30 to maintain the total weight of specimen at 1.00. The compressive strength was performed at 7, 14 and 28 days age, and the specimens were cured at ambient temperature. The depth penetration test was performed by vicat test. The results show that the highest strength value of 16.38 MPa was achieved by fly ash/mud-volcanic geopolymer with ratio of limestone to fly ash of 0.30:0.40. The initial setting time of 90 minutes indicates that the early reaction of fly ash/mud-volcanic geopolymer shows a comparable setting time to the normal concrete with initial setting time of 45 minutes. These results indicate that fly ash and mud-volcanic can be used as cement replacement materials.

1. Introduction

The volcanic mudflow disaster since 2006 in Sidoarjo Indonesia has caused material losses for the affected environment. The volcanic mudflow continues to occur to this day and it has reached the area of impact affected around 10 square kilometers and a depth of 30 meter [1-3]. The use of volcanic mudflow as a basic material for construction materials has been carried out to overcome the large volume of mud-volcanic [4-6]. The investigation results of previous studies indicate that the mud-volcanic has a high silicate, aluminate and ferrite content [7].

The use of cement, also known as Portland Cement (PC), as concrete-based material has been widely used as construction materials. However, the cement production process has a significant issue to the environment [8, 9]. The production of 1 ton of cement also delivers 0.7 to 1 ton of carbon dioxide (CO_2) gas, which is one of the main cause of global warming problems [10]. However, the global cement production has grown rapidly in recent years. It is one of the largest emissions source of anthropogenic CO_2 gas [11]. The use of cement replacement materials is required to be encouraged to cope with this issue. One alternative solution is to utilize fly ash, a waste material from coal combustion process, as cement replacement materials [12, 13].

The utilization of fly ash as a fully cement replacement material has been extensively investigated. Fly ash as a precursor material is combined with an alkali activator solution, a mix of sodium hydroxide

(NaOH) and sodium silicate, and it is known as geopolymer [14, 15]. The basic reaction of geopolymer is a polymer reaction between silicate and aluminate formed a Si-O-Al- matrix through a geopolymerization process. However, the main issue of using geopolymer is the requirement of high curing temperature to accelerate the geopolymer reactions due to the low calcium content in fly ash material [16, 17].

This paper reports the utilization of mud-volcanic and fly ash as waste materials in the development of fly ash-based geopolymer mortar. Fly ash with high calcium compound was used as the primary material of geopolymer specimens. The mud-volcanic from the volcanic mudflow disaster in Sidoarjo, Indonesia, was used for this research. The variations of limestone was implemented to investigate the fly ash/mud-volcanic geopolymer behavior. The addition of limestone was applied to overcome the high temperature issue during the geopolymer curing process. The strength properties of geopolymer specimen was identified by a compressive strength at the age of 28 days. The depth penetration test (Vicat test) was performed to investigate the setting times of geopolymer specimens.

2. Research methodology

2.1. Materials

Fly ash from a coal power plant Indonesia was used as a geopolymer specimen primary material. It has a high ferrite (54.67%), low silicate (13.28%) and aluminate (4.82%) contents. It is categorized class C fly ash with high CaO content of 16.24% (> 10%) and the total content of SiO_2+Al_2O_3+ Fe_2O_3 is 72.77% (> 50%) in accordance with ASTM C618 [18]. Mud-volcanic materials were collected from the disaster of mud eruption in Sidoarjo, Indonesia, while limestone material was available from the market material distributor. All materials chemical compositions were determined by X-Ray Fluorescence (XRF) test using PANalytical test equipment. The chemical compositions of fly ash, mud-volcanic, and limestone are listed in Table 1.

Table 1. Chemical composition of fly ash, mud-volcanic, and limestone (mass %).

Materials	SiO_2	Al_2O_3	Fe_2O_3	CaO	K_2O	TiO_2	Mn_2O_3
Fly ash	13.28	4.82	54.67	16.24	1.38	0.86	0.51
Mud-volcanic	26.64	8.02	46.17	8.57	3.35	2.25	0.61
Limestone	-	-	0.34	96.12	-	-	-

The solution of alkaline activator was prepared by mixing sodium silicate and 10 Molar sodium hydroxide (NaOH). The 10 Molar NaOH solution was produced by dissolving 400 grams of NaOH pellet to 1 liter of water. The sodium silicate to NaOH ratio of 1.5 was applied for this research.

2.2. Mix design

The details of fly ash/mud-volcanic geopolymer (FAMVG) mortar mix design are listed in Table 2.

Table 2. Mix proportions of fly ash/mud-volcanic geopolymer (FAMVG) mortars

Mixture	Portland Cement	Lime stone	Fly ash	Mud-Volcanic	Fine aggregate	Sodium silicate	NaOH 10 M	Water
PC	1	-	-	-	3.0	-	-	0.45
FAMVG1	-	0.10	0.60	0.30	3.0	0.212	0.141	-
FAMVG2	-	0.20	0.50	0.30	3.0	0.212	0.141	-
FAMVG3	-	0.30	0.40	0.30	3.0	0.212	0.141	-
FAMVG4	-	0.40	0.30	0.30	3.0	0.212	0.141	-
FAMVG5	-	0.50	0.20	0.30	3.0	0.212	0.141	-

The total fine aggregate in geopolymer mortars was maintain to 75% by volume for all mixtures, while the total weight of binder (fly ash, limestone, mud-volcanic, and activator solution) was kept to 25.3%. The density of fine aggregate, fly ash, and mud-volcano was 2.604, 2.895, and 2.533, respectively, while the water solid ratio (w/s) of 0.21 was applied.

2.3. Curing models
Control mortar specimen was cured at ambient temperature (24°C - 30°C) before tested. In general, a geopolymer specimen is required to be cured at high temperature, however due to the inclusion of limestone, a room temperature curing process was applied for FAMVG mortars.

2.4. Testing mortar specimens
A cube steel mold of 5 x 5 x 5 cm3 is used to make the FAMVG mortar specimens. The strength performance of FAMVG specimens was performed by compressive strength and vicat tests. The test of compressive strength was carried out at 7, 14, and 28 days in accordance with ASTM C109 [19]. The depth penetration test (Vicat test) was undertaken in accordance with ASTM C191 [20] to identify the setting time of FAMVG paste as the results of mud-volcanic and limestone variations.

3. Results and discussion
The strength performance of FAMVG specimens for all mixtures are shown in Table 3 and Figure 1. All FAMVG mortar specimens demonstrate a better strength performance compared to PC control except for FAMVG4 and FAMVG5 throughout 28 days.

Table 3. Compressive strength of FAMVG mortars

Mixture	Compressive strength (MPa)		
	7 days	14 days	28 days
PC	7.30	9.60	9.30
FAMVG1	9.10	10.24	12.58
FAMVG2	9.45	11.04	13.10
FAMVG3	12.64	16.20	16.38
FAMVG4	7.10	6.30	4.60
FAMVG5	5.20	4.50	3.80

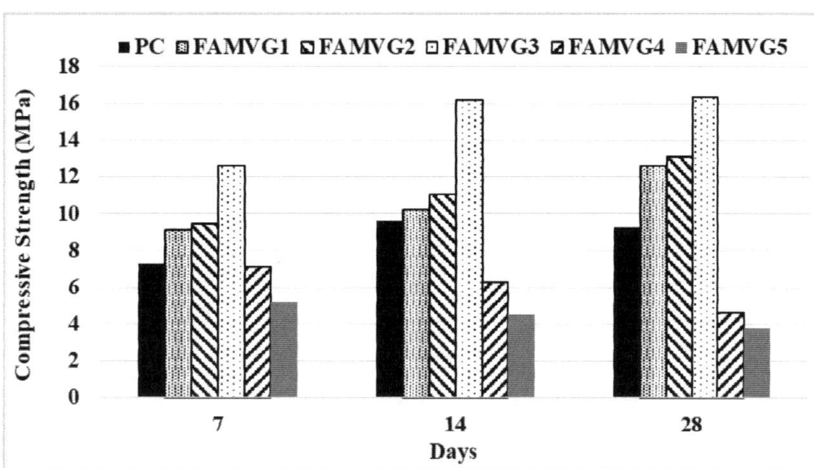

Figure 1. The strength performance of fly ash/mud-volcanic geopolymer mortars

The highest early strength (7 days) of 12.64 MPa was shown by FAMVG3 (30% limestone) compared to the PC control with the strength merely of 7.30 MPa. The FAMVG3 also exhibits the highest strength throughout 28 days with the strength of 16.20 MPa and 16.38 MPa at 14 and 28 days, respectively. A comparable compressive strength was also performed by FAMVG2 with the initial and final strength of 9.45 MPa and 13.10 MPa, respectively. Furthermore, FAMVG1, FAMVG2, and FAMVG3 geopolymer mixtures demonstrates a significant improvement in strength along with time.

In addition, inclusion of limestone more than 30% tends to lessen the strength of FAMVG specimens. FAMVG4 (limestone 40%) and FAMVG5 (limestone 50%) mortar specimens exhibit a lower strength performance compared to all mixtures. FAMVG5 specimen with the strength of 5.20 MPa and 3.80 MPa achieves the lowest compressive strength at 7 days and 28 days, respectively. Moreover, the higher addition of limestone tends to hinder and decrease the strength development of FAMVG specimens. The addition of 40% limestone decreases the strength of FAMVG4 specimen by 35.2% from 7.10 MPa at 7 days to 4.60 MPa at final strength. Further addition of limestone by 50% in FAMVG5 decreases the strength by 26.9% from 5.20 MPa to 3.80 MPa. Similar finding was found by [6], the author found that the addition of limestone more than 30% on fly ash-based geopolymer specimen using mud-volcanic as sand replacement material tends to reduce the strength performance.

Indeed, the addition of limestone improves the strength performance of fly ash/mud-volcanic geopolymer mortars as shown in FAMVG1, FAMVG2, and FAMVG3. The geopolymer reaction involves the primary reaction between silicate and aluminate to form geopolymer matrix [21]. However, due to the low aluminate compound in fly ash raw material, it causes un-reacted silicate during the geopolymerization reaction [22]. The addition of limestone increases the calcium content and drive a reaction between calcium and silicate. This reaction produces a calcium-silicate-hydrate (C-S-H) compound similar to the traditional concrete, which improves the strength performance of FAMVG specimens.

However, a high limestone inclusion tends to reduce the strength performance as shown in FAMVG4 and FAMVG5 specimens. It might attributes to the excess calcium compound, which does not react with silicate. This excess calcium forms a calcium hydroxide compound which leads to a concrete crack due to the material expansion [23]. This leads to the strength reduction throughout the time. Furthermore, the inclusion of limestone significantly governs the setting time rate of fly ash/mud-volcanic geopolymer specimens as shown in Figure 2.

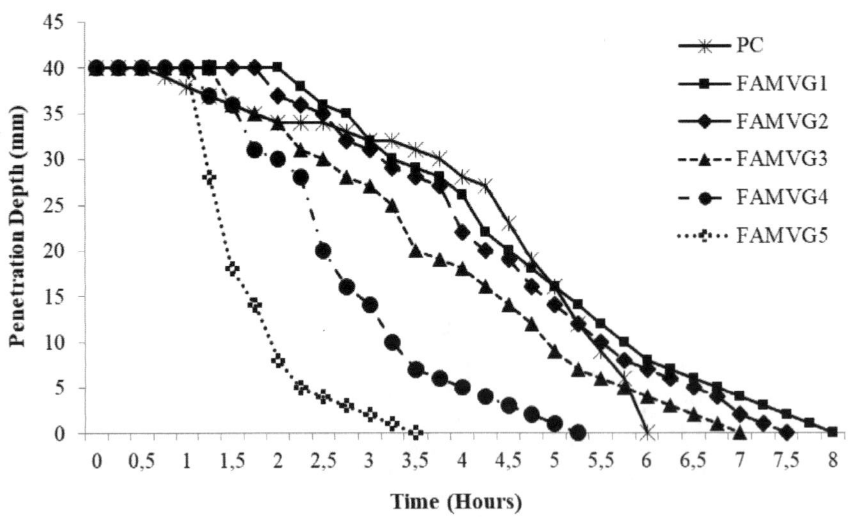

Figure 2. The depth penetration test fly ash/mud-volcanic geopolymer mortars

Based on the results of vicat test, the standard setting time (initial and final) rate of normal concrete (PC control) was 0.75 hours and 6 hours, respectively. The fly ash/mud-volcanic geopolymer tends to have a longer initial and final setting rate. The FAMVG1 with 0% limestone addition has initial and final setting time rate of 2.25 hours and 8 hours, respectively. Despite fly ash/mud-volcanic geopolymer can be made at ambient temperature, it requires a longer time to achieve its structural integrity. The inclusion of limestone significantly affects the setting time rate. The inclusion of 10% limestone affected the initial setting time rate from 2.25 hours to 1.25 hours and 8 hours to 7.5 hours for the final setting time. FAMVG3 specimen demonstrates a similar setting time rate with the initial and final setting time of 1.5 hours and 7 hours. The fastest setting time rate shows by FAMVG5 with limestone addition of 50%. It has initial setting time of 1.25 hours and final setting time of 3.5 hours. The accelerate setting time rate might be caused by two reaction, a reaction between silicate and aluminate (Si-O-Al-) and reaction between calcium and silicate (C-S-H).

4. Conclusion

This research deals with the strength properties of high calcium class C fly ash/mud-volcanic geopolymer with variations of limestone. Based on the findings, it can be concluded that:

- A mix of high calcium, mud-volcanic, and limestone can be used as cement replacement material in geopolymer specimen due to the similarity compositions of silicate and aluminate.
- The addition of limestone in fly ash-mud volcanic geopolymer significantly affects the strength performance. The 30% inclusion of limestone gives the highest strength of 16.38 MPa at the final 28 days.
- High addition of limestone significantly decreases the strength performance of fly ash-mud volcanic geopolymer throughout the time due to the un-reacted excess calcium.
- The setting time rate of fly ash-mud volcanic geopolymer governs by the inclusion of limestone. It accelerates the setting time (initial and final) rates, which affects the hardening in room curing temperature.
- High inclusion of limestone at 50% significantly increases the setting time rate of fly ash-mud volcanic geopolymer in 60 minutes at final setting time. However, it tends to decrease the strength due to the difficulty during the casting process.
- The use of fly ash and mud-volcanic is one alternative material for cement replacement in construction material. It also helps to overcome the high volume of mud volcanic problem and global warming issue cause by cement production.

References

[1] Drake P 2018 *Environmenal Communication* 12 (2) 261-273
[2] Mohsin A 2017 *Arcadia* Spring 2017 (5)
[3] Krisnayanti B D and Agustawijaya D S 2014 *Journal of Degraded and Mining Lands Management* 1 (4) 207-210
[4] Antoni et al 2013 *Advanced Materials Research* 626 224-228
[5] Razak R A, et al. 2015 *International Journal of Molecular Sciences* 16 11629-11647
[6] Wardhono A, et al 2018 *Materials Science and Engineering (IOP Conference Series)* 434 (1)
[7] Handoko L, et al. 2015 *Procedia Engineering* 125 324-330
[8] Meyer C 2009 *Cement and Concrete Composites* 31 (8) 601-605
[9] Berry M, et al 2009 in: *The World of Coal Ash (WOCA) Conference* (Lexington, USA).
[10] Li C, et al. 2011 *Materials Science Forum* 685 181-187
[11] Andrew R M 2018 *Earth System Science* Data 10 195-217
[12] Okoye F N 2017 *Materials Today* 4 (4 Part E) 5599-5604
[13] Singh B, et al 2015 *Construction and Building Materials* 85 78-90
[14] Davidovits J 1994 in: First *International Conference on Alkali Cements and Concretes* (Kiev, Ukraine)
[15] Hardjito D, et al. 2005 *Australian Journal of Structural Engineering* 6 (1)
[16] Kong D L Y and Sanjayan J G 2010 *Cement and Concrete Research*

[17] Wardhono A 2018 *Journal of Physics (IOP Conference Series)* 1108 (1)
[18] ASTM Standard 2004 ASTM C618-03
[19] ASTM Standard 2004 ASTM C109-02
[20] ASTM Standard 2004 ASTM C191-04
[21] Pacheco-Torgal F, et al 2008 *Construction & Building Materials* 22 1308-1314
[22] Collins F and Sanjayan J G 2001 *Cement and Concrete Composites* 23 345-352
[23] Sinsiri T, et al 2010 *International Journal of Minerals, Metallurgy and Materials* 17 683-690

Acknowledgments
The authors would like to express their sincere thanks to Kementerian Riset Teknologi dan Pendidikan Tinggi Contract No. 193/SP2H/LT/DRPM/2019 for the funding and Universitas Negeri Surabaya for the support.

IOP Conf. Series: Materials Science and Engineering **669** (2019) 012017 doi:10.1088/1757-899X/669/1/012017

Shifting cropping shifts by efficient irrigation water to produce maximum rice productivity

S Azis[1]

[1] Civil Engineering Department, Nationale Institute of Technology Malang, East Java, Indonesia

cup.subandiyah@gmail.com

Abstract. Agriculture is the main support of the economy in Indonesia so that the sustainability of the productivity of paddy fields must be maintained properly. Rice fields rely heavily on irrigation water supplies from technical rigs. Continuity of water flow must be maintained in accordance with water requirements and cropping patterns in each paddy field. In connection with changes in conditions lately there has been a shift in the rainy season and the dry season which results in erratic discharge of surface water both in quantity and on time so that the use of irrigation water must be done wisely and efficiently to maintain the productivity of the yield of paddy fields. From the results of the implementation of the cropping pattern set by the relevant agency there has been a deviation of 13.81%, which means that it is necessary to improve the implementation of cropping patterns and adjust the time of rain so that rice productivity remains maximal.

1. Introduction

The provision of irrigation water for agriculture needs to be managed in a wise and sustainable manner so that its existence and functions are maintained, one of which is to provide benefits in agriculture. The Irrigation Area (DI) Molek is one of 44 Irrigation Areas in the Malang-Indonesia Regency area which has 3,983 hectares of paddy fields. Maintenance of irrigation network is also very important in order to ensure continuity of irrigation water allocation. [1]

Planting planning is outlined in the planting document called the Global Planting Plan (RTTG). From year to year, the method of preparing this Global Planting Plan has not changed even though the current situation and conditions are different.

The duration of the rainy and dry seasons is not fixed, sometimes the rainy season lasts less than 6 (six) months and sometimes more, as well as the dry season [2]. This condition causes the availability of surface water to become erratic and if the surface water is affected, the water discharge conditions are also affected so that it needs to be done by adjusting the surface water adaptation to climate change [3]. Climate change is projected to have a significant impact on conditions affecting agriculture [4].

From the results of the evaluation of the Global Planting Plan presented in Table 1 attached below, it can be concluded that there has been a deviation in the pattern of cropping arrangements of 13.81%, meaning that the implementation of the Global Planting Plan is not in accordance with the plan and needs to develop a new Global Planting Plan applicative and dynamic and keep up with the progress of the rain discharge conditions.

Table 1. Evaluation of the global planting plan

Irrigation / Raw Rice Field Area (Ha)	RTTG (Ha)	Realization of Planting Patterns Rice Special (Ha)	Compliance with RTTG (%)	Deviation (%)
DI Molek 3983	3974	3425	86.11	13.81

From the problems that occur, make the background and rationale for doing research to produce an appropriate analysis of water needs as a basis for determining the Global Planting Plan.

All stakeholders involved in the planning and implementation of the allocation of irrigation water understand climate change situations and conditions that are very influential on surface water and irrigation water so they argue that the Global Planting Plan arrangement must be adapted to climate change conditions to avoid the possibility of crop failure. [5].

More and more research shows that the world is warming and will continue to warm up when greenhouse gas concentrations increase in the future [6]. Productivity is influenced by a number of climate change variables including rainfall patterns, temperature increases, changes in sowing and harvesting, water availability and land suitability [7]. One of the most important impacts of climate change is the change in water available at the regional and local levels [8].

2. Materials and methode

Administratively, Molek Irrigation Area is located in Malang Regency. The average annual rainfall is 1,500 mm. Water sources Molek Irrigation Area originates from Mount Arjuno which flows through Brantas River which is channeled through the 12.44 Km Molek canal. The research framework is as in Table 2.

Table 2. Study framework

No	Stages of Calculation / Analysis	Data source	Calculation / analysis method	Results
1	Data Quality Test	Historical rainfall and discharge data	Rescaled Adjusted Partial Sums Method	Analysis requirements are accepted (still within consistent limits) if the values of Q / (n0.5) and R / (n0.5) count smaller than the values of Q / (n0.5) and R / (n0.5)
3	Mainstay Debit	Debit Data	Weibull Method Pr = m / (n+1) * 100 %	Debit of Wet Season , Normal Season, Dry Season
4	Mainstay Rainfall	Historical rainfall data	$R_{80} = \frac{n}{5} + 1$ (reliability 80 %) $R_{50} = \frac{n}{2} + 1$ (reliability 50 %)	Mainstay Rain Pattern
5	Effective Rainfall	Historical rainfall data	Frequency Analysis Formula	rice : R_e = (0,70 x R_{80})/day secondary crops : R_e= R_{50}/day
6	Evapotranspirati on	Climatology Data	Modified Penman	Eto , mm/day
7	Water Needs for Soil Processing and Nursery	The duration of the tillage	Determined	Land preparation = 30 days Soil processing + Nursery = 250 mm/day

Table 2. Study framework (cont.)

No	Stages of Calculation / Analysis	Data source	Calculation / analysis method	Results
8	Percolation	Soil sample data and map of soil type distribution	Soil Test in the Laboratory to determine permeability numbers	Percolation Numbers
9	Water Requirements for Land Preparation	Potential Evapotranspiration of the Penmann Modification Method, Percolation Calculation	Suyono + Takeda C_{PL}= Cland preparation x Extensive planting ratio	IR, mm/day
10	Water Level Requirement	Estimated soil fertility	Determined	Change of Water Layer = 50 mm/month
11	Irrigation Efficiency	Debit Intake	Water loss in Primary, Secondary and Tertiary channels	Irrigation Efficiency Rainy Season Irrigation EfficiencyDry Season 1 Irrigation EfficiencyDry Season 2
12	Balance sheet needs and availability of water	Regional Irrigation of Irrigation Water Needs Molek	Water balance concept	Comparison of the value of the amount of availability to needs
13	Planting Pattern Simulation	Data and Simulation	Relative Crops Factor - Area of Relative Palawija	Irrigation Water Needs and Extensive Planting Area are optimal for Wet Season Year, Normal Season Year and Dry Season Year.
14	Preparation of a Global Planting Plan	Simulation Results	Rainy Season Early Planting Dry Season 1 Early Planting Dry Season 2 Early Planting	Wet Season Year Dry Season Year Normal Season Year

3. Results and discussion

The results of the analysis of the data and observations can produce the following result as shown in Table 3, Table 4, Table 5 and Table 6.

Table 3. Rainfall data consistency results

Station Name	Number		Coefficient Determination (%)
	Station	Stasiun Uji	
Sumber Pucung	36	Station 1	99.89
Kepanjen	39	Station 2	99.59
Kali Pare	40	Station 3	99.80
Pagak	64	Station 4	99.21

Table 4. Analysis of availability of irrigation intake discharge

Opportunity (80%)	Jan	Feb	Mar	Apr	May	Jun	Jul	Aug	Sep	Oct	Nov	Dec
80	4.80	5.97	6.47	6.47	6.42	6.15	5.85	5.04	4.89	4.54	5.69	5.69
90	3.52	3.63	5.00	3.65	3.49	3.22	3.09	2.98	2.88	2.91	4.35	3.14
25	7.26	7.38	7.43	7.38	7.40	7.46	7.42	7.37	7.16	6.99	6.71	7.35

Table 5. Mainstay rainfall 80% and 50%

Month	Period	R 80 (mm)	R 50 (mm)	Month	Period	R 80 (mm)	R 50 (mm)
January	1	72.09	77.66	July	1	0.00	1.41
	2	29.32	55.80		2	0.00	0.00
	3	98.04	95.02		3	0.00	0.00
February	1	71.77	121.63	August	1	0.00	0.00
	2	122.96	61.68		2	0.00	0.00
	3	71.19	102.42		3	0.00	0.00
March	1	123.02	37.64	September	1	0.00	0.00
	2	87.31	84.56		2	0.00	0.06
	3	26.13	192.76		3	0.00	2.94
April	1	82.52	119.91	October	1	0.00	20.56
	2	63.62	99.42		2	0.00	41.37
	3	46.64	0.01		3	24.31	43.02
May	1	24.43	47.22	November	1	11.54	73.98
	2	32.68	31.43		2	41.77	58.50
	3	0.00	0.49		3	72.52	73.59
June	1	0.00	5.39	December	1	30.26	210.53
	2	0.00	14.12		2	43.95	15.88
	3	0.00	0.49		3	122.86	92.94

Table 6. Effective rainfall analysis

Month	Period	R 80 (mm)	Re rice			Re Secondary Plant		
			(mm)	(mm/day	(lt/sec/ha)	(mm)	(mm/day)	(lt/sec/ha)
January	1	72.09	50.47	5.05	0.59	77.66	7.77	0.90
	2	29.32	20.52	2.05	0.24	55.80	5.58	0.65
	3	98.04	68.63	6.86	0.80	95.02	9.50	1.10
February	1	71.77	50.24	5.02	0.58	121.63	12.16	1.41
	2	122.96	86.07	8.61	1.00	61.68	6.17	0.72
	3	71.19	49.83	4.98	0.58	102.42	10.24	1.19
March	1	123.02	86.11	8.61	1.00	37.64	3.76	0.44
	2	87.31	61.12	6.11	0.71	84.56	8.46	0.98
	3	26.13	18.29	1.83	0.21	192.76	19.28	2.24
April	1	82.52	57.77	5.78	0.67	119.91	11.99	1.39
	2	63.62	44.53	4.45	0.52	99.42	9.94	1.15
	3	46.64	32.65	3.26	0.38	0.01	0.00	0.00
May	1	24.43	17.10	1.71	0.20	47.22	4.72	0.55
	2	32.68	22.87	2.29	0.27	31.43	3.14	0.36
	3	0.00	0.00	0.00	0.00	0.49	0.05	0.01
June	1	0.00	0.00	0.00	0.00	5.39	0.54	0.06
	2	0.00	0.00	0.00	0.00	14.12	1.41	0.16
	3	0.00	0.00	0.00	0.00	0.49	0.05	0.01
July	1	0.00	0.00	0.00	0.00	1.41	0.14	0.02
	2	0.00	0.00	0.00	0.00	0.00	0.00	0.00
	3	0.00	0.00	0.00	0.00	0.00	0.00	0.00

Table 6. Effective rainfall analysis (cont.)

Month	Period	R 80 (mm)	Re rice			Re Secondary Plant		
			(mm)	(mm/day	(lt/sec/ha)	(mm)	(mm/day)	(lt/sec/ha)
August	1	0.00	0.00	0.00	0.00	0.00	0.00	0.00
	2	0.00	0.00	0.00	0.00	0.00	0.00	0.00
	3	0.00	0.00	0.00	0.00	0.00	0.00	0.00
September	1	0.00	0.00	0.00	0.00	0.00	0.00	0.00
	2	0.00	0.00	0.00	0.00	0.06	0.01	0.00
	3	0.00	0.00	0.00	0.00	2.94	0.29	0.03
October	1	0.00	0.00	0.00	0.00	20.56	2.06	0.24
	2	0.00	0.00	0.00	0.00	41.37	4.14	0.48
	3	24.31	17.02	1.70	0.20	43.02	4.30	0.50
November	1	11.54	8.08	0.81	0.09	73.98	7.40	0.86
	2	41.77	29.24	2.92	0.34	58.50	5.85	0.68
	3	72.52	50.76	5.08	0.59	73.59	7.36	0.85
December	1	30.26	21.19	2.12	0.25	210.53	21.05	2.44
	2	43.95	30.76	3.08	0.36	15.88	1.59	0.18
	3	122.86	86.00	8.60	1.00	92.94	9.29	1.08
Total		1741.25	909.27	90.93	10.55	1782.42	178.24	20.68

Climatology data and Potential Evaporation for calculating irrigation water requirements is shown in Table 7.

Table 7. Climatology data and ETo (Potential Evaporation)

Month	Average Temperature °C	Humidity%	Wind velocity km/day	Duration of solar radiation hour	Radiation MJ/m²/day	ETo (Potential Evaporationl mm/day
January	22.9	76	95	6.1	19.3	3.81
February	23.2	76	94	5.5	18.5	3.73
March	23.2	76	76	6.3	19.3	3.77
April	23.4	71	83	6.6	18.5	3.66
May	23.1	68	91	7.8	18.6	3.58
June	23.3	69	98	7.8	17.7	3.42
July	22.4	69	102	7.9	18.2	3.46
August	22.2	70	102	8.3	20.2	3.78
September	22.7	72	102	8.6	22.3	4.20
October	23.0	70	94	8.2	22.5	4.34
November	23.4	68	91	8.2	22.5	4.43
December	22.8	73	89	7.0	20.5	4.02

The duration of land preparation is 30 days. Water requirements for tillage and nursery are 250 mm which consists of 200 mm used for saturation and 50 mm for nursery. To estimate the water requirements for land preparation, the percolation value in the study area must be known. The percolation value is shown in Table 8.

Table 8. Magnitude of Percolation Value

No	Kinds of Land	Vertical percolation (mm/day)
1.	Sandy loam	3 – 6
2.	Loam	2 - 3
3.	Clay	1 – 2

Then the water requirements for land preparation can be calculated which the results whichare shown in Table 9.

Table 9. Water requirements for land preparation

Month	Eto mmday	Eo mm/day	P mm/day	M mm/day	S mm	T days	K	IR mm/day
January	3.810	4.191	4.5	8.691	250	30	1.0429	13.42
February	3.730	4.103	4.5	8.603	250	30	1.0324	13.36
March	3.770	4.147	4.5	8.647	250	30	1.0376	13.39
April	3.660	4.026	4.5	8.526	250	30	1.0231	13.31
May	3.580	3.938	4.5	8.438	250	30	1.0126	13.25
June	3.420	3.762	4.5	8.262	250	30	0.9914	13.14
July	3.460	3.806	4.5	8.306	250	30	0.9967	13.17
August	3.780	4.158	4.5	8.658	250	30	1.0390	13.40
September	4.200	4.620	4.5	9.120	250	30	1.0944	13.71
October	4.340	4.774	4.5	9.274	250	30	1.1129	13.81
November	4.430	4.873	4.5	9.373	250	30	1.1248	13.88
December	4.020	4.422	4.5	8.922	250	30	1.0706	13.58

Water Level Requirement is estimated at 50 mm. When using a 10 day period, the Water Level Requirement (WLR) of 50 mm is divided into 30 days, which is equal to 1.67 mm / day.

Irrigation efficiency is a comparison between the water discharge that arrives at the agricultural land and the water discharge coming out of the take-up gate. Before arriving at the paddy field, water must be flowed from the source through the main, secondary and tertiary canals Irrigation Efficiency.

Irrigation efficiency is calculated based on water loss during the drainage process along the channel. Irrigation efficiency can be shown in Table 10. Recapitulation of water balance conditions for existing planting can be shown in Table 11.

Table 10. Calculation of Irrigation Efficiency

Season	Month	Primary Loss (m3/sec)	Secondary Loss (m3/sec)	Irrigation Efficiency (%)	average (%)
Rainy Season	November	28.05	17.64	66.38	68.813
	December	20.30	16.65	71.26	
	January	20.82	21.33	68.22	
	February	19.36	20.74	69.39	
Dry Season 1	March	22.35	16.42	70.21	69.057
	April	22.42	16.64	70.03	
	May	21.83	19.06	68.94	
	June	25.28	19.05	67.05	
Dry Season 2	July	25.74	19.29	66.67	64.275
	August	28.80	18.91	65.29	
	September	30.17	19.64	64.21	
	October	33.35	23.08	60.93	

Table 11. Recapitulation of Water Balance Conditions for Existing Planting

Month	Period	Irrigation Needs Intake (m3/sect)	Dry Season River Discharge (m3sect)	Information	Normal Season River Discharge (m3/sec)	Information	Wet Season River Discharge (m3sec)	Information
January	1	7.06	3.52	Not enough	4.80	Not enough	7.26	Enough
	2	7.93	3.52	Not enough	4.80	Not enough	7.26	Not enough
	3	5.35	3.52	Not enough	4.80	Not enough	7.26	Enough
February	1	5.32	3.63	Not enough	5.97	Enough	7.38	Enough
	2	3.30	3.63	Enough	5.97	Enough	7.38	Enough
	3	4.91	3.63	Not enough	5.97	Enough	7.38	Enough
March	1	3.41	5.00	Enough	6.47	Enough	7.43	Enough
	2	4.39	5.00	Enough	6.47	Enough	7.43	Enough
	3	6.31	5.00	Not enough	6.47	Enough	7.43	Enough

Table 11. Recapitulation of Water Balance Conditions for Existing Planting (cont.)

Month	Period	Irrigation Needs Intake (m3/sect)	Dry Season River Discharge (m3sect)	Information	Normal Season River Discharge (m3/sec)	Information	Wet Season River Discharge (m3sec)	Information
April	1	4.54	3.65	Not enough	6.47	Enough	7.38	Enough
	2	5.32	3.65	Not enough	6.47	Enough	7.38	Enough
	3	6.32	3.65	Not enough	6.47	Enough	7.38	Enough
May	1	6.39	3.49	Not enough	6.42	Enough	7.40	Enough
	2	6.08	3.49	Not enough	6.42	Enough	7.40	Enough
	3	6.61	3.49	Not enough	6.42	Not enough	7.40	Enough
June	1	5.98	3.22	Not enough	6.99	Enough	7.46	Enough
	2	5.58	3.22	Not enough	6.99	Enough	7.46	Enough
	3	5.77	3.22	Not enough	6.99	Enough	7.46	Enough
July	1	6.65	3.09	Not enough	5.85	Not enough	7.42	Enough
	2	6.67	3.09	Not enough	5.85	Not enough	7.42	Enough
	3	6.67	3.09	Not enough	5.85	Not enough	7.42	Enough
August	1	6.60	2.98	Not enough	5.04	Not enough	7.37	Enough
	2	7.04	2.98	Not enough	5.04	Not enough	7.37	Enough
	3	7.86	2.98	Not enough	5.04	Not enough	7.37	Not enough
September	1	8.61	2.88	Not enough	4.89	Not enough	7.16	Not enough
	2	8.48	2.88	Not enough	4.89	Not enough	7.16	Not enough
	3	7.66	2.88	Not enough	4.89	Not enough	7.16	Not enough
October	1	6.85	2.91	Not enough	4.54	Not enough	6.99	Enough
	2	6.22	2.91	Not enough	4.54	Not enough	6.99	Enough
	3	5.64	2.91	Not enough	4.54	Not enough	6.99	Enough
November	1	4.68	4.35	Not enough	5.69	Enough	6.71	Enough
	2	4.16	4.35	Enough	5.69	Enough	6.71	Enough
	3	3.56	4.35	Enough	5.69	Enough	6.71	Enough
December	1	4.52	3.14	Not enough	5.69	Enough	7.35	Enough
	2	5.32	3.14	Not enough	5.69	Enough	7.35	Enough
	3	4.37	3.14	Not enough	5.69	Enough	7.35	Enough

Considering that the planting in one year is divided into 3 (three) seasons, namely the Rainy Season (MH), Dry Season 1 (MK1) and Dry Season 2 (MK2), and based on the effective rainfall pattern, the rainy month is November to April. Good months of start the planting is as the follows.

- Schedule for early planting of November period I
- Schedule for early planting of November period II
- Schedule for early planting of November period II
- Schedule for early planting of December period I
- Schedule for early planting of December period II
- Schedule for early planting of December period III
- Schedule for early planting of January period I
- Schedule for early planting of January period II
- Schedule for early planting of January period III
- Schedule for early planting of February period I
- Schedule for early planting of February period II
- Schedule for early planting of February period III

4. Conclusion

This shift in planting time was generated from simulations of rise planting area and secondary crops which produced the most optimal planting area.
Based on the research, shifting planting schedule that produces maximum production is as follows.

IOP Conf. Series: Materials Science and Engineering **669** (2019) 012017 doi:10.1088/1757-899X/669/1/012017

Table 12. Conclusion

Season	Global Planting Plan of Eksisting	New Global Planting Plan of Wet Season Year	New Global Planting Plan of Normal Season Year	New Global Planting Plan of Dry Season Year
Rainy Season				
Group I	10 December	1 December	20 February	20 February
Group II	15 December	10 December	1 March	1 March
Group III	20 December	20 December	10 March	10 March
Dry Season 1				
Group I	21 March	1 April	20 June	20 June
Group II	25 March	10 April	1 July	1 July
Group III	30 March	20 April	10 July	10 July
Dry Season 2				
Group I	20 August	1 August	20 October	20 October
Group II	25 August	10 August	1 November	1 November
Group III	30 August	20 August	10 November	10 November

References

[1] Azis S and Widodo H 2016 *International Journal of Agricultural Research, Innovation and Technology* (IJARIT) 6(2) .61-68

[2] Azis S 2011 *Journal of Basic and Applied Scientific Research (JBASR)*. Text Road Journals Publications. www.textroad.com 1(10) 1709-1714

[3] Azis S 2016 *Australian Journal of Basic and Applied Sciences (AJBAS)* 10(14) 207-216.

[4] Ziad A M and Sireen 2011 *African Journal of Environmental Science and Technology* 4(3) 183-192

[5] Azis S 2018 *International Journal of Civil Engineering and Technology (IJCIET)* 9(10) 1371–1381

[6] Seo S N, et al 2008 *Journal of Agricultural Research* 68(1) 69-79

[7] Shakoor U, et al 2011 *Journal of Agricultural and Sciences* 4(2) 327-333

[8] Kazem J 2013 *American Journal of Climate Change* 2(4) 296-305

IOP Conf. Series: Materials Science and Engineering **669** (2019) 012018 doi:10.1088/1757-899X/669/1/012018

The comparative study of flood modelling with the unsteady and the steady flow on Ngotok river

G Idfi [1], I D Wahyono[1], A Yulistyorini[1], and N L Khomsiati [1]

[1] Departement of Civil Engineering, State University of Malang, Indonesia

gilang.idfi.ft@um.ac.id

Abstract. The Flood modeling in a river is usually modeled with the one-dimensional flow. The Flood is one form of unsteady flow that can be simulated with the HEC-RAS program. The unsteady flow modeling results often occur error due to unstable program analysis. The stability of the model is influenced by the river channel, the slope of the river bed and the cross-sectional of the river. In flood forecasting required by maximum discharge data and maximum flood elevation, then the steady flow model is often used as an alternative to the calculation of flood forecasting. This study aims to compare the advantages and disadvantages of flood modeling with a steady flow and unsteady flow. The location of study in Ngotok River, Mojokerto Regency, Indonesia. Hydrological modeling uses HEC-HMS and hydraulic modeling uses HEC-RAS. The resulting modeling shows that the water elevation profile of the steady flow model has a higher than the unsteady flow model. The time required for the simulation of the unsteady flow model is longer than the steady flow model. The average water level elevation produced by the unsteady flow model is +27.17 m while the steady flow model is +29.62 m.

1. Introduction

The full bank capacity of a river is expected to accommodate the flood discharge. To find the optimal of flood modeling, the method must be approaching the existing condition. The simulation model with the software is often used in flood modeling planning on the steady or the unsteady flow. HEC-RAS is one of the software to analyze the flood problem. The unsteady flow model creates by the component flow according to the function of distance and time. While the steady flow is the modeling with the same component flow in every place and time [1]. In the unsteady modeling is often found the stable results. The input data on the unsteady model is more complicated like a cross-section of the river, the initial condition of the river, the upstream and downstream boundary condition. When the unsteady modeling on HEC-RAS occurs constraints, the engineers can replace it with the steady flow model. Flood routing methods on the rivers need to simulate with the unstable component flow [2]. If the flood modeling generated by the unsteady flow is unstable, then the simulation can be diverted to steady flow [3]. In the unsteady flow must be changed the discrete finite difference approximation to solve it, so there are some forms of error.

2. Location of Study

Area of this study on Ngotok River in Mojokerto, Indonesia. The length of this river is ± 26 km and the width is 60 m. The Ngotok river has the downstream area in Brantas River [5]. This river has the mild slope from the upstream area until the downstream area. The upstream boundary condition is the hydrograph entered on cross-section number 267. The downstream boundary condition is water surface, rating curve and stage hydrograph entered on a first cross-section. Location of study can be shown in figure 1.

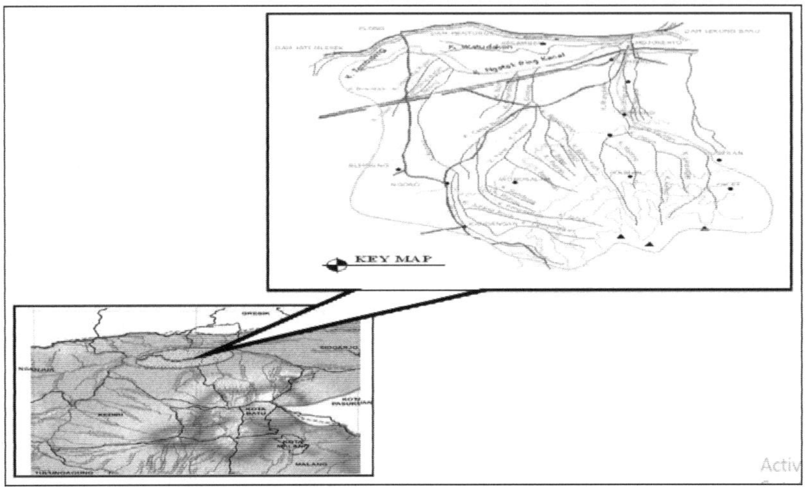

Figure 1. Location of the Ngotok river

3. Methodology

A drainage basin with rain gauge data is not uniformly distributed [6]. To determine the maximum precipitation is used The Thiessen Polygon Method [7]. This method using the rainfall data from Balai Besar Wilayah Sungai (BBWS) Brantas for 20 years (1995-2004). To create the Thiessen Polygon, draw lines connecting the stations and drag a straight line from the middle position from the line connection [8]. The hydrology analysis using the Soil Conservation Service (SCS) with HEC-HMS 4.1 [9]. The SCS methods determine the runoff discharge based on the cumulative of rainfall, land covers, soil characteristic and humidity [10]. The model equations can be shown below :

$$P_e = \frac{(P-I_a)^m}{P-I_a+S} \quad (1)$$

Where Pe: The Cumulative Precipitation, P: Time Concentrate (hours), Ia: Initial loss, S: Storage capacity

$$S = \frac{1000-10CN}{CN}\ (English\ Unit) \quad (2)$$

The storage capacity depends on by Curve Number (CN) [11]. It can be shown below

$$S = \frac{25400-254CN}{CN}\ (Metric\ Unit) \quad (3)$$

The hydraulics model using the one dimension steady and unsteady flow with HEC-RAS 4.1. To calculate the water surface elevation, this software using the cross section and long section data from the rivers. The unsteady model can be done by the St. Venant equation [4]. This equation develops from the continuity equation and the momentum equations. The flow chart of this study can be shown in figure 2 below

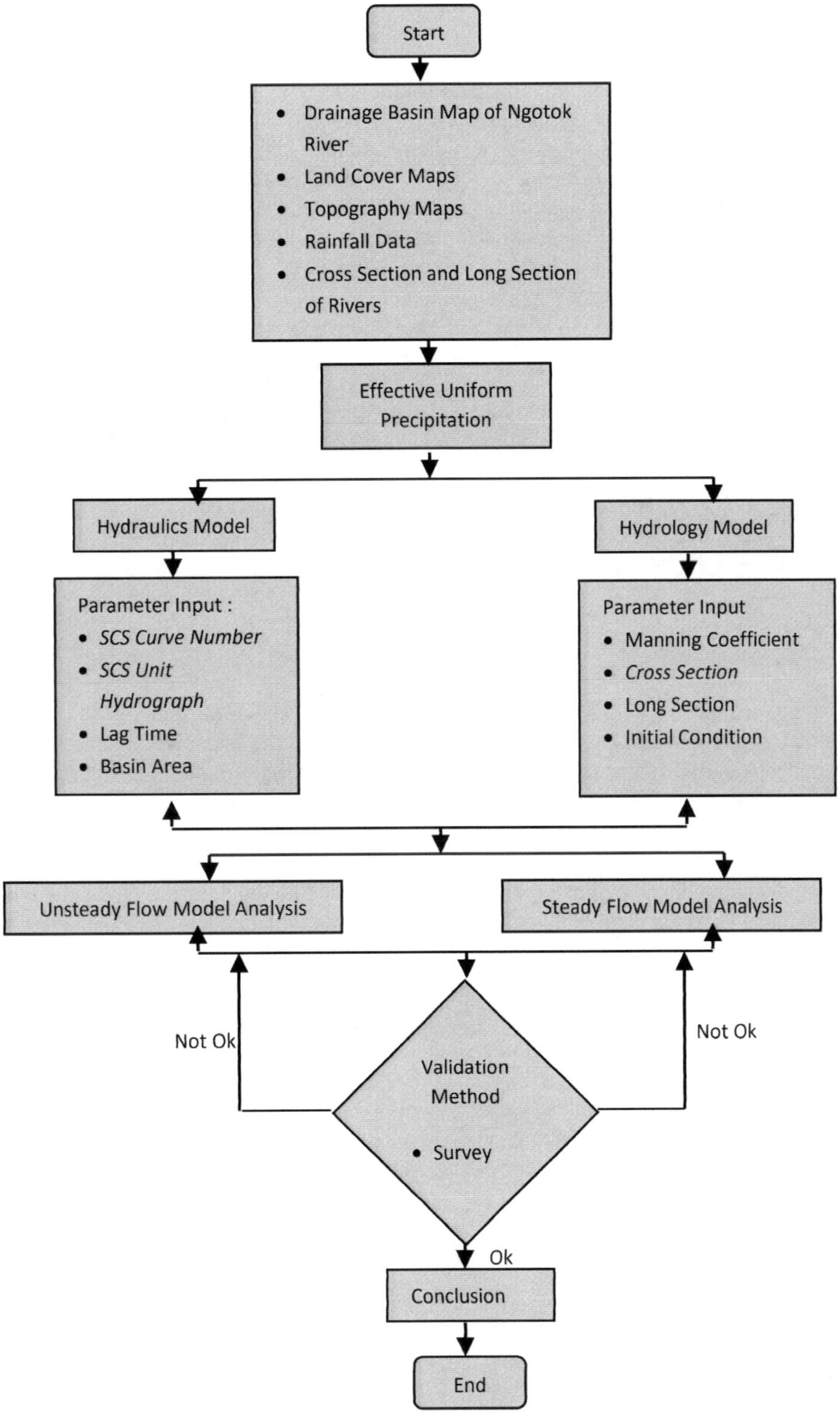

Figure 2. Flow chart

IOP Conf. Series: Materials Science and Engineering **669** (2019) 012018 doi:10.1088/1757-899X/669/1/012018

4. Analysis and discussion

4.1. Hydrology model

The result of the modeling must be calibrated to know whether the modeling results in accordance with the existing. Due to the existing discharge data not recorded by the Automatic Water Level Recorder (AWLR), so the calibration method using the survey by the residents around the Ngotok River to determine the behavior of this river. The flow characteristic like a flood surface elevation and the inundation area. The survey was conducted on 29 October 2016. From this survey can be known when the Ngotok overtopping, flood inundates the resident and the rice field. The depth of inundation is 20-30 cm. The survey method can be shown at the figure 3

Figure 3. The survey method

In this study, the return period of discharge using 25 years. The input data of the hydrology model is catchment area, the land cover, curve number, lag time and the effective precipitation. The all of them is inputted by HEC-HMS model. The result of the hydrology model is the flood hydrograph. It can be shown at figure 4

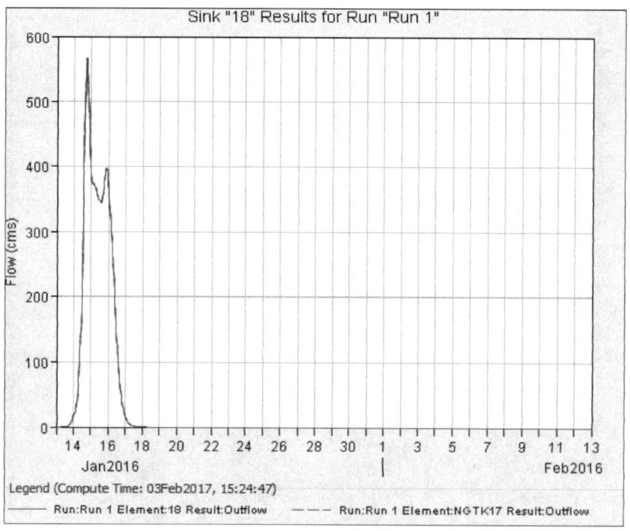

Figure 4. Hydrograph modelling

From the hydrograph on figure.4, the maximum discharge model on The Ngotok River is 566.72 m^3/s. the result of the hydrology model used the upstream boundary condition for the hydraulics model.

4.2. Hydraulic model

The input data on this model is maximum flood discharge for the steady flow model and the hydrograph data for the unsteady flow [12]. The result of the modeling must be calibrated to know whether the modeling results in accordance with the existing. Due to the existing discharge data not recorded by the Automatic Water Level Recorder (AWLR) [13], so the calibration method using the survey by the residents around the Ngotok River to determine the behavior of this river. Flood is modeled by the steady flow and the unsteady flow. The water surface of the steady flow and the unsteady flow can be shown in figure. 5 and 6

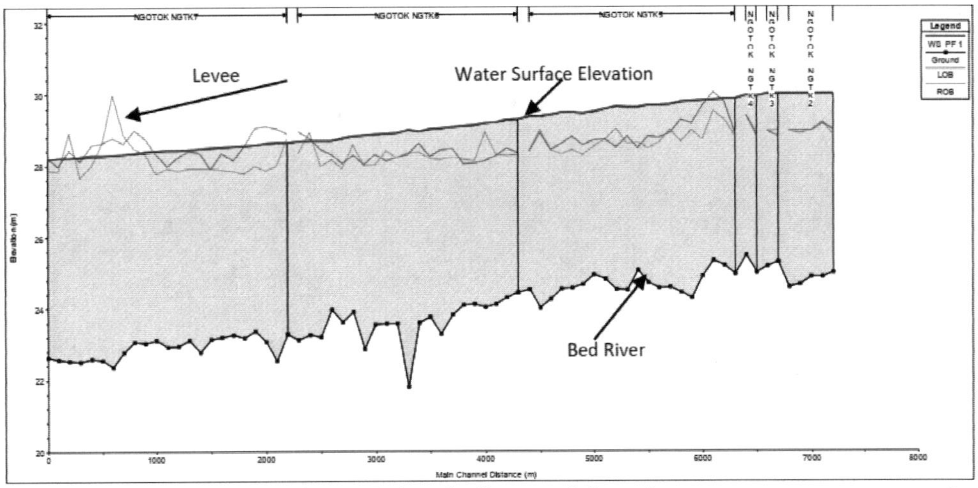

Figure 5. Steady flow model result

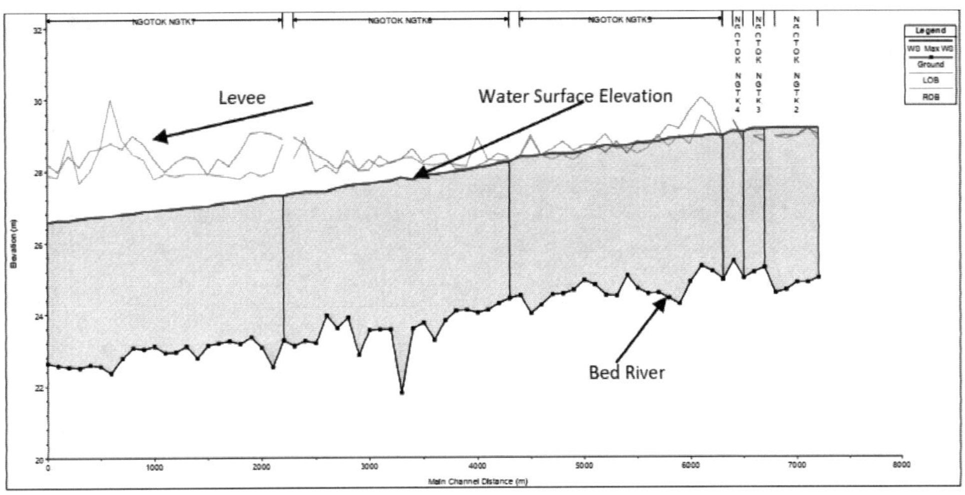

Figure 6. Unsteady flow model result

From both of the figureure, the upstream section of the Ngotok River is overtopping condition on the steady flow and the unsteady flow modeling. This condition can be explained that the water surface elevation (blue line) higher than the right and left levee of river elevation (green and red line). This result also conforms with the survey data that the inundation area at the upstream section of Ngotok River. The steady flow model result (black line) has a higher water surface elevation than the unsteady flow model (blue line). The overlay long section of the Ngotok river with the both of modeling can be shown at figure.7

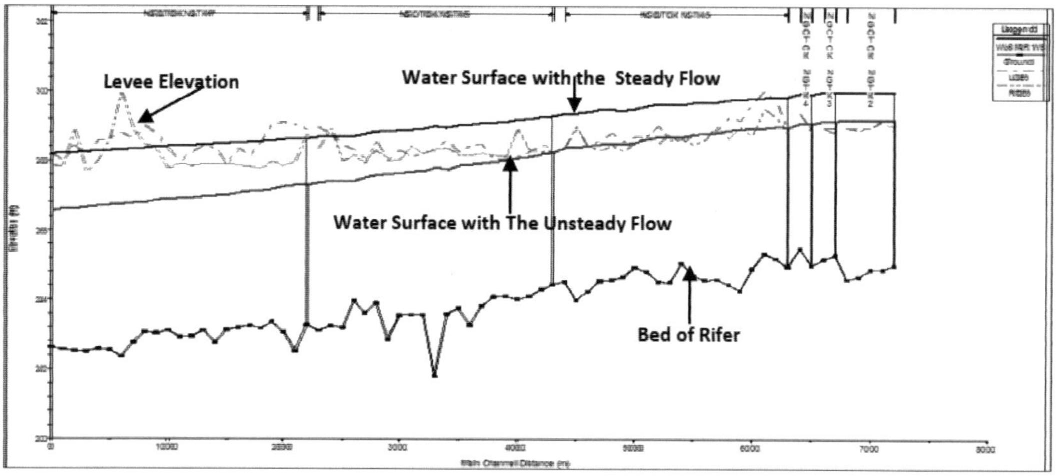

Figure 7. The comparation result

5. Results and conclusion

The comparison of the steady flow and the unsteady flow one-dimensional flow models using HEC-RAS (Table 1)

Table 1. The result comparation

Discussion	The Unsteady Flow Model	The Steady Flow Model
Method	The Gradually Varied Flow, except at culvert, siphon, pier of bridge and weir. The equation used is momentum equation and energy conservation	The uniform flow that not depend by the time and distance
Hydraulic Input Data	The geometric and the cross section of river with a much the cross section to eliminate the errors	The geometric and the cross section of river
Hydrology Input Data	Upstream Boudary : Flood Hydrograph Downstream Boundary : tide of water, rating curve	Upstream Boudary : Maximum Discharge Downstream Boundary: tide of water, rating curve
Model Stability	Discharge,River Geometric sungai	None
Water Elevation	Average = +27.17 m	Average = +29.62 m

References

[1] Istiarto 2014 The Flow Simulation of 1-D using HEC-RAS (istiarto.staff.ugm.ac.id)
[2] Istiarto 2014 HEC-RAS Steady or Unsteady Flow Analysis (istiarto.staff.ugm.ac.id)
[3] Anggrahini 1997 Hidrolika Saluran Terbuka (CV. Citra Media, Surabaya)
[4] V. T. Chow 1992 Open Channel Flow., (Erlangga, Jakarta)
[5] Sri Harto Br 1993 Analysis of Hydrology. (PT Gramedia Utama, Jakarta)
[6] Lensley, Ray K, Franzini, Joseph B. 1991 Teknik Sumber Daya Air Jilid II, (CV. Citra Media, Surabaya)
[7] C.D. Soemarto 1987 Hydrology Engineering, (Erlangga, Jakarta.)
[8] E. Suhartanto 2008 The Manual Book of HEC-HMS and Application at Water Resource

Management, (CV Citra, Malang)
[9] B. Triatmojo 2008 Hidrologi Terapan (Beta Offset, Yogyakarta)
[10] USACE 2000 Hydrologic Modelling System HEC-HMS Technical Reference Manual (http//www.hec.usace.army.mil. Maret)
[11] US Army Corp of Enginneering 2008 HEC-RAS User Manual (Davis, California)
[12] Van Rijn, C. Leo 1990 Principles of Fluid Flow and Surface Waves in Rivers, Estuaries,Seas and Ocean (Delft Hydraulic)

IOP Conf. Series: Materials Science and Engineering **669** (2019) 012019 doi:10.1088/1757-899X/669/1/012019

The effect of reinforcement bridging on the elastic fracture energy of concrete

A H Patty [1]

[1] Widya Karya Catholic University, Civil Department, Jl. Bondowoso 2 - Malang, 65115

agneshpatty@gmail.com

Abstract. Failure of structures may be analyzed based on strength criterion or fracture mechanics criterion. Strength criterion deals with material resistance beyond the critical conditions such as yield, and ultimate strength. In fracture mechanics the critical condition is defined by toughness which can be stress intensity factor (K) or fracture energy (G). This study explores the critical point during loading up to the limit of elasticity based on linear elastic fracture mechanics – LEFM. However, nonlinearity frequently appears due to the existence of a relatively large fracture process zone – FPZ located at the crack-tip. The assumption of LEFM in quasi – brittle material such as concrete is therefore limited to large size structures only. The well-known approach to obtain the fracture energy Gf for infinite large structures is size effect law – SEL. Gf is defined as the specific energy, i. e. energy per unit crack plane area. This is elastic energy which is linear. To answer a question whether reinforcement affect the elastic fracture energy, a research was conducted following the principles of work-of-fracture – WOR using the RILEM Specification Test Method. Three-point-bend beam specimens were made of normal concrete, subjected to monotonic load P, until they failed. P–u relationship prevails the real work. The elastic fracture energy based on WOR that is G_{F-el}^{R} was found by divided the area under P–u curve of elastic range with ligament. The result is, G_{F-el}^{R} equal to 397.87 N/m for beam A1 (bending failure l/h=5.5), 133.6 N/m for beam B1 (Shear failure, l/h=5.0). 101.93 N/m for beam B2 (shear failure, l/h=5.0), 153.75 N/m for beam B3 (shear failure, l/h=5.0). The greater value G_{F-el}^{R} comparing to Gf implies that reinforcement affect significantly the elastic fracture energy.

1. Introduction

It is well known that failure behaviour of structures is greatly influenced by the behaviour of the material used. Concrete as a discrete material tends to be full of cracks even under service loads. According to fracture mechanics specialist, it appears natural that concrete structures should be designed based on fracture mechanics [1]. This approach is to find out the maximum critical conditions during the failure process. Fracture parameters which mainly affect the failure capacity are FPZ (contributed by 'bridging') and fracture toughness.

1.1. Fracture Process Zone - FPZ

Failure of quasi-brittle structures primarily is resulted from cracking of concrete, which leads to softening type-load deformation behaviour. Hence, it makes sense to accept that failure process of quasi-brittle mechanism is controlled by a propagation of the fracture zone.

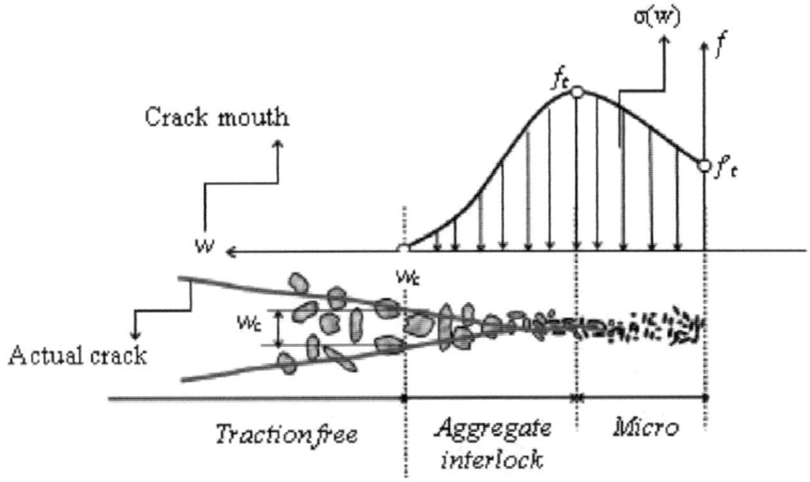

Figure 1. Phenomenon of bridging mechanism

This zone exists at the tip of the actual crack and plays a role as crack arrester. Illustrated in Figure 1, one can see that this zone occurs as result of aggregate interlock. If these micro cracks solid significantly, then it may be having as yield or plastic zone under a constant stress – yield stress. According to the idea of Dugdale and Barentblatt [2][3], this yield stress will maintain as long as the forming of a plastic zone with a certain length, to meet the singularity condition.

The yield stress is defined as the lowest stress in which the asymptote of elastic stress distribution begins to infinite. Taken from [4] concrete has yield stress of 25 MPa. By applying the modified crack closure and the J-integral principle simultaneously, Patty (2004)[5] proposed a yield stress for fiber reinforced lightweight concrete with fiber volume fraction, of 0.5%, 1%, and 1.5% to be 14.051 MPa, 15.002 MPa and 17.104 MPa respectively.

1.2. Fracture Process Zone – FPZ

Fracture toughness can be clarified as both critical strain energy release rate denoted by parameter G_c, and critical stress intensity factor denoted by parameter K_c. These two parameters meet the following relationship

$$K_c = \sqrt{EG_c} \tag{1}$$

One of the fundamental principles of fracture mechanics is that unstable crack growth (brittle fracture) occurs when the stress intensity factor K reaches its critical value K_c. In case of reinforced concrete beams, the stress intensity factor due to applied load can be reduced by traction due to bridging effect between reinforcing steel and the concrete itself. So, to prevent structural materials from brittle fracture keep the calculated toughness below its critical value. Concrete is considered as a quasi-brittle material whose failure mechanism can be described based on energy principles as follow

$$G_q = G_c + G_\sigma \tag{2}$$

where G_q is energy release rate at the tip of the effective quasi-brittle crack. G_c is the material surface energy, needed to create surface and G_σ is a part of energy to overcome the cohesive pressure $\sigma(w)$ to separate the crack surfaces, G_c is dealing with LEFM while G_σ is dealing with NLFM

2. Literature review

Formula 2 describing a phenomenon that concrete may fail in two manners; brittle fracture, if G_c is overcome, prevails

$$G_q = G_c \tag{3}$$

but if G_σ is overcome it prevails

$$G_q = G_\sigma \tag{4}$$

The question in this case, does reinforcement-bridging affects G_c?. In case of quasi-brittle concrete, bridging occurs when the crack has advanced beyond on aggregate that continues to transmit stresses across the crack until it ruptures or pulled out [6]. Bridging due reinforcement works as a composite action between steel bars and matrix, in case of reinforced concrete.

2.1. *Researches of G_c on plain concrete*

As mentioned before, G_c is energy consumed to create cracks. When the length of the cracks reaches their critical value ($a = a_c$) then failure will occur rapidly. This condition happens, because of the no existence of micro cracks, consequently LEFM shall prevails. However, non-linearity frequently appears due to the existence of a relatively large fracture process zone located at the crack-tip. The assumption of LEFM in quasi-brittle materials is therefore limited to large size structures only. Figure 2 illustrates the load-deflection curves of geometrically similar structures [7]

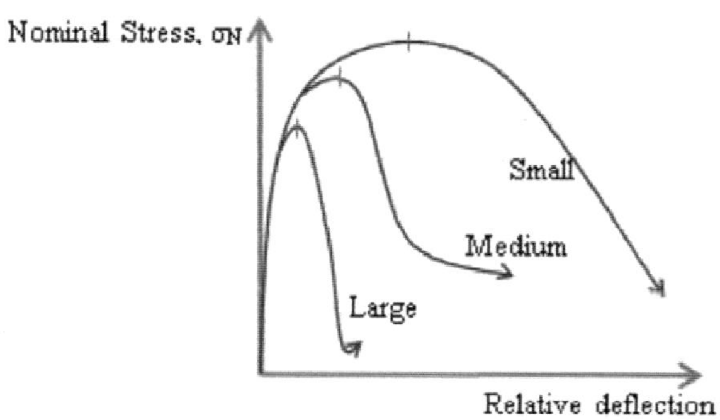

Figure 2. Load deflection diagram of geometrically similar structures of different size [7]

G_c is material property which has to be modeled for infinite large structures.

2.2. *Size Effect Law (SEL)*

The well-known approach to obtain the fracture energy G_f for infinity large structures was proposed by Bazant (1987) based on size effect [8]. G_f is defined as the specific energy, i. e. energy per unit crack plane area. For mode I fracture with three-point-bend loading (Figure 3) G_f may be computed as follow

$$G_f = \frac{g(\alpha_0)}{EA_B} \tag{5}$$

where

E = modulus of elasticity
α_0 = a_0/d
a_0 = length of the initial crack
d = height of specimen
A_B = constant to be determined

$g(\alpha_0)$ is a geometric factor, given by

$$g(\alpha_0) = \left(\frac{S}{b}\right)^2 \pi\alpha_0[1.5g_1(\alpha_0)]^2 \tag{6}$$

where S = span of beam specimen, $\alpha_0 = a_0/b$ and $g_1(\alpha_0)$ is the geometric function for stress intensity factor, as a function of α_0 for certain ratio of $\frac{S}{b}$.

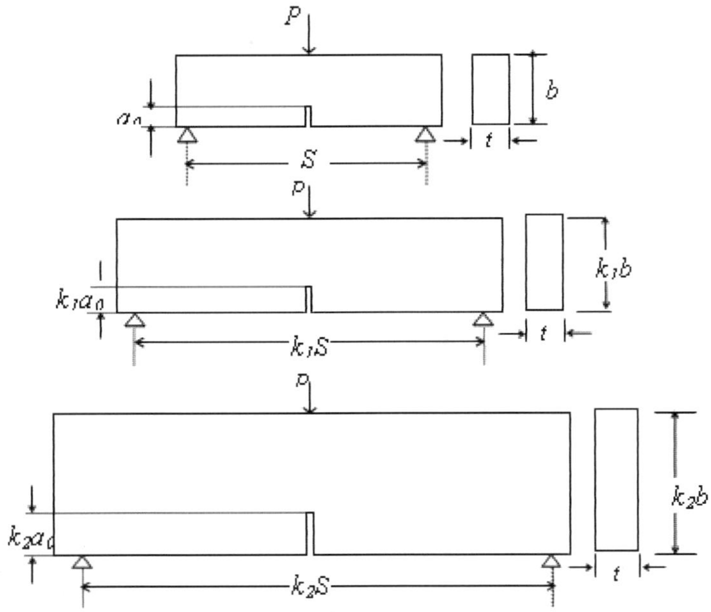

Figure 3. Series of geometrically similar structures (k_1 and k_2 are constant, two-dimensional similarity) [8]

2.3. Reported SEL-Based researches

Nallathambi and Karihaloo [9] used geometrically similar beam to determine G_f of normal concrete with maximum aggregate size $da = 20$ mm, that resulted $G_f = 21.600$ N/m.

Bazant and Pfeiffer [8][10] have reported $G_f = 36.60$ N/m for concrete with $E = 27.7$ GPa, and $da = 13$ mm. Concrete with $E = 32.9$ GPa and $da = 5$ mm, $G_f = 23.67$ N/m. Further, $G_f = 38.35$ N/m for concrete with $da = 12.7$ mm with $E = 31.077$ GPa.

Alexander [11] has found $G_f = 71.06$ N/m for concrete with $da = 19$ mm with $E = 32.5$ GPa.

Chan and Shieh [12] calculated G_f of lightweight concrete with $E = 19.7$ GPa; based on SEL, he found $G_f = 34.42$ N/m. Also reported for $E = 22.3$ GPa, $G_f = 37.2$ N/m.

The same approach used by Chan and Wang to determine fibre reinforced concrete with 1% fibre volume fraction resulting in $G_f = 37.4$ N/m.

2.4. G_c of reinforced concrete based on 'Work-of-Fracture'

Work-of-fracture based on real work *P–u* which is obtained from a three-point-bend beam testing. Real work is the first energy method. The term 'real work' implies that the force and displacement considered are real [13]. Shown in Figure 4, *P* is external load increased monotonically and *u* is load line displacement 'work' is defined as

$$w = \int_0^{u1} P d_u \quad (7)$$

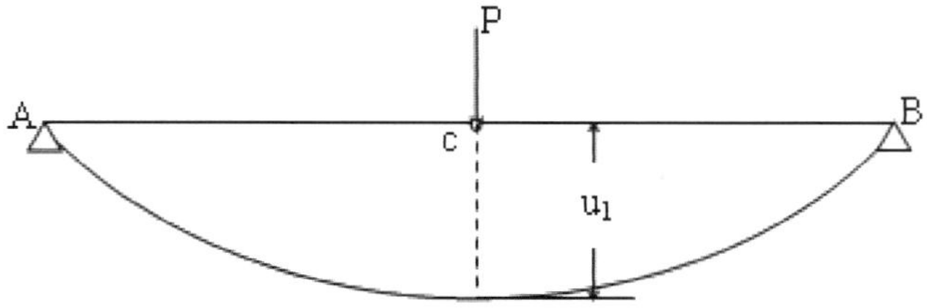

Figure 4. Beam under elastic bending

The relationship described by formula 7 represents the work needed to affect the considered point (point-c) to a value *u*

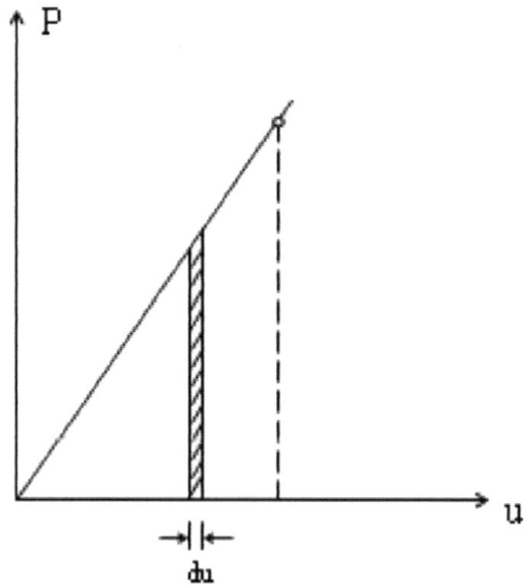

Figure 5. Real work principle [13]

Further, the principle of real work applied as work-of-fracture is written as

$$W_f = \int_0^{u1} P d_u \tag{8}$$

Where u_l = final displacement at which the load is reduced to zero.
The fracture energy according to the RILEM definition, G_F^R, represents the average fracture energy in the ligament [14][15] obtained as

$$G_F^R = \frac{W_f}{bl}, l = (l - \alpha_0)d \tag{9}$$

where, b = thickness of beam
l = length of ligament
d = depth of beam
$\alpha_0 = a_0/d$, a_0 = length of notch
as shown in Figure 6a, followed by P-u curve described by Figure 6b.

In this 'case study', the part A of P-u relationship which is the elastic energy, may be calculated using the principle of 'work-of-fracture' under the following basic consideration: The elastic peak load P_c will drop to zero instantly at the moment $u=u_c$, if there is no bridging (due to reinforcing or aggregate interlock). This means that u_c corresponding to P_c is similar to u_l corresponding to P=0.

Hence, the determination of elastic fracture energy for certain size of beams may be defined based on 'work-of-fracture' by using the formula 9 with $P=P_c$ and $u=u_c$

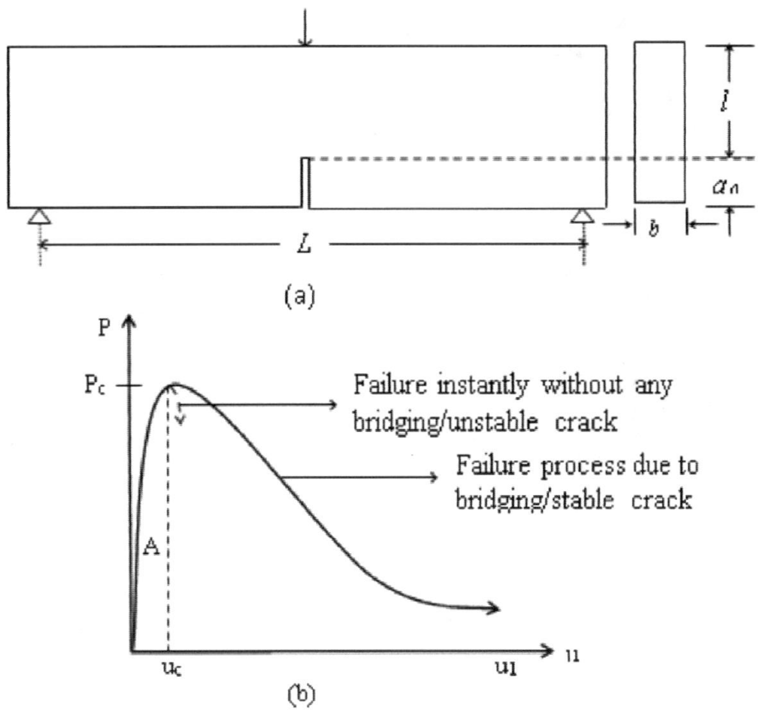

Figure 6. Three point bend beam (a) Specimen geometry, (b) *P-u* curve

3. A case of Work-of-Fracture – WOR

Three-point-bending reinforced concrete beams were tested as shown in Figure 6. There were four specimens made of normal concrete with f'c=30 MPa and reinforced by 2 #12 mm just to avoid them from instantly failure, as shown Figure 7 and Figure 8. All of them, has (200x100) mm dimension with *l/h*=5.5 for beam A1; 5.0 for beam B1, B2, and B3. The dimension met the requirement of RILEM [16] regarding the maximum size of aggregate. Following are the result of loading where *P* was increased monotonically until it fails.

4. Result and Discussion

Basically, concrete fails in two manners, brittle fracture which is LEFM based or quasi-brittle fracture which is nonlinear fracture mechanics – NLFM based. Brittle fracture deals with the energy consumed to create cracks; failure occurs at the moment when a critical length of the crack (a_c) is obtained. This study deals with brittle elastic energy performed by plain concrete and reinforced concrete as well. Plain concrete used SEL to find G_f and reinforced concrete used WOR to find $G_{F}^{R}\_el$. The result is, G_{F-el}^{R} equal to 397.87 N/m for beam A1 (bending failure *l/h=5.5*), 133.6 N/m for beam B1 (Shear failure, *l/h=5.0*), 101.93 N/m for beam B2 (shear failure, *l/h=5.0*), 153.75 N/m for beam B3 (shear failure, *l/h=5.0*). The greater value of G_{F-el}^{R} comparing to G_f implies that reinforcement significantly affect the elastic fracture energy. The result of the analysis shows that due reinforcement, the elastic energy for bending failure is greater than for shear failure. In addition, the critical length (a_c) for bending failure must be longer than that of shear failure. In bending failure plastic hinge might be developed due to wedge forces between reinforcement and concrete. On the contrary plastic hinge is not to be expected to occur in reinforced concrete with shear failure.

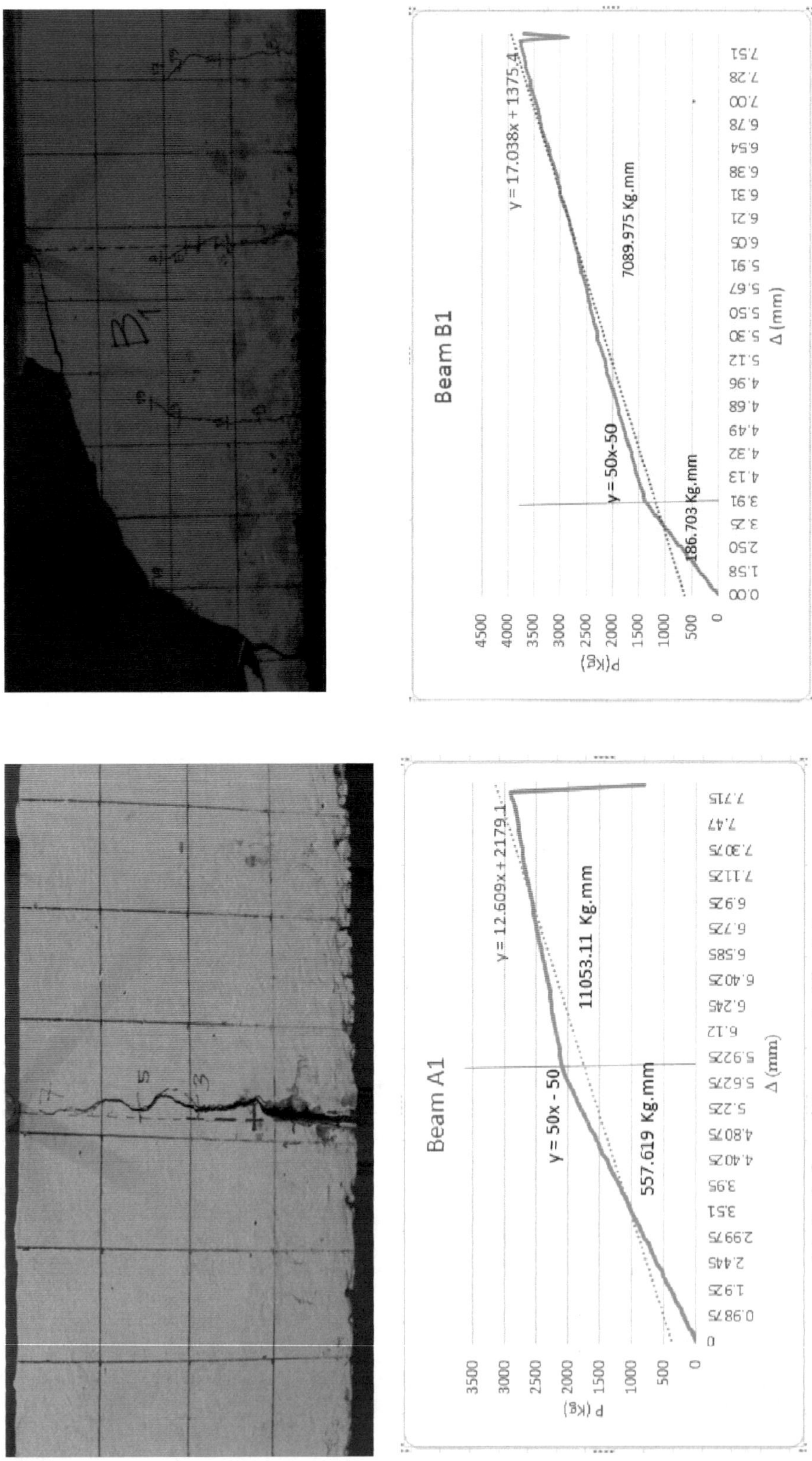

(a) Reinforced Beam – 1, l/h=5.5

(b) Reinforced Beam – 2, l/h=5.0

Figure 7. Results of three-point loading beam (A1 and B1)

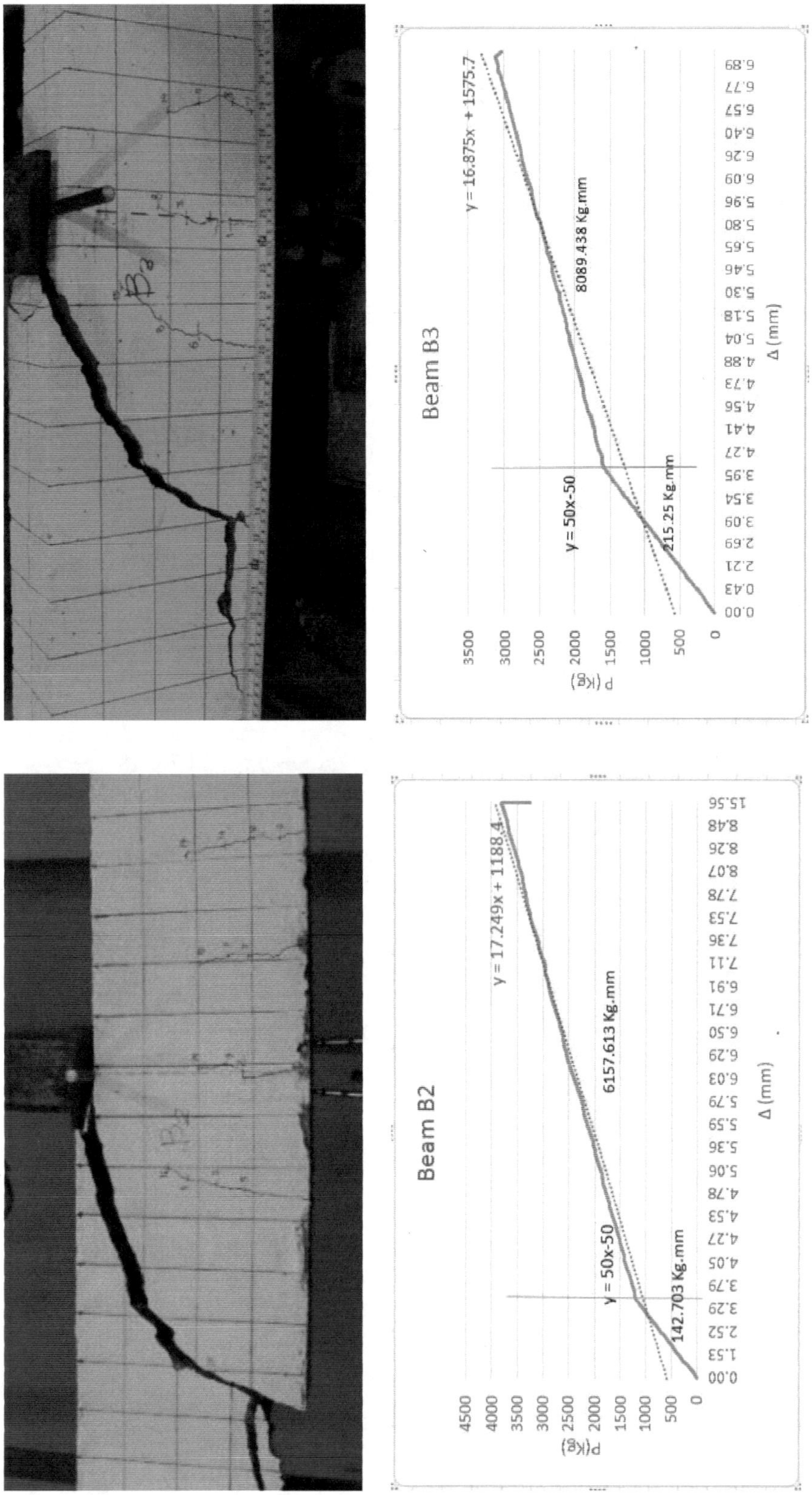

(a) Reinforced Beam – 3, *l/h=5.0*

(b) Reinforced Beam – 4, l/h=5.0

Figure 8. Results of three-point loading beam (B2 and B3)

5. Conclusion

- Failure of structures catastrophically can be avoided by creating bridging effect, which is leading to perform micro crack at the tip of main crack.
- Slenderness of bending element may affect the mode of fracture, i. e. bending or shear.
- Presence of reinforcement roles in increasing elastic fracture energy, not only in energy dissipation (after the limit of elastic fracture energy is reached).

References

[1] Bazant Z P 1985 *Mechanics of Fracture and Progressive Cracking in Concrete Structures* in Fractures Mechanics of Concrete: Structural Application and Numerical Calculation, Martinus Nijhoff Publishers, Dordrnecht /Boston/ Cancaster
[2] Barentblatt G I 1962 The Mathematical Theory of Equilibrium Cracks in Brittle Fracture *Advance Applied Mechanics* vol 7, pp.55-129.
[3] Dugdale D S 1960 Yielding of Steel Sheets Containing Slits *Journal of Mechanics, Physics, and Solids* vol 8, pp. 252-258.
[4] Roylance D 1996 *Mechanics of Materials* John Wiley & Sons (INC, New York)
[5] Patty A H 2004 *Analisis Mekanika Fraktur Diimplementasikan Pada Beton Ringan Serat Baja dengan Bukaan Tarik Tunggal* Ph. D. thesis (Bandung Institute of Technology)
[6] Shah S P, et al 1995 *Fracture Mechanics of Concrete* John Wiley & Sons. (INC, New York)
[7] ACI Committee 1989 Fracture Mechanics of Concrete: Concepts Models and Determination of Material Properties *ACI Committee* 446, ACI 446.1R-91
[8] Bazant Z P and Pfeiffer P A 1987 Determination of Fracture Energy from Size Effect and Brittleness Number *ACI Materials Journal* vol. 84 No 6 463-480.
[9] Nallathambi P and Karihaloo B L 1986 Determination of Specimen – size Independent Fracture Toughness of Plain Concrete *Magazine of Concrete Research* vol 38 76 -79
[10] Foote R.M.L, et al 1986 Crack Growth Resistance Curves in strain softening Materials," *Journal of Mechanics and Physics of solids* vol. 34, No 6 pp 593-607.
[11] Karihaloo B L and Nallathambi P 1991 *Notched Beam Test: Mode I Fracture Toughness* in Fractures Mechanics Test Methods for Concrete, RILEM Report 5, edited by Shah S P and Carpinteri (A Champ & Hall, London) pp. 1-86.
[12] Chang T P and Shieh M M 1995 Fracture Properties of Lightweight Concrete *Cement and Concrete Research* vol. 26, No. 2 181-188
[13] Harold I Laursen 1988 *Structural Analysis* McGraw-Hill Book Company (New York)
[14] Nakayama J 1965 *Journal of American Ceramics Society* vol. 48 No 11, pp 583-87
[15] Moavenzadeh F and Kuguel R 1969 *Journal of Materials* vol 4 No 3, pp 497-519.
[16] RILEM Committee 1985 on Fracture Mechanics of Concrete – Test Methods Determination of the Fracture Energy of Mortar and Concrete by means of Three-Point Bend Test on Notched Beams *Materials and Structures* vol. 18 No 106, pp. 285 – 290.

IOP Conf. Series: Materials Science and Engineering **669** (2019) 012020 doi:10.1088/1757-899X/669/1/012020

Life cycle assessment of low costs housing in Indonesia

S Zuraida[1], Y S Wahyuni[2] and D Larasati[3]

[1] Civil Engineering Department, InstitutTeknologiSains Bandung, Indonesia
[2] Civil Engineering Department, Universitas Muhammadiyah Sukabumi, Indonesia
[3] Architecture Department, Institut Teknologi Bandung, Indonesia

siswanti.zuraida@itsb.ac.id

Abstract. Housing demands in Indonesia cause an increase in the supply needs of building materials where some of these materials have a negative impact on the environment. Life Cycle Assessment (LCA) method is used to assess the potential environmental impact caused by housing demand of low-costs housing in Indonesia which its types are standardized to types of 21, 36 and 45. The data of building materials were collected from some developers that involved in development of low-costs housing. Embodied Energy (EE) and Carbon Emission (CE) were used as indicators of LCA. The result of the study showed that EE and CE value is increasing linearly by the addition of housing area. It also found out that material volume has significant value to rise EE and CE. Based on the result, with the same volume of square meters, the amount of energy used to build two types of houses 21 was higher than type of 45 houses.

1. Introduction

Indonesia as a developing country has a population growth rate of 1.4% per year. This condition leads to the fulfillment of housing demand at around 800 thousand units per year. The capacity of stakeholders to fulfill housing demand over the past five years is only 400 to 500 thousand units, while housing backlogs are currently at around 11.4 million units [1]. As regards to this situation, the Indonesian government has proclaimed a program to provide affordable housing for low-cost income [2]. The fulfillment of housing demand is then implemented in two types of housing, landed housing and vertical housing. On one hand, the program has a positive effect to fulfill affordable housing needs for society, but on the other hand, housing construction that conducts continuously and massively does require a very large amount of material supply.

The housing development constantly needs a supply of building materials in large quantities. The procurement of building materials in large quantities will lead to the exploitation of natural resources. There is a large potential environmental impact due to the procurement of building materials for the fulfillment of housing demand. The building material production process follows the building material lifecycle, from the extraction, processing, manufacturing to ready-to-use building materials [3].

Several studies that assess the environmental impact of building material and construction process used Life Cycle Assessment (LCA) as a method. LCA based on ISO 14040 is one of the environmental management techniques for risk assessment, environmental performance evaluation, environmental audits, and analysis of environmental impacts [4]. LCA is usually not used to assess social or economic aspects, but its approach and methodology can be applied to other aspects. In several LCA-based studies, several indicators regarding environmental impact were measured by embodied energy and carbon emissions [5-16]. Some even measure the high environmental costs that must be incurred due to these emissions, known as eco-costs [6, 7, 17-23].

In response to the issue of housing development that occurred in Indonesia, this study analysed the potential environmental impacts of building material used for low costs housing. Embodied Energy (EE) and Carbon Emission (CE) are used as indicators of impact assessment in this study. The focus of the research will be to assess the life cycle of the planning process, which is in the stage of selecting building materials to be used. The objective is to find out earlier potential environmental impacts that will be caused by design decisions and the selection of building materials. This decision can be used as a step to disaster mitigation. Meanwhile, the types of housing that measured in this study are type 21, type 36 and type 45 based on previous research that had been conducted [12][13] in measuring the EE of the three types of housing that used Building Material Analysis (WUPA).

2. Literature review

2.1. Low Costs Housing Concept in Indonesia

Low costs housing is referred to as the calculation of minimum space standard per person (m2) that has been determined to be occupied by around 3 to 4 persons. There are several provision about low costs housing such as is showed on the table 1.

Table 1. Minimum size requirement for Low-Cost Housing [24]

Standard per-person (m^2)	Areas (m^2) for 3 persons	Areas (m^2) for 4persons
Minimum Threshold 7,2	21,6	28,8
IndonesiaStandard 9,0	27,0	36,0
InternationalStandard 12,0	36,0	48,8

The observation of the building typology of low costs housing units that built by the developer, starting from the Aceh region to Papua, shows a typology that is not much different in terms of floor area, appearance and selection of building materials used that generally refer to type of floor area; 21, 36 and 45 [12][13]. The design of the building looks simple units generally called minimalist modern. The technical specifications of the building materials used by simple housing units in all regions turned out to be almost the same.

The same procedure is used to collect the use of building materials used by simple housing units. The data is compiled based on information from each developer, both through the study of information presented in housing brochures, as well as direct interviews with developers that showed on the table [12][13].

Table 2. Building components that used for low costs housing in Indonesia

Building Components	Developer 1	Developer 2	Developer n
Foundation	Stone	Stone	Stone
Floor	Concrete	Concrete	Concrete
Column	Ceramic	Ceramic	Ceramic
Wall	Brick	Brick	Brick
Window	Wood	Aluminium	Wood
Door	Wood Panel	Wood Panel	Wood Panel

Table 2. Building components that used for low costs housing in Indonesia (cont.)

Building Components	Developer 1	Developer 2	Developer n
Roof	Wood frame	Steel frame	Wood frame
Roof Cover	Aluminium	Clay Tile	Clay Tile

2.2. Embodied energy and carbon emission of building material

EE building materials in the building material lifecycle are aimed at the total energy needed to produce building materials including the procurement of raw materials, fuel for the procurement of raw materials, the process of producing building materials and the types of transportation to move raw materials to be ready on site. This is referred to as from cradle to the grave [25]. Whereas EE which is calculated from the initial process to the outgoing manufacturing products, transportation to the location, installation or construction, maintenance, and destruction or recycling is referred to as from cradle to the grave [3].

The calculation of the EE and CE value of building materials will indicate the amount of energy that needed and the magnitude of the environmental impact that might occur depending on the selected material. Designers or architects will get the prediction impact of the building materials that will help to decide suitable material for the design. It based on review and reconsideration of the EE and CE amount that embedded in building materials. Improvement decisions may involve building materials, building methods, access to material resources, and efficient use of transportation in the delivery of materials to the location. The definite goal is to make efficient use of energy based on the choice of building materials in the design.

Several studies on the value of EE that have been carried out were discussed about:

- Calculation of the amount of EE and CE value to calculate the energy-saving potential that can be done on various building objects [26-28]
- Development of methods and calculation models for EE values that will be used, to obtain methods and models that are seen to be closest to research needs that are mostly done in research is the development of EE measurement methods. The measurement of EE value is part of the Lifecycle Assumption (LCA) assessment on the building cycle (lifecycle building). However, until now there are no specific calculation standards for EE so that special studies are needed to produce specific EE calculation standards that can be referred to globally [29]

2.3. Work Unit Prices Analysis (WUPA) and Material Unit Analysis (MUA)

Analysis of Work Unit Prices (WUPA) is a method of calculating job descriptions both for the needs of the Bill of Quantity (BQ) and the total calculation of work in terms of volume and price of work [12,13]. WUPA is provided by Indonesia Standardization Board. In each component of the work consists of numbers of coefficients, units, description of employment, the volume of work and price of work. each part of the component selected only calculates that it relates to building materials so that the price of work is not counted in this study. Resources in the management of construction projects calculated in the WUPA consist of building material components, supporting equipment, and labor.

This research specifically only discusses building materials as a fixed component that provides an overview of the greatest needs of a project including the needs of the types of simple houses that are mass-built by continuous development. The results of the selection of components in the WUPA consist of the number of coefficients, units, job descriptions and volume of building materials needed and becomes a new term called Material Unit Analysis (MUA) [12,13].

3. Methodology

Key steps on doing this research:

- Determine the object of research: low costs housing of type 21, type 36 and type 45 as the models set out in this study.
- Formulation of the Material Unit Analysis (MUA) method that developed from the Analysis of Work Unit Prices (WUPA) that has been commonly used by development stakeholders in Indonesia for building/project planning
- Through the MUA process on the object of research, the inventory data of building materials are referred to as Carbon and Energy (ICE) Inventory, Bath, United Kingdom.
- The calculation of the total EE and CE value of the building to map the use of building materials and types of materials that produce the highest environmental impact value.
- Conducting a comparative analysis of environmental impact results of each low costs housing types and significant aspects that lead to high environmental impact values on the research object
- Developing environmental impact mitigation strategies through consideration of environmental impact values of building materials.

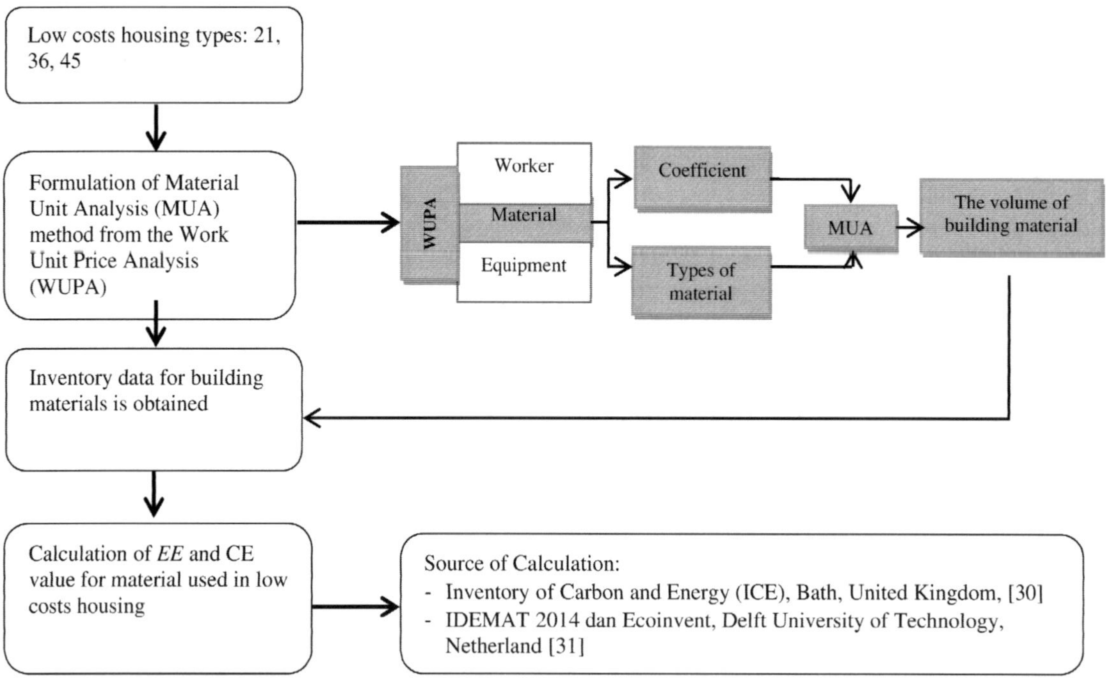

Figure 1. Steps of research methodology

3.1. Development of Material Unit Analysis (MUA) from Work Unit Prices (WUPA)

The table shows the initial format of WUPA based on SNI 6897 2008 for wall construction per 1 m2.

Table 3. Sample for work unit prices for 1 m^2 wall construction

Code	Coef	Unit	Work Detail	Vol.
1	*2*	*3*	*4*	*5*
D	WALL CONSTRUCTION			74,41
D.1	1M^2 Construction of bricks: Thick 1/2 brick, 1 Pc: 2 Ps			
	70	Pc	Bricks 5 x 11 x 22	5.208

Table 3. Sample for work unit prices for 1 m^2 wall construction (cont.)

Code	Coef	Unit	Work Detail	Vol.
1	*2*	*3*	*4*	*5*
	18,95	Kg	Portland Cement (Pc)	1.410
	0,0380	m^3	Sand River (Ps)	2
	0,3000	OH	Worker	22
	0,1000	OH	Stonemason	7
	0,0100	OH	Foreman	0,7
	0,0150	OH	Overseer	1,1162

Then the selection process is carried out where labor requirements are removed from the table and only show the need for building materials consisting of specifications of building materials in the form of red bricks, sand, and cement.

Table 4. Elimination process sample to get Material Unit Analysis (MUA) from Work UnitPrices Analysis (WUPA) for 1 m^2 wall construction

Code	Coef	Unit	Work Detail	Vol.
1	*2*	*3*	*4*	*5*
D	WALL CONSTRUCTION			74,41
D.1	1M^{2}Construction of bricks: Thick 1/2 brick, 1 Pc: 2 Ps			
	70	Pc	Bricks 5 x 11 x 22	5.208
	18,95	Kg	Portland Cement (Pc)	1.410
	0,0380	m^3	Sand River (Ps)	2
	~~0,3000~~	~~OH~~	~~Worker~~	~~22~~
	~~0,1000~~	~~OH~~	~~Stonemason~~	~~7~~
	~~0,0100~~	~~OH~~	~~Foreman~~	~~0,7~~
	~~0,0150~~	~~OH~~	~~Overseer~~	~~1,1162~~

The results of the selection process are shown in the table. The same process is done for other types of work.

Table 5. Material Unit Analysis (MUA) for 1 m^2 wall construction

Code	Coef	Unit	Work Detail	Vol.
1	*2*	*3*	*4*	*5*
D	WALL CONSTRUCTION			74,41
D.1	1 M^{2}Construction of bricks: Thick 1/2 brick, 1 Pc: 2 Ps			
	70	Pc	Bricks 5 x 11 x 22	5.208
	18,95	Kg	Portland Cement (Pc)	1.410
	0,0380	m^3	Sand River (Ps)	2

3.2. Conversion of building material unit

One thing that stands out from the results of the MUA calculation is that there are variations in operational units that are not following the inventory data from ICE Bath and Eco-Invent TU Delft. For example iron in kilograms (kg), concrete sand in cubic (m3) while for energy conversion kg units is needed. This variation of material units requires the next step to facilitate calculations of energy [12-15].

IOP Conf. Series: Materials Science and Engineering **669** (2019) 012020 doi:10.1088/1757-899X/669/1/012020

Table 6. Conversion process of Building Material Unit

Code	Coef	Unit	Coef. Conversion	Unit Conversion	Work detail
1	*2*	*3*	*4*	*5*	*6*
K	70	Pc	70 x 3 kg = 210	Kg	Bricks 5 x 11 x 22
	18,95	Kg	18,95	Kg	Portland Cement (Pc)
	0,0380	m^3	p = m/V, so, m = p x V	Kg	Sand River (Ps)

4. Results and Discussions

4.1. Embodied Energy (EE) and Carbon Emission (CE) value-based on the use of building material of low costs housing

The data had been collected in previous studies showed that 9 types of material have a large percentage in each unit [12-15].

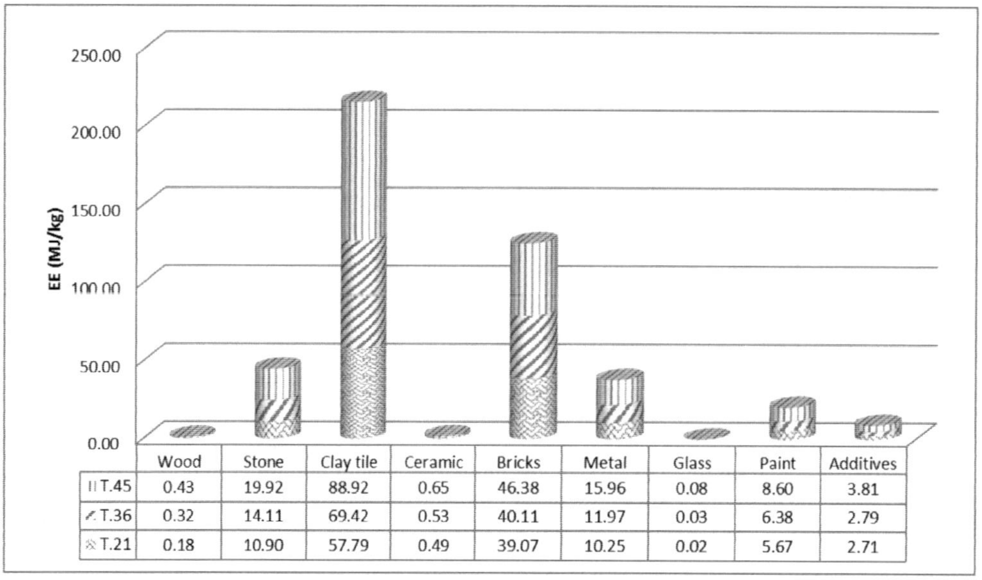

Figure 2. Building material Embodied Energy (EE) value of low costs housing

Figure 2. shows that the high EE value of building material is not determined by the EE unit value, but due to the use of material volume. For example, EE value for brick is 2.5 MJ / Kg, while for glass is 15 MJ / Kg. However, the calculation shows that EE value for brick higher than glass due to the use of brick volume is larger than glass. The picture also shows data from 9 types of building materials identified as having the highest EE value of building materials among 46 other types of building materials that had been collected by [12,13], only one type of material which includes material that is renewable resources. Whereas 8 other types of building materials are included in the category of non-renewable material resources.

Figure 3. Above also shows the value of potential environmental impacts both physically and in the form of Carbon Emission (CE) that is contributed by building materials used. Non-renewable natural resources are not following the principles required by the commitment of sustainable development. This fact should be a concern by all parties in how to suppress the use of large volumes of building materials derived from non-renewable natural resources.

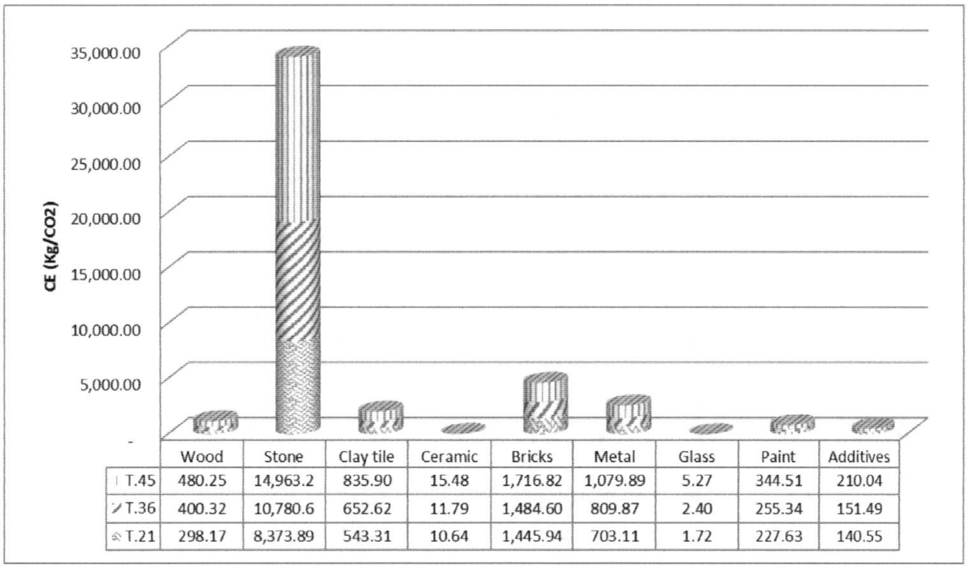

	Wood	Stone	Clay tile	Ceramic	Bricks	Metal	Glass	Paint	Additives
T.45	480.25	14,963.2	835.90	15.48	1,716.82	1,079.89	5.27	344.51	210.04
T.36	400.32	10,780.6	652.62	11.79	1,484.60	809.87	2.40	255.34	151.49
T.21	298.17	8,373.89	543.31	10.64	1,445.94	703.11	1.72	227.63	140.55

Figure 3. Building material Carbon Emission (CE) value of low costs housing

4.2. Embodied Energy (EE) and Carbon Emission (CE) value-based on the types low costs housing

All low costs housing units of type 21, type 36 and type 45 are compared to the results of their calculations to find out the comparison of the total EE and CE value of building materials as shown below. The studies of EE value had been carried out [12-15].

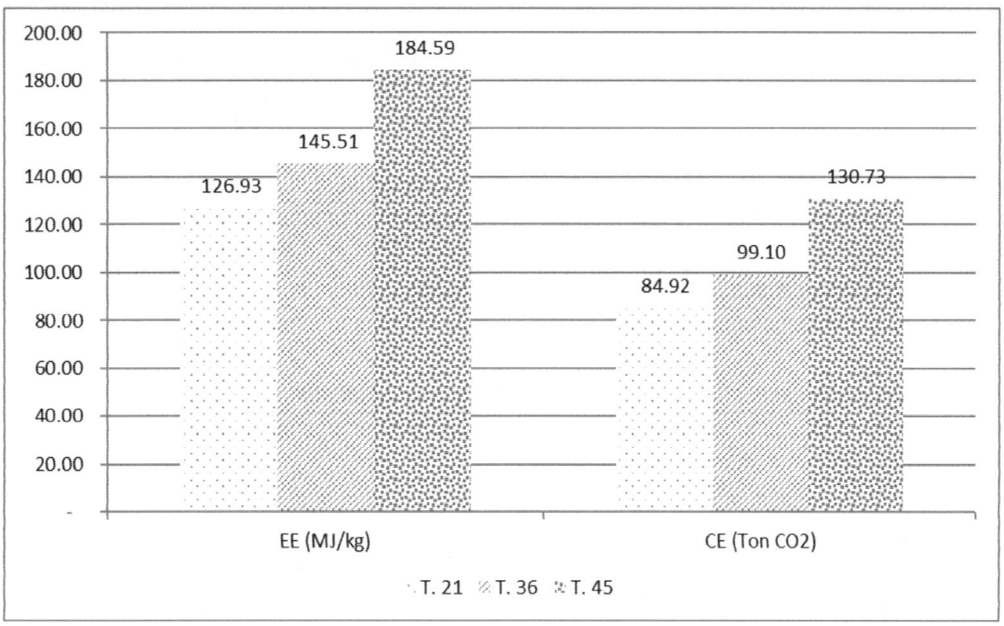

Figure 4. Embodied Energy (EE) and Carbon Emission (CE) value of low costs housing

The figure above shows the comparison of the total EE and EC value of building material between types of low costs housing unit types 21, 36 and 45 has a linear increase. This is due to the addition of space functions that have an impact on increasing the need for the volume of building materials. Thus, significant factors remain in the volume of material used in each type of housing unit.

Table 7. The Comparison of Increasing Value of EE, CE, and Area for Low Costs Housing

	T.21	T.36	T.45
T.21		EE= 1,15times CE =1,16 times	EE = 1,45times CE = 1,54 times
T.36	Area = 1,71 times		EE = 1,27times CE = 1,32 times
T.45	Area = 2,14 times	Area= 1,25 times	

It found out the increasing of total EE and CE value of building materials for type 21, type 36 and type 45. The data shows that to build 2 units of type 21 with an area of only 42 m^2, the EE is still below the area of type 45 while it contained a total EE value of 1,45 and CE value of 1,54 times more than EE and CE material contained in 1 unit type 45. This means that there is a waste in the use of EE and CE material if type 21 is built more. The reason is the high volume of material used by type 21 as the smallest unit, which has an impact on the increase of EE and CE material.

5. Conclusion

The EE and CE value of the processed unit material will be higher than the EE and CE value of the non-processed unit material. This is because the processing to produce building materials will require energy and it included in calculation EE and CE value of the material. All processed materials also have material EE and CE values that are different, so the significant factor of the EE and CE value of the material is the volume of material used. A significant factor in determining the high value of the EE and CE of a low costs housing type depends on its volume. If the volume of a material used is higher than others, then the type of material will contribute to the high EE and CE value. Otherwise, if certain types of material even have large EE and CE material values but the volume of material used is low, the EE and CE value is decreasing. In conclusion, with the same volume of square meters, the amount of energy used to build two types of houses 21 was higher than type of 45 houses.

References

[1] Badan Pusat Statistik 2016 Laporan Badan Pusat Statistik – Backlog unit rumah atas dasar kepemilikan rumah *Badan Pusat Statistik BPS*

[2] Direktorat Jendral Anggaran 2015 Peranan APBN dalam Mengatasi Backlog bagi Masyarakat Berpenghasilan Rendah (MBR) *Kementrian Keuangan Republik Indonesia* Dirjen Kemenkeu RI 16

[3] Thormark C 2002 A Low Energy Building In A Life Cycle—Its Embodied Energy Energy Need For Operation And Recycling Potential. Building and Environment, 37, 429-435

[4] ISO 14040 2006 *Environmental Management* — Life Cycle Assessment — Principles And Framework

[5] Monahan J and Powell J C 2011. An Embodied Carbon and Energy Analysis of Modern Methods of Construction in Housing: A Case Study Using A Lifecycle Assessment Framework. *Journal of Energy and Buildings* 43 179–188

[6] Pellegrino, et al. 2011. Energy-Efficient Choice of Brick Façade in Kolkata India *Procedia Engineering* 21 737 – 744

[7] DongHun et al 2011 Sustainable Design of Reinforced Concrete Structures Through Embodied Energy Optimization. Energy and Buildings 43 2028–2033

[8] Crishna N, et al 2011 Embodied Energy and CO2 in UK Dimension Stone *Resources*

Conservation and Recycling 55 1265– 1273.
[9] Shukla A, et al 2009 *Embodied Energy Analysis Of Adobe House* Renewable Energy 34 755–761
[10] Tingley D D and Davison B 2012 Developing an LCA Methodology to Account for The Environmental Benefits of Design for Deconstruction *Journal Building and Environment* 57 387-395
[11] Utama N, et al 2012 Embodied Impacts Of Traditional Clay Versus Modern Concrete Houses In A Tropical Regime *Building and Environment* 57 362-369
[12] Wahyuni Y S 2017. Pengembangan Strategi Mitigasi Dampak Lingkungan Melalui Pertimbangan Nilai Embodied Energy Material Pada Pembangunan Rumah Sederhana. *Disertasi*: Insititut Teknologi Bandung
[13] Larasati D, et al 2017. Embodied Energy Calculation in Mitigating Environmental Impact of Low-Cost Housing Construction. Proceeding of The 6th International *Conference of Euro Asia Civil Engineering Forum (EACEF)* vol. 138.
[14] WahyuniY et al 2016. Application of Embodied Energy Calculationfor Low-Cost Housing in Indonesia. Proceeding of Sriwijaya International *Conference on Engineering, Science and Technology (SICEST)* 2016
[15] Wahyuni Y, et al. 2016. Embodied Energy Value Data Development Of ConstructionMaterial In Sustainable Planning And Partnership. Proceeding of *International Seminar on Partnership for Sustainable Housing and Urban Development*
[16] Wahyuni Y, et al 2016 Model Of Embodied Energy Calculation For Low-Cost Housing In Indonesia. Proceeding of *International Conference on Dwelling Form (I-dwell)*
[17] Vogtländer J, et al 2010 The Sustainability Of Bambu Products For Local And Western European Applications: LCAs and Land-Use *Journal of Cleaner Production* 18 1260-1269
[18] van der Lugt P, van den Dobbelsteen A, Abrahams R 2003 Bamboo as a building material alternative for Western Europe? A study of the environmental performance, costs and bottlenecks of the use of bamboo (products) in Western Europe. *Journal of Bamboo and Rattan* 2(3): 205-223
[19] van der Lugt P, et al 2006 An environmental, economic and practical assessment of bamboo as a building material for supporting structures. Construction and Building Materials 20 648-656
[20] Zuraida S and Larasati D 2015 The Assessment Of Marginal Prevention Costs On Bamboo Preservation Method*Procedia Environmental Sciences*vol 28 789-798
[21] Zuraida S and Larasati D 2016. Economic Analysis of Bamboo Preservation Method in Indonesia*Proceeding of ASEAN Bamboo Symposium*
[22] Zuraida S and Larasati D 2016 Eco-Costs Life Cycle Assessment of Traditional Housing in Indonesia: The Caseof West Java*Proceeding of International Symposium of Architectural Interchanges in Asia (ISAIA)* 2016
[23] Zuraida S and Larasati D 2015 Environmental Impacts Of Bamboo Preservation Methods As An Assessment Of Sustainable Material*Proceeding of International Seminar On Partnership For Sustainable Housing And Urban*
[24] Kementrian Pekerjaan Umum dan Perumahan Rakyat 2016 *Kebijakan dan Strategi Penyediaan Perumahan Tahun Anggaran 2015-2019* Kementrian Pekerjaan Umum dan Perumahan Rakyat Kemenpu-Pera, 1, 3
[25] Chang Y, et al 2010 The embodied energy and environmental emissions of construction

projects in China: An economic input–output LCA modelEnergy and Building 39 6321-6330.

[26] ChenT.Y, et al 2001 Analysis of embodied energy use in the residential building of Hong Kong Energy 26 323-340.

[27] Hana M.Y, et al 2013 Embodied Energy Consumption of Building Construction Engineering : Case study in E-town Beijing. Energy and Building 64 62-72.

[28] Wang E, et al 2012 An AHP-weighted aggregated data quality indicator (AWADQI) approach for estimating embodied energyof building materials *International Journal Lifecycles Assess* 17 76-773

[29] Dixit, et al 2010 *Identification of parameters for embodied energy measurement* A literature review Energy and Buildings 42 1238–1247.

[30] Inventory of Carbon and Energy (ICE) (University of Bath, United Kingdom)

[31] IDEMAT 2014 and Ecoinvent Delft (University of Technology, Netherland)

IOP Conf. Series: Materials Science and Engineering **669** (2019) 012021 doi:10.1088/1757-899X/669/1/012021

Performance of aerobic and anaerobic landfill bioreactor for municipal solid waste treatment in Jatibarang landfill, Semarang

M A Budihardjo[1], A R Rizaldianto[1] and F I Muhammad[1]

[1] Department of Environmental Engineering, Faculty of Engineering, Diponegoro University, Indonesia

m.budihardjo@ft.undip.ac.id

Abstract. An investigation on the use of landfill bioreactors was conducted to elaborate an alternative solution on the municipal solid waste management in Jatibarang Landfill, Semarang. In addition, this study is also trying to develop a prevention on leachate and greenhouse gases pollution which are potentially emits from a landfilling process. There are six landfill bioreactors that were constructed for this study, which consist of two anaerobic bioreactors (R1 and R2) and two aerobic bioreactors (R3 and R4). Leachate recirculation was applied on R1 and R2 since the leachate production began while R3 and R4 was given air injection with 2 l/min flow rate. R1 and R2 maintained its thermophilic phase longer and generated more methane gases than aerobic bioreactors. Meanwhile, both aerobic bioreactors reached the thermophilic phase on day-7, faster than the anaerobic bioreactors.

1. Introduction

Landfilling has become the most widely final treatment of municipal solid waste management (MSW) in the world [1]. In Semarang, majority of waste that composed by organic and inorganic matter, will be managed in Jatibarang Landfill [2]. As a result of a landfilling process, highly concentrated organic and inorganic materials called leachate were formed [3]–[5]. There are various factor that determine the composition of leachate, such as water content, density, and characteristic of the solid waste [6]. It is categorized as a wastewater which has the potential to pollute surface and ground water [7], [8]. Other than leachate, methane gases are often formed as a product of landfilling process. Methane gases have a potential to trigger global warming 25 times higher than carbon dioxide [9]. Hence, both leachate and methane gases need to be managed proper to avoid the pollution on environment.

To overcome these problems, a mechanical biological treatment (MBT) is needed, such as landfill bioreactor due to its ability in reducing organic contents in leachate, accelerating biodegradation process, and recovering methane gases [8]. By making a biodegradation process become faster, it helps the solid waste to be stabilized quickly [7]. In addition, the recovered methane gases are likely to become an economical landfill gas (LFG) so that it can be used as an alternative source of electricity for housing [9], [10]. The accelerated biodegradation process and improved LFG generation can be achieved through in-situ treatment, for instance leachate recirculation or water addition to control the moisture content in the landfill bioreactor, as well as air injection [10]–[14]. Other parameters can also be controlled by utilizing this method, namely pH, redox, nutrient distribution, and bacteria [15].

This study aims to investigate the use of landfill bioreactor in improving the quality and amount of methane gases from the biodegradation process of organic waste in Jatibarang Landfill. Several parameters like temperature, pH, COD, and CH4 gas volume are monitored. During the research, two types of landfill bioreactor are used, such as aerobic and anaerobic.

2. Methodology

Four (4) landfill bioreactors (2 anaerobic bioreactors and 2 aerobic bioreactors) were built for this study. Figure 1 shows parts of the bioreactor. Two liters of water were given every day to all bioreactors by pumping, while leachate recirculation was given to anaerobic bioreactor by flowing it through the top of the bioreactor. Leachate recirculation was given to anaerobic bioreactor since the beginning of its leachate generation. Meanwhile, aerobic bioreactors are given 2 l/min air injection.

Leaf waste was given in R1and R3 while R2 and R4 were filled with vegetable waste, all of the bioreactors has a weight of 10 kg each. The study was conducted for 60 days at room temperature. Parameters such as temperature, leachate quality (pH and COD), and methane gas (CH4) were examined. The temperature and pH were monitored every day, while COD tests were carried out every 5 days. To measure the methane gases generation, a test were carried out every 2-3 days. Bioreactor was built from plastic-drum with a diameter of 50 cm and 150 cm in high. The body part of the bioreactor was given 3 layers, namely PE (polyethylene), aluminum foil, and stretch film to prevent any influences from external factors that possibly enter the bioreactor. The lid of the bioreactor was also made of plastic added with iron crates to strengthen the lid and prevent unwanted influx of air. The organic wastes used in this study were 1-week old leaf and vegetable from Jatibarang Landfill, Semarang.

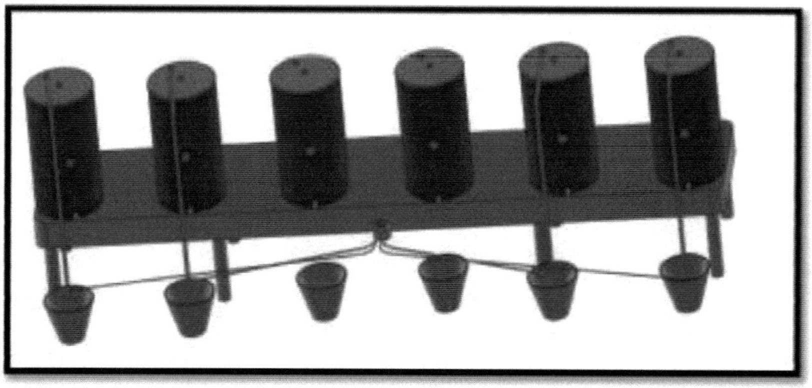

Figure 1. Arrangement of Landfill bioreactor

3. Result

3.1. Temperature

Temperature fluctuations during the research process is shown in Figure 2. The initial temperature in the bioreactor was around 28-30oC. In R1 and, the temperature was increasing until it reached thermophilic temperatures. R1 and R2 showed that the thermophilic temperature was reached on day-23 (R1) and day-18 (R2). The thermophilic phase is a phase where bacterial metabolic activity occurs optimally so that the biodegradation of organic materials occurs faster [16]. The temperature continued to rise until it reached the peak temperature, which occurred on day-35 for R1 (60oC) and day-37 for R2 (61oC). After reaching its peak temperature, the temperature in all of bioreactors gradually dropped until reached the room temperature.

Meanwhile, the aerobic bioreactor experienced a significant temperature increase in the first 2 weeks, which was caused by aeration. R3 and R4 reached thermophilic phase starting from day-7, which was maintained until day-21. The peak temperature of R3 and R4 was reached on the day-12 and day-16 respectively, with a temperature of 55oC, which then decreases the temperature to near room temperature. In the other hand, vegetable waste can maintain thermophilic temperatures longer than leaf waste.

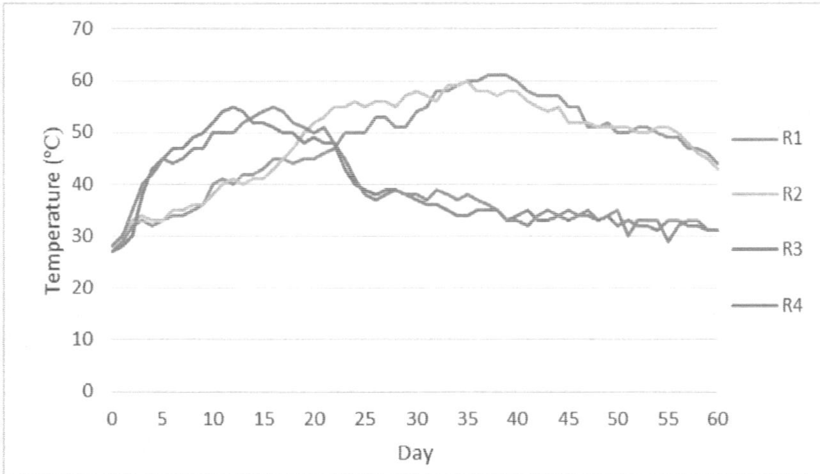

Figure 2. Variation of temperature in landfill bioreactor.

3.2. Leachate pH

The pH value is an important parameter to note because it describes the quality of the leachate produced. The fluctuations of the pH value every day are illustrated in Figure 3. Optimum pH value has the range from 6.5-8.0 to provide a good environment for bacteria to metabolize [17]. Initially, R1, and R2 had pH value above 5.0 and decreased until day-7, which had a pH value ranging from 4.7-4.9. The fall of pH value could be caused by acid accumulation due to the hydrolysis process under anaerobic conditions because microbial organisms need environments with different acidity to do their activities [18], [19]. The pH value of the bioreactor R1 and R2 began to rise until it reached a value of 7, which is a normal pH, on day-34 and day-37 respectively and continued with a gradual increase to reach the value of 7.0 in end of the study. In the bioreactor R3 and R4, a decrease in pH did not occur. This is likely due to the time for a short accumulation of acids so that the pH value increases and aeration is reducing fermentation process so that acidic compounds are not formed much [20], [21]. The pH value increases significantly until it reached the optimum value (6.5) on the day-3 for R3 and day-4 for R4. Fluctuations in pH values continued with a slow increase until the end of the study. The pH values of all bioreactors were still in the range of effective values, but the temperature parameter indicated that bacterial activity has decreased.

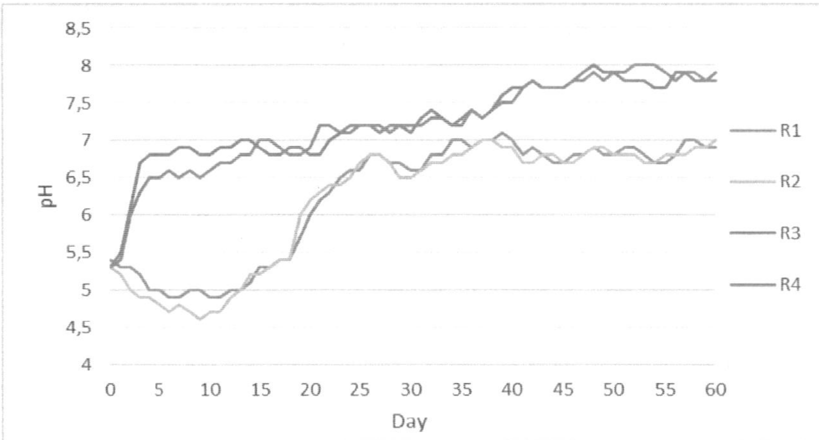

Figure 3. Variation of pH value in landfill bioreactor

3.3. Leachate COD

Beside the pH value, COD also determines leachate quality, which indicates the organic content contained in leachate. The COD value tends to decrease from the start of the study, as illustrated in Figure 4. Initially, the COD values of the waste were about 1,300 mg/l. R1 and R2 were decreasing until it dropped to about 1,100 mg/l in day-30. After day-30, more significant decrease happened because of the temperature entered thermophilic phase. R2 showed a lower COD value in the end of study, which had longer thermophilic period. COD removal for R1 reached 31.3% and 34.1% for R2.

Meanwhile, R3 and R4 decreased rapidly in the early of the study because it reached thermophilic phase in the first 7 day. It continued to decrease until the end of the study with 38.7% COD removed in R3 and 37.7% COD removed in R4, which R3 showed slightly lower COD value from day-15.

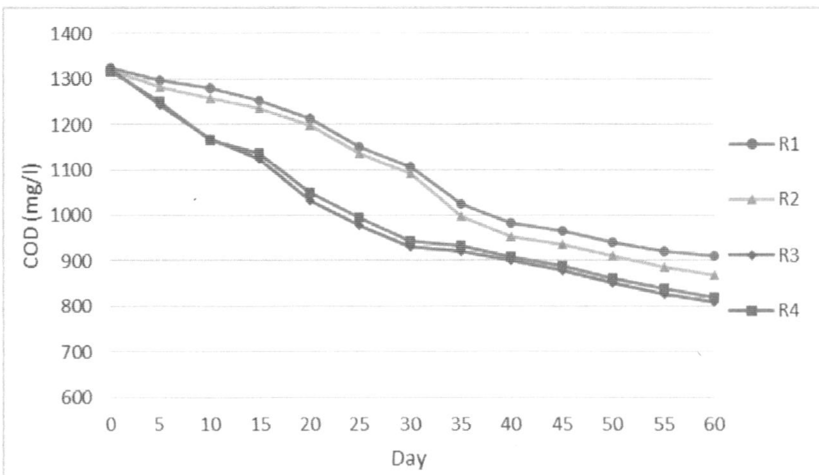

Figure 4. Variation of COD in landfill bioreactor

3.4. Methane production

Figure 5 illustrates the changes of methane gas generated from each landfill bioreactor. Initially, R1 and R2, which reached the thermophilic phase on day-23 and day-18, produced methane gas of 2.1 µg/m3 and 4.8 µg/m3 on day-5. The production of methane gas continues to rise until the temperature of the bioreactor reached its peak, with the amount of 32. μg/m3 for R1 and 35.2 μg/m3 for R2 on the day-40 respectively.

On the other hand, R3 and R4 began producing methane gas on the day-5, which were about 13,1 μg/m3 and 14,7 μg/m3. This value continues to increase until day-10 with methane gas produced at 21.1 μg/m3 and 22.1 μg/m3. After the temperature in the bioreactor gradually decreases, the production of methane gas also decreases following bioreactor temperature. The total production of methane for each bioreactors were 199.7 μg/m3, 228.7 μg/m3, 100.4 μg/m3, and 110.6 μg/m3 for R1, R2, R3, and R4 respectively. Methane gas in the anaerobic landfill bioreactor was generated more than aerobic landfill because of in the anaerobic landfill the stable condition for methane production was reached [22].

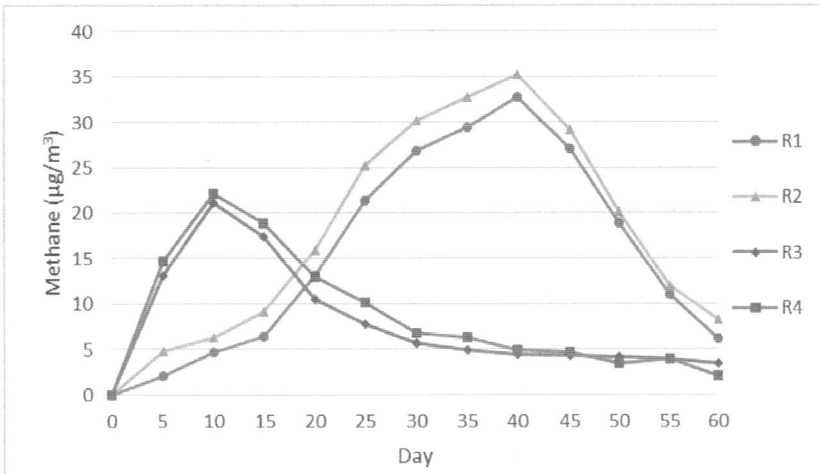

Figure 5. Profiles of methane generation

4. Conclusion

This study showed different results of aerobic and anaerobic landfill bioreactors performance. R1 and R2 were given leachate recirculation from it started generating leachate, whereas R3 and R4 given air injection. These conditions were given to accelerate the biodegradation process, which can shorten the stabilization of the waste. Aerobic bioreactor has faster biodegradation process because it reached thermophilic phase and optimum pH faster than anaerobic, because the suitable environment of bacteria are affected by these parameters. Meanwhile, anaerobic bioreactor produced more methane gas than aerobic because it reached the optimum condition for methane gas production. In the other hand, vegetable waste can maintain thermophilic phase better than leaf waste but it contains more organic materials than leaf. Nevertheless, further investigation is needed in a larger scale to ensure the method is suitable to improve Jatibarang Landfill.

References

[1] Giusti L 2009 *A review of waste management practices and their impact on human health* Waste Manag vol 29 no. 8, pp. 2227–2239
[2] Budihardjo M A and Wahyuningrum I F S 2018 Recovery Practice of Unsorted Solid Waste: from Landfill towards Economic Benefits in Semarang, Indonesia *MATEC Web Conf* vol 159, p. 02019
[3] Maria F D, et al 2013 *Leachate purification of mechanically sorted organic waste in a simulated bioreactor landfill* Waste Manag Res vol 31 No. 10, pp. 1070–1074
[4] Maria F D, et al 2013 *Experimental and life cycle assessment analysis of gas emission from*

mechanically-biologically pretreated waste in a landfill with energy recovery Waste Manag vol 33 No 11, pp. 2557–2567

[5] Maria F D and Micale C 2014 *What is the acceptable margin of error for the oxygen uptake method in evaluating the reactivity of organic waste* Waste Manag vol 34 No. 8, pp. 1356–1361

[6] MochamadA. B, et al 2018 Characterization of Leachate from the Integrated Solid Waste Treatment Plant at Diponegoro University Indonesia E3S *Web Conf* vol 73, p. 07017

[7] Yang Q, et al. 2015 A novel pretreatment process of mature landfill leachate with ultrasonic activated persulfate: Optimization using integrated Taguchi method and response surface methodology Process Saf. Environ. Prot vol 98, pp. 268–275

[8] Wei H 2019 Strategy of rapid start-up and the mechanism of de-nitrogen in landfill bioreactor J. *Environ. Manage* vol 240 no. January, pp. 126–135

[9] Broun R and Sattler M 2014 A comparison of greenhouse gas emissions and potential electricity recovery from conventional and bioreactor landfills *J. Clean. Prod* vol. 112 no. July 2014, pp. 2664–2673, 2016.

[10] Reinhart D R and Townsend T G Landfill Bioreactor Design and Operation. 1998 Boca Raton, FL (USA): CRC Press LLC.

[11] Byun B, et al 2019 Stability of bioreactor landfills with leachate injection configuration and landfill material condition Comput *Geotech* vol. 108 no. April 2018, pp. 234–243

[12] Maria F D, et al 2016 Treatment of mechanically sorted organic waste by bioreactor landfill: Experimental results and preliminary comparative impact assessment with biostabilization and conventional landfill Waste Manag vol. 55, pp. 49–60

[13] Townsend T 2015 *Sustainable Practices for Landfill Design and Operation* 2015

[14] Ritzkowski M, et al 2016 Aeration of the teuftal landfill: Field scale concept and lab scale simulation Waste Manag vol. 55, pp. 99–107

[15] Norbu T 2005 Pretreatment of municipal solid waste prior to landfillingWaste Manag vol. 25 no. 10, pp. 997–1003

[16] Kumar S 2011 Bioreactor landfill technology in municipal solid waste treatment: An overview Crit. *Rev. Biotechnol* vol. 31 no. 1, pp. 77–97

[17] Sekman E, et al 2011 Pilot-scale investigation of aeration rate effect on leachate characteristics in landfills Fresenius Environ. Bull vol. 20 no. 7 A, pp. 1841–1852

[18] Budihardjo M A, et al 2018 Leaves Composting Process and the Influence of Rumen Content and Bran Addition *MATEC Web Conf* vol. 159, p. 02033

[19] Long Y, et al 2009 *Nitrogen transformation in the hybrid bioreactor landfill* Bioresour. Technol vol. 100 no. 9, pp. 2527–2533

[20] Qiu Z, et al 2010 Effect of Aerating Frequency on Stabilizing Process of Aerobic Bioreactor Landfillin 2010 4th *International Conference on Bioinformatics and Biomedical Engineering*, pp. 1–4.

[21] Borglin S E, et al 2004 Comparison of aerobic and anaerobic biotreatment of municipal solid waste *J. Air Waste Manag. Assoc* vol. 54 no. 7, pp. 815–822

[22] Xu Q, et al 2014 Methane production in simulated hybrid bioreactor landfill Bioresour Technol vol. 168, pp. 92–96

IOP Conf. Series: Materials Science and Engineering **669** (2019) 012022 doi:10.1088/1757-899X/669/1/012022

Planned investigation of UTES potential in Algeria

S Amara[1], T Baghdadli[1], and B Nordell[2]

[1] Materials and Renewable Energy Research Unit (URMER), University of Tlemcen, BP 119 Tlemcen, Algeria
[2] Department of Civil, Environmental and Natural Resources Engineering Luleå University of Technology (LTU), SE-97187 Luleå, Sweden

sofiane.amara@yahoo.fr

Abstract. The unbalance between supply and demand of heat can be managed by thermal energy storage (TES). For large-scale systems the underground is used as storage medium or storage volume. Aquifer storage (ATES) is most suitable for very large applications, Borehole storage (BTES) the most general system in all scales and the rock cavern storage (CTES) is best suited for extremely high loading/extraction loads. The construction of any of these systems requires knowledge about site-specific properties of the ground i.e. geology and groundwater conditions. Current paper gives a brief review of the potential and advantage of Underground Thermal Energy Storage (UTES) technology utilization in buildings for the hard climate.

1. Overview

Energy is the backbone of an economy and solving the energy trilemma of sustainability, affordability, and adaptation to climate change is a huge challenge that the world faces daily. The major part of countries with this problem are located in Sub Saharan Africa. With most of the countries having electricity access rates less 50% and high dependence on traditional biomass for cooking, the need for clean and affordable energy becomes imperative. This, therefore, prompts a need for thorough multidisciplinary research and development in clean and sustainable energy sectors. Consequently, renewable energy resources in Africa must be exploited. These sustainable sources of energy can supply energy demands of different sectors in residential, commercial, and industrial sectors with minimal impact on the environment.

Africa is endowed with various renewable energy resources which the solar resource is more evenly distributed across the continent when compared to others. With issues of deforestation and need for production of food crops which impedes the full exploitation of biomass and intermittency of wind and solar energy, this leaves geothermal energy resource as the one of the least intermittent renewable sources for heating and/or cooling buildings, electricity production and other uses.

Generally, geothermal energy entails the harnessing of thermodynamic quantities which have equal ability of a physical system to generate work or heat; from the earth crust's underneath. Through the means of conduction and convention, geothermal heat that exists deep inside the earth's interior emerges close to the earth's surface. The heat coming from the earth crust is caused by the decomposition of radioactive isotopes of uranium, thorium, potassium and earth crust core which has a temperature of 1000-4500°C [1].

Besides from volcanoes, geothermal energy also appears at the surface in the form of hot springs, steam vents, hot soil, boiling mud, geysers, hydrothermal craters, structures ascribed circulation of hot fluids moving to the surface from the underground. These fluids contain the convective component of the natural heat flow from the earth interior. There is also a conductive component to the natural heat flow of the earth, which arises because of the temperature gap between the deep underground and the soil surface, and is controlled by the thermal conductivity property of the rocks [2].

2. Utilization of geothermal energy

One can harness steam/water heated by the geothermal heat by the mean of different technologies and also use it for several purposes. The use of geothermal fluid relies considerably on its thermodynamic characteristics and chemistry. The geothermal system from which the fluid originated determines these factors. Geothermal fluids have been categorized in a different way by number of authors. Some have done so by using temperatures while others have used enthalpy [3]. The most mutual criterion is based upon enthalpy. The resources are divided into low, medium and high enthalpy resources. Table 1 depicts the classifications proposed by several authors.

Table 1. Classification of geothermal resources (°C) [3]

Classification of resources	Muffler and Cataldi [4]	Hochstein [5]	Benderitter and Cormy[6]	Nicholson [7]	Mburu [8]
Low Enthalpy Resources	<90	<125	<100	≤150	≤190
Intermediate/ Medium Enthalpy Resources	190-150	125-255	100-200	-	-
High Enthalpy Resource	>150	>255	>200	>150	>190

Depending on the enthalpy, geothermal fluid can be used either to generate electricity or for direct applications. Electricity generation is the common and dominant form of utilization of high-temperature geothermal resources while low to medium resources are better suited for non-electric (direct) application (Table 2).

Table 2. Basic technology commonly used for geothermal energy.[9]

Reservoir temperature	Reservoir fluid	Common use	Technology commonly chosen
High temperature, >220°C	Water or Steam	Power generation and Direct use	Flash steam; Combined (flash and binary) cycle Direct fluid use; Heat exchangers; Heat pumps
Intermediate temperature, 100-220°C	Water	Power generation and Direct use	Binary cycle Direct fluid use; Heat exchangers; Heat pumps
Low temperature, 30-150°C	Water	Direct use	Direct fluid use; Heat exchangers; Heat pumps

Table 3. Applications of geothermal energy.[9]

Electricityproduction	Wells drilled into a geothermal reservoir produce hot water and steam from depths of up to 3 km. The geothermal energy is converted at a power plant into electric energy, or electricity. Hot water and steam are the carriers of the
Directuse	Applications that use hot water from geothermal resources directly. Examples: space heating, crop & lumber drying, food preparation, aquaculture, industrial processes, greenhouses etc. Historical traces back to ancient Roman times, e.g.
Geothermalheat pumps	Taking advantage of relatively constant earth temperature as the source and sink of heat for heating and cooling, as well as hot water provision. It is one of the most efficient heating and cooling systems available.

3. Underground Thermal Energy Storage (UTES)

Energy is needed to heat or cool space, but mostly, its availability is always a constraint. The gap between energy supply and demand issues is solved by the energy storage. The potential of using renewable energy sources as geothermal energy, solar heat and waste heat has been in escalate thanks to the energy storage system. Underground Thermal Energy Storage (UTES) is the most often used storage technology for heat and "cold". It has been proven that the ground is the best candidate of storing heat or cool in great quantity and for long period of time [10]. The use of bedrock, sediment and groundwater in the subsurface as a huge UTES system allows for storage of surplus heat (regardless of whether it is solar in origin, radiogenic (geothermal), or simply waste summer heat from buildings) until a time when it is needed. To put it another way, the ground can be turned into a heat store; it can preserve summer heat until winter and winter cold until summer, letting us to correspond supply and demand for minimum one year. Ground-Source Heat Pumps (GSHP) are frequently used to 'pump' exess heat into the ground and to drawn out heat when needed for space heating. For the most optimum, it can be used to procure a balanced combination of both heating and cooling, with surplus summer heat being stored in the ground for next extraction in winter.

Many different UTES systems have been created and tested. Two types of system, Aquifer (ATES) and Borehole (BTES) storage.

- Aquifer Thermal Energy Storage (ATES), where heat transfer and storage are by warm or cold groundwater.
- Borehole Thermal Energy Storage (BTES), where boreholes are used as heat exchangers for transfer of thermal energy to/from the ground. The volumetric heat capacity of the rock is used to store the heat.

4. Research design and methodology.

The objective of this research is to estimate the geothermal resource potential in several regions in Algeria and to estimate the applicable UTES in the said regions. This means identifying the geothermal resource potential, in terms of rock and fluid properties, surface extent, volume, and collect necessary information for decision making on developing geothermal power plants and also direct use of available resources. To achieve this, the following surveys should be taken in consideration according to [2]:

4.1. Geological survey

The geological survey will include the study of surface rocks in terms of mineral content, fluid inclusions, as well as studying the hydrology of the area, aiming at:

- • Mapping surface geology.
- • Identifying the age of rocks.
- • Establishing local stratigraphy and possible permeable deep zones for geothermal exploitation.

- • Identifying volcanic formations of recent geological age, which indicate the presence of hot magma chambers close to the surface.
- • Identifying hydrothermally altered rocks, which are a product of a fossil or active hydrothermal system.

The data of this survey will be gotten from literature as wells as from the oil exploration companies in Algeria if available.

4.2. Mapping thermal manifestations

Thermal manifestations (steam vents, geysers, hot springs, hot ground, boiling mud, hydrothermal craters) are the outlet of rising hot fluids that prove the presence of an active hydrothermal system in an area, which may be the target for geothermal power generation. They may provide useful clues concerning the location, the temperature, the chemistry and the depth of the deep geothermal resource. In addition, the flow of each thermal manifestation is measured, which is necessary for calculating the convective heat flow from the area.

4.3. Geochemical survey

It includes sampling of surface and near surface waters, chemical analysis and application of geo-thermometers. The latter presumes that the chemical equilibrium of dissolved species stays unchanged during their ascent from depth towards the surface, and as chemical equilibrium is a function of temperature, the deep temperature of the fluids can be obtained. An experienced geochemist will be enquired to assist in the derivation of credible estimations pertaining the temperature and chemistry of the deep fluids.

4.4. Thermal gradient measurements

It consists of drilling different shallow bores, of depths 50-100m, measuring downhole temperature, and defining the distribution of the geothermal gradient in the area, which, along with the thermal conductivity of the rocks, are the factors of conductive heat flow from the area.

4.5. Gravity survey

It implies measuring the local gravity field, which depends on the density of subsurface layers. In volcanic environments where there is a large difference in the density between the upper volcanic formations and the deeper basement of metamorphic rocks, the gravity survey provides useful information concerning the thickness of the volcanic formations.

4.6. Deep exploration drilling and testing

A geothermal resource is proven only if there is at least one deep geothermal production well. In addition to proving the resource, deep exploratory wells give important information on subsurface geology, properties of the rock, minerals present, fluid inclusions, as well distribution of key reservoir parameters, such as temperature, pressure, porosity and permeability, which are essential to evaluate both quantity and quality of the geothermal resources. Related tests include sampling of rocks (cores), and fluid, wire line logging, production tests, pressure drawdown and build-up tests, injection tests and interference tests.

After data collection, several statistical and mathematical models would be applied to estimate the natural heat flow, stored heat, reservoir volume, qualitative thermal recovery factor for each area. Also the Kehle temperature correction to downhole temperatures would be applied.

4.7. Thermal response test (TRT)

To find out the suitability of the UTES especially the BTES, the thermal response test (TRT) will be used to determine the ground thermal properties. The TRT is an indirect (in-situ) measurement approach, which is the easiest and most accurate way to determine the effective thermal properties. The methodology of TRT include [11]:

- Drilling of a test borehole and inserting with U-tube of a particular size.
- The test borehole is filled with (ground)water or some grouting
- The tube and fluid is left to stabilize for several days.
- The initial temperature along the borehole will be noted.
- A circulation pump is used to circulate the heated fluid through the borehole
- Heat is supplied by some kind of heater usually a heat pump
- The inlet outlet temperature is monitored and the ground thermal conductivity and borehole resistance are estimated from this temperature response
- Other information such as the initial ground temperatures and indications of groundwater flow would be noted and taken into consideration.

4.8. Data Analysis using GIS

Geographic Information System (GIS) Model for Geothermal Resource Exploration (GM-GRE) toolbox will be used for data integration and analyses to identify the suitability of a particular geothermal resource type and the extent of areas that warrant further detailed investigation. Digital data layers and maps can be used in the GIS environment to develop a geothermal favorability map. It integrates three different data sources, namely geological and geochemistry sources and the geothermal gradient, are applicable to the tool to assess the geothermal potential on a regional scale [12]. All these systems and steps took a lot of time and are very expensive and also the difficulty to obtain authorization from the government.

5. CTES potential in Algeria

CTES still have a big advantage – the powers of injection and extraction are very high. In the Rock Cavern Thermal Energy Store (CTES) energy is stored at a constant temperature during the year in an underground cavern. In such a system with a large volume of air or water it is of great importance to maintain a stratified temperature profile in the cavern that cannot be obtained.

Figure 1 and 2 show the potential of different existing caverns in Algeria country that are characterized by a constant temperature. This system (Caverns) means available underground duct systems that could be used as a source for heating and cooling. Such a system would be made simply by pumping air through the Cavern and conducting this air to the air conditioned building. For example the BeniAad Cavern at Tlemcen in North West Algeria is characterized by a temperature about 13°C during all the year where the ambient temperature varies from -6°C in winter to 40°C in summer. The ambient air then approaches the temperature of the Cavern (~13°C) during the passage through the ground before used for ventilation, heating, and cooling of buildings that can reduce energy consumption inside the building [14].

(a) (b)

Figure 1. Potential of different existing caverns in Algeria, (a) Grotte de Cervantes at Algiers, and (b) Grotte de Ziama Mansouriah at Jijel

IOP Conf. Series: Materials Science and Engineering **669** (2019) 012022 doi:10.1088/1757-899X/669/1/012022

(a) (b)

Figure 2. Potential of different existing caverns in Algeria, (a) Grotte de Beni Aad Tlemcen, and (b) Ghar Boumaaza Tlemcen

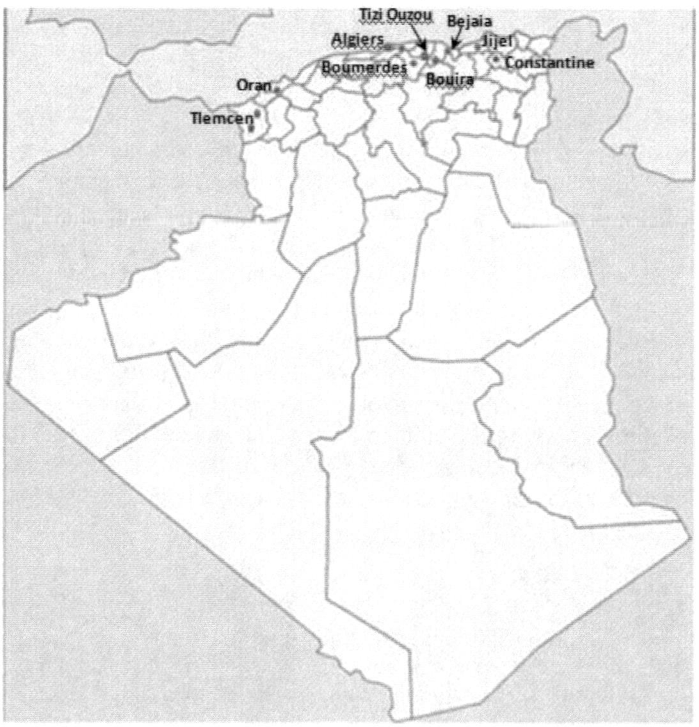

Figure 3. Potential of CTES that can be used for heating and cooling

Another old existing system has been studied called Fouggara situated in the desert of Algeria [15]. This study shows the feasibility of using this system for heating and cooling buildings where the temperature inside the system is constant about 21°C during the year Figure 4

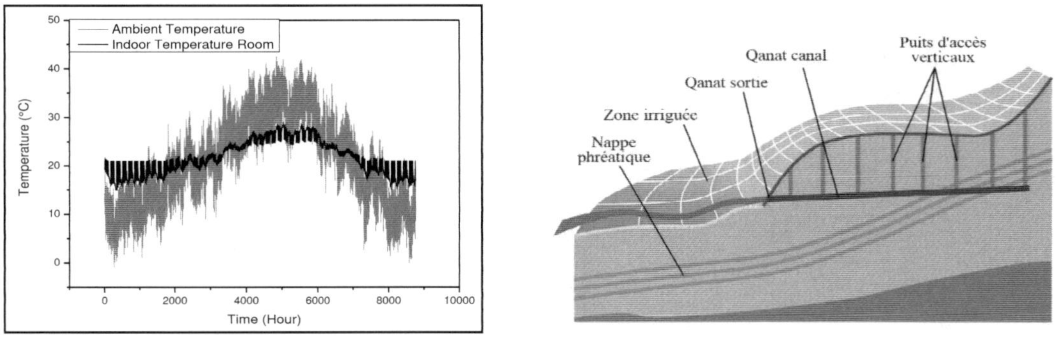

Figure 4. Ambient and indoor temperature variation

Figure 5. Fouggara in Adrar

6. Conclusion

As geothermal energy production and use become more prominent in today's renewable energy. Many countries are actively exploiting geothermal energy for both electricity generation and other uses but only a limited number of countries use geothermal energy directly for heat production. In Algeria, balneotherapy remains practically the only direct use of this energy with more than two hundred hot springs spread over the northern part of the country. About one third (33%) of them have temperatures above 45°C. There are even high temperature sources up to 118°C in Biskra. Algeria being the largest country in Africa in terms of area, it necessarily has a large potential in Albian nappes (80% of Algerian territory). Geothermal energy remains under exploited in Algeria. Aware of this significant potential of geothermal energy in the energy sector, it is necessary to focus its interest on geothermal energy for the development of the agricultural sector. This is why in this paper we describe different steps to investigate the underground thermal energy storage that can be planned for the near future.

References

As geothermal energy production and use become more prominent in today's renewable energy. Many countries

[1] Ikechukwu O, Agbidi D C and BamisileO O 2015 *Exploration And Application of Geothermal Energy in Nigeria* **6** 2 726–732
[2] Mendrinos D,Karytsas C and Georgilakis P S 2008 *Assessment of Geothermal Resources for Power Generation* **10** 51262–1267
[3] Dickson M and Fanelli 2004 *What Is Geothermal Energy* Inst. di Geosci. e Georisorse, CNR,

Pisa, Italy
[4] Muffler R, P and Cataldi 1978 *Methods for Regional Assessment of Geothermal Resources*
[5] Hochstein M P 1990 *Classification and Assessment Of Geothermal Resources* in: Dickson, M.H. and Fanelli, M., eds., Small Geothermal Resources: A Guide to Development and Utilization," UNITAR New York 31—57
[6] Benderitter Y and Cormy Y 1990 *Possible Approach To Geothermal Research And Relative Costs* in Dickson, M.H. and Fanelli MSmall Geothermal Resources: A Guide to Development and Utilization, New York: UNITAR 59–69
[7] Nicholson K 1993 *Geothermal Fluids*(Berlin: Springer Verlag) vol. **18** 264
[8] HUTTRER G W*The Status of World Geothermal Power Generation 1995 – 2000* Geothermics 2001 **30** 7–27
[9] Mburu M 2012 *Geothermal Energy Utilisation*
[10] Midttømme K, Banks D, Ramstad R K, Sæther O M and Skarphagen H 2008 *Ground-Source Heat Pumps and Underground Thermal Energy Storage — Energy for the future*93–98
[11] Gehlin S 2002 *Doctoral Thesis Thermal Response Tes* 1402–1544
[12] Noorollahi Y,Ghasempour R and Jalilinasrabady S 2015 *A GIS Based Integration Method for Geothermal Resources Exploration And Site Selection* **33** no. 2243–257
[13] Wikipedia, https://fr.wikipedia.org/wiki/Cat%C3%A9gorie:Grotte_en_Alg%C3%A9rie
[14] Nordell B 2000 *Large-scale Thermal Energy Storage, WinterCities'2000, Energy and Environment*t (Sweden : Luleå)
[15] Amara S, Nordell B and Benyoucef B 2011 *Using Fouggara for Heating and Cooling Buildings in Sahara*, Energy Procedia, Science Direct Elsevier **6** 55 - 64

IOP Conf. Series: Materials Science and Engineering **669** (2019) 012023 doi:10.1088/1757-899X/669/1/012023

Pervious low-calcium fly ash geopolymer concrete using gap-graded limestone coarse aggregates in ambient curing

P H Simatupang[1], M A J Adu[1], Y M I Sadukh[1] and Jonbi[2]

[1] Civil Engineering Department, Faculty of Science and Engineering, University of Nusa Cendana, Kupang, Indonesia
[2] Civil Engineering Department, Faculty of Engineering, University of Pancasila, Jakarta, Indonesia

partogihsimatupang@gmail.com

Abstract. This paper describes the research of pervious geopolymer concrete using limestone from Timor Island in Indonesia as coarse aggregate. No sand was used to make this pervious geopolymer concrete. Two types of crushed limestone aggregates used were coral limestone (organic limestone) and limestone (calcilutite). Abrasion losses of the aggregates using Los Angeles Machine test were 18.9% for limestone (calcilutite) and 34.98% for coral limestone. Those aggregates had many characteristics such as 10-20 mm of diameter length, angular shapes and apparent specific density 2.74-2.75. The mass ratios of aggregate to geopolymer paste were varied from 2.86 to 6.91. To make the geopolymer paste, it was used Class F or Low-Calcium Fly Ash from Bolok-Kupang Coal Fired Power Plant as a rich silica-alumina of precursor material. Meanwhile, the mass ratio of natrium silicate solution to natrium hydroxide solution 10 M was 1.0 and the mass ratio of fly ash to alkaline solution was 2.0. Ambient curing was applied to all specimens. Mechanical characteristics obtained at the age of 28 days of hardened pervious geopolymer concrete were: (1) compression strength, (2) splitting tensile strength, and (3) abrasion losses using Los Angeles Machine Test. Besides, density of hardened specimens were also obtained. In addition, the comparisons of mechanical properties of pervious geopolymer concretes using coral limestone and limestone (calcilutite) as coarse aggregate were also conducted. According to the results of this research, it is feasible to make pervious geopolymer concrete using Timor limestone as coarse aggregate due to its mechanical characteristics.

1. Introduction

Geopolymer concrete is a relatively new concrete. This is the most promising concrete to replace the conventional or Portland Cement based concrete. The difference between geopolymer concrete and portland cementbased concrete is the paste i.e geopolymer paste and cement paste. Meanwhile, all types of aggregates as a filler in cementbased concrete can be used to make geopolymer concrete. Comprehensive researches and applications of geopolymer paste, mortar or concrete have been available in three well-known text books about geopolymer [1-3]. The dominant geopolymer paste used was fly ash based geopolymer paste. The reason was that fly ash was an abundant waste material so that usaged of it was a solve of environmental problem. Many types of fly ash in Indonesia have been used to make the durable geopolymer paste, mortar or concrete [4], [5], [6]. From previous results, Indonesian fly ash based geopolymer material is comparable to cement based material according to its mechanical characteristics.

Most application of geopolymer concrete is well known except pervious geopolymer concrete. In the last seven years, very few papers or reports of researches were available according to pervious geopolymer concrete.The most cited paper of it was the work of Khon Kaen University of Thailand [7]. They used limestone as coarse aggregate. Their pervious geopolymer concretes were made of Class C fly ash or lignite fly ash with CaO 19.4% and dry curing 60OC. It was contrast to papers or research report of pervious cement-based concrete by [8]. In Indonesia, pervious cement-based concrete are also well known and applied material. Then, locally available materials can actually be used to produce High Performance Concrete [9,10].

ACI 522R-10 has recommended the range of composition per m3 of pervious cement-based concrete to be applied as road pavement i.e cement (270-415 kg), coarse aggregate (1190-1480 kg), and mass ratio of water to cement (0.27-0.34). Meanwhile, in Indonesia, the volume ratio of aggregate to paste is preferred and the range of it is 2:1 – 12:1 with range of compressive strength 2 MPa – 35 MPa [9],[11].

Therefore, this study focused on pervious geopolymer concrete using Class F or Low-Calcium fly ash with ambient curing. Coarse aggregates used were crushed coral limestone and crushed limestone (calcilutite) from Timor Island Nusa Tenggara Timur Province-Indonesia. The mass ratios of aggregate to geopolymer paste were varied from 2.86 to 6.91. To make the geopolymer paste, it was used Class F Fly Ash from Bolok-Kupang Coal Fired Power Plant as a rich silica-alumina of precursor material. Meanwhile, the mass ratio of natrium silicate solution to natrium hydroxide solution 10 M was 1.0 and the mass ratio of fly ash to alkaline solution was 2.0. The physical and mechanical characteristics of the pervious geopolymer concrete were obtained.

2. Experimental programme

This subsection will present raw material used, sample made, and testing conducted. There was no treatment to limestone aggregates and fly ash as raw materials. Testing conducted were: (1) the characteristics of aggregates including specific gravity and abrasion losses of aggregates in The Los Angeles Machine, (2) compression and splitting tensile strength, (3) Abrasion Losses of concrete samples in The Los Angeles Machine.

2.1. Materials

Two types of crushed limestone aggregates used were coral limestone (organic limestone) and limestone (calcilutite). The aggregates were crushed by hand with diameter 10-20 mm and angular shapes as shown in Figure 1.

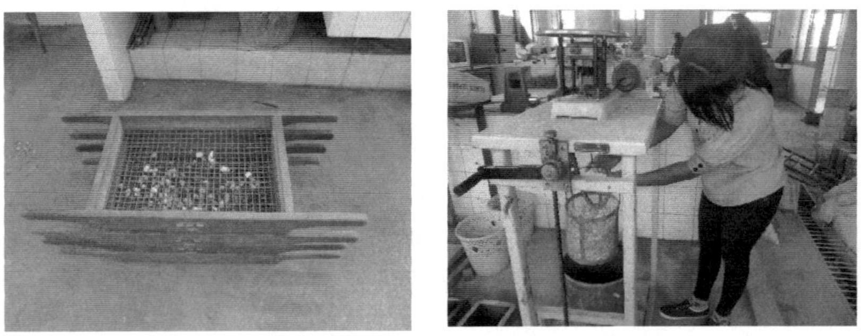

Figure 1. The gap-graded limestone coarse aggregates

The characteristics of coarse aggregates are given in Table 1. Apparent specific gravity of the coral limestone (organic limestone) and limestone (calcilutite) are almost the same. Meanwhile, there is a significant difference of abrasion losses between the coral limestone and limestone (calcilutite). The coral limestone has higher value of abrasion losses so that this aggregate has weaker abrasion strength.

The gap-graded limestone coarse aggregates used is given in Figure 2. This gradation was not comply to any standard of gradation of coarse aggregate in Indonesian Standard (SNI). The size of coarse aggregate used was dominantly in the range 12.7 – 15.8 mm.

The mixture composition of geopolymer paste was fly ash and alkaline activator solution with certain mass ratios. Alkaline activator solution was a mixture of 10 M sodium hydroxide solution and natrium silicate with mass ratio = 1:1. Class F or Low-Calcium fly ash from Power Plant Bolok Kupang was used as a precursor material to make geopolymer. This fly ash had CaO 8.48%, SiO2 43.18 % and Al2O3 17.78 %. More detail oxide compositions and XRD (X-Ray Diffraction) of fly ash can be referred in Simatupang et al [6].

Table 1. Characteristics of coarse aggregates

Characteristics	Coral Limestone	Limestone (Calcilutite)
Dry Specific Gravity	2,57	2,68
SSD Specific Gravity	2,62	2,70
Apparent Specific Gravity	2,74	2,75
Absorption	1,74	0,75
Abrasion Losses by Los Angeles Machine Test	34,98	18,90

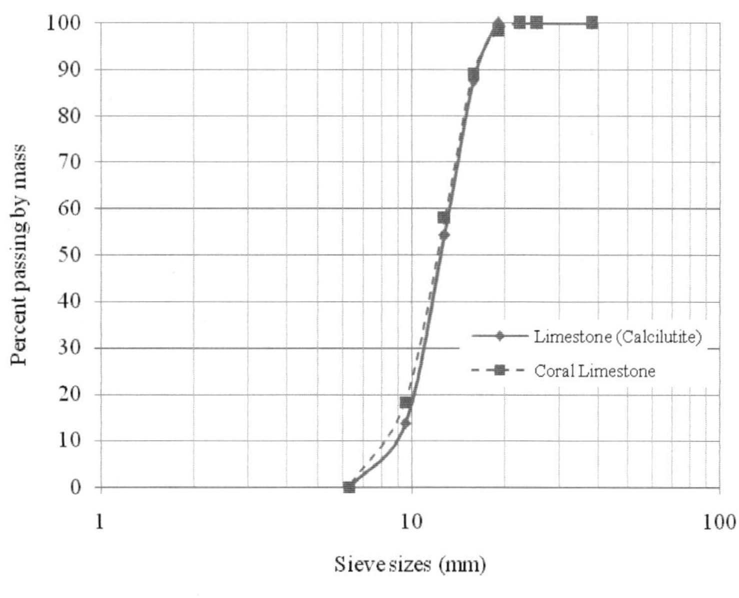

Figure 2. The gap-graded limestone coarse aggregates.

2.2. Mixture compositions, mixing, and casting

Sample used were cylindric concrete with diameter 10 cm and 20 cm length. After fresh pervious concrete casted, the samples were cured in ambient curing. Composition of pervious geopolymer concrete for each m3 used in this paper is given in Table 2. Every mixture was prepared i.e 3 samples for compression testing, 3 samples for splitting tensile strength, 2 samples for abrasion losses. Results presented in this paper are the mean of those samples.

The mixing and casting of pervious geopolymer concrete is shown in Figure 3. To make geopolymer paste, it was mixed fly ash and alkali activator solution by using Paint Mixer with 1800 rpm for 2-3 minutes. After, geopolymer paste was ready, it was poured in concrete mixer with coarse aggregate. Concrete mixer run for 5 minutes. Fresh geopolymer concrete was casted in plastic formwork with 3 layers. Each layer was compacted by steel rod. Then, samples were cured with ambient curing. After 28 days, the formworks were released

Table 2. Composition of pervious geopolymer concrete (per m^3).

Mixture	FA[a] (kg)	NS[b] (kg)	NH[c] (10 M) (kg)	Aggregate (kg)	Type of Aggregate	Mass ratio of Aggregate/Geopolymer Paste
A1	360	90	90	1548	Coral limestone	2,86
A2	180	45	45	1548	Coral limestone	5,73
B1	360	90	90	1865	Limestone (Calcilutite)	3,45
B2	180	45	45	1865	Limestone (Calcilutite)	6,91

[a]Fly Ash

[b]Natrium Silicate

[c]Natrium Hydroxide

Figure 3. Mixing and casting pervious geopolymer concrete

2.3. Testing

2.3.1. Compression and splitting tensile strength. The compressive strength was tested at the age of 28 days using ASTM C 39/C 39M-01 standard or SNI 03-1974-1990 (Indonesian standard). Meanwhile, the splitting tensile strength was tested at the age of 28 days using ASTM C 496-96 standard or SNI-03-2847-2002 (Indonesian standard). The reported values of strength were the average of three samples. Compression testing samples were capped at top and bottom side with a gypsum capping compound. The compression and splitting tensile strength testing is shown in Figure 4.

Figure 4. The testing of compression and splitting tensile strength of pervious geopolymer concrete.

2.3.2. Abrasion Losses in The Los Angeles Machine. The abrasion losses and impact testing of aggregates was tested using ASTM C 131-01 standard or SNI 2417-2008 (Indonesian standard). The abrasion losses was calculated using Eq. (1).

$$1.1.1.\ N = (a - b)/a$$

Where N is the abrasion and impact value, a is mass of original sample (gram) and b is mass of sample retained in No.12 (gram). The reported values of strength were the average of two samples. Mass of each sample was 5000 gram and comply to gradation B (in Indonesian standard). Steel ball bearing used was 11 balls with total mass 4584±25 gram. The Los Angeles machine rotated at 30 to 33 r/min for 500 revolutions.

In addition, the abrasion and impact testing with similar procedure was also applied to pervious geopolymer concrete. The origin samples of abrasion and impact testing of pervious geopolymer concrete were from the broken splitting tensile samples.

3. Results and discussion

3.1. Compressive strength

The results of testing are summarized in Table. 3. The compressive strength of samples A2 and B2 (higher ratio of aggregate to geopolymer pastes) were higher than the strength of samples Al and B1 (lower ratio of aggregate to geopolymer pastes). It means excessive geopolymer paste that trapped at the bottom of samples gives negative effect as shown in Figure 5. This is very different manner with pervious cement-based concrete that the higher ratio volume of aggregate to cement paste the higher compressive strength of pervious cement-based concrete [9],[11]. An important consideration to make pervious geopolymer concrete is the different cohesive and rheology characteristics between geopolymer paste and cement paste.

The results of this research compared to Tho-in et al [7] are shown in Figure 6. The compressive strength of pervious geopolymer concretes in this research are lower than the concretes strength of

Tho-in et al [7]. There is no good relationship between the results of this research and the results of Tho-in et al [7]. This result indicates that the compressive strength of the geopolymer paste in this research was not optimally contributed.

Table 3. Results of testing.

Mixture	Density (kg/m^3)	Compressive Strength (MPa)	Tension strength (MPa)	Abrasion Losses (%)
A1	1777	1.66	0.35	36.60
A2	1645	2.55	0.48	35.14
B1	1925	1.95	0.42	23.06
B2	1748	3.01	0.61	22.50

Figure 5. The excessive geopolymer pasteswere trapped at the bottom of samples.

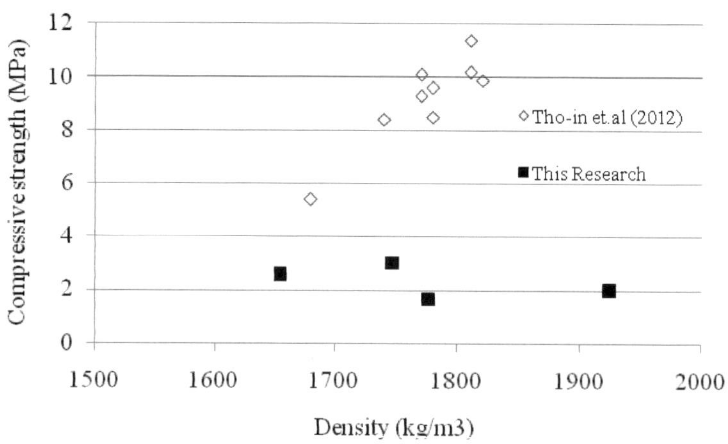

Figure 6. Relationship between compressive strength and density of pervious geopolymer concrete.

3.2. Splitting tensile strength.

The splitting tensile strength of samples A2 and B2 (higher ratio of aggregate to geopolymer paste) were higher than the strength of samples Al and B1 (lower ratio of aggregate to geopolymer paste). The splitting tensile strength tends to have the same trend with the compressive strength. Figure 7 shows good linear relationship between compressive strength and splitting tensile strength of pervious geopolymer concrete with R2=0.86.

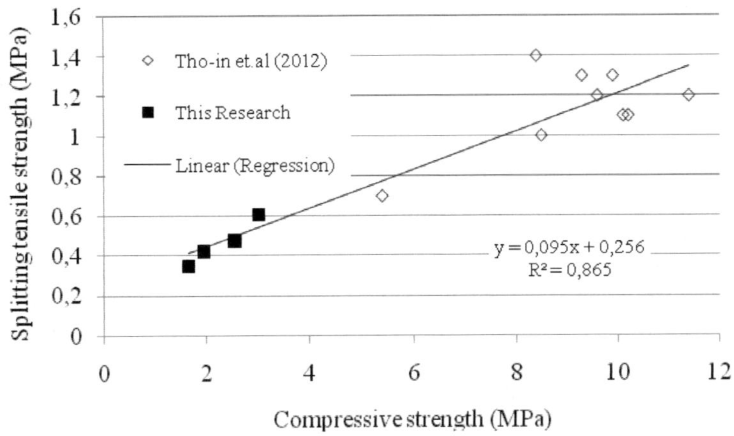

Figure 7. Relationship between compressive strength and splitting tensile strength of pervious geopolymer concrete.

3.3. Abrasion losses of samples.

The abrasion and impact strength of aggregate and pervious geopolymer concrete is shown in Figure 8. The abrasion and impact strength of pervious geopolymer concrete is lower than its aggregate regardless the type of aggregate. It is indicated with higher abrasion and impact value of pervious geopolymer concrete compared to the aggregate. There is a significant abrasion and impact strength of pervious geopolymer concrete according to their aggregate. It depends closely to the abrasion and impact strength of aggregate.

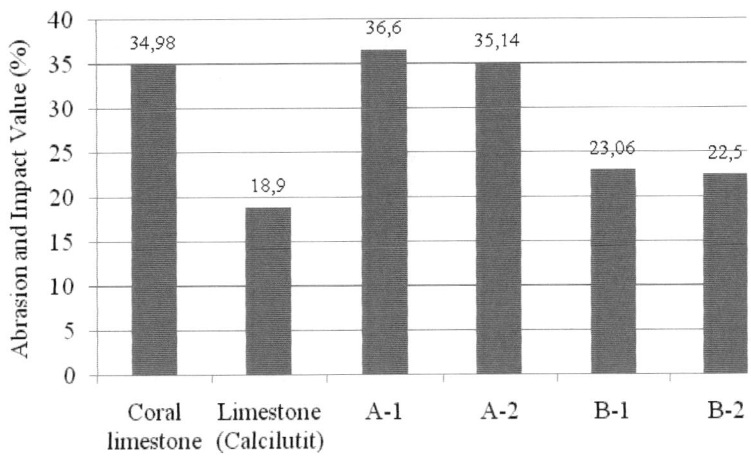

Figure 8. Abrasion and impact values of aggregate and pervious geopolymer concrete.

4. Conclusions

It has been shown mechanical characteristics of pervious geopolymer concrete using two types of Timor limestone as coarse aggregate. The pervious geopolymer concrete using crushed coral limestone aggregate is comparable to pervious geopolymer concrete using crushed limestone (calcilutite) according to its compression strength and splitting tensile strength. The abrasion and impact strength of pervious geopolymer concrete closely depends on the strength of its aggregate.

An important consideration to make pervious geopolymer concrete is cohesive and rheology characteristic of geopolymer paste. Furthermore, the excessive geopolymer paste must be limited to make reliable pervious geopolymer concrete. It is recommended that the strength of this pervious geopolymer concrete need to be increased by using different curing method, continuous grading of coarse aggregate and finding the optimal compositions of the pervious geopolymer concrete.

References

[1] Shi C, Krivenko P V, Roy D 2006 *Alkali-activated cements and concrete* (New York: Taylor and Francis).
[2] Provis J L, Deventer van J S J 2009 *Geopolymers Structure, processing, properties and industrial applications* (United Kingdom: Woodhead Publishing Limited and CRC Press).
[3] Davidovits J 2008 *Geopolymers:Chemistry and Applications* (Saint-Quentin : Institut Geopolymer).
[4] Simatupang P H, Imran I, Pane I, and Sunendar B J 2015 *Eng Technol Sci* **47** No 32 31-249.
[5] Simatupang P H 2017 AIP Conference Proceedings 1887 020028 doi 10.1063/1.5003511.
[6] Simatupang P H, Lubalu, Sianturi H L, and Bunganaen 2019 *Matec Web of Conferences* 276, 01031; https//doi.org/10.1051/matecconf/201927601031.
[7] Tho-in T, Sata V, Chindaprasirt P, Jaturapitakkul C 2012 *Construction and Building Materials* **30** 366–371.
[8] ACI Committee 522 2010 *Pervious Concrete* (Detroit USA : Report No. 522R-10, American Concrete Institute).
[9] Kardiyono T 2010 *Non-Sand Concrete* (Gadjah Mada University Press).
[10] Jonbi J, Hariandja B, Imran I, and PaneI 2012 *Applied Mechanics and Materials* **175–177** 1067–1071.
[11] Trisnoyuwono D 2014 *Proceedings of The 2nd National Conference on Science and Engineering* (Sainstek2) (University of Nusa Cendana).

Acknowledgments

Thanks to Ir. Leonard Simatupang as a manager of Bolok-Kupang Coal Fired Power Plant for providing fly ash. Also thanks to Dr. Herry Z. Kotta for explaining about Timor limestone.

IOP Conf. Series: Materials Science and Engineering **669** (2019) 012024 doi:10.1088/1757-899X/669/1/012024

Cooling treatment of alkali solution to prolong the setting time of high calcium fly ash based geopolymers

O K Wattimena[1], A Antoni[1], R T Surja[1], R Mintura[1] and D Hardjito[1]

[1]Civil Engineering Department, Petra Christian University, Indonesia

antoni@petra.ac.id

Abstract. The problem of rapid or short setting time has been reported by many researchers when utilizing high calcium fly ash as an aluminosilicate source material in the synthesis of geopolymers. Meanwhile, there are very few studies carried out to overcome this obstacle. This study was conducted to evaluate the cooling treatment of the alkaline solution to prolong the setting time of the geopolymer. Alkaline solutions with five different temperatures were used to synthesize the geopolymer paste. Each set of fresh geopolymer paste was tested using the Vicat apparatus according to ASTM C191 to measure its setting time. The results showed that a decrease in the temperature of the alkaline solutions prolonged both the initial and the final setting time of the geopolymer paste. An alkaline solution with a temperature of 9°C could double the initial setting time compared to the control mixture (31°C). In the mortar phase, the cooling treatment also increased the compressive strength of the geopolymer mortar besides longer setting time.

1. Introduction

Concrete has become a major requirement in almost all construction industries because of its strength, durability and also because of its ease to be formed as desired. The main ingredient of a concrete mixture is cement, which binds all other materials together. Portland cement is one of the most popular cement, which processed from the natural limestone. However, as globally known, the production process of Portland cement releases an amount of carbon dioxide that causes the greenhouse effect in the atmosphere. Therefore, since the global warming issue becoming world concern, numerous researches were done to develop alternatives to Portland cement [1-3].

Geopolymer concrete is one of the most promising alternative binders in concrete production. Geopolymer concrete is a synthesis product from the aluminosilicate materials (which are in the amorphous to semi-crystalline phases) [4]. Metakaolin, slag, and fly ash are widely known as some of the aluminosilicate source materials that can be used for the synthesis of geopolymer concrete. Nowadays, the use of fly ash as an aluminosilicate source material becomes more beneficial because fly ash is an abundant waste material that harmful to the environment if not used. As the waste material derived from the coal combustion, fly ash give various properties both chemically and physically, which significantly influences the synthesis of the geopolymers [5].

According to ASTM C168, fly ash was divided into two classes, class F (low calcium) and class C (high calcium). When working with class F fly ash, the geopolymer concrete produced an exceptional performance both in compressive strength and durability against severe environment [6,7]. However, when dealing with class C fly ash or fly ash with high calcium content, the fast setting time problem arises due to the presence of calcium oxide [8]. While class F fly ash-based geopolymer concrete generally needs heat to start the geopolymerization process, the class C fly ash-based geopolymer set rapidly in minutes after the alkali solution introduced into the solid mixtures, even in the room temperature (±25°C).

A previous study showed that setting time of the high calcium fly ash-based geopolymer concrete influenced by both physical and chemical composition of the fly ash. It also showed that the initial temperature of the fly ash affects the setting time of the mixture, where a higher temperature of the fly ash leads to faster setting time [9]. Some research also shows that the short setting time problem could occur because of the significant amount of calcium compound in the fly ash. The presence of calcium compound makes the setting time of the geopolymer governed by the formation of CSH or CASH-like phases [10]. This rapid or short setting time problem could be a challenging problem when the high calcium fly ash-based geopolymer concrete is applied in the field.

Several studies have been conducted to provide alternatives solution addressing these setting-time problems. One study shows that increasing NaOH concentration could prolong the setting time of the geopolymer paste [11]. Another study shows a partial replacement of fly ash with diatomite or diatomaceous earth clay is beneficial as it prolongs the setting time. However, the compressive strength and the modulus of elasticity of the paste are reduced [12]. Another attempts to overcome this problem is using chemical additives like sucrose [13], borax [14], acid solutions [15]. However, some of the alternatives have a negative side effect on the other properties of the geopolymers, such as decreasing compressive strength.

Depart from the fact that the chemical reaction involved in the polymerization process is an exothermic reaction, in addition, using cold water in conventional concrete, that use Portland cement, has also been generally applied to reduce the rate of hydration and extend the setting time of the concrete, especially when working with mass concrete. This study was done to evaluate the cooling effect of both the solid material and the alkaline solution to the setting time of the geopolymer paste and also the compressive strength of the hardened mortar.

2. Materials, mix-design, and methods

2.1. Materials and mix design

All geopolymer made in this study use fly ash obtained from Paiton steam power plant in East Java, Indonesia as the source of aluminosilicate material. An XRF test (showed in Table 1) was done to find out the chemical composition of the fly ash and the Loss on Ignition (LoI) value is also shown. A pH value of 11.2 (tested using a pH meter immersed in a solution consisting of 20 grams of fly ash and 80 grams of distilled water) also confirmed the high calcium content in the fly ash. The physical properties of the fly ash given by PSA (Particle Size Analysis) showed that 10% of the fly ash particle having a diameter smaller than 0.67 μm, 50% of the fly ash particle having diameter smaller than 6.72 μm, and 90% of the fly ash particle having diameter smaller than 35.95 μm. The fly ash had a specific surface area (SSA) of 570.14 cm²/g and the specific gravity of 2.753.

Table 1. Chemical composition of the fly ash from XRF test

Oxide	SiO_2	Al_2O_3	Fe_2O_3	CaO	MgO	K_2O	Na_2O	SO_3	LoI
% weight	34.29	16.62	15.38	18.18	7.52	1.35	2.97	1.63	0.36

Alkali solution was prepared using sodium hydroxide (NaOH) and type 52 sodium silicate (Na_2SiO_3). 8M sodium hydroxide solution was made by dissolving 24 grams of sodium hydroxide flakes into distilled water until the solution volume reaches 75ml, while the sodium silicate used was 60 grams, so that the mass ratio of 2.5 between sodium silicate liquid and sodium hydroxide solid was obtained. All the alkali solution was kept until reaching room temperature before further process. Then, for some alkali solution was kept in the freezer for a specific time to decrease its temperature before mixed with the fly ash. For the geopolymer paste, the mixture consists of 300 grams of fly ash, 75 ml of sodium hydroxide, and 60 ml of sodium silicate. Sand from Lumajang quarry, East Java, Indonesia was used to make geopolymer mortar with fly ash and sand mass ratio of 1:2.

2.2. Methods and testing

Alkali solutions with five different temperatures were used to make five sets of geopolymer paste. Right before the alkali solution come into contact with the fly ash, the temperature of the solution was measured using a digital thermometer. Both the initial and the final setting time of each geopolymer paste was tested using the Vicat apparatus as described by ASTM C191 [16]. For compressive strength test, geopolymer mortar was made and cast into 5×5×5 cm cube molding. The compressive strength test was carried out by crushing three cube samples at testing age of 3, 7, 14, and 28 days.

3. Results and discussions

The initial alkali solution temperature was 31°C in the room temperature and before entered the freezer for further cooling. This alkali solution was also used to make the control geopolymer paste. The initial setting time of the control geopolymer paste was about 14 minutes, while the final setting time was 26 minutes. The initial setting time was similar to the previous study for a similar pH of fly ash [17]. This result confirmed that the setting time of high calcium fly ash-based geopolymer is strongly affected by the pH value of the fly ash.

Table 2. Initial and final setting times for different activator solution temperature

Activator Solution Temperature (°C)	Initial Set (mins)	Final Set (mins)	Set Temperature of the paste (°C)	
			Outside	Inside
31	14	26	-	-
22	20	-	28	30.7
19	25	48	-	-
9	28	42	-	-
7	36	-	25	28.8

By cooling the alkali activator solution, it was shown that both the initial and the final setting times of the geopolymer paste could be prolonged. Figure 1 shows the strong correlation between the setting time of the geopolymer paste and the alkali activator solution temperature. The cooling treatment might prolong the setting time of the geopolymer paste probably because, in a lower temperature, the chemical reaction runs slowly, especially in the general polymerization. However, polymerization reactions, especially in the condensation phase, require heat to be able to take place. Figure 2 confirmed that a geopolymer paste was not able to harden when it stored in the freezer, while the sample stored in the room temperature starts to set after the geopolymer paste reaches 25°C at about 40 minutes.

Two geopolymer mortars were made to investigate the effect of cooling treatment towards the geopolymer compressive strength. The control sample (FAG-CT) was synthesized using materials in the room temperature condition. For the second sample (FAG-CL), all materials were pre-treated by storing in the freezer until the temperature of the material close to 0°C before finally being synthesized.

Table 3 shows the compressive strength for each testing age and also the average density of each sample. The higher density on the FAG-CL sample indicates that the fresh mortar has more workability compared to the control sample. The 28-days compressive strength of the FAG-CL sample was 65.20 MPa, and it was about 20 percent higher when compared to FAG-CT sample, which has the compressive strength of 54.27 MPa. Higher compressive strength of the FAG-CL sample could be either because of the increased density or because of the better chemical reaction processes. Not only at 28-days compressive strength, but the cooling treatment also provides higher compressive strength at an early age of 3 days.

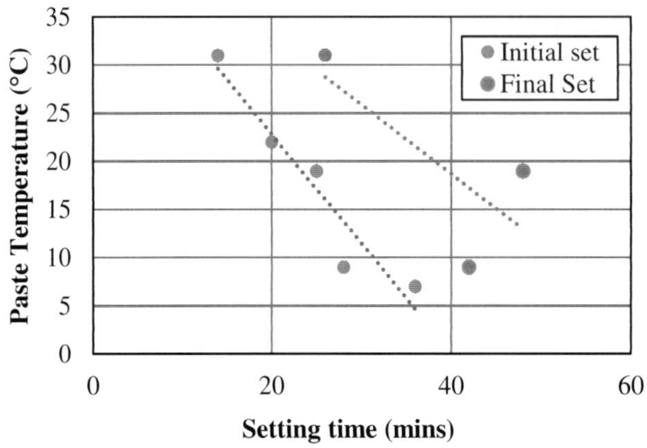

Figure 1. Setting time of geopolymer pastes

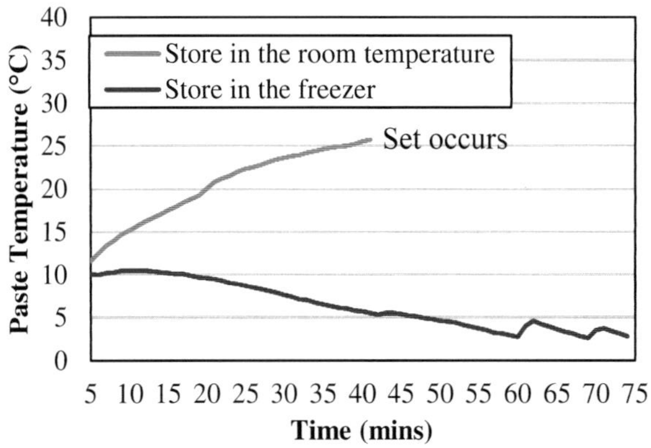

Figure 2. The internal temperature of the geopolymer paste

Table 3. Compressive strength & density of geopolymer mortar

Sample	Compressive Strength (MPa)				Density (kg/m³)
	3 days	7 days	14 days	28 days	
FAG-CT	30.80	37.33	47.60	54.27	2.344
FAG-CL	38.67	47.47	53.33	65.20	2.417

In contrast to the low calcium fly ash-based geopolymers whose the compressive strength of the concrete is not influenced by age [6], high calcium fly ash-based geopolymers show the development of strength along with age. As shown in Figure 3, both the FAG-CT samples and the FAG-CL samples showed strength development as well as conventional concrete (which use Portland cement). For FAG-CT sample, the compressive strength development for 3, 7, and 14 days was 56.75%, 68.78%, and 87.71% respectively. While for FAG-CL sample, the compressive strength development for 3, 7, and 14 days was 59.31%, 72.81%, and 81.79%. The strength development indicates that the chemical reaction occurred in high calcium fly ash-based geopolymer is not only the polymerization but also the hydration reaction due to the presence of calcium compound in the fly ash.

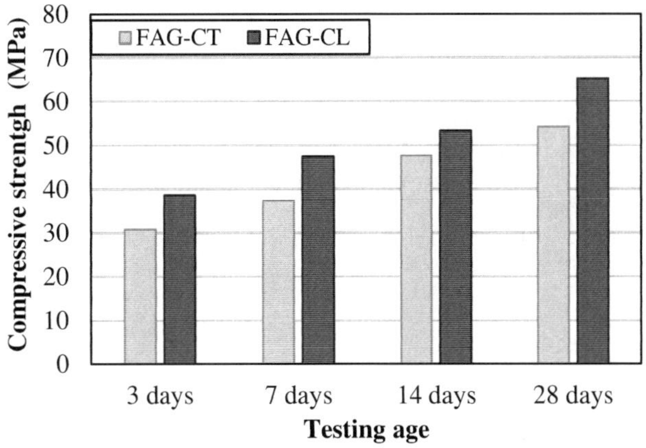

Figure 3. Strength development of geopolymer mortar

4. Conclusions

From the above study, it can be concluded that the cooling treatment of alkali solutions and the constituting materials can prolong the setting time of high calcium fly ash-based geopolymer. The lower temperature of the material would prolong the initial and final setting time of the geopolymer. However, it should be noted that high calcium fly ash-based geopolymer cannot harden under freezing conditions i.e., below 0°C. The cooling treatment can also increase the compressive strength of the geopolymer mortar both at an early age and at 28-days compared to the room temperature specimens.

In an industrial perspective, using high calcium fly ash as geopolymer precursors could give more benefit both in saving time and cost because it does not need the application of heat to start the geopolymerization reaction, and the casting process can be shortened.

Regarding the rapid setting time problem of high calcium fly ash-based geopolymer concrete, by using cooling treatment, it is more likely to be applied in the precast concrete industry or for repair material, where short setting time and high early age compressive strength are expected. In addition, the cooling treatment is also a more promising method to prolong the setting time of high calcium fly ash-based geopolymer concrete compared to the use of chemical additive as it did not reduce the compressive strength of the concrete.

References

[1] Davidovits J 1991 *J Therm Anal* **37** no 8 1633–1656
[2] Palomo A, Grutzeck M W, and Blanco M T 1999 *Cem Concr Res* **29** no 8 1323–1329
[3] Berry M, Cross D, and Stephens J 2009 World of Coal Ash Conference
[4] Davidovits J 2015 Geopolymer Chemistry & Application (Morrisville: Institute Geopolymere)
[5] Wattimena O K, Antoni, and Hardjito D 2017 *AIP Conference Proceedings* **1887** 020041
[6] Hardjito D and Rangan B V 2005 *Development and properties of low-calcium fly ash-based geopolymer concrete* (Perth, Australia: Curtin University of Technology)
[7] Song X J, Marosszeky M, Brungs M, and Munn R 2005 *Int. Conf. Durab. Build. Mater. Components* LYON 17–20
[8] Topark-ngarm P, Chindaprasirt P, and Sata V, 2014 *J. Mater. Civ. Eng* **27** no. 2011 1–7
[9] Antoni, Wijaya S W, and Hardjito D 2016 Mater. Sci. Forum 841, 90–97
[10] Chindaprasirt P, De Silva P, Sagoe-Crentsil K, and Hanjitsuwan S 2012 *J. Mater. Sci* **47** no. 12 4876–4883
[11] Hanjitsuwan S, Hunpratub S, Thongbai P, Maensiri S, Sata V, and Chindaprasirt P 2014 Cem. Concr. Compos **45** 9–14
[12] Phoo-Ngernkham T, Chindaprasirt P, Sata V, and Sinsiri T 2013 Indian J. Eng. Mater. Sci **20** no.4 310–318

[13] Kusbiantoro A, Ibrahim M S, Muthusamy K, and Alias A 2013 Procedia Environ. Sci **17** 596–602
[14] Antoni, Wijaya S W, Satria J, Sugiarto A, and Hardjito D 2016 *Mater. Sci. Forum* **857** 416–420.
[15] Antoni, Herianto J G, Anastasia E, and Hardjito D 2017 *AIP Conf Proc* **1887** 020042
[16] ASTM C191-18a 2018 *ASTM International*
[17] Antoni, Satria J, Sugiarto A, and Hardjito D 2017 *MATEC Web Conf* **97**, 01026

Setting time and compression strength of low-calcium fly ash geopolymer paste with non-calcined red soil substitution in ambient curing

P H Simatupang[1], V Nine[1] and A E Sinaga[1]

[1] Civil Engineering Department, Faculty of Science and Engineering, University of Nusa Cendana, Kupang, Indonesia

partogihsimatupang@gmail.com

Abstract. This paper describes the research of the effects of non-calcined red soil substitution in low-calcium fly ash geopolymer paste in ambient curing. Mechanical characteristics observed were setting time and compression strength. The non-calcined red soil substitutions were varied i.e 0%, 5%, 25% and 50% of mass of fly ash. The alkaline solutions used were natrium silicate solution and natrium hydroxide solution 10 M with mass ratios 1:1 and 2:1. The results show that the higher the substitution of non-calcined red soil the longer the initial and final time setting of the geopolymer paste. Meanwhile, the higher the substitution of non-calcined red soil the lower the compression strength of the geopolymer paste. Furthermore, geopolymer paste with ratio of Na_2SiO_3 / NaOH = 2.0 will has longer initial and final setting than geopolymer paste with ratio of Na_2SiO_3 / NaOH = 1.0. Geopolymer paste with ratio of Na_2SiO_3 / NaOH = 2.0 will has higher compression strength than geopolymer paste with ratio of Na_2SiO_3 / NaOH = 1.0. All the relationships between the substitution of non-calcined red soil and setting time or compression strength of geopolymer paste can be drawn in exponential function.

1. Introduction

Geopolymer paste is a relatively new binder to make a solid material. This is the most promising binder to replace the cement paste. Low-calcium fly ash and other industrial by product based geopolymer material in Indonesia have been studied i.e: compression strength of geopolymer mortar [1], characteristics of geopolymer paste in sulphuric acid solution [2], pigment in geopolymer paste [3] and environmental aspects of geopolymer material [4].

Clay or soils based geopolymer materials studied were dominantly using calcination of the materials to get the amorphous silica-alumina precursor [5,6]. The calcination of raw material produced good geopolymer solidification. However, the researches on non-calcined clay or soil (including dredged soil and red soil/laterite soil) geopolymer material were relatively rare [7-11]. Several papers revealed that clays or soils have enough SiO_2 and Al_2O_3 to make geopolymer material i.e SiO_2 47.49-70.23% and Al_2O_3 11.79-35.69% [5], SiO_2 31.7-46.7% and Al_2O_3 10.6-23.04% [6] and SiO_2 61.56% and Al_2O_3 22.53% [7]. Meanwhile, in Nusa Tenggara Timur Province Indonesia, red soil is a raw material to make red fired brick. Red fired brick is always produced by burning the red soils in the furnace.

Several studies on the non-calcined red soils geopolymer mortar have presented sufficient compression strength i.e 7.73 MPa [7], 1.4-7 MPa [10]. However to increase its strength and durability of non-calcined red soil geopolymer material, it was recommended to use fly ash as a main precursor.

IOP Conf. Series: Materials Science and Engineering **669** (2019) 012025 doi:10.1088/1757-899X/669/1/012025

Until now, there were very few papers conducting research on the characteristics of setting time and compressive strength of low-calcium fly ash geopolymer paste with non-calcined red soils substitution.

Therefore, this study focused on setting time and compressive strength of low-calcium fly ash geopolymer paste with non-calcined red soil substitution. Red soil used as solid substitution was non-treatment soil from Desa Obelo Kupang District, Nusa Tenggara Timur Province, Indonesia. To make the geopolymer paste, it was used Low-Calcium or Class F Fly Ash from Bolok-Kupang Coal Fired Power Plant as a rich silica-alumina of precursor material. Meanwhile, the mass ratio of natrium silicate solution to natrium hydroxide solution 10 M was 1:1 and 2:1. The mass ratio of solid (fly ash+non-calcined red soil) to liquid (alkaline activator solution) was 2.0. The characteristics of setting time and compressive strength of low-calcium fly ash geopolymer paste with non-calcined red soils substitution were obtained.

2. Experimental details

This subsection will present raw material used, sample made, and testing conducted. There was no treatment to fly ash and non-calcined red solid as raw materials. Testing conducted were: (1) the setting time testing and (2) compression strength testing.

2.1. Materials

Low-calcium fly ash from Power Plant Bolok Kupang was used as precursor material to make geopolymer paste. To substitute fly ash, it was used non-calcined red soil from Desa Obelo Tanah Merah in Kupang District-Nusa Tenggara Timur Province, Indonesia. These materials are shown in Figure 1. There is no treatment to fly ash. Before usage of non-calcined red soil, this soil should be dried in room temperature until 24 hours.

Figure 1. Raw material used: Low-Calcium Fly Ash (left) and Non-Calcined Clay (right)

The mixture composition of geopolymer paste was low-calcium fly ash, red soil and alkaline activator solution with certain mass ratios. Alkaline activator solution was a mixture of natrium silicate and 10 M sodium hydroxide solution with mass ratio 1: 1 and 2:1. Class F or low-calcium fly ash from Power Plant Bolok Kupang was used as a precursor material to make geopolymer. This fly ash had CaO 3.64%, SiO_2 57.47 % and Al_2O_3 21.88 %. More detail oxide compositions by XRF (X-Ray Fluorescence) testing is given in Table 1. Meanwhile, XRD (X-Ray Diffraction) pattern of fly ash can be shown in Figure 2. Many minerals found in the fly ash were Quartz, Gypsum, Hematite and Magnitite. It was not conducted microscopic characteristics on non-calcined red soil.
Some text.

Table 1. Oxide Compositions of Fly Ash by XRF (X-Ray Fluorescence) Testing.

Oxide	MgO	Al_2O_3	SiO_2	K_2O	Na_2O	CaO	TiO_2	MgO	Fe_2O_3	LOI
Wt (%)	1,41	21,88	57,47	1,59	2,69	3,64	0,81	1,41	3,61	1,77

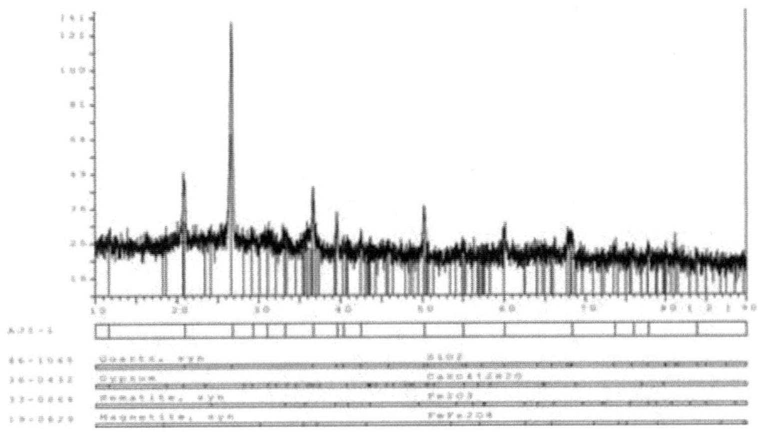

Figure 2. X-Ray Diffraction pattern of fly ash

2.2. Mixture compositions, mixing, and casting

Sample used were cylindric paste with diameter 2.5 cm and 5 cm height. After fresh geopolymer paste casted, the samples were cured in ambient curing. Composition of geopolymer paste studied is given in Table 2. The substitutions of non-calcined red soil to fly ash are 0%, 5%, 25% and 50%. Every mixture was prepared i.e 9 cylindric pastes (three samples for compression testing on 14 days, three samples for testing on 28 days and others for laboratory assets) and 1 sample of setting time.

The mixing and casting of geopolymer paste is shown in Figure 3. To make geopolymer paste, it was mixed fly ash, non-calcined red soil and alkali activator solution by using Paint Mixer with 1800 rpm for 2-3 minutes. After, geopolymer paste was ready, it was poured in plastic formwork. Then, samples were cured with ambient curing.

Table 2. Composition of geopolymer paste.

Mixture	FA[a] (gr)	Non-Calcined Clay (gr)	NS[b] (gr)	NH [c] (10 M) (gr)	Ratio FA/ (NS+NH)	Ratio NS/NH	Substitution of Non-Calcined Clay to FA
A-AC-1	650	0	162.5	162.5	2	1	0 %
A-AC-2	600	0	216.7	108.3	2	2	0 %
B-AC-1	570	30	162.5	162.5	2	1	5 %
B-AC-2	570	30	216.7	108.3	2	2	5 %
C-AC-1	450	150	162.5	162.5	2	1	25 %
C-AC-2	450	150	216.7	108.3	2	2	25 %
D-AC-1	300	300	162.5	162.5	2	1	50 %
D-AC-1	300	300	216.7	108.3	2	2	50 %

[a] Fly Ash
[b] Natrium Silicate
[c] Natrium Hydroxide

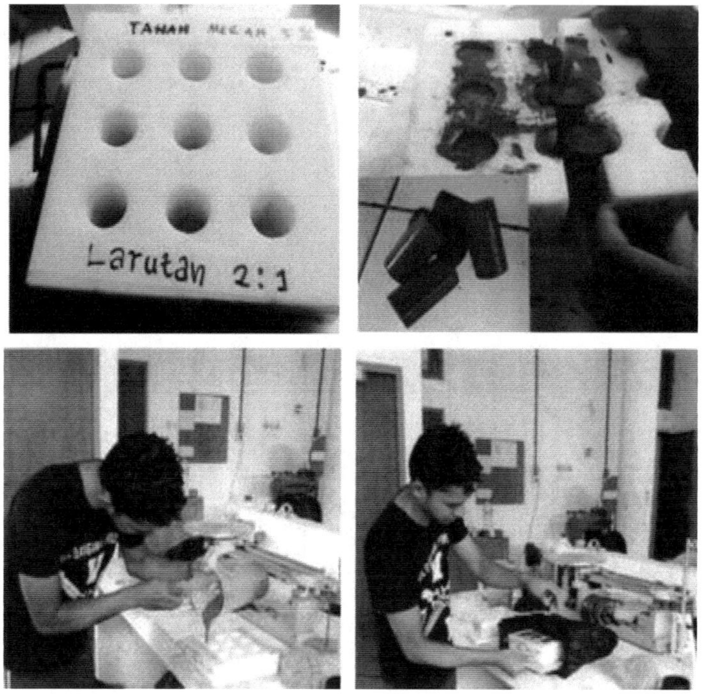

Figure 3. Mixing and casting of geopolymer paste.

2.3. Testing

2.3.1. Setting time testing. The setting time of geopolymer paste was conducted using Vicat apparatus like setting time testing on cement paste in ASTM C191-13. The setting time testing is given in Figure 4. Every 15 minutes penetration test was taken using 1-mm Vicat needle until the final setting occurred. Then initial setting time could be found by determining 25 mm of penetration. The results of setting time were the mean of two observations of penetration depth on two different locations.

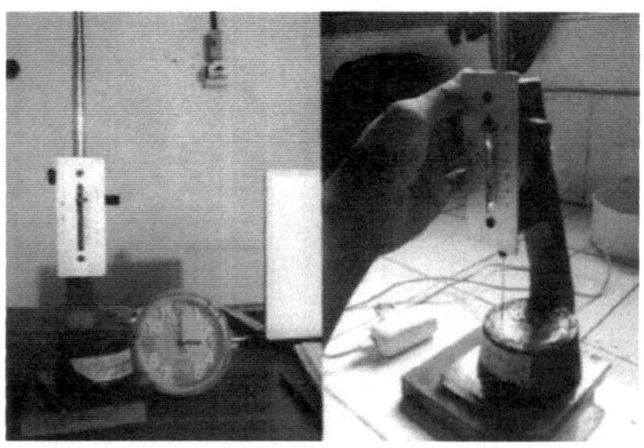

Figure 4. Setting time testing of geopolymer paste

2.3.2. Compression strength testing. Hardened geopolymer pastes were tested using Compression Testing Machine with capacity 312.5 KN. The compression strength testing is given in Figure 5. Compression testings were conducted for 14 days and 28 days age of samples. The results of compressive strength were the mean of three samples for every variable.

Figure 5. Compression strength testing of geopolymer paste.

3. Results and discussion

3.1. *Setting Time*

The results of setting time testing are given in Figure 6. Setting time testings were completely conducted except sample D-AC-2. The setting time of geopolymer paste D-AC-2 was not finished until 84 hours. So, the setting time testing was stopped and the end value of setting time was extrapolated. The results of initial and final setting time of geopolymer paste are shown in Figure 7. It is shown that the more the substitution of non-calcined red soil the high the initial and final setting time of geopolymer paste. The initial and final setting time depend on the ratio of Na_2SiO_3 to NaOH solution. Geopolymer paste with ratio of Na_2SiO_3 / NaOH =2.0 (A-AC-2, B-AC-2, C-AC-2 and D-AC-2) will has longer initial and final setting than geopolymer paste with ratio of Na_2SiO_3 / NaOH =1.0 (A-AC-1, B-AC-1, C-AC-1 and D-AC-1).

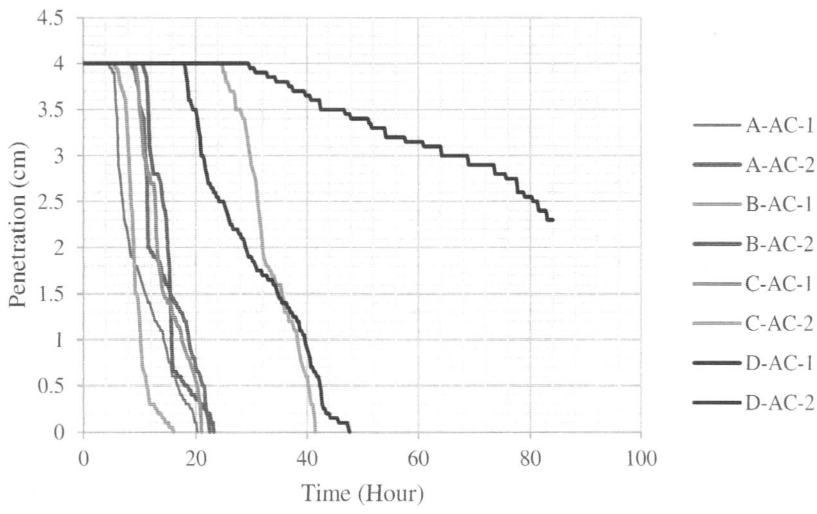

Figure 6. Setting time of geopolymer paste

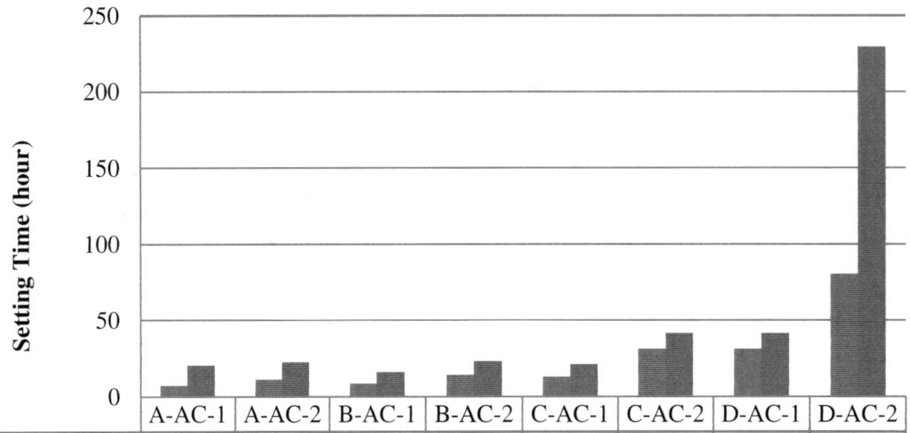

	A-AC-1	A-AC-2	B-AC-1	B-AC-2	C-AC-1	C-AC-2	D-AC-1	D-AC-2
■ initial setting time	7	11.3	8.5	14	12.75	31	31	80.5
■ final setting time	20.25	22.5	16	23.25	21	41.5	41.5	229.31

Figure 7. Initial and Final setting time of geopolymer paste

The relationships of setting time and substitution of non-calcined red soil are given in Figure 8. and Figure 9. Those relationships can be drawn as the exponential functions.

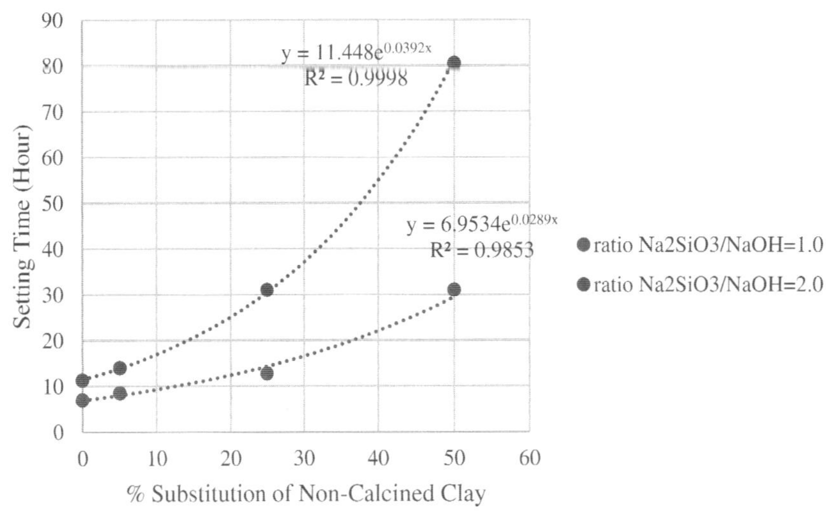

Figure 8. Relationships of Initial Setting and substitution of non-calcined red soil.

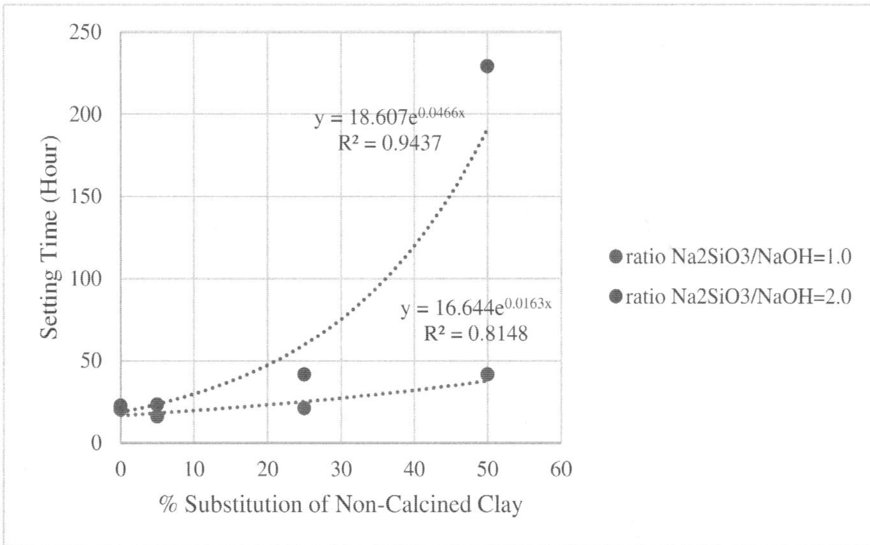

Figure 9. Relationships of Final Setting and substitution of non-calcined red soil

3.2. Compression Strength

The results of compression strength testing are given in Figure 10 and Figure 11. It is shown that the more the substitution of non-calcined red soil the low the compression strength of geopolymer paste. The compression strengths also depend on the ratio of Na_2SiO_3 to NaOH solution. Geopolymer paste with ratio of Na_2SiO_3 / NaOH =2.0 (A-AC-2, B-AC-2, C-AC-2 and D-AC-2) will has higher compression strength than geopolymer paste with ratio of Na_2SiO_3 / NaOH =1.0 (A-AC-1, B-AC-1, C-AC-1 and D-AC-1).

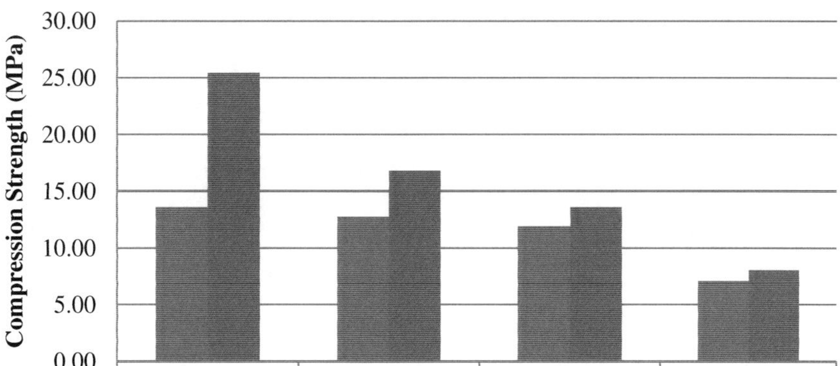

	A-AC-1	B-AC-1	C-AC-1	D-AC-1
■ 14 days	13.58	12.73	11.88	7.08
■ 28 days	25.45	16.80	13.58	8.06

Figure 10. Compression strength of geopolymer paste with ratio $Na_2SiO_3/NaOH$ = 1.0.

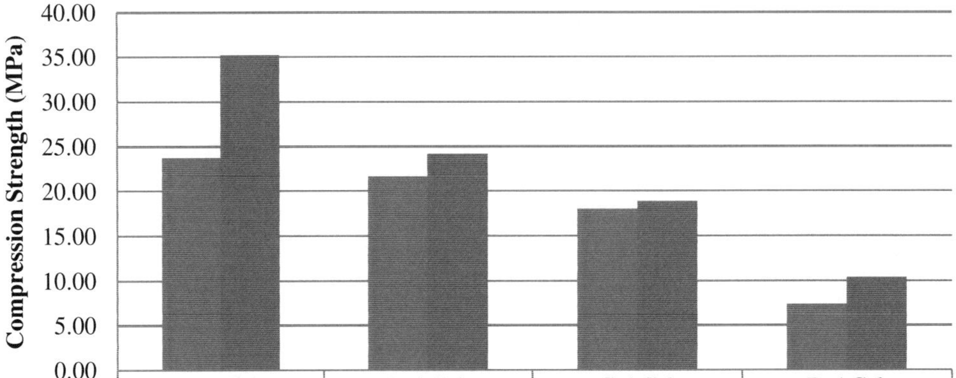

	A-AC-2	B-AC-2	C-AC-2	D-AC-2
■ 14 days	23.76	21.64	18.03	7.38
■ 28 days	35.21	24.18	18.88	10.39

Figure 11. Compression strength of geopolymer paste with ratio $Na_2SiO_3/NaOH = 2.0$.

The relationships of compression strength and substitution of non-calcined clay are given in Figure 12 and Figure 13. Those relationships can be also drawn as the exponential functions like the relationships between setting time and substitution of non-calcined clay.

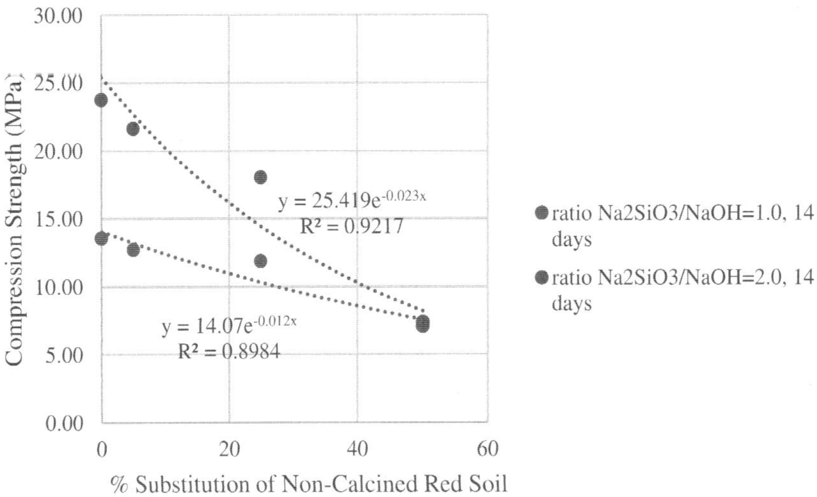

Figure 12. Relationship between compressive strength and substitution of non-calcined red soil for age 14 days

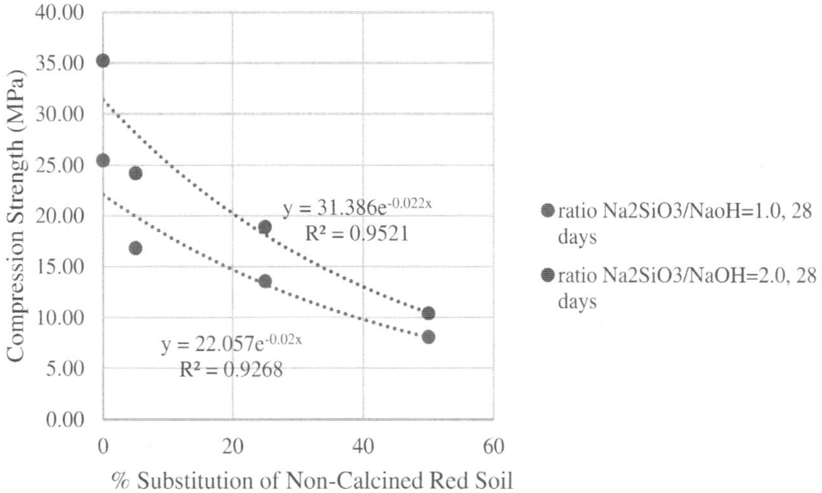

Figure 13. Relationship between compressive strength and substitution of non-calcined red soil for age 28 days.

The correlation between setting time and compressive strength of those mixtures can be described as follows. The longer the initial setting time of the paste, the lower the compressive strength of the paste. Similarly, the longer the final setting time of the paste, the lower the compressive strength of the paste. This correlation does not depend on the ratio of natrium silicate solution to natrium hydroxide solution.

4. Conclusions

It has been shown characteristics of setting time and compression strength of low-calcium fly ash geopolymer paste with non-calcined red soils substitution. The results show that the higher the substitution of non-calcined red soil the longer the initial and final time setting of the geopolymer paste and the higher the substitution of non-calcined red soil the lower the compression strength of the geopolymer paste. Furthermore, geopolymer paste with ratio of Na_2SiO_3 / NaOH =2.0 will has longer initial and final setting than geopolymer paste with ratio of Na_2SiO_3 / NaOH =1.0. Geopolymer paste with ratio of Na_2SiO_3 / NaOH =2.0 will has higher compression strength than geopolymer paste with ratio of Na_2SiO_3 / NaOH =1.0. All the relationships between substitution of non-calcined red soil and setting time or compression strength can be drawn in exponential function.

References

[1] Simatupang P H, Imran I, Pane I, and Sunendar B 2015 J.Eng.Technol.Sci **47** No.3 231-249.
[2] Simatupang P H 2017 *AIP Conference Proceedings* 1887 020028 doi **10**.1063/1.5003511.
[3] Simatupang P H, Hanafi R, Sunendar B, Imran I and Pane I 2012 *Advanced Materials Research* **450-451** 281-285.
[4] Simatupang P H 2018 *CSID Journal of Sustainable City and Urban Development* **1** Issue 1 23-30.
[5] Mohsen Q, Mostafa N Y 2010 *Ceramic Silikaty*. **54** (2), 160–168.
[6] Obonyo E A, Kamseu E, Lemougna P N, Tchamba A B, Melo U C, and Leonelli C 2014, *Sustainability*. **6** 5535–5553.
[7] Uddin M N, Saraswathy V 2018 *Civil Engineering and Architecture*. **6** (1), 34–39`.
[8] Maskell D, Heath A, Walker P J 2014 *Key Engineering Materials*. **600**175–185.
[9] Dungca J R, Ang K D, Isaac A M L, Joven J J R, Sollano M B T 2019 *International Journal of GEOMATE* **16**, Issue 57 9–14.

[10] Uddin M N, Saraswathy V, Elumalai P 2015 *International Journal of Scientific and Engineering* Research **6** Issue 5.
[11] Sreelakshmi S, Kalayan Kumar G, Krishnaiah E 2017 *Proceedings of Indian Geotechnical Conference/ GeoNEst* (India : IIT Guwahati).

The use of natural fiber from oil palm empty fruit bunches for soft soil stabilization

Y F Arifin[1], Misnawati[1] and E Normelani[2]

[1] Civil Engineering Department, University of Lambung Mangkurat, Banjarbaru
[2] Geography Education Department, University of Lambung Mangkurat, Banjarmasin

y.arifin@ulm.ac.id

Abstract. The use of natural fiber as a construction material, especially soil stabilization continues to grow. This paper focuses on the use of natural fiber to increase shear strength and bearing capacity of soft soil. The fiber used is obtained from oil palm empty fruit bunches (EFB) as a by-product of palm oil mills. Soft soil mixed with fiber with a composition of 5, 6, 7, and 8% fiber in the mixture. Some tests such as the standard compaction test, unconfined compression test, laboratory vane test, and California Bearing Ratio test were carried out. The results show that the soft soil used in this study can be compacted at fiber content higher than 5%. The maximum density obtained is 0.92 g/cm^3 at 7% fiber content. The compacted soil-EFB mixtures successfully increase the shear strength and bearing capacity of the soft soil shown by the results of the UCT, laboratory vane, and CBR tests. The soil consistency changes from soft to medium soil. The maximum q_u, s_u, and CBR obtained are 0.8 kg/cm^2, 0.65 kg/cm^2, and 6%, respectively at optimum fiber content of 6 to 7%.

1. Introduction

The use of fiber for construction materials has been carried out, especially for increasing the strength of concrete [1, 2, 3]. Addition of steel fiber to concrete increases strength and maximum displacement, and also reduces the number of cracks in concrete [2]. Fiber is also developed for soil stabilization using either synthetic fibers such as tire shredders [4], Nylon fiber [5], polypropylene fiber [6, 7], glass fiber [8], basalt fiber [9], as well as natural fibers such as coir fiber [10], wheat straw, barley straw, and wood shaving [11], and bamboo fiber [12].

Ignoring whether fiber is used synthetic or natural, soil stabilization with fiber is influenced by fiber content [4, 5, 6, 7, 8, 9, 11, 12, 14,16], fiber length [7, 9, 12, 16], water content [14], fiber characteristics [15, 8, 11], fiber diameter [12], soil properties [5, 15], and stress conditions [4, 10, 15].

The interaction between soil and fiber is interesting to learn in improving the nature of soil engineering. [13] stated that fiber-reinforced soil shear strength has two components, including the shear strength of the soil matrix and the tensile stress acting on the fibers. Besides that, [14] stated that the bonding of soil and fiber caused the contribution of fiber to the increase in shear strength to the pull out the mechanism and the tensile strength of the fiber itself. This mechanism explains the interaction of soil and fiber in general, where other possible interactions occur between soil and fiber, especially for natural fibers. [11] found that natural fiber absorbs more water than the soil. This behavior is needed, especially for the stabilization of soft soils that have high water content.

Some other advantages of using natural fiber are environmentally friendly alternatives, locally available, can create composites with cement/lime, inexpensive, and biodegradable [5, 17]. Therefore, this paper discusses the alternative use of natural fiber from oil palm empty fruit bunches (EFB) for stabilizing soft clay soils. The fibers are a byproduct of the palm oil industry. The use of this fiber as an alternative to soft soil stabilization is rarely discussed. [18] has succeeded in increasing the strength of laterite brick slightly by adding 3% EFB in the mixture of laterite, sand, and cement.

For applications in the field, fiber reinforcement is possible to repair slopes and strengthen thin layers of soil where synthetic materials such as geotextiles and geogrids are challenging to implement [13]. Fiber can also be used to stabilize subgrade soil from sand to high plasticity clay [20]. Therefore, this paper focuses on providing new information about the possibility of using natural fiber from an oil palm empty fruit bunch as soft clay reinforcement material.

2. Material used

2.1. Soil

The soft soil used was taken near the city of Banjarmasin, the capital of South Kalimantan. The soft soil characteristics are summarized in Table 1. The physical properties were determined using the tests based on ASTM standards [21]. The soft clay was classified as an organic soil with high plasticity (OH) based on USCS classification system.

Table 1. Summary of soft soil characteristics used

Properties		Soft soil
Specific gravity		2.31
Water content	%	106.07
Liquid limit	%	61.00
Plastic limit	%	34.87
Plasticity index	%	26.13
Fine content	%	95.12
Clay content	%	56.32
Soil type	USCS	OH

2.2. Fiber

The fiber used is oil palm empty fruit bunches (EFB) as a byproduct of palm oil processing in PT. Perkebunan Nusantara XIII, Pleihari (Figure 1 (a)). The fiber has a water content of 9.8% and the density of 0.45 g/cm^3, as shown in Table 2. The fiber density is smaller than that reported by [18]. However, the value is very close to the realistic density of natural fibers of coil and sisal (i.e., 0.67-1.07 g/cm^3) as reported by [19]. The fiber has a diameter of 200-500 μm measured from the SEM image (Figure 1(b)) and rough surface, as shown in Figure 1 (c).

A test was carried out to determine the water absorption of EFB used by soaking it for one, two, and three days. The results are shown in Table 2. After soaking, the fiber moisture content increases to 384.99% and does not change significantly on the second and third days with water levels of 422.71% and 415.87%, respectively. The result confirmed previous findings that the fibers absorb more water than soil and continue to increase with increasing fiber content [11].

3. Techniques and procedures

In sample preparation, the size of fiber used varies depending on the test performed. For compaction, CBR, and vane tests, the fiber length used was a maximum of 10 cm or smaller than the mold size used (i.e., 11.51 cm). For the UCT test, a 1 cm fibers size was used to adjust the diameter of the sample used (i.e., 4.77 cm).

Because the soft soil used had very high water content, the initial test was performed to determine the percentage of fiber at which compaction can be carried out. Based on trial tests, the mixture can be

IOP Conf. Series: Materials Science and Engineering **669** (2019) 012026 doi:10.1088/1757-899X/669/1/012026

compacted at the fiber content of 5%. The compositions of fiber used in this study were 5, 6, 7, and 8%. Moreover, four types of specimens were prepared with different EFB contents (on a dry mass base); namely, 5% EFB, 6% EFB, 7% EFB, and 8% EFB. The specimens were dynamically and statically compacted depend on the tests performed.

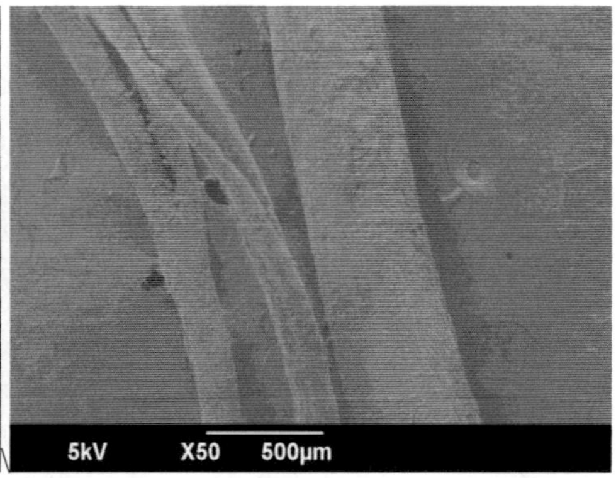

(a) (b)

(c)

Figure 1. EFB fiber (a) raw material, and SEM image with a magnification of (b) 500×, and (c) 1500×

Table 2. Summary of fiber from empty fruit bunches (EFB) characteristics used

Properties		EFB fiber
Water content	%	9.8
Density	g/cm^3	0.45
Water absorption:		
one day	%	384.99
two days	%	422.71
three days	%	415.87

Some test techniques used include standard Proctor compaction test, California bearing capacity (CBR) test, and laboratory vane tests (using dynamically compacted samples) and unconfined compression tests (using statically compacted samples).

To ensure fiber distribution in the sample is evenly distributed, a trial test was conducted on the sample by dividing it into three parts, namely top, middle, and bottom. The soil and samples were separated by soaking in water. The dry weight of fiber and soil were compared. The results are shown in Table 3. The fiber contents in the sample were evenly distributed with an average of 5.07% and 5.99% for samples of 5% EFB and 6% EFB, respectively.

Table 3. Fiber distribution in the samples

Fiber percentage plan	5%	6%
Top	5.12	5.96
Middle	5.02	6.02
Bottom	5.08	5.99
Average	5.07	5.99

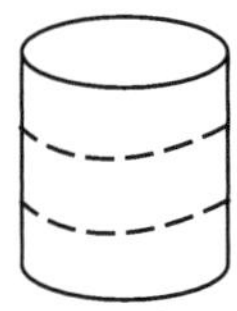

4. Results and discussions

4.1. Compaction test

Table 4 shows compaction test data using standard Proctor for samples with fibers content of 5, 6, 7, and 8%. Compaction was performed by three layers and 25 blows per layer. There are two fundamental things obtained from Table 4; there are changes in water content and sample density. The water content of the sample decreases with increasing fiber content to a 7% EFB, as shown in Figure 2 (a). This reduction in water content was due to the addition of fiber that fills the soil pores. The figure also shows that the sample that can be compacted has moisture content smaller than the liquid limit. At high water content, the pore water pressure increases in compaction proses so that fiber does not contribute much to this condition. This agrees well with the findings reported by [14].

Table 4. Optimum fiber content determination

Fiber content	%	5	6	7	8
No. of layers		3	3	3	3
No. of blows		25	25	25	25
Density	g/cm^3	1.365	1.392	1.382	1.307
Water content	%	62.40	56.98	51.05	51.51
Dry density	g/cm^3	0.84	0.88	0.92	0.87

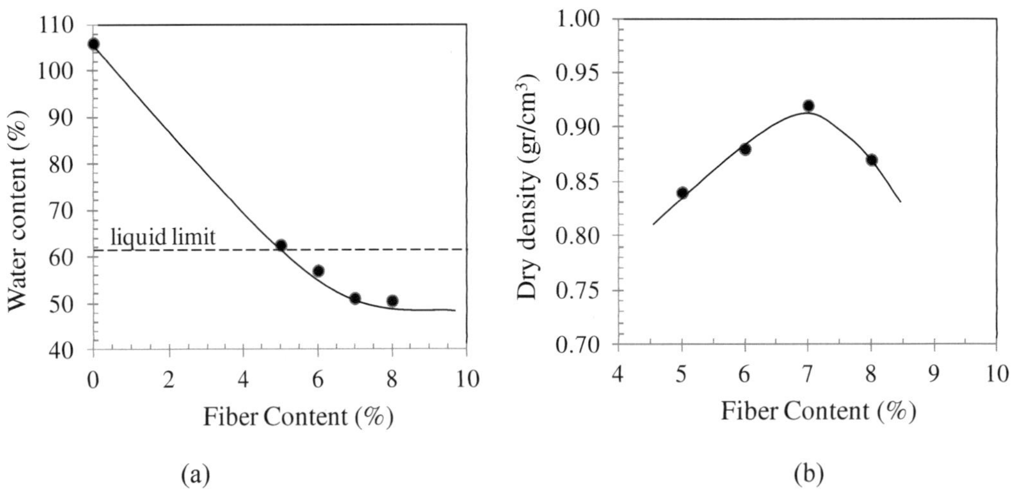

Figure 2. (a) Water content and (b) dry density of samples as a function of fiber contents

Figure 2 (b) shows sample density as a function of fiber content. Soil density increases to 7% fiber content with a density of 0.92 g/cm^3. The smaller density was obtained at 8% fiber content. When the soil has begun to be plastic (i.e., water content less than liquid limits) , the bond between fiber and soil has begun to form. The more fiber, the higher the soil-fiber bonding due to pull-out. While at 8% fiber content, the interaction of soil and fiber decreases because the amount of fiber reduces the bonds between soil and fiber. In this condition, compaction is difficult to perform, and the large pores appear in the sample. Visually, Figure 3 shows pictures of compacted samples for each percentage of fiber.

Figure 3. The photograph of samples compacted at different fiber content (a) 5% EFB, (b) 6% EFB, (c) 7% EFB, and (d) 8% EFB

4.2. Unconfined compression test

The unconfined test was performed to a statically compaction specimen due to the presence of fiber. The sample with a diameter of 4.75 cm and 9.24 cm high were used. The result obtained from the test was unconfined compression strength (q_u). Figure 4 shows the q_u as a function fiber content. Consistent with compaction result, the q_u increases by increasing fiber content. The peak of q_u (i.e., 0.8 kg/cm^2) was obtained at the fiber content of 7%. The value decreases at 8% fiber content. The result reveals that the presence of fiber changes the soil consistency from very soft to medium soil based on [22] (i.e., medium soil 0.48-0.96 kg/cm^2).

4.3. California bearing ratio test (CBR)

Figure 5 shows CBR as a function of fiber content. Different from compaction and q_u results, the maximum CBR of 6% was obtained in between 6 to 7% fiber content (i.e., 6.4%). Random distribution of fiber may result in optimum fiber content shifted between 6 and 7%. Consistent with UCT, the presence of fiber and compaction results in increasing bearing capacity of soil from very soft to CBR of 6%.

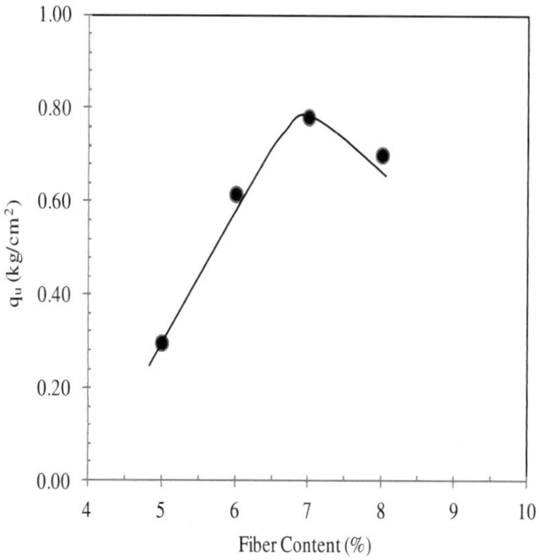

Figure 4. q_u as a function of the fiber content

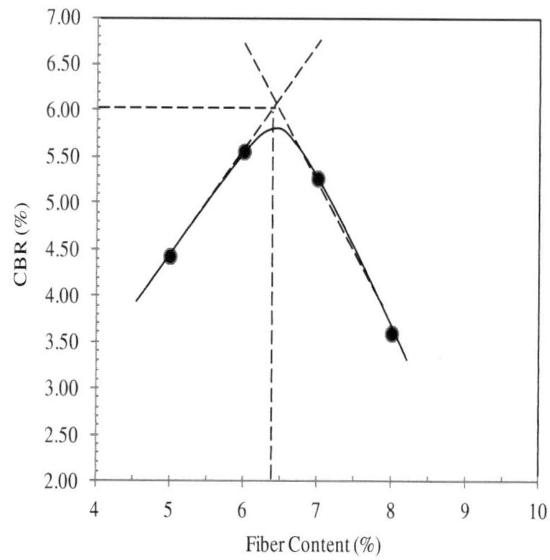

Figure 5. CBR as a function fiber content

4.4. Laboratory vane test

To avoid the vane being resisted by fiber as reported by [17], the small vane of 1.85 cm was used. Figure 6 shows undrained shear strength (s_u) obtained from laboratory vane test at different compaction energy. As shown in the figure, the maximum s_u of soil compacted by 10 and 25 blows are 0.46 kg/cm^2 and 0.65 kg/cm^2 placed at the fiber content of 6%. The energy of 25 blows is a standard Proctor compaction energy. The result reveals that the optimum fiber content obtained from laboratory vane test gives the smallest optimum fiber content. The smaller vane used in this study may result in reducing the fiber contribution in the mixtures. The test was measured the shear strength of soil between fiber; it was entirely the shear strength of soil-fiber mixtures. By increasing the compaction energy (i.e., 56 blows), the s_u increases to 0.71 kg/cm^2 at the fiber content of 6.7% approximately.

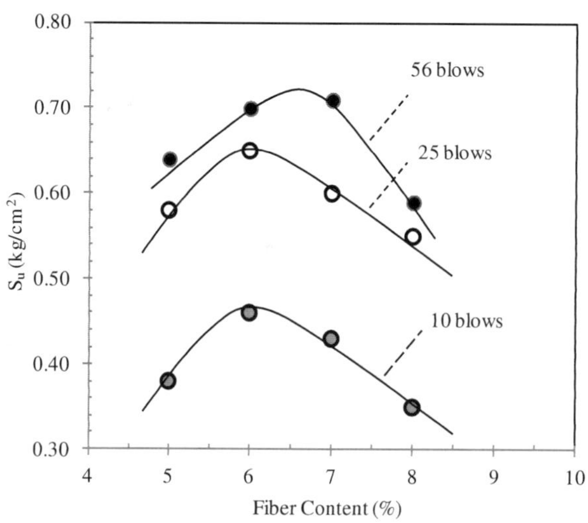

Figure 6. Undrained shear strength as a function fiber content

5. Conclusion

The effects of natural fiber made from oil palm EFB in enhancing soft soil shear strength were presented. Several important findings obtained from the study are:

- EFB function is not only as fiber to increase shear strength but also absorbs water, allowing soft soil to be compacted.
- The soft soil used in this study can be compacted at fiber content higher than 5%. The maximum density obtained is 0.92 g/cm^3 at 7% fiber content.
- The compacted soil-EFB mixtures successfully increase the shear strength and bearing capacity of the soft soil shown by the results of the UCT, laboratory vane, and CBR tests. The soil consistency changes from soft to medium soil.
- The maximum q_u, s_u, and CBR obtained are 0.8 kg/cm^2, 0.65 kg/cm^2, and 6%, respectively at optimum fiber content of 6 to 7%. Other paragraphs are indented (BodytextIndented style).

References

[1] Kayali O. 2004. Effect of high volume fly ash on mechanical properties of fiber reinforced concrete. *Materials and structures.* **37** pp 318-327

[2] Ranjbaran F, Rezayfar O, Mirzababai R. 2018. Experimental investigation of steel fiber-reinforced concrete beams under cyclic loading. *International Journal of Advanced Structural Engineering.* **10** pp. 49–60

[3] Iqbal S, Ali I, Room S, Khan SA, Ali A. 2019. Enhanced mechanical properties of fiber reinforced concrete using closed steel fibers. *Materials and Structures* **52**:56

[4] Ahmed A, El-Naggar M.H. 2018. Effect of cyclic loading on the compressive strength of soil stabilized with bassanite–tire mixture. *J Mater Cycles Waste Manag.* **20** pp 525–532

[5] Brahmachary T K, Ahsan M K, Rokonuzzaman M. 2019. Impact of rice husk ash (RHA) and nylon fiber on the bearing capacity of organic soil. *SN Applied Sciences* **1**:273

[6] Chore H S and Vaidya M K. 2015. Strength Characterization of Fiber Reinforced Cement–Fly Ash Mixes. *Int. J. of Geosynth. and Ground Eng.* **1**:30

[7] Gupta D and Kumar A. 2016. Strength Characterization of Cement Stabilized and Fiber Reinforced Clay–Pond Ash Mixes. *Int. J. of Geosynth. and Ground Eng.* **2**:32

[8] Khelifi H, Lecompte T, Perrot A, Ausias G. 2016. Mechanical enhancement of cement-stabilized soil by flax fibre reinforcement and extrusion processing. *Materials and Structures.* **49**. pp. 1143–1156

[9] Gao L, Hu G, Xu N, Fu J, Xiang C, Yang C. 2015. Experimental Study on Unconfined Compressive Strength of Basalt Fiber Reinforced Clay Soil. *Advances in Materials Science and Engineering.* Vol. 2015. Article ID 561293

[10] Anggraini V, Asadi A, Huat B B K, Nahazanan H. 2015. Performance of Chemically Treated Natural Fibres and Lime in Soft Soil for the Utilisation as Pile-Supported Earth Platform. *Int. J. of Geosynth. and Ground Eng.* **1**:28

[11] Ashour T and Wu W. 2010. The influence of natural reinforcement fibers on erosion properties of earth plaster materials for straw bale buildings. *Journal of Building Appraisal.* **5**, 4, 329–340

[12] Brahmachary T K and Rokonuzzaman M. 2018. Investigation of random inclusion of bamboo fiber on ordinary soil and its effect CBR value. *Geo-Engineering.* **9**:10

[13] Li C and Zornberg J G. 2013. Mobilization of Reinforcement Forces in Fiber-Reinforced Soil. *Journal of Geotechnical and Geoenvironmental Engineering.* **139** (1) pp 107-115

[14] Diambra A and Ibraim E. 2014. Modelling of fibre–cohesive soil mixtures. *Acta Geotechnica.* **9**:1029–1043

[15] Chou J-S and Ngo N-T. 2018. Engineering strength of fiber-reinforced soil estimated by swarm intelligence optimized regression system. *Neural Comput & Applic.* **30**:2129–2144

[16] Deb K and Narnaware Y K. 2015. Strength and Compressibility Characteristics of Fiber-

Reinforced Subgrade and their Effects on Response of Granular Fill-Subgrade System. *Transp. in Dev. Econ.* **1**:1–9

[17] Spritzer J M, Khachan M M, Bhatia S K. 2015. Influence of Synthetic and Natural Fibers on Dewatering Rate and Shear Strength of Slurries in Geotextile Tube Applications. *Int. J. of Geosynth. and Ground Eng.* **1**:26

[18] Ismail S and Yaacob Z. 2011. Properties of Laterite Brick Reinforced with Oil Palm Empty Fruit Bunch Fibres. *Pertanika J. Sci. & Technol.* **19**(1): 33 – 43

[19] Ali M. 2012. Natural fibres as construction materials. *Journal of Civil Engineering and Construction Technology.* **3**(3), pp. 80-89

[20] Santoni L, Tingle S, Webster L. 2001. Engineering properties of sand–fiber mixtures for road construction. *J.Geotech.Geoenviron.Eng.* **127**(3) 258-268

[21] ASTM. 1997. *Annual Book of Standards*. 04.08 and 04.09, Soil and rock, ASTM International. West Conshohocken. PA

[22] Terzaghi K and Peck R B. 1967. *Soil mechanics in engineering practice*. John Wiley & Sons. New York

IOP Conf. Series: Materials Science and Engineering **669** (2019) 012027 doi:10.1088/1757-899X/669/1/012027

Modal parameters identification of light rail bridges

E Zulkifli[1] and D R Widarda[2]

[1] Civil Engineering Department, Institut Teknologi Bandung, Indonesia
[2] Civil Engineering Department, Parahyangan Catholic University, Indonesia

dinarw@unpar.ac.id

Abstract. One of important aspect in Structural Health Monitoring System (SHMS) is the initial condition of the structure. This paper studies the initial condition of light rail train (LRT) bridges before officially operated. There are four bridges to be evaluated, consists of three simple supported bridges and one continuous bridge. Data measurement was using tri-axial accelerometer EZDYN with a train-set as the dynamic loading source. The modal parameters are obtained by means of operating modal analysis (OMA) in time domain. Three methods are simulated, that are ARMA, ITD and SSI. The data processing applied shows that all three methods give satisfactory results that confirm each other, for both natural frequencies and vibration modes.

1. Introduction

To assure the best performance of structure during its life service time, the health of structure should be monitored continuously. Structure can undergo low performance due to natural aging, lack of maintenance or changing of load. Structural health monitoring (SHM) is intended to detect the damage of material or geometric properties of structures.

One of important aspect in SHM is the initial condition of the structure. The initial condition is taken as a datum of considered health structure. Take an advantage of evaluation of modal parameters as the structure constructed, the structural design can be evaluated, as well. It was done by comparing the actual modal parameter from field measurements to modal parameters from structural design.

This paper discusses the investigation of modal parameters of a new constructed light rail bridges by means of dynamic vibration. The investigation is carried out to yield the initial modal parameter of the bridge before it officially operates. Periodic evaluation of modal parameter will give an early detection of damage and prevent a bigger loss. For completeness, numerical model using finite element method is performed and compared to the field measurement results.

The modal parameters identification of bridge has been reported in some papers. Vu et al. [1] have reported the identification of modal parameters of a bridge before and after rehabilitation. Modal parameters were extracted using Operating Modal Analysis (OMA) in four methods, that are the pick picking method (PP), the Least Square Complex Exponential method (LSCE), the Autoregressive method (AR) and the Autoregressive Moving Average method (ARMA). Cabboi et al. [2] conducted FDD and covariance-based SSI schemes to identify the modal parameters of a centenary iron-arch bridge. The paper discussed natural frequencies, vibration modes and furthermore the possible damage or structural anomaly. OMA method has been applied in general structure as reported in Ediansjah et al. [3], that evaluate modal parameters in platform structure.

2. Operating modal analysis

Operating Modal Analysis (OMA) is one of promising method to obtain structural modal parameters. The method required no input excitation and applicable to wide structure. OMA is highly applicable in evaluating the real structure compares to FRF method that need certain and measurable laboratory excitation. Some methods of OMA are based on time domain and the others on frequency domain. This paper focuses on time domain OMA, that are Auto regressive moving average (ARMA), Ibrahim Time Domain (ITD) and Stochastic Subspace Identification (SSI). Every methods intend to gather information related to physical behaviour of structural system from correlation function from fields measurement [4,5].

2.1. *Auto regressive moving average (ARMA).*

ARMA method identifies a system and predict the output based on previous input and output. The complete discussion about ARMA can be found in Modal Analysis book by He and Fu [5]. Only short explanation is presented below.

The ARMA (n,m) model is characterized by

$$x_t = \sum_{r=1}^{n} \phi_r x_{t-r} - \sum_{s=1}^{m} \theta_s a_{t-s} + a_t, \quad n > m \tag{1}$$

The order of auto regressive model number of previous responses denoted by n. m denotes the order of moving average. x_t is the response that correlates with previous response x_{t-1}, x_{t-2}, $\cdots x_{t-n}$, input of time t, a_t, and its previous input a_{t-1}, a_{t-2}, $\cdots$, a_{t-m}. At the time t, the response x_t is decomposed into two parts that are deterministic and dependent part $\sum_{r=1}^{n} \phi_r x_{t-r} - \sum_{s=1}^{m} \theta_s a_{t-s}$ that happened in the past; and random and independent part, a_t, that represent a current input. ϕ_r denotes auto regressive coefficients and θ_s denotes moving average coefficients.

The ARMA(n,m) model for N DOF system follows

$$x_t = \{-x_{t-1} \quad \cdots \quad -x_{t-2N} \quad a_{t-1} \quad \cdots \quad a_{t-2N+1}\} \begin{Bmatrix} \phi_1 \\ \vdots \\ \phi_{2N} \\ \theta_1 \\ \vdots \\ \theta_{2N-1} \end{Bmatrix} \tag{2}$$

The transfer function for N-DOF system is

$$\alpha_{ij}(s) = \frac{x_i(s)}{a_j(s)} = \frac{b_0 + b_1 s + \ldots + b_{2N-2} s^{2N-2}}{a_0 + a_1 s + \ldots + a_{2N} s^{2N}} \tag{3}$$

Applying z-transform to transfer function results

$$\alpha_{ij}(z) = \frac{x_i(z)}{a_j(z)} = \frac{\theta_0 + \theta_1 z^{-1} + \ldots + \theta_{2N-1} z^{-2N+1}}{1 + \phi_1 z^{-1} + \ldots + \phi_{2N} z^{-2N}} \tag{4}$$

Then, solving the characteristic equation of form

$$1 + \phi_1 z^{-1} + \ldots + \phi_{2N} z^{-2N} = 0 \tag{5}$$

And taking relationship between z-transform and Laplace transform results

$$z_r = e^{s_r \Delta} \tag{6}$$

Then the natural frequency, ω and damping ratio, ξ_r of the r-th mode can be obtained from

$$\omega_r = \frac{1}{\Delta}\sqrt{z_r z_r^*} \quad \text{and} \quad \xi_r = \frac{\ln(z_r z_r^*)}{2\omega_r \Delta} \tag{7}$$

where Δ is sampling time resolution. Order of ARMA model takes a significant role in estimating the modal parameter [5]

2.2. *Ibrahim Time Domain (ITD)*

ITD method obtains modal parameter using free vibration of displacement, velocity or acceleration responses. The decaying free vibration is obtained by sampling the displacement data with a selected time interval. Brief description of the scheme is displayed here, the comprehensive explanation can be found in He and Fu [5] and some papers [6,7,8].

From the dynamic equation

$$\mathbf{M\ddot{x}} + \mathbf{C\dot{x}} + \mathbf{Kx} = \mathbf{F(t)}, \tag{8}$$

define a new vector

$$y = \begin{bmatrix} x \\ \dot{x} \end{bmatrix}. \tag{9}$$

Rewrite Equation (8)

$$\mathbf{\dot{Y}} = \mathbf{AY} \text{ and } \mathbf{A} = \mathbf{\dot{Y}Y}^{-1} \tag{10}$$

where

$$\mathbf{Y} = \begin{bmatrix} x(t_1) & x(t_2) & \dots & x(t_{2N}) \\ \dot{x}(t_1) & \dot{x}(t_2) & \dots & \dot{x}(t_{2N}) \end{bmatrix}; \quad \mathbf{\dot{Y}} = \begin{bmatrix} \dot{x}(t_1) & \dot{x}(t_2) & \dots & \dot{x}(t_{2N}) \\ \ddot{x}(t_1) & \ddot{x}(t_2) & \dots & \ddot{x}(t_{2N}) \end{bmatrix}. \tag{11}$$

And modal parameters can be found from Equation (11) and (12)

$$\mathbf{A} = \begin{bmatrix} \mathbf{0} & \mathbf{I} \\ -\mathbf{M}^{-1}\mathbf{K} & -\mathbf{M}^{-1}\mathbf{C} \end{bmatrix} \tag{12}$$

2.3. *Stochastic Subspace Identification (SSI)*

Simple description of SSI method is displayed here. Complete explanation can be found in Overschee and De Moor [9]. Consider a dynamic system under the stochastic random excitation or random noises in state space form

$$\begin{aligned} \mathbf{z}(k+1) &= \mathbf{Az}(k) + \mathbf{w}(k) \\ \mathbf{y}(k+1) &= \mathbf{Cz}(k) + \mathbf{v}(k) \end{aligned} \tag{13}$$

w(k) and *v(k)* are white noises vector representing the process and measured noises, respectively. *w(k)* and *v(k)* are statistically uncorrelated. The cross-correlation function is determined using Equation (13)

$$\mathbf{R}(k) = E\left[\mathbf{y}(k+m)\mathbf{y}(m)^T\right] = \mathbf{CA}^{k-1}E\left[\mathbf{z}(m+1)\mathbf{y}(m)^T\right] = \mathbf{CA}^{k-1}\mathbf{G} \tag{14}$$

where $G = E[z(m+1)y(m)^T]$Then, block Hankel matrix can be constructed using cross correlation matrix

$$\mathbf{H}_{n_1,n_2} = \begin{bmatrix} \mathbf{R}_1 & \cdots & \mathbf{R}_{n_2} \\ \vdots & \ddots & \vdots \\ \mathbf{R}_{n_1} & \cdots & \mathbf{R}_{n_1+n_2-1} \end{bmatrix} = \begin{bmatrix} \mathbf{CG} & \cdots & \mathbf{CA}^{n_2-1}\mathbf{G} \\ \vdots & \ddots & \vdots \\ \mathbf{CA}^{n_1-1}\mathbf{G} & \cdots & \mathbf{CA}^{n_1+n_2-2}\mathbf{G} \end{bmatrix} = \mathrm{O}_{n_1} B_{n_2} \tag{15}$$

where $\mathrm{O}_{n_1} = \begin{bmatrix} \mathbf{C} \\ \vdots \\ \mathbf{CA}^{n_1-1} \end{bmatrix}$ is observability matrix and $B_{n_2} = \begin{bmatrix} \mathbf{G} & \cdots & \mathbf{A}^{n_2-1}\mathbf{G} \end{bmatrix}$ is extended controllability matrix.

Multiplying weighting matrices, W_1 and W_2, to Hankel matrix and decomposing it will form

$$\mathbf{W}_1\mathbf{H}_{n_1,n_2}\mathbf{W}_2 = \begin{bmatrix} \mathbf{U}_1 & \mathbf{U}_2 \end{bmatrix} \begin{bmatrix} \mathbf{S}_1 & \mathbf{0} \\ \mathbf{0} & \mathbf{0} \end{bmatrix} \begin{bmatrix} \mathbf{V}_1^T \\ \mathbf{V}_2^T \end{bmatrix} = \mathbf{U}_1\mathbf{S}_1\mathbf{V}_1^T \tag{16}$$

Equilibrate (3) and (4) and decomposing into $W_1\ O_{n1}$ and $B_{n2}\ W_2$ results possible solution

$$\mathrm{O}_{n_1} = \mathbf{W}_1^{-1}\mathbf{U}_1\mathbf{S}_1^{1/2} \tag{17}$$

Then the matrices A and C can be obtained using (3) and known observability matrix in (5)

$$\mathrm{O}_{n_1-1}^{\uparrow} = \mathrm{O}_{n_1-1}\mathbf{A} \tag{18}$$

where $\mathrm{O}_{n_1-1}^{\uparrow} = \begin{bmatrix} \mathbf{CA} \\ \vdots \\ \mathbf{CA}^{n_1-1} \end{bmatrix}$ and $\mathrm{O}_{n_1} = \begin{bmatrix} \mathbf{C} \\ \vdots \\ \mathbf{CA}^{n_1-2} \end{bmatrix}$

The system matrix $\mathbf{C}$ is the first row of O_{n_1}.

SSI can be determined as balanced realization (BR) if no weighting matrices are employed. Otherwise, by maximizing the correlation between the measured time history data the weighting matrices governs

$$\mathbf{W}_1 = (\mathbf{L}^+)^{-1} \text{ and } \mathbf{W}_2 = (\mathbf{L}^-)^{-1} \tag{19}$$

where

$$\mathbf{R}^+ = \begin{bmatrix} \mathbf{R}_0 & \cdots & \mathbf{R}_{n_1-1} \\ \vdots & \cdots & \vdots \\ \mathbf{R}_{n_1-1} & \cdots & \mathbf{R}_0 \end{bmatrix} = \mathbf{L}^+\mathbf{L}^{+\mathrm{T}} \text{ and } \begin{bmatrix} \mathbf{R}_0 & \cdots & \mathbf{R}_{n_2-1} \\ \vdots & \cdots & \vdots \\ \mathbf{R}_{n_2-1} & \cdots & \mathbf{R}_0 \end{bmatrix} = \mathbf{L}^-\mathbf{L}^{-\mathrm{T}} \tag{20}$$

3. Case studies

Four light rail bridges are taken as study cases. Three bridges are simply supported bridges and the other is continuous and curved bridge. The length of bridges is displayed at Table 1.

Table 1. Type, section and length of bridges.

	Type of bridge	Section	Length [m]
PC30M	Simple support	PC-I Girder	30
PC40M	Simple support	PC-I Girder	40
SB60M	Simple support	Composite Steel Box Girder	60
SB60L	Curved-continuous	Composite Steel Box Girder	60

Data acquisition was using 6 units digital accelerometer EZDYN as shown in Figure 1, each unit consists of tri-axial accelerometer transducer and built-in data logger. Acceleration data was time history that has been recorded offline with time synchronized to all sensors. Time synchronizing is important to guarantee that certain responses from 18 sensors relate to the same certain time. Sampling rate is set to 100Hz with vary time duration depends on the situation on site.

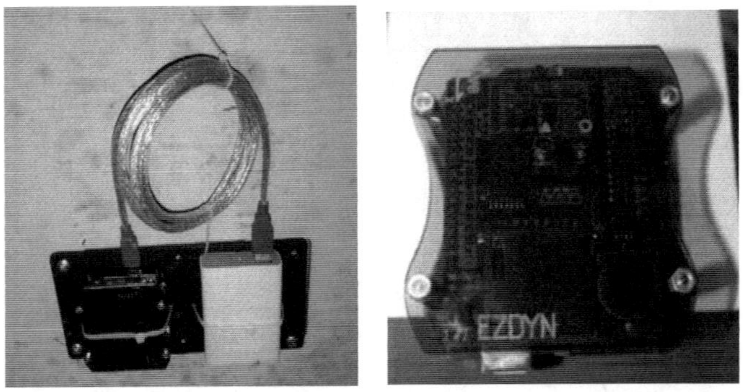

Figure 1. Tri-axial accelerometer EDZYN

Accelerometer is placed in ¼, ½ and ¾ bridge span, as displayed in Figure 2, at the bottom side of bridge deck as shown in Figure 3.

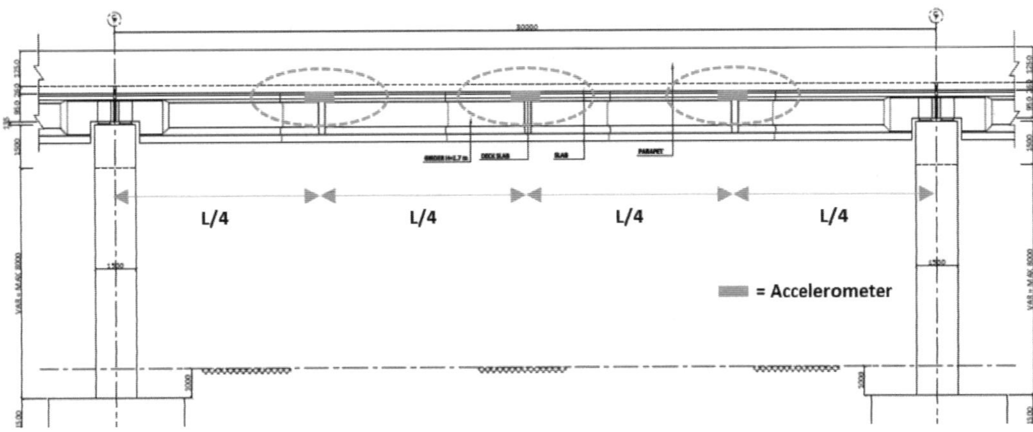

Figure 2. Typical accelerometer placements along longitudinal span.

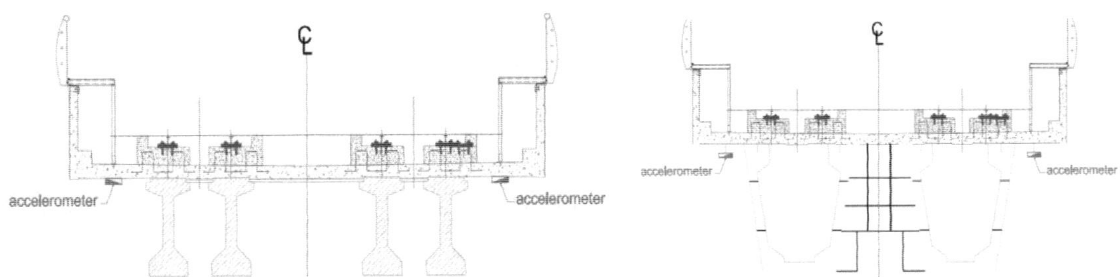

a) I-girders b) Composite steel box girder

Figure 3. Typical bridge section of I-girder and composite steel girder

Dynamic vibration of bridge sourced from running train of 51860mm length, 2650 width, 12 ton axle load, 100ton of net weight (without payload) and 25 km/h speed. Dimension and loading set of train is displayed in Figure 4.

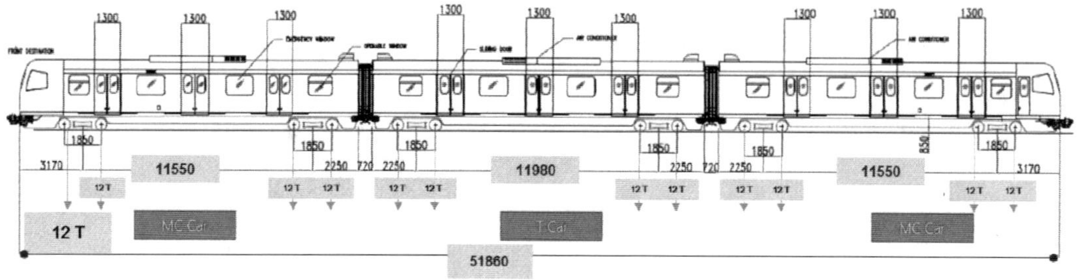

Figure 4. Train dimension and loading set

4. Discussion

Raw measurement data from accelerometer is processed to obtain responses: acceleration, velocity and displacement from each channel. Typical responses in time domain and frequency domain are shown in Figure 5 - Figure 8, for four bridges being evaluated. As an example, the accelerometer BD records acceleration in z-direction (vertical). The figure informs the maximum responses (MAX), peak to peak (P2P) and root mean square (RMS) of each time domain responses; while frequency domain displays maximum responses and its corresponding frequency.

Figure 5 shows time history response of 30m bridge with simple support (PC30M). The maximum acceleration, velocity and displacement are 959mm/s^2, 21.73mm/s and 3.033mm, respectively.

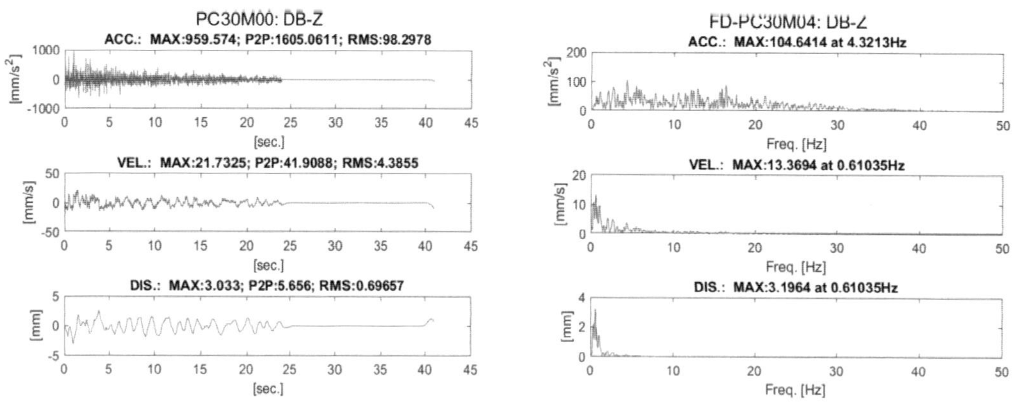

a) responses in time domain b) responses in frequency domain

Figure 5. Typical output of structural responses of bridge PC30M: displacement, velocity and acceleration in time domain and frequency domain

The dynamic responses, for example the Figure 5, are further processed using Operational Modal Analysis to extract the modal parameters. For comparison, OMA is done in 3-time domain methods that are ARMA, ITD and SSI. Parameters to be obtained are structural natural frequencies, mode shapes and mass participation ratio.

Table 2 shows the natural frequency estimation using several data sets of 30m simply supported bridge (PC30M). Three methods of OMA (ARMA, ITS and SSI) is employed to obtain estimated natural frequency from each data set. In P30M bridge, the ARMA method gives lower value compare to two other methods. The estimated natural frequency is taken from average natural frequency or three analysis methods.

Table 2. Natural frequency estimation of PC-I girder bridge PC30M

Data set	ARMA	ITD	SSI
0	3.621	3.775	3.708
1	3.519	3.798	3.725
2	3.541	3.723	3.784
3	3.546	3.722	3.785
4	3.245	3.22	3.297
5	3.499	3.88	3.625
Average Natural Frequency [Hz]	3.495	3.686	3.654
Estimated Natural Frequency [Hz]		± 3.61	

Measurement of 40m simply supported bridge gives maximum acceleration 450mm/s^2, velocity 9.8mm/s and displacement 1.5313mm, as shown in Figure 6. The estimated natural frequency is 2.66Hz, evaluated from 2 data sets as shown in Table 3. In this case, the SSI method gives the lowest natural frequency.

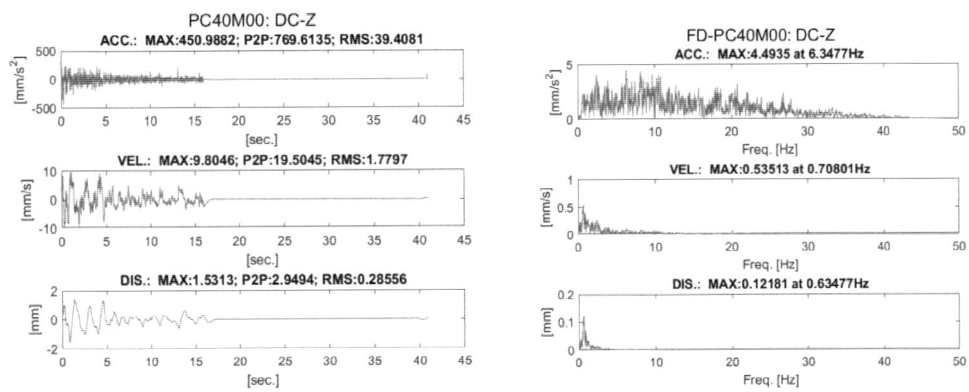

Figure 6. Typical output of structural responses of bridge girder PC40M: displacement, velocity and acceleration in time domain and frequency domain

Table 3. Natural frequency estimation of bridge girder PC40M

Data set	ARMA	ITD	SSI
1	2.813	2.818	2.378
2	2.822	2.826	2.329
Average Natural Frequency [Hz]	2.818	2.822	2.354
Estimated Natural Frequency [Hz]		± 2.66	

Estimated natural frequency of SB60M (simply supported, 60m length composite steel box) bridge is 2.14Hz, as shown in Table 4. The SSI method results the lowest natural frequency, compared to other method. Time history responses in Figure 7 displays the maximum acceleration, velocity and displacement are 1813 mm/s^2, 60.061m/s and 7.3577mm respectively.

IOP Conf. Series: Materials Science and Engineering **669** (2019) 012027 doi:10.1088/1757-899X/669/1/012027

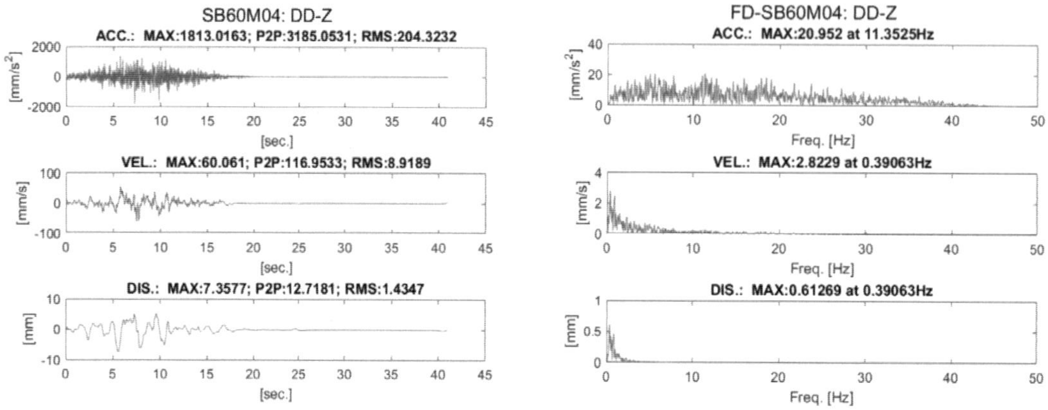

Figure 7. Typical output of structural responses of bridge girder SB60M: displacement, velocity and acceleration in time domain and frequency domain

Table 4. Natural frequency estimation of bridge girder SB60M

Data set	ARMA	ITD	SSI
1	2.178	2.081	1.9911
2	2.162	2.164	2.1076
3	2.150	2.1507	2.2718
Average Natural Frequency [Hz]	2.163	2.132	2.124
Estimated Natural Frequency [Hz]		± 2.14	

Structural response of 60m steel composite with continuous support (SB60L) bridge is shown in Figure 8. Maximum responses are 1415mm/s^2 acceleration, 51.6719mm/s velocity and 7.3268mm displacement. The estimated natural frequency of SB60L is 1.72Hz, as shown in Table 5.

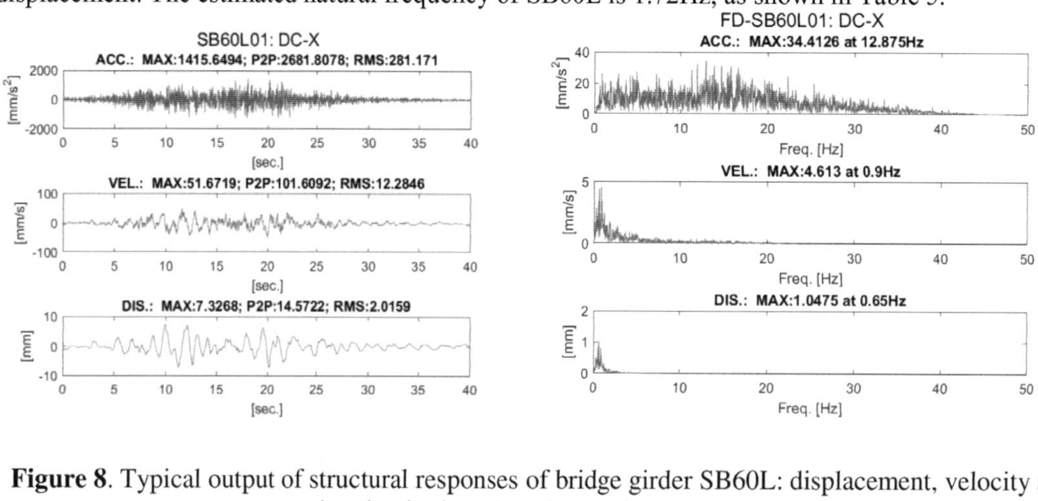

Figure 8. Typical output of structural responses of bridge girder SB60L: displacement, velocity and acceleration in time domain and frequency domain

Table 5. Natural frequency estimation of bridge girder SB60L

Data set	ARMA	ITD	SSI
1	1.718	1.718	1.715
Estimated Natural Frequency [Hz]		± 1.72	

Mode shape extracted from measurement data meets the mode shape of finite element analysis. For example, the PC30M case shown in Figure 9. At natural frequency 3.52Hz, finite element model gives torsional mode (Figure 9a) and the measurement with ARMA (Figure 9b) yields the torsional mode, as well.

a) mode shape from finite element model b) mode shape from measurement

Figure 9. Mode shape of simple supported bridge 30m

5. Conclusions

Some remarks in this contribution are:

- The three OMA methods, namely ARMA, ITD and SSI gives convergence results one to another,
- The mode shape from measurement meet the mode shape from finite element analysis,
- The estimated natural frequencies meet the natural frequencies from finite element analysis

References

[1] Vu V Hung, Thomas M, Lakis A and Marcouiller L 2007 Identification of modal parameters by operational modal analysis for the assessment of bridge rehabilitation Proc. 2nd Int. Operational Modal Analysis Conf. (Copenhagen)

[2] Cabboi A, Gentile C and Saisi A 2014 Vibration-based SHM of a centenary bridge: a comparative study between two different automated OMA techniques Proc. 9th International Conf. on Structural Dynamics (Porto)

[3] Ediansjah and Widarda D 2017 Dynamic properties evaluation of a platform structure Procedia Engineering 171 pp 1222-27

[4] Ewins D 2000 Modal testing: theory, practice and application (Baldock: Research Studies Press Ltd)

[5] He J and Zhi Z F 2001 Modal Analysis (Butterworth-Heinemann)

[6] Ibrahim S R and Milkulcik E 1977 A method for direct identification of vibration parameters from the free response Shock Vibr. Bull. 47 pp 183–96

[7] Ibrahim S R 1977 Random decrement technique for modal identification of structures. J. Spacecr. Rockets 14 pp 696–700

[8] Ibrahim S R 1978 Modal confidence factor in vibration testing J. Spacecr. Rockets 15 pp 313–16

[9] Overschee P and De Moor B 1996 Subspace identification for linear systems, theory, implementation, application (Kluwer Academic Publishers)

Acknowledgement

The authors highly appreciate the PT Drei Meta Dinamika for providing measurement data.

IOP Conf. Series: Materials Science and Engineering **669** (2019) 012028 doi:10.1088/1757-899X/669/1/012028

Application of steady flow model to overcome inundation based on eco drainage system in the Port Area of Dobo City, Aru Islands Regency

B E Prasetyo [1], G Idfi[1], Mujiyono[1] and E Suwarno[1]

[1] Civil Engineering Department, State University of Malang, Indonesia

gilang.idfi.ft@um.ac.id

Abstract. This research was carried out in inundation that occurred in the Port area of Dobo City, Aru Islands Regency. Inundation that occurs is due to the unavailability of adequate and well-arranged drainage. This research aims to plan a drainage system in the Port area of Dobo City that can overcome the inundation that arises and the influence of tides. Data processing is carried out with two analyzes including hydrological analysis and hydraulic analysis. For hydraulics analysis, this includes cross section planning and drainage system modeling with HEC-RAS 4.0. In conventional drainage in the case of high tides, the results of the modeling show that the overflow of water on the channel that will occur is at an elevation of +2.48 m. After adding a long storage with dimensions of 1.2 m x 3.5 m along 101.6 m and sea walls as high as + 2.48 m. The results of the modeling show that the inundation height that will occur is at an elevation of +0.61 m. Of the several scenarios that are performed for the most optimal two-door openings, which are opened by 30 cm which later the water will be at + 12 cm elevation. Whereas the most optimal one-door openings are opened by 70 cm which later the water will be at + 6 cm elevation.

1. Introduction

Inundation that occur are mainly caused by the absence of a drainage system that is able to accommodate runoff discharge that occurs [2]. Drainage originating from the verb 'to drain' which means drying or draining water, is the terminology used to express systems related to handling the problem of excess water both above and below the surface of the land. Generally, the drainage system in the Port area of Dobo City is a natural drainage channel that has not been properly arranged to channel inundation that occurs in the area. This is exacerbated by the amount of sedimentation and garbage that covers the canal so that it reduces the capacity of existing channels.

An alternative solution that can be done to solve the inundation problem in the Dobo City Port area and fulfill the zero delta Q policy principle is to use a drainage system based on Eco Drainage. Eco drainage is defined as an effort to manage excess water (rainwater) with various methods including by accommodating through water reservoirs to be used directly, accommodating in artificial reservoirs, permeating and flowing to nearby rivers without increasing the burden on the river concerned and always maintain the system so that it can be sustainably efficient [4]. The main priority in realizing this concept is by managing surface runoff through developing facilities to endure rainwater (rainfall retention fascilities) so there is no inundation on the trunk chanel [3]. Based on its function, rainwater detention facilities can be in the form of storage types and infiltration types.

2. Methodology

In data processing carried out with two analyzes including hydrological analysis and hydraulic analysis [5]. Hydrological analysis carried out included analysis of probability distribution, probability distribution test, planned rainfall calculation, analysis of time concentrate, rainfall intensity analysis, and the last is calculation of planned flood discharge by dividing the planning area into several sub-catchments (sub catchments), all hydrology analysis calculations is carried out using Microsoft Excel 2010. For hydraulics analysis, includes cross section channel planning and drainage network modeling using HEC-RAS 4.0 to determine the happening of back water on water level elevation profile [6].

Furthermore, eco drainage system planning is carried out by adding a long storage to the planned drainage channel [7]. Planning is done to determine the dimensions of the long storage also to know the amount of discharge that can be accommodated in the long storage. After knowing the dimensions of the long storage and obtained technical data about the height and time of tide occurring, following by the outlet door scenario planning in anticipating back water caused by tides. Furthermore, after the planning phase is completed, proceed with making conclusions and suggestions [8].

3. Analysis and Discussion

The flood discharge plan is the most important part and is the basis for determining the capacity and dimensions of water buildings. In this plan, the determination of the amount of planned flood discharge carried out using a rational method where the discharge used in the planning is the peak flood discharge [9].

Table 1. Amount of plan discharge for each channel

Channel	Discharge m^3/s	Channel	Discharge m^3/s	Channel	Discharge m^3/s
1.a	0.080	8a	0.048	11.a	0.028
2.a	0.038	BC 5	0.644	BC 9	0.827
BC 1	0.118	8b	0.064	11.b	0.002
4a	0.190	9b	0.799	12.b	0.048
1.b	0.051	9.a	0.178	12.a	0.042
4b	0.143	10.a.a	0.002	12.c.a	0.002
2.b	0.026	BC 6	0.180	12.c	0.003
3a	0.071	10.a.b	0.007	BC 10	0.051
BC 2	0.071	10.a	0.219	12.d.a	0.002
3b	0.049	10.a.c	0.014	12.d	0.003
5a	0.177	10.a.d	0.016	13.a	0.307
BC 3	0.367	BC 7	0.234	13.b	0.885
6.a.a	0.004	10.b.a	0.002	14.a	0.313
6.a	0.065	10.b	0.266	14.b	0.890
BC 4	0.065	10.b.b	0.013	BC 11	0.931
6.b	0.054	10.b.c	0.014	15.b	0.087
5b	0.169	BC 8	0.280	BC 12	0.087
7	0.597	11.a.a	0.016	15.a	1.046

Hydraulics analysis carried out to plan the dimensions of the channel that is able to pass the planned flood discharge. Hydraulics analysis includes calculation of channel capacity, flow velocity, and the amount of discharge that can be flowed by the channel.

Table 2. Calculation of hydraulics analysis

Channel	b	h	Qhydraulics	Channel	b	h	Qhydraulics
	m	m	m^3/second		m	m	m^3/second
1.a	0.5	0.4	0.089	10.a.c	0.3	0.3	0.030
2.a	0.4	0.3	0.045	10.a.d	0.3	0.3	0.030
BC 1	0.6	0.5	0.152	BC 7	0.7	0.7	0.288
4a	0.6	0.6	0.191	10.b.a	0.3	0.2	0.018
1.b	0.4	0.4	0.065	10.b	0.7	0.7	0.288
4b	0.6	0.5	0.152	10.b.b	0.3	0.3	0.030
2.b	0.3	0.3	0.030	10.b.c	0.3	0.3	0.030
3a	0.5	0.4	0.089	BC 8	0.8	0.7	0.348
BC 2	0.4	0.4	0.065	11.a.a	0.3	0.3	0.030
3b	0.5	0.4	0.089	11.a	0.3	0.3	0.030
5a	0.6	0.6	0.191	BC 9	1.1	1.1	0.960
BC 3	0.8	0.8	0.411	11.b	0.3	0.2	0.018
6.a.a	0.3	0.2	0.018	12.b	0.4	0.4	0.065
6.a	0.5	0.4	0.089	12.a	0.4	0.3	0.045
BC 4	0.5	0.4	0.089	12.c.a	0.3	0.2	0.018
6.b	0.4	0.4	0.065	12.c	0.3	0.2	0.018
5b	0.6	0.6	0.191	BC 10	0.4	0.4	0.065
7	1.0	0.9	0.654	12.d.a	0.3	0.2	0.018
8a	0.4	0.4	0.065	12.d	0.3	0.2	0.018
BC5	1.0	0.9	0.654	13.a	0.8	0.7	0.348
8b	0.4	0.4	0.065	13.b	1.1	1.1	0.960
9b	1.1	1.0	0.854	14.a	0.8	0.7	0.348
9.a	0.6	0.6	0.191	14.b	1.1	1.1	0.960
10.a.a	0.3	0.2	0.018	BC 11	1.2	1.1	1.088
BC 6	0.6	0.6	0.191	15.b	0.5	0.4	0.089
10.a.b	0.3	0.2	0.018	BC 12	0.5	0.4	0.089
10.a	0.7	0.6	0.238	15.a	1.2	1.1	1.088

After all the dimensions used for calculating the hydraulic discharge are able to accommodate the planned flood discharge, it is then added with a free board of 10 cm on each channel. Material used in the channel using precast concrete (U-ditch). The dimensions of the channel used on each channel showed on table 3.

Table 3. Channel dimension

Channel	Dimension	Channel	Dimension
	mm		mm
1.a	500 x 500	10.a.c	300 x 400
2.a	400 x 400	10.a.d	300 x 400
BC 1	600 x 600	BC 7	700 x 800
4a	600 x 700	10.b.a	300 x 300
1.b	400 x 500	10.b	700 x 800
4b	600 x 600	10.b.b	300 x 400
2.b	300 x 400	10.b.c	300 x 400
3a	500 x 500	BC 8	800 x 800
BC 2	400 x 500	11.a.a	300 x 400
3b	500 x 500	11.a	300 x 400
5a	600 x 700	BC 9	1100 x 1200

Table 3. Channel dimension (cont.)

Channel	Dimension mm	Channel	Dimension mm
BC 3	800 x 900	11.b	300 x 300
6.a.a	300 x 300	12.b	400 x 500
6.a	500 x 500	12.a	400 x 400
BC 4	400 x 500	12.c.a	300 x 300
6.b	400 x 500	12.c	300 x 300
5b	600 x 700	BC 10	400 x 500
7	1000 x 1000	12.d.a	300 x 300
8a	400 x 500	12.d	300 x 300
BC 5	1000 x 1000	13.a	800 x 800
8b	400 x 500	13.b	1100 x 1200
9b	1100 x 1100	14.a	800 x 800
9.a	600 x 700	14.b	1100 x 1200
10.a.a	300 x 300	BC 11	1200 x 1200
BC 6	600 x 700	15.b	500 x 500
10.a.b	300 x 300	BC 12	500 x 500
10.a	700 x 700	15.a	1200 x 1200

The eco drainage modeling is carried out by adding long storage to the downstream of the planned drainage system combined with the sea wall. The following are the results of modeling on eco drainage:

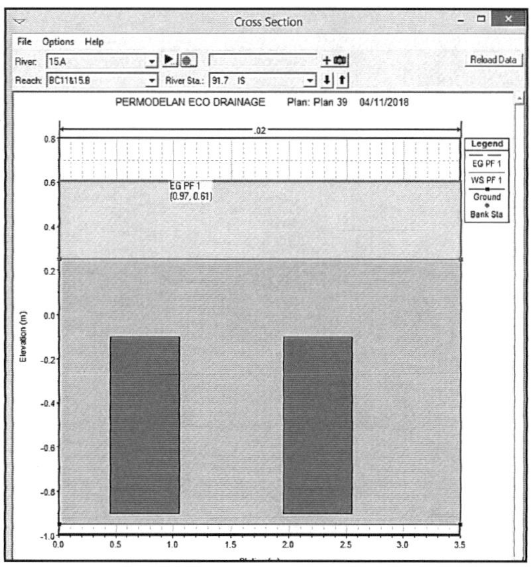

Figure 1. Eco drainage modeling results

Gate opening scenarios and their effect on decreasing water level on the channel. Modeling carried out with 7 openings scenarios.

- Two Gate Opened 20 cm, in this scenario modeling results show that there is still overflow around the overflow channel which occurs as high as 14 cm. From the results of this modeling, the reduction results obtained by 13.7%.

- Two Gate Opened 30 cm, in this scenario modeling results show that there has been no overflow occurs around the channel. From the results of this modeling, the reduction results obtained by 30.4%.
- Two Gate Opened 50 cm, in this scenario modeling results show that there has been no overflow occurs around the channel. From the results of this modeling, the reduction results obtained by 60.2%
- One Gate Opened 20 cm, Another Gate Closed, in modeling this scenario, there is still an overflow occurs around the overflow channel which occurs as high as 26 cm. From the results of this modeling, the reduction in water level obtained by 6.2%.
- One Gate Opened 30 cm, Another Gate Closed, in modeling this scenario, there is still an overflow occurs around the overflow channel which occurs as high as 20 cm. From the results of this modeling the results of the reduction in water level obtained by 9.9%.
- One Gate Opened 50 cm, Another Gate Closed, in modeling this scenario, there is still an overflow occurs around the overflow channel which occurs as high as 7 cm. From the results of this modeling the results of the reduction in water level obtained by 18%.
- One Door Gate 70 cm, Another Gate Closed, In modeling this scenario is found that there is no overflow occurs around the channel. From the results of this modeling the results of the reduction in water level obtained by 41.6%.

The results of gate opening scenarios can be informs at graphics below

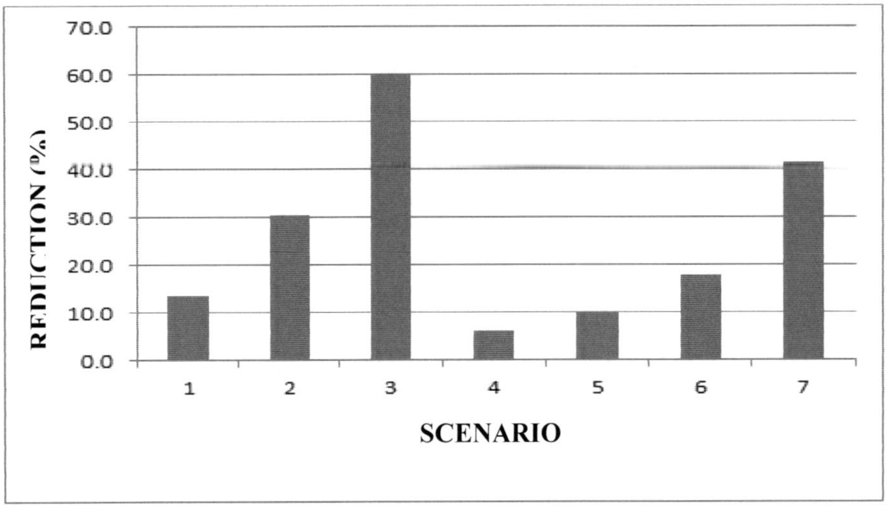

Figure 2. Reduction of discharge based on scenario

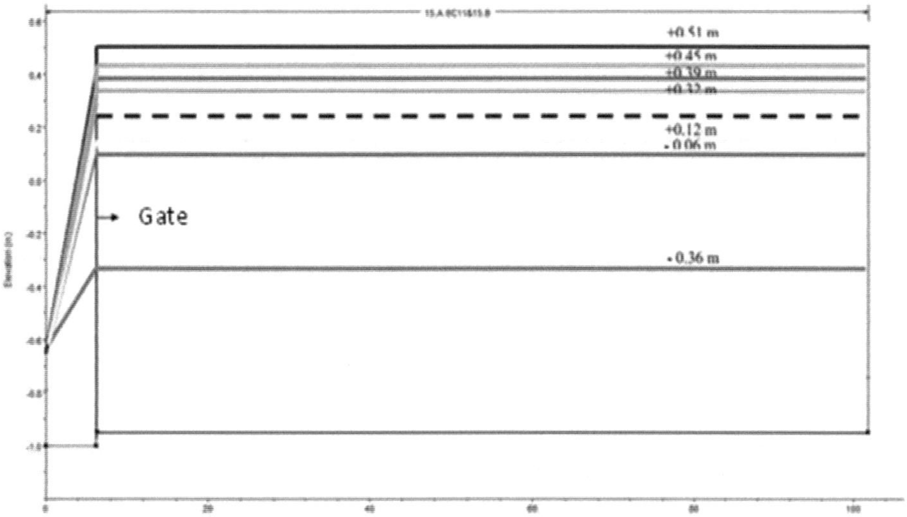

Figure 3. Comparation of water surface

= Scenario 1	= Scenario 4	= Scenario 7
= Scenario 2	= Scenario 5	= Levee
= Scenario 3	= Scenario 6	

4. Results and Conclusion

Based on the results of planning data analysis and modeling results in the HEC-RAS 4.0, it can be concluded that:

- From the results of the calculation of the plan flood discharge obtained for the downstream discharge of the drainage system is 1.046 m3 / second.
- In conventional drainage when sea tides occur, the results of the modeling show that water overflows on the channel that will occur are at an elevation of +2.48 m.
- After adding long storage with dimensions of 1.2 m x 3.5 m along 101.6 m and sea walls as high as + 2.48 m. The results of the modeling show that the inundation height that will occur is at an elevation of +0.61 m. With an overflow height of 36 cm. So that the construction of longitudinal reservoirs in the drainage system will reduce the inundation height by 1.87 m
- From several scenarios carried out for the most optimal two-door openings by opening 30 cm which later the water will be at an elevation of +0.12 m. Whereas the most optimal one-door openings are opened by 70 cm which later the water will be at an elevation of +0.06 m.

References

[1] Idfi G 2017 Perbandingan Model Aliran Banjir Unsteady Flow dan Steady Flow pada Sungai Ngotok Ring Kanal. Jurnal Bangunan, 22(2), 31-40.
[2] Istiarto 2014 HEC-RAS Steady or Unsteady Flow Analysis (istiarto.staff.ugm.ac.id)
[3] Anggrahini 1997 Hidrolika Saluran Terbuka (CV. Citra Media, Surabaya)
[4] Chow V T 1992 Open Channel Flow., (Erlangga, Jakarta)
[5] Harto S Br 1993 Analysis of Hydrology. (PT Gramedia Utama, Jakarta)
[6] Lensley, Ray K, Franzini and Joseph B. 1991 Teknik Sumber Daya Air Jilid II, (CV. Citra Media, Surabaya)
[7] Soemarto C D 1987 Hydrology Engineering, (Erlangga, Jakarta.)
[8] Suhartanto E 2008 The Manual Book of HEC-HMS and Application at Water Resource Management, (CV Citra, Malang)
[9] Triatmojo B 2008 Hidrologi Terapan (Beta Offset, Yogyakarta)

IOP Conf. Series: Materials Science and Engineering **669** (2019) 012029 doi:10.1088/1757-899X/669/1/012029

Barriers and critical success factors towards sustainable hazardous waste management in electronic industries – A review

Y H Chin[1] and H N A Yong[2]

[1] Specialist I, Universiti Tunku Abdul Rahman, Kampar, Malaysia
[2] Assistant Professor, Universiti Tunku Abdul Rahman, Kampar, Malaysia

yhchin@utar.edu.my

Abstract. This study reviews existing literature to determine the barriers and critical success factors (CSFs) towards sustainable hazardous waste management in electronic industries. Sustainable hazardous waste management can considerably reduce the volume of wastes going to landfill sites and can therefore reduce the burden to the environment in terms of pollution prevention and greenhouse gaseous reduction. This goal is achievable through efforts, such as waste recovery, waste-to-energy technology and waste-to-raw material conversion. Hazardous waste in Malaysia mainly comes from recovery facilities, metal refinery, metal finishing and coating, cement manufacturing, petroleum refinery and chemical, electrical and electronic industry. The electrical and electronic industry is a major contributor of hazardous waste generated in Malaysia. Some hazardous wastes, such as heavy metal sludge, generated by electrical and electronic industries can be utilised as alternative fuel for cement industries. The residue produced subsequently can be utilised as raw material for cement manufacturing. Despite the economic and environmental benefits of sustainable hazardous waste management efforts amongst industries, they remain relatively weak. The identified barriers include (1) complexity of legal requirements, (2) lack of systems and tools, (3) hazardous waste quantity and quality, (4) lack of awareness about environmental and sustainability issues, (5) poor partnership/networking, (6) public resistance, (7) financial constraints, (8) lack of information and (9) insufficient environmental resources. Nevertheless, the benefits gained from adopting sustainable hazardous waste management will continuously encourage an increasing number of organisations to devote efforts onto it. The success of this initiative depends on a few critical factors, including (1) readiness of hazardous waste beneficiary, (2) economic and environmental benefits, (3) stakeholders' pressure, (4) adoption of ISO14001, (5) effects on corporate image, (6) commitment from the top management and (7) incentives and reward.

1. Introduction

Hazardous waste, also known as scheduled waste in Malaysia, is any waste that possesses hazardous characteristics and can adversely affect the public health and the environment. First Schedule of Environmental Quality (Scheduled Wastes) Regulations in 2005 listed 77 types of scheduled wastes and introduced provisions for the management of wastes. Regulation 7 of Environmental Quality (Scheduled Wastes) Regulations in 2005 allows hazardous waste producers to manage their respective hazardous wastes sustainably by converting the hazardous waste into energy or raw materials for other interested parties. However, sustainable hazardous waste management efforts amongst industries remain relatively weak.

Malaysia is one of the 12 mega biodiversity countries in the world. However, the environmental performance of the country has worsened in recent years. Malaysia ranks 63rd in 2016 compared with 51st in 2014 in environmental performance index [1]. By contrast, its neighbouring country the Philippines has improved 48 ranks in the same period [2].

Hazardous waste in Malaysia mainly comes from recovery facilities, metal refinery, metal finishing and coating, cement manufacturing, petroleum refinery and chemical, electrical and electronic industry [3]. Amongst these industries, the scheduled waste generated from the electrical and electronic industry contains sludge composed of chromium, nickel, copper, lead, zinc, cadmium, tin, aluminium, beryllium and vanadium, which are categorised under code SW204 in the First Schedule of the Environment Quality (Scheduled Waste) Regulation in 2005.

The statistics published by Department of Environment (DOE) Malaysia (Table 1) showed that the total scheduled wastes generated in Malaysia in 2017 was approximately 2,017,281 metric tons (MT), of which 226,748 MT (11.24%) was SW204 (sludge containing heavy metals). SW204 is the third highest scheduled waste generated by various industries.

Table 1. Quantity of scheduled waste generated by waste code [4]

Name of Waste	Waste Code	MT/Year	Percentage (%)
Dross/slag/clinker/ash	SW 104	706,750.00	35.03
Gypsum	SW 205	510,724.45	25.32
Heavy metal sludge	SW 204	226,747.90	11.24
Spend lubricating oil	SW 305	94,528.12	4.69
Spent acid	SW 206	86,664.98	4.30
Contaminated container	SW 409	47,509.28	2.36
Mixture of scheduled waste and non-scheduled waste	SW 422	31,962.60	1.58
E-waste	SW 110	28,604.15	1.42
Pathogen clinical waste	SW 404	28,375.24	1.41
Others	-	255,360.04	12.65
TOTAL		2,017,280.76	100

Source: [4]

One of the major generators of SW204 is the electrical and electronic industry (Table 2). SW204 is the heavy metal sludge produced from industrial wastewater treatment plants. These amounts of sludge are potential solid fuel (coal substitute) for cement industries and other energy generation facilities, such as power plants and steam boilers.

Table 2. Quantity of scheduled waste generated by industry [4]

Type of Industry	Quantity of Waste	
	(MT/Year)	Percentage (%)
Scheduled Waste Treatment and Disposal Facilities	510,724.45	25.32
Power Plant	363,087.02	18.00
Chemical Industry	276,242.51	13.69
Electrical and Electronic Industry	223,897.54	11.10
Metal Refinery	166,618.59	8.26
Workshops	63,094.96	3.13
Others	116,2948.38	20.50
TOTAL	2,766,613.45	100.00

Source: [4]

However, out of 226,748 MT of SW204 generated, only 30,516 MT (7.43%) was reused as raw material for product manufacturing through special management under Regulation 7 of Environmental Quality (Scheduled Wastes) Regulations in 2005 (Table 3). By contrast, the remaining 196,232 MT (92.57%) of SW204 eventually ended up in sanitary landfill sites and thus had no economic benefit and added burden to the environment.

Table 3. Special management of scheduled waste under Regulation 7 [4]

Waste Code	Source	Metric Ton	%	Special Management Method
SW 104	Power Plant	363,087.02	71.16	Reuse as raw material for product manufacturing
SW 104	Industry	18,515.50	3.63	Reuse as raw material for product manufacturing
SW 204	Drinking water treatment plant	26,032.33	5.10	Sanitary landfill
SW 204	Industry	63,822.59	12.51	Sanitary landfill
SW 204	Industry	30,515.88	5.98	Reuse as raw material for product manufacturing
SW 406	Industry	4,824.78	0.95	Sanitary landfill
SW 427	Industry	1,954.05	0.38	Reuse as neutralisation agent
SW 320, 325, 418	Industry	885.78	0.17	Sanitary landfill
SW 421	Industry	501.35	0.10	Reuse as raw material for product manufacturing
SW409	Industry	23.40	0.00	Reuse as raw material for product manufacturing
SW431	Industry	10.39	0.00	Slow burning method
SW 405, 429	Industry	10.00	0.00	Sanitary landfill

Source: [4]

The figures indicate that electrical and electronic industries can reduce the amount of hazardous waste going to landfill sites through sustainable waste management efforts. Rather than sending the scheduled wastes to landfills, which incurs a large cost, the heavy metal sludge generated from electronic industries can be used as alternative solid fuel for cement industries due to its high calorific value [5]. The fuels extracted from hazardous waste or the thermal use of solid recovered fuels in the clinker can eliminate the hazardous waste without producing any harmful gaseous by-products in the incinerator whilst providing cost savings for electronic and cement industries. Moreover, the residue left after the incineration of scheduled waste can be recycled as additive material to improve cement properties, such as reduced permeability, high resistance to sulphates and harsh environment condition, enhanced workability, or high-quality finish [3].

Table 4. Approximate calorific value of materials

Type of Material	CV (MJ/kg)	CV (kCal/kg)
Industrial and hazardous waste	22–40	5257–9558
Medical waste	19–24	4540–5735
Domestic waste (without recycling)	7–16	1673–3823
Domestic waste (after recycling)	10–14	2389–3345

Source: [6]

This paper provides an overview of common barriers and critical success factors (CFSs) towards sustainable hazardous waste management in electronic industries. The outcome of this literature review is expected to inform the scheduled waste generator and the energy producer or user about alternative ways of managing the scheduled waste in a sustainable and cost-effective manner through waste recovery and waste-to-energy efforts.

2. Methods

This study aims to identify the barriers and critical success factors (CSFs) that influence sustainable hazardous waste management efforts amongst electronic industries. Data and information were extracted from secondary data resources, such as government agency statistical reports, environmental reports and journal publications.

3. Importance of Sustainable Hazardous Waste Management

Global environmental problems, such as climate change, pollution and depletion of natural resources, have adversely affected human health and caused diseases. Sustainable hazardous waste management has become an essential global agenda that leads to the emergence of environmental protection legislations related to the control of hazardous substances and wastes. The awareness of consumers about the environmental impact resulting from their respective demand has motivated voluntary efforts to reduce the ecological footprint [7] and has therefore created new market opportunities for manufacturers of green products. Furthermore, the emergence of highly stringent requirements penalising the environmentally harmful behaviours manufacturers and the proactive governance of Non-Governmental Organizations (NGOs) have motivated manufacturers to better control their activities to reduce reputation risk and prevent additional cost.

To date, very limited studies have been conducted on sustainable hazardous waste management in Malaysia. Investigating the potential barriers to sustainable hazardous waste management in Malaysia is important due to the drastic increase in manufacturing activities and hazardous wastes. The barriers to sustainable hazardous waste management should be addressed immediately to tackle the environmental issues at present [8]

The impact caused by mishandling of hazardous waste, even in small amounts, is sometimes intolerable for some small countries with fragile environment by nature [9]. Many developed countries now have effective hazardous waste management systems in place. By contrast, countries with a long-term industrial base, such as China [10] and India [11], have not yet developed their hazardous waste management systems to the same extent.

Hazardous waste, which has toxicity, ignitability, corrosivity and reactivity, must be managed carefully to avoid unintentional release to the environment, which can lead to adverse effects to humans and biodiversity. Thus, the Ninth Malaysia Plan empowered the relevant agencies in adopting the global harmonised system to monitor the movements and safe handling of hazardous waste. In addition, the utilisation of technologies to treat and convert hazardous waste into a resource for reuse will be facilitated and implemented [12]. DOE Malaysia is also aggressively promoting a shift from 'cradle-to-grave' approach, in which wastes are treated as unwanted by-products, to 'cradle-to-cradle' approach, in which hazardous wastes are regarded as alternative resources and potential for energy recovery [3].

4. Barriers to Sustainable Hazardous Waste Management

'Sustainability is a difficult term to be fully understood by many [13]. Sustainable hazardous waste management is achievable through waste recovery and waste-to-energy efforts that promote efficient use of energy and natural resources, reduce volume of waste generated and prevent pollution. However, the adoption of sustainable hazardous waste management remains subject to some barriers.

4.1. Complexity of legal requirements

DOE Malaysia encourages sustainable scheduled waste management under the provision of Regulation 7 of Environmental Quality (Scheduled Wastes) Regulations in 2005. However, the extensive evaluation on the characteristic of the hazardous waste requires various testing on the waste and can thus be a potential factor discouraging the industry to proceed.

4.2. Lack of systems and tools

The availability of systems and tools is a crucial element of hazardous waste management because utilisation or disposal of the wastes is unattainable within the precondition on a long period. The management issues become increasingly complex because the waste has heterogeneous characteristics in terms of bioavailability, chemistry and toxicity. Thus, these hazardous wastes require special collection, treatment and recycling [14]. Li et al. [15] reported that incineration is the main refuse treatment in most cities in China with the aim to reduce waste volume, create less harm to humans and the environment and recycle the energy. However, recovering energy through incineration only works when the calorific value of the waste reaches a target value.

4.3. Hazardous waste quantity and quality

One of the potential beneficiaries of hazardous waste is the cement industry. A huge cement plant produces several thousand tons of clinker per day. Therefore, the demand for energy is high and must be met continuously. To cater to such demand, not only standard fuels, such as coal, fuel oil or natural gas, but also alternative fuels (solid hazardous wastes) must be available on site in sufficient quantity at any time [16]. The procurement of alternative fuels, including solid hazardous wastes, must fulfil several critical criteria, such as fuel properties (meet the specification required by the cement plant), availability (consistent quantity), price (cheaper than existing fuels used), process compatibility (will not damage the kiln), cement compatibility (no effect on cement quality), environmental compatibility (comply to emission limit) and approvability (comply to hazardous substance limit).

4.4. Lack of awareness about environmental and sustainability issues

One of the critical barriers for implementation of sustainable hazardous waste management is the lack of awareness about the environmental and sustainability issue. According to Siebenhuner and Arnold [17], for a company to become more sustainability orientated, it should make changes that include the introduction of resource-efficient technologies, sustainability reporting schemes and by providing sustainable products, services and product-service combinations.'

4.5. Poor partnership/networking

Statistical analysis done by Gunawardana [18] showed that most hazardous industries in Sri Lanka do not have proper integration amongst the parties that are responsible in hazardous waste management processes. This situation may be due to that the hazardous management area does not acquire sufficient concern and lack awareness on the importance of correct handling of hazardous waste. Moreover, some hazardous industries can have insufficient support and cooperation from hazardous waste management parties. Gunawardana [18] argued that many countries' hazardous waste transport organisations and agencies of law enforcement are cold and detached towards hazardous waste management; this situation leads to restriction in this matter. Finding external partners to invest in sustainable hazardous waste technologies is difficult [8].

4.6. Public resistance towards hazardous industries

Van der Sloot and Kosson [19] pointed out that national awareness of liquid and solid hazardous waste issues grew exponentially in the middle of 1970s with extensive concern over mishandled waste, miscellaneous dumping of liquid waste and rare but visible transportation accidents. In Sir Lanka, some hazardous industries have been forced to close down because of harsh public resistance. Gunawardana [18] found that most industries generating hazardous waste have dealt with public resistance because of inappropriate management of hazardous waste.

4.7. Financial constraints
The success of sustainable hazardous waste management depends on the amount of investment spent on it. Various waste reduction projects change the standard production progression to assist recovery of hazardous waste material and involve high cost [18]. Thus, the lack of adequate finance for supporting eco-efficiency initiatives acts as a barrier.

4.8. Lack of information
The lack of technology information and market availability will prevent a company from adopting sustainable practices because the market demand is uncertain [8]. Sustainable hazardous waste management is a newer subject than other environmental management parameters. Information is an essential point for the electronic industry that applies modern technologies, and local and international information are necessary to investigate various compositions of hazardous waste. With the correct information, various kinds of hazardous waste can be identified efficiently and contribute massively towards successful hazardous waste management practices. Hazardous waste is always produced as a part of a structure. Therefore, identification of waste with the reference of updated information is important [18].

4.9. Insufficient environmental resources
Environmental management knowledge is essential to provide environmental management skilled personnel in the particular field for constructing a sustainable organisation. Lack of expertise in this area leads to less effort in sustainable hazardous waste management. Furthermore, findings showed that trained employees do not prefer to stay long in an organisation to assist in sustainable management [8]. The organisation that focuses on short-term tasks has low emphasis on long-term sustainable management practices [20]. This situation will further obstruct the adoption of these practices because the related works will bring burden for other employees and result in low success rate. Insufficient qualified staff can be a critical constraint to the development of sustainable hazardous waste management in an organisation.

5. Critical Success Factors of Sustainable Hazardous Waste Management
The amount of electronic wastes (e-wastes) disposed in 2017 was approximately 28,604 MT. E-waste may contain hazardous substances, such as heavy metals, carcinogens and ozone depleting substances, that are risky to human health and the environment; nevertheless, it contains precious materials that can be recovered as secondary resources for reuse and can thus minimize the exploitation of natural resources and reduce greenhouse gas emissions [21]. In Malaysia, 146 e-waste recovery facilities have the total capacity to handle more than 24,000 MT of e-waste per month. Amongst the 146 e-waste recovery facilities, 128 are partial recovery facilities engaged in physical or manual segregation of e-wastes for further processing; on the contrary, the remaining 18 are full recovery facilities that can process e-wastes to recover precious metals [4].

5.1. Readiness of hazardous waste beneficiary
Manufacturing of cement consumes a large portion of raw material and energy including heat and electricity. The process uses approximately 3.2 GJ to 6.3 GJ of energy and 1.7 t of raw materials in cement production. This energy intensive industry can spend thermal energy of up to 20%–25% of the production cost and consume 110–120 kWh of electrical energy per ton of cement [3]. Thomanetz [5] indicated that hazardous waste can be used as alternative fuels in cement plants, but the waste must first be processed in a special preconditioning plant. Alternative fuels made from several liquids and pasty and solid hazardous wastes can yield up to 120 000 t/year. A fine conveyable calorific fuel is created by blending the waste with saw dust. Converting the hazardous waste into cement fuels can save the conventional fuel and raw material whilst preventing landfilling and underground storage of waste. The mineral residues from the process are embedded within the clinker rather than dumped as waste.

Operational cost can be decreased because this process has lower price than the conventional incineration of hazardous waste.

5.2. Economic and environmental benefits
Reduction in cost and environmental load is achievable through an appropriate management of waste incineration with decreased primary energy consumption [22]. Anett Zajáros et al. [23] showed that solvent recovery in manufacturing technology of EVOH-based adsorbent can reduce the amount of hazardous waste water from 265.2 t/year to 5.5 t/year (98% reduction).

Alternative fuels as thermal recycling use industrial wastes, such as waste oils, plastics, auto shredded residues and waste tyres, and clay substitutes as material recycling use sewage sludge and industrial wastewater treatment sludge with alumina and silica [24-25]. Turning hazardous waste into raw material or energy (solid fuel) has the following benefits towards the environment, public health and economy:

- High temperature with long residence time and intensive turbulence within the rotary kiln can destroy most of the hazardous organic matter.
- Mineral residues will be embedded within the clinker and become part of the product; no waste will be disposed.
- Reduce the usage of conventional fuel (such as coal, fuel oil or natural gas).
- Reduce the amount of waste going to landfill and underground storage.
- Reduce the cost of sending waste to conventional hazardous waste incineration.
- Reduce the rate of natural resource depletion (coal and petroleum).
- Reduce potential health and environmental impact caused by hazardous waste.

5.3. Stakeholders' pressure
Stakeholders refer to the individuals and groups of people who can influence the performance of the company or who are influenced by the actions of the company [26]. Environmental consciousness of a company indicates harmonising environmental performance with the expectations of stakeholders [27]. Thus, the performance of the company is affected by the pressures of its perception from stakeholders and importance for the progress of proactive environmental strategies. Stakeholders can motivate the acts of proactive environmental strategies. For example, the development of relationship with stakeholders and even the probability of affecting some of them are advantages contributing to environmental proactivity [28].

5.4. Adoption of ISO14001
As stipulated under Clause 6.1.1 of ISO 14001:2015, an organization shall determine the environmental aspects of its activities, products and services under its control and influence and their associated environmental impacts from a life cycle perspective. Such requirement intends to reduce the generation of waste by adopting 'cradle-to-cradle' approach that requires sustainable waste management strategies for its success. Therefore, adopting ISO14001 will directly motivate an organisation to make much effort in managing their waste sustainably. Quazi et al. [29] argued that the top management of companies that have been practising ISO14001 standard will eventually be more concerned on the natural environment than other companies that have not.

5.5. Effects on corporate image
The environmental effort of large companies can imply positive effect on many customers [28]. King and Lenox figured that large organisations have higher success rate of probability in adopting the ISO 14001 standard than their small counterparts.

5.6. Commitment from the top management
Del Brio et al. observed that the organisation has high attention on sustainable environmental practices when the involvement of the top management and their understanding of the benefits, disadvantages and tools of the environmental management is high. Priorities of the top management strongly influence the

organisational culture with respect to environment management, corporate responsibility and sustainability management [30]

5.7. Incentives and reward

Incentives and reward can stimulate great efforts in some cases. Such initiatives make people feel that their extra efforts are being appreciated and thus motivate them to progress further. Thus, the Malaysia government provides incentives in the form of investment tax allowance for the purchase of green technology assets and income tax exemption for the use of green technology services and systems. Incentives are also offered for establishing waste ecological parks.

6. Conclusions

The increasing generation of hazardous waste calls the need for action in hazardous waste management. Sustainable hazardous waste management can considerably reduce the volume of wastes going to landfill sites and can therefore reduce the burden to the environment in terms of pollution prevention and greenhouse gaseous reduction. Adopting these strategies has many barriers, such as complexity of legal requirements, lack of systems and tools, waste quantity and quality, lack of awareness about environmental and sustainability issues, poor partnership/networking, public resistance, financial constraints, lack of information and insufficient environmental resources, that need to be overcome

The benefits gained from adopting sustainable hazardous waste management will continuously encourage an increasing number of organisations to exert much effort onto it. The success of this initiative depends on a few factors, including readiness of hazardous waste beneficiary, economic and environmental benefits, stakeholders' pressure, adoption of ISO14001, effects on corporate image, commitment from the top management and incentives and reward.

In the future work, a few experienced hazardous waste handlers from electronic companies will be interviewed to help us identify the overlooked barriers and critical success factors of sustainable scheduled waste management. The primary data will be collected from personnel involved in the management of scheduled wastes amongst electronic manufacturers through the distribution of self-administered questionnaires. This study will use close-ended question method to collect data and measure items by using a five-point Likert scale. The collected data will be analysed using principal component analysis to identify the most important CSFs and barriers for the respondents.

References

[1] Malaysia Productivity Corporation (MPC) 2016 Malaysia's Performance in Environmental Performance Index.
[2] Yale University 2016 Environmental Performance Index (EPI) 2016.
[3] Che Jamin N and Mahmood N 2015 A case study on cement manufacturing. Hazardous waste management in Malaysia, 8, pp.2-7.
[4] Department of Environment 2017 Environmental Quality Report 2017.
[5] Thomanetz E 2012 Solid recovered fuels in the cement industry with special respect to hazardous waste. *Waste Management & Research*, 30(4), pp.404-412.
[6] Calorific value of waste. (2018, February 8). Retrieved from http://www.igniss.com/calorific-value-waste
[7] Harrison R, Newholm T, Shaw D (eds) 2005 The ethical consumer. Sage, Thousand Oaks
[8] Abdullah M, Zailani S, Iranmanesh M, Jayaraman K 2015 Barriers to green innovation initiatives among manufacturers: the Malaysian case. *Review of Managerial Science.* 10(4):683–709
[9] Mallak S K and Ishak M K 2012 Waste minimization as sustainable waste management strategy for Malaysian industries. UMT *11th International Annual Symposium on Sustainability Science and Management*
[10] Duan H, Huang Q, Wang Q, Zhou B and Li J 2008 Hazardous waste generation and management in China: A review. *Journal of Hazardous Materials.* Vol 158(2), pp. 221-227.

[11] Dutta S K, Pupadhyay V and Sridharan U 2006 Environmental management of industrial hazardous waste in India. *Journal of Environ Science & Engg.* Vol 48, No 2, pp.143-150.
[12] Jaminl N C and Mahmood N Z 2015 Scheduled Waste Management in Malaysia: An Overview. Advanced Materials Research, ISSN: 1662-8985, Vol. 1113, pp 841-846
[13] Lozano R 2013 A Holistic Perspective on Corporate Sustainability Drivers. *Corporate Social Responsibility and Environmental Management*, 22(1), 32–44.
[14] Rout R P, Bhunia P, Ramakrishnan A, Surampalli Y R, Zhang C T and Tyagi D R 2016 Sustainable Hazardous Waste Management/Treatment: Framework and Adjustments to Meet Grand Challenges. *Sustainable Solid Waste Management*, 46.
[15] Li X, Zhang C, Li Y and Zhi Q 2016 The Status of Municipal Solid Waste Incineration (MSWI) in China and its Clean Development. Energy Procedia, 104, pp.498-503.
[16] Thomanetz E 2012 Solid recovered fuels in the cement industry with special respect to hazardous waste. *Sage Journal: Waste Management & Research* 30(4), 404–412
[17] Siebenhuner B and Arnold M 2007 Organizational learning to manage sustainable development. *Business Strategy and the Environment* 16: 339–353
[18] Gunawardana K 2017 Safety Issues in Hazardous Waste Management in Heavy Industries. *SSRN Electronic Journal*, 64.
[19] Van der Sloot H and Kosson D 2012 Use of characterisation leaching tests and associated modelling tools in assessing the hazardous nature of wastes. *Journal of Hazardous Materials*, 207-208, pp.36-43.
[20] Ylinenpaa H 1998 Measures to overcome barriers to innovation in Sweden-fits and misfits. In: European small business seminar in Vienna, Austria.
[21] Suja F, Rahman R A, Yusof A and Masdar M S 2014 E-Waste Management Scenarios in Malaysia. *Journal of Waste Management*, 2014, 1-7.
[22] Abaecherli M L, Cap on-García E, Szijjarto A and Hungerbühler K 2017 Optimized energy use through systematic short-term management of industrial waste incineration. Comput. Chem. Eng. 104, 241-258.
[23] Zajáros A, Szita K, Matolcsy K and Horváth D 2018 Life Cycle Sustainability Assessment of DMSO Solvent Recovery from Hazardous Waste Water. *Periodica Polytechnica Chemical Engineering.* 62(3), pp. 305-309
[24] Nee A Y H 2009 Motivation to Adoption of ISO 14001 Environment Management System: Environment Performance and Competitive Advantage. *14th International Conference on Thinking* (2009 Malaysia)
[25] Nee A Y H 2010 Developing An Occupational Safety And Health Management System In An Electronics Manufacturer. The 13th Conference & Exhibition On Occupational Safety And Health 1-3rd August 2010, Kuala Lumpur Convention Centre, Malaysia.
[26] Freeman R E 1984 Strategic Management: a Stakeholder Approach. Pitman–Ballinger: Boston.
[27] Gupta M C 1994 Environmental management and its impact on the operations function. *International Journal of Operations and Production Management* 15(8): 34–51.
[28] González-Benito J and González-Benito Ó 2006 A review of determinant factors of environmental proactivity. *Business Strategy and the Environment*, 15(2), pp.87-102.
[29] Quazi H A, Khoo Y, Tan C and Wong P 2001 Motivation for ISO14000 certification: development of a predictive model. Omega. 29(6): 525–542.
[30] Schaefer A 2004 Corporate sustainability-integrating environmental and social concerns? Corp. Soc. Responsib. *Environ. Manag.* 11 (4), 179-187.

IOP Conf. Series: Materials Science and Engineering **669** (2019) 012030 doi:10.1088/1757-899X/669/1/012030

Base flow identification using conceptual hydrology model

N Suryoputro[1], A Yulistyorini[1], G Idfi[1], and N E B Alias[2]

[1] Department of Civil Engineering, Faculty of Engineering, Universitas Negeri Malang, Malang, Jawa Timur, Indonesia.

[2] Department of Hydraulics and Hydrology, Faculty of Civil Engineering, Universiti Teknologi Malaysia, Johor, Malaysia

nugrohosuryoputro@gmail.com

Abstract. Base flow was a vital component in controlling the flow of a river. Base flow came from the groundwater flow. Base flow was the river flow that occurred during the rainless period. Conceptual hydrology model was a model that displays the hydrology process in mathematical formulation and separating the production and routing functions. The tank model was one of the conceptual models. The tank model simulated the watershed by replacing some reservoirs with several tanks. This research aimed to identify the base flow using conceptual hydrology model. The study was conducted in the Bango sub-watershed.. The results exhibited that the tank model was dissatisfactory to simulate the surface flow compared to base flow. Surface flow model tended to be higher (overestimate) whereas the base flow tended to be lower (underestimate) compared to the observation data. The statistic test result displayed that the tank model could describe the base flow well.

1. Introduction

Base flow is a vital component in controlling the flow of a river. Base flow comes from the groundwater flow. According to Harto [1], base flow is the river flow that occurred during the rainless period. Factually, this definition was difficult to understand because, during the rainless period, the river contains interflow element. In his research, Harto stated the three watershed parameters that influenced the base flow: source frequency (ratio between total segment one rivers to total river from all segments), drainage density (total length of rivers in all segments per unit of watershed area), and drainage frequency (total rivers in all segments per unit of watershed area).

Limantara [2] defined base flow as the drainage result of the aquifer and is a minimum discharge. Similar argument by Viessman et al. [3] stated that base flow is the result of percolation water that cascades to groundwater reservoir and flows to the river as groundwater flow.

Base flow separation technique from the river flow hydrograph has a long history in the hydrology field and can predict low-flow and water quality. There are various methods to separate the base flow. There are three categories in the separation method: analytic, empiric, and chemical/isotopic or tracer [4]. The base groundwater and surface flow theory (for example, the analytical solution from Boussinesq approximation, unit hydrograph model, and reservoir yield-related theories from aquifer) is the base for the analytical method. The empirical method covers (1) recession analysis, (2) graphic method, (3) low-pass filter, and (4) using groundwater elevation to measure the base flow contribution based on the correlation between groundwater elevation and river flow. The tracer method uses dissolved chemical and/or stable isotope to separate the hydrograph into components based on water mass balance and tracer.

Recursive Digital Filter (RDF) is a method of the low-pass filter, adapted from the signal-processing theory [5]. RDF method observes the surface discharge as a high-frequency signal, whereas base flow is a low-frequency signal. By filtering the high-frequency signal (surface discharge) from the flow, the method identifies the low-frequency signal (base flow) [6]. The filtered result depends on several specific parameters of watershed. The measured data in the base flow as the calibration base for these parameters. However, measurable data rarely available, and in practice, this parameter is often arbitrarily determined [7].

Many factors influences the base flow amount, for example, fluvial geomorphology, land, land use, and climate [8],[9]. Soil topography and properties have a relatively small influence on the base flow in a short period. Meanwhile, the changes in land use have various effects, especially the changes in forest land [10],[11]. Changes in forest land and others correlate with the changes in evapotranspiration, infiltration, and refilling the groundwater deposits in a watershed; and all influence base flow [9],[12]. Planning and managing watershed required practical knowledge about the correlation between changes in forest land on other land uses and base flow process. The most significant adverse effect on base flow change is by changes in forest land followed by agriculture and habitation. Grasslands, barren land, and shrubs did not have a significant impact [13].

Conceptual hydrology model is a model that displays the hydrology process in a mathematical formulation and separating the production and routing functions [1]. Some conceptual models of rainfall-runoff are NAM model [14], tank model [15], and HBV model [16].

The conceptual model position is between the physical-based model and the black box model. The conceptual model generally represents essential components that connected hydrology input and output. Usually, the term conceptual describes a simple arranged model from a small number of conceptual elements that are interrelated, and each represents the stage of the hydrological cycle in a part of the land. Most used element in the conceptual model is the storage component. Each storage has one input and one or more output to represent the watershed storage such as surface detention, soil moisture, and others. Routing uses storage and linear channel. The basic conceptual model consists of a series of moisture flow from one element to the next [17].

The conceptual model began for small and homogenous watershed modelling. However, this model could be in use in watershed with diverse topography and vegetations, and watershed with thousands of square kilometres. Input data for this model is simple and easy to obtain [17]. The conceptual model developed started from the 1950s when hydrology experts obtained a hydrology approach technique [18]. Then, in the 1960s, a conceptual model of lumped rainfall-runoff rapidly developed with a clear understanding of the physical hydrology cycle component (conceptual element). They connected each conceptual element and obtained rainfall-runoff correlation in a subsystem. The development of the hydrology model in 1980s led to a complicated model to predict the changes in land use and input-output spatial effect. That stage also developed the distributed parameter model with two and three dimensions displays. The hydrology model started to use the elevation model from a map or satellite image. After the 1980s, hydrology model evolved on a global scale that was macro-scale hydrology model.

The purpose of this study was to identify base flow using another alternative approach, the conceptual hydrological model. The conceptual hydrological model that will be used is the tank model. Hydrological tank model is a conceptual model of rainfall-runoff. Sugawara and Funiyuki introduced this model in 1956 [22]. This model consists of a series of linear tanks arranged in a series or parallel with output holes on the side and bottom side of the tank.

2. Research method

2.1. Research location

The study was conducted in the Bango sub-watershed. (Figures 1 and 2). The coordinates were between 112.56668^0 - 112,94181^0 East Longitude and 7.76549^0 - 7.87407^0 South Latitude with the elevation from

+500 m up to 1,800 m. The area of the Bango sub-watershed was 239.71 Km². The length of the main river was 12.5 Km. The average land slope in Bango sub-watershed was ±2.0% (sloping)

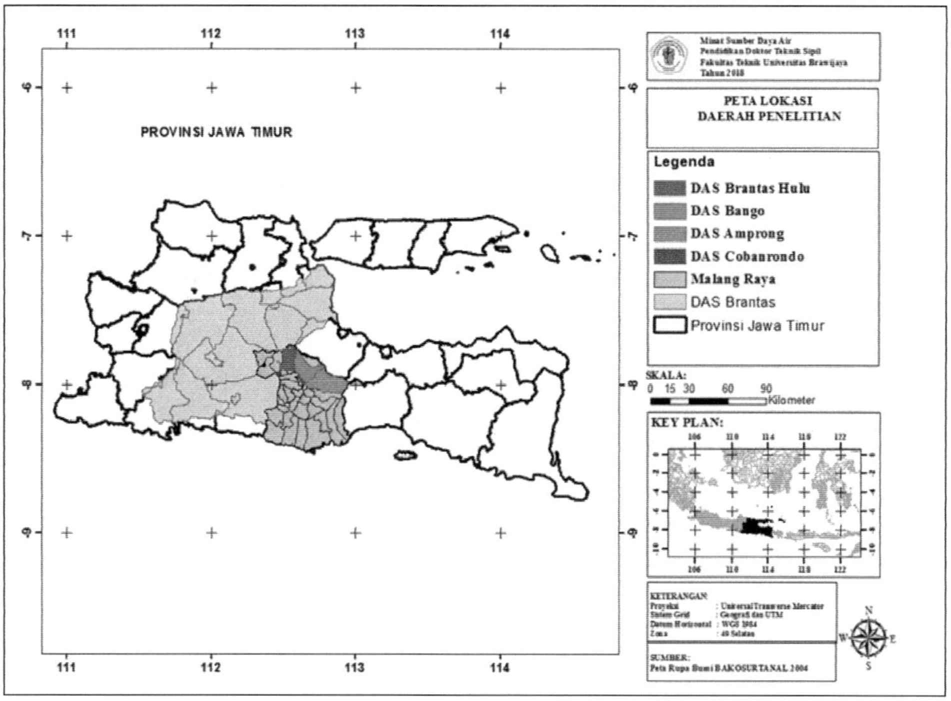

Figure 1. Research location.

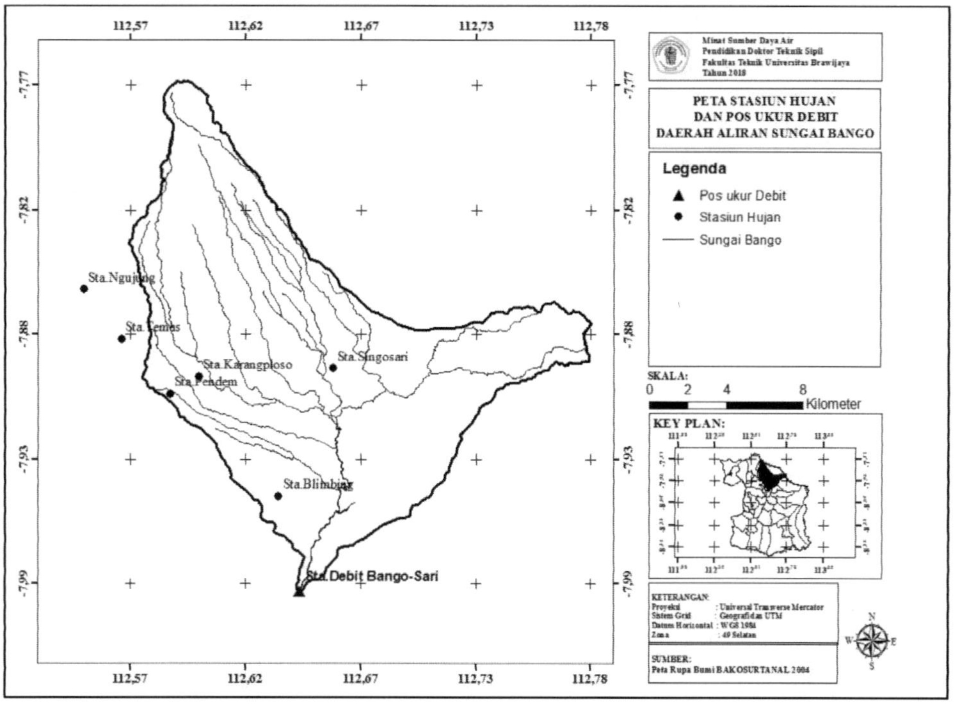

Figure 2. Bango sub-watershed

IOP Conf. Series: Materials Science and Engineering **669** (2019) 012030 doi:10.1088/1757-899X/669/1/012030

Land use in the Bango sub-watershed consists of: 25% settlements, 43% plantations, 14% paddy fields and 18% forests. The land fraction in the location generally consisted of 26% sand, 51% silt, and 23% clay. Therefore, it belonged to a silt loam texture category with infiltration rate between 7.5–15 mm/hour [19] with moderate-slow classification [20]. This information affected low surface runoff potential [21].

2.2. Tank model
Hydrological tank model was a rainfall-runoff conceptual model. Sugawara and Funiyuki introduced this model in 1956 [22]. This model consisted of a series of linear tanks arranged in a series or parallel with output holes on the side and bottom side of the tank. The tank model connected discharge as the effect of rain, evaporation, and groundwater storage from the previous duration so that the developed conceptual model was deterministic non-linear. The tank model simulated the watershed by replacing some reservoirs with several tanks. There are two parameters in the tank model: 1) outlet coefficient at the side and the bottom of the tank parameter, 2) groundwater storage parameter. The parameters were calibrated automatically using the algorithmic genetic method. The excel macro program was developed for the development of genetic algorithms.

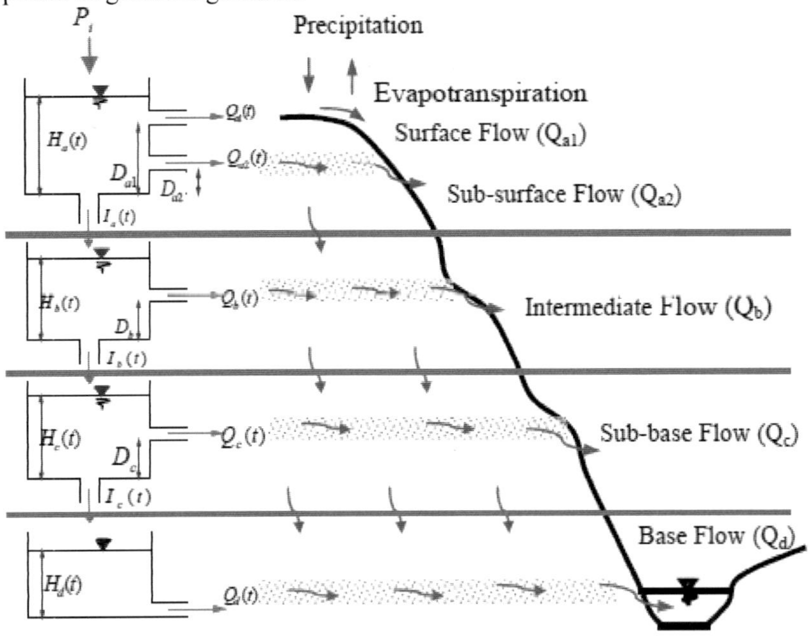

Figure 3. Design tank model scheme.

The total outflow from the side outlet (Q) of each tank, supposedly the water flow accumulation from the system in watershed, formulated as below:

$$Total_{(t)} = Q\ Surface\ flow_{(t)} + Q\ Total\ base\ flow_{(t)} \quad (1)$$

$$Q\ Total\ base\ flow_{(t)} = Q\ Sub\text{-}surface\ flow_{(t)} + Q\ Intermediate\ flow_{(t)} + Q\ Sub\text{-}base\ flow_{(t)} + Q\ Base\ flow_{(t)} \quad (2)$$

The water balance equation in the tank model:

$$\frac{d}{dt} H_{(t)} = P_{(t)} - Q_{(t)} \quad (3)$$

Where P is rainfall (mm day^{-1}), Q is total discharge (mm day^{-1}), H is water storage height (mm), t is time (day). The start ($t=1$) determined the initial water level in tank A storage ($Ha(1)$), tank B ($Hb(1)$), tank C ($Hc(1)$), and tank D ($Hd(1)$). Next, at ($t+1$), the renewed storage formula is as below:

$$Ha_{(t+1)} = Ha_{(t)} + P_{(t)} - Qa1_{(t)} - Qa2_{(t)} - Ia_{(t)} \quad (4)$$

$$Hb\,(t+1) = Hb\,(t) + Ia\,(t) - Qb\,(t) - Ib\,(t) \quad (5)$$

$$Hc\,(t+1) = Hc\,(t) + Ib\,(t) - Qc\,(t) - Ic\,(t) \quad (6)$$

$$Hd\,(t+1) = Hd\,(t) + Ic\,(t) - Qd\,(t) \quad (7)$$

2.3. *Base flow separation method*

This research used Recursive Digital Filter (RDF) with One Parameter Algorithm filter to separate the base flow. RDF method used Hydro Office BFI+ 3.0 software [23]. The equation of RDF with One Parameter Algorithm filter is:

$$q_{b(t)} = \frac{k}{2-k}\, q_{b(t-1)} + \frac{1-k}{2-k}\, q_t \quad (8)$$

Where:

q(t)	: The real river discharge value at t-day
qb(t)	: The actual base flow value at t-day
qb(t-1)	: The base flow value before t-day
k	: Filter parameter with recession constant
t	: Daily time interval

2.4. *Model evaluation*

Assessing reliability in a model to reproduce the nature phenomenon required some criteria. Principally, there are two criteria here: visual (graphics) and using a statistical analytical tool (Indarto, 2010 in [24])

2.4.1. Visual or graphic criterion. Visually, reliability observes the coherency/similarity between measured and calculated output, for example, the scatter-plot result between measured and calculated discharge, making a difference residual graphic between measured and calculated discharge, and FDC graphic comparison between measured and calculated discharge (Podger, 2004 in [24]).

2.4.2. Statistical criteria. Quantitatively, model reliability to reproduce the natural phenomenon determined by a statistic with various parameters. One or several objective functions usually measured errors between the measured and calculated quantitatively. This research used several criteria:

2.4.2.1 The determination coefficient (r^2). Formulated as below:

$$r^2 = \left(\frac{\sum_{i=1}^{n}(O_i-\bar{O})(P_i-\bar{P})}{\sqrt{\sum_{i=1}^{n}(O_i-\bar{O})^2}\sqrt{\sum_{i=1}^{n}(P_i-\bar{P})^2}} \right)^2 \quad (9)$$

the total observation data during routing period, *Oi* is the i-observed data, $\bar{O}$ is the average observation data, *Pi* is the i-value of the surface model, $\bar{P}$ is the average surface model. The range of r2 is between 0 and 1 (perfect suitability). The value of 1 means the distribution of the surface model value is the same with observation.

2.4.2.2 *Nash-Sutcliffe Efficiency (NSE) Coefficient.* Nash and Sutcliffe (1970) suggested NSE coefficient and formulated as below:

$$NSE = 1 - \frac{\sum_{i=1}^{n}(O_i-P_i)^2}{\sum_{i=1}^{n}(O_i-\bar{O})^2} \quad (10)$$

The range of NSE is between 1.0 and -∞. The NSE value = 1.0 shows that the model reliability is excellent.

2.4.2.3 *Root Mean Square Error (RMSE).* RMSE equation is as below:

$$RMSE = \sqrt{\frac{\sum_{i=1}^{n}(P_i - O_i)^2}{n}} \quad (11)$$

The range of RMSE values is between 0 and ∞. The RMSE value = 0 shows that the model reliability is excellent.

2.4.2.4 *Percent Bias (PBIAS).* PBIAS equation is as below:

$$PBIAS = \frac{\sum_{i=1}^{n}(O_i - P_i).100}{\sum_{i=1}^{n} O_i} \quad (12)$$

The value of PBIAS < ±10 shows that model reliability is excellent.

2.4.2.5 *Mean Absolute Error (MAE).* MAE equation is as below:

$$MAE = \frac{\sum_{i=1}^{n}|O_i - P_i|}{n} \quad (13)$$

The range of MAE value is between 0 and ∞. The value of MAE = 0 shows that the model reliability is excellent.
To summarize, Table 1 displays the statistical criteria to assess model reliability [25].

Table 1. Criteria to Assess the Hydrology Model Performance.

Statistical Criteria	Value	Performance Classification
Determination coefficient (r^2)	$0{,}00 \leq r^2 \leq 0{,}50$	Dissatisfactory
	$0{,}50 < r^2 \leq 1{,}00$	Satisfactory
Nash-Sutcliffe Efficiency (NSE) Coefficient	0,75 < NSE < 1,00	Very good
	0,65 < NSE ≤ 0,75	Well
	0,50 < NSE ≤ 0,65	Satisfactory
	0,40 < NSE ≤ 0,50	Acceptable
	NSE ≤ 0,40	Dissatisfactory
Root Mean Square Error (RMSE)	A value below half the standard deviation of the observation data	Satisfactory
Percent Bias (PBIAS)	PBIAS < ± 10	Very good
	± 10 ≤ PBIAS < ± 15	Well
	± 15 ≤ PBIAS < ± 25	Satisfactory
	PBIAS ≥ ± 25	Dissatisfactory
Mean Absolute Error (MAE)	A value below half the standard deviation of the observation data	Satisfactory

Source: [25]

3. Results and discussion

The tank model represented the surface flow in the top tank and the base flow in the tank below it consecutively. In this research, the surface flow and base flow models then compared to the results of the separation flow observation analysis using Recursive Digital Filter (RDF). Comparison between total flow, surface flow and base flow results from observations and tank model can be seen in Figures 4 to 6.

The results exhibited that the tank model was dissatisfactory to simulate the surface flow compared to base flow. Surface flow from tank model tended to be higher (overestimate) whereas the base flow tended to be lower (underestimate) compared to the observation data (Figures 5 and 6).

The analysis obtained a conclusion that tank model had dissatisfactory performance in simulating surface flow or extreme discharge, meanwhile it could give a satisfactory performance to simulate low-flow. Phien et al. [26] also stated a similar statement, that tank model could not simulate flood hydrograph (surface flow or surface runoff).

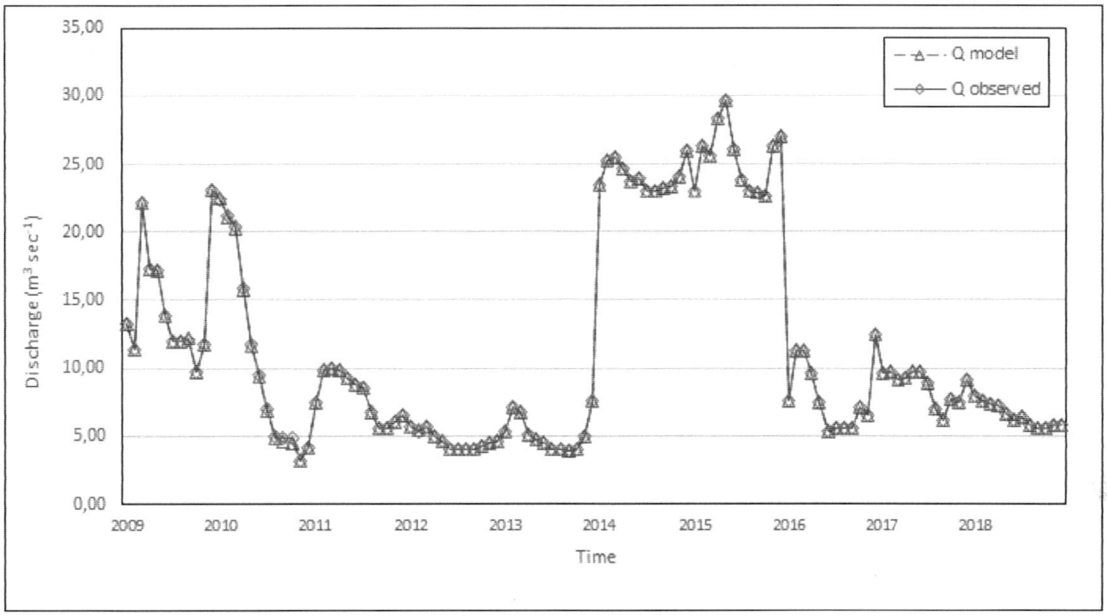

r^2 =	1,000	RMSE =	0,041	MAE=	0,012
NSE =	1,000	PBIAS =	0,103		

Figure 4. Comparison of observed total flow (surface flow + base flow) compared to the results of the tank model

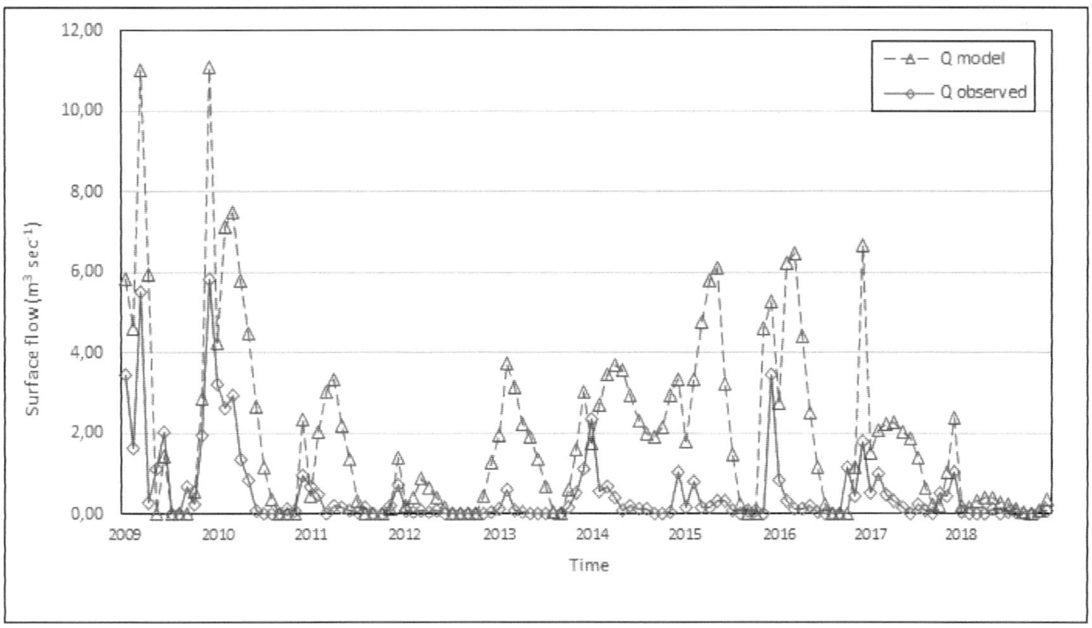

r^2 =	0,483	RMSE =	2,263	MAE=	1,563
NSE =	-4,214	PBIAS =	300,181		

Figure 5. Comparison of observed surface flow compared to the results of the tank model.

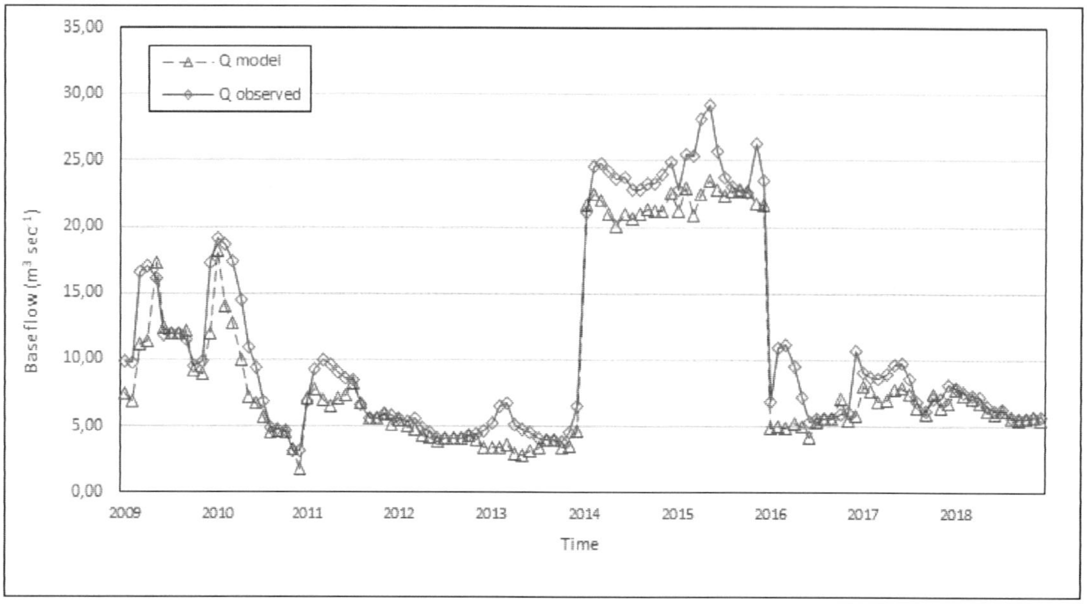

r^2 =	0,951	RMSE =	2,271	MAE=	1,570
NSE =	0,906	PBIAS =	14,203		

Figure 6. Comparison of observed base flow compared to the results of the tank model.

The inaccuracy of the results of the surface and base flow models in this study was predicted due to the model not yet considering the level of soil moisture before the infiltration and surface flow processes in the first tank {27], [28].

Overall, tank model performance could simulate rainfall-runoff well (Figure 4), in line with the research result of Suryoputro et al. [29].

4. Conclusions

The conclusion from this research was that conceptual hydrology model could simulate base flow well. Based on the statement, conceptual hydrology model could be developed to predict the effect of the change in land use on base flow changes.

References

[1] Harto S 1993 Analisis Hidrologi (Hydrological Analysis) (Jakarta: Gramedia Pustaka)
[2] Limantara L M 2010 Hidrologi Praktis (Practical Hydrology) (Bandung: Lubuk Agung)
[3] Viessman W, Knapp J W, Lewis G L and Harbaugh T E 1977 Introduction to Hydrology (Newyork: Harper and Row)
[4] Stewart M K 2015 Hydrol. Earth Syst. Sci. 19 2587–2603
[5] Zhang R, Li Q, Chow T L, Li S and Danielescu S 2013 Hydrol. Process 27 2659–2665
[6] Nathan R J and McMahon T A 1990 Water Resources Research 26 1465–1473
[7] Nejadhashemi A P, Shirmohammadi A and Montas H J 2003 Evaluation of Streamflow Partitioning Methods, Pap. No. 032183 in ASAE Annual International Meeting, edited by St Joseph M (Las Vegas, Nevada, USA)
[8] Knox J C 2006 Geomorphology 79 286–310
[9] Price K 2011 Prog. Phys. Geogr 35 465–492
[10] Varis O and Vakkilainen P 2001 Geomorphology 41 93–104
[11] Zhuang D F, Deng X Z, Zhan J Y and Zhao T 2002 Geogr. Res 21 667–674
[12] Dou X, Chen B, Black T A, Jassal R S and Che M 2015 Forests 6 1897–1921

[13] Huang X D, Shi Z H, Fang N F and Li X Forests 7(1) 16
[14] Danish Hydraulic Institute 1982 NAM Model Users Guide (Denmark: Danish Hydraulic Institute)
[15] Sugawara M, Watanabe T, Ozaki E and Katsuyame Y 1983 Tank model (Tsukuba: National Research Center for Disaster Prevention)
[16] Harlin J 1991 Hydrol 22 15-36
[17] Jain S K 1993 Sciences -Journal- des Sciences Hydrologiques 38(5) 431-441
[18] Xu C 2002 Hydrologic Models (Uppsala: Uppsala University Department of Earth Sciences Hydrology)
[19] Suripin 2004 Sistem Drainase yang Berkelanjutan (Sustainable Drainage System) (Yogyakarta: Andi Offset)
[20] Vargas T B, Uribe C M, Aristizábal E F G and Martínez J F E 2017 Revista Facultad de Ingeniería 85 70-85
[21] SNI 03-2415-1991 2004 Tata Cara Pengendalian Debit Banjir (Procedures for Control of Flood) (Jakarta: National Standardization Agency)
[22] Kuok K K, Harun S, and Chiu PC 2011 Int. J. Advance. Soft Comput. Appl. 3 (3) 1-17
[23] Gregor M 2010 BFI +3.0 User's Manual (Bratislava: Hydro Office)
[24] Sutapa I W 2013 Pengaruh Perubahan Iklim terhadap Pemodelan Debit (Effects of Climate Change on Discharge Modeling) Dissertation (unpublished) (Malang: Universitas Brawijaya)
[25] Silva M G, Netto A O A, Neves R J J, Vasco A N V, Almeida C and Faccioli G G 2015 Journal of Environmental Protection 6(8) 837-850
[26] Phien, H N and Pradhan P S 1983 Water SA 9 (1) 93-102
[27] Takegami S 1980 JSIDRE 86 36-41,a1
[28] Tsutsumi A, Jinno K and Berndtsson R 2004 Hydrological Sciences 49(2) 205-226
[29] Suryoputro N, Suhardjono, Soetopo W, Suhartanto E, Limantara L M and Utaya S 2018 Journal of Environmental Research, Engineering and Management 74(3) 34-41

IOP Conf. Series: Materials Science and Engineering **669** (2019) 012031 doi:10.1088/1757-899X/669/1/012031

The effect of white soil on geopolymer mortar porosity

Y A Priastiwi [1], Muhrozi[1], R Y Adi[1], D Daryanto[1] and Z Salamsyah[1]

[1]Department of Civil Engineering, Diponegoro University, Indonesia

dwidaryanto23@gmail.com

Abstract. This study analysed the effect of white soil as a filler on geopolymer mortar. The materials used were fly ash, white soil from Kupang as fillers, sand, water, and NaOH as alkaline activator. The comparison of geopolymer mortar mixture was 1 Pca: 3 Sand with water factor 0.5. The Pca was composed by white soil and fly ash with alkaline activator and water. The percentage of white soil used were 5, 10, 15, 20, 25 and 30%. The white soil chosen passed the number 200 sieve. Moreover, the dimension of mortar used was 5 x 5 x 5 cm. To remove the water content from the mortar, the oven was set at 60 °C for 24 hours until the mortar weight did not change. The result shows that porosity decreases with the increase of white soil percentage. Increase in the white soil percentage by 5, 10, 15 and 20% can reduce porosity by 0.9, 2.7, 2.5 and 2.9%. Nevertheless, the white soil, with a percentage of 25 and 30, increased porosity by 3.7 and 3.8%, respectively. This happened because in these percentages, the white soil hardens faster than the other percentages, impacting the compaction process on the mould.

1. Introduction

The geopolymers are materials that are produced from the reaction of materials containing a lot of silicon and aluminium with alkaline activators [1]. Many of these elements are found in fly ash, a waste material from the Electric power plant centre (PLTU), which is the remaining after the coal combustion. The use of geopolymers can reduce gas dioxide emissions, which is one of the causes of the greenhouse effect [2]. The use of geopolymer technology can reduce about 80% of carbon dioxide gas emissions into the atmosphere, caused by the cement and aggregate industries [3]. The other advantages are: it can reduce carbon dioxide gas emissions by reducing fly ash (the environmental pollutant waste) and can be used as building materials [4].

In the Kupang area of East Nusa Tenggara, there are natural ingredients called white soil. The white soil is a rock that has undergone changes in chemical composition, caused by weathering and the influence of underground water conditions. The white soil has brownish-white features; it is compact and rather difficult to dig with simple tools. It has chemical constituents, namely SiO_2, P_2O_5, Al_2O_3, Fe_2O_3, FeO, MnO, MgO, CaO, Na_2O, K_2O, TiO_2 and SO_3 [5]. People in Kupang use this white soil to construct pavement and light concrete by mixing it with cement [6]. The availability of white soil in Kupang area is quite abundant. Therefore, other utilisations of white soil must be done. Thus, the white soil can be used more effectively. In Kupang, there is also a Bolok PLTU, which produces fly ash from the remaining of coal combustion. However, the waste fly ash, if not properly treated will potentially be an environmental pollutant. Therefore, a study is needed to overcome this potential environmental pollution in Bolok PLTU.

To overcome the potential problem of pollution by fly ash, it can be used as a geopolymer material with the white soil as its filler. The white soil was chosen as a geopolymer filler because the majority of the white soil constituent material is calcium (Ca), which was carried out by the research [7] [8]. The calcium content can strengthen geopolymer specimens so that, it is expected to produce smaller porosity than the normal geopolymer mortar.

2. Research Methods

2.1. Mortar Geopolymer Composite Material

The composition of the geopolymer mortar are fly ash, white soil, fine aggregate, alkaline activator (NaOH) and water. The function of fly ash is to work as a binder and white soil as a filler. These materials, mixed with NaOH and water, formed geopolymer binders. The material used must comply with the requirements to produce good quality mortar.

2.1.1. White soil from Kupang, NTT.

The white soil is a natural product from Kupang, East Nusa Tenggara. The white soil has several chemical compounds, including calcium, magnesium, silica, alumina and iron, expressed in the form of oxides (CaO, MgO, SiO_2, Al_2O_3, Fe_2O_3 [9]. The white soil used was particles of soil with a size smaller than 0.074 mm or passed the number 200 sieve, as shown in Figure 1(a).

2.1.2. Fly Ash

The fly ash is a mineral residue in fine grains, produced from combustion of refined coal at a power plant center. The fly ash used was class F from Tanjung Jati B Jepara PLTU. The fly ash used had the particle size smaller than 0.074 mm or passed the number 200 sieve, as shown in Figure 1(b).

2.1.3. Fine Aggregate

The fine aggregate or sand used in this study was Muntilan sand. According to fine aggregate requirements based on American Standard Testing and Materials (ASTM) C33[10], the maximum diameter of sand was 4.76 mm and minimum diameter was 0.074, as shown in Figure 1(c).

2.1.4. Alkaline Activator

The alkaline activator is a compound that is used to facilitate the condensation polymerization reaction on the geopolymer mortar. The alkali activator used was sodium hydroxide (NaOH), commonly known as caustic soda, as shown in Figure 1(d)

(a) White soil **(b)** Fly Ash **(c)** Sand **(d)** NaOH

Figure 1. Constituent material of geopolymer mortar

2.2. Making Mortar Test Objects

In conducting this research, several tools were needed to support the work process and testing, which are as follows:

2.2.1. Tools

- Mortar molds with dimensions of 5x5x5 cm,
- Measuring Cup
- Pounder
- Porcelain bowl
- Rubber Hammer
- Tray
- Scales
- Mixer
- Small spade
- Covenant, and others

2.2.2. Composition of Mortar Geopolymer

In this study, the composition of the geopolymer mortar were fly ash, white soil, fine aggregate, alkaline activator (NaOH) and water. The comparison of geopolymer mortar mixture was 1Pca: 3Ps with 0.5 water factor. The Pca is white soil and fly ash, and Ps is sand. There are 7 variations of test object with different compositions according to the percentage of white soil filler, which varies from 0% to 30%. This percentage is based on trial error methods to get the best composition. In each variation, 6 object were made. The composition of mortar mixtures as shown in the Table 1.

Tabel 1. Composition of mortar geopolymer mixtures

Test Object	White soil filler (kg)	Fly ash (kg)	Sand (kg)	Water (kg)	(% White soil to binder)
1.	0	3	9	1.5	0% white soil filler
2.	0.150	2.850	9	1.5	5% white soil filler
3	0.300	2.700	9	1.5	10% white soil filler
4.	0.450	2.550	9	1.5	15% white soil filler
5.	0.600	2.400	9	1.5	20% white soil filler
6.	0.750	2.250	9	1.5	25% white soil filler
7.	0.300	2.100	9	1.5	30% white soil filler

Usage of 8 M NaOH

Calculation of NaOH molarity with 8 M NaOH

M = 8 M

Water = 1.5 kg = 1.5 liters = 1500 ml

Mr NaOH = 40

$$M = \frac{NaOH\ weight}{Mr} \text{ x } \frac{1000}{Water\ Volume}$$

$$8 = \frac{NaOH\ Weight}{40} \text{ x } \frac{1000}{1500}$$

NaOH Weight = 480 gram

2.2.3. Method for Making Test Objects

- White soil and fly ash used were particle that pass sieve no. 200. The white soil had water content of 0%, as shown in Figure 2(a) , Figure 2(b), Figure 3(a) and Figure 3(b).

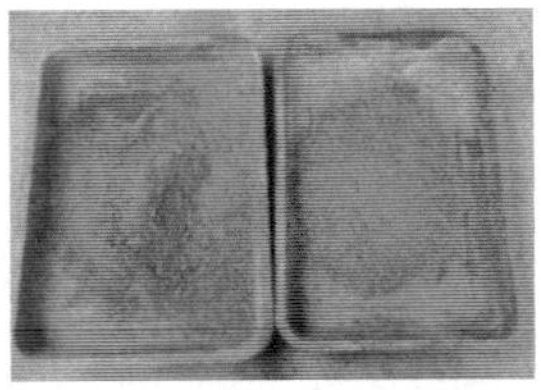

(a) (b)

Figure 2. (a) Fine fly ash and **(b)** Original fly ash

(a) (b)

Figure 3. (a) Original condition and (b) fine condition of white soil

- The white soil, fly ash, sand, water, and NaOH were weighed as planned.
- The water and NaOH were mixed. Because of the heat produced during the mixing process, some waiting time was needed for the sample to be cold.
- The white soil, fly ash and sand were taken into the mixer.
- The engine was turned on, so that all the previous ingredients were evenly mixed for 2 minutes, then the mixer was turned off.

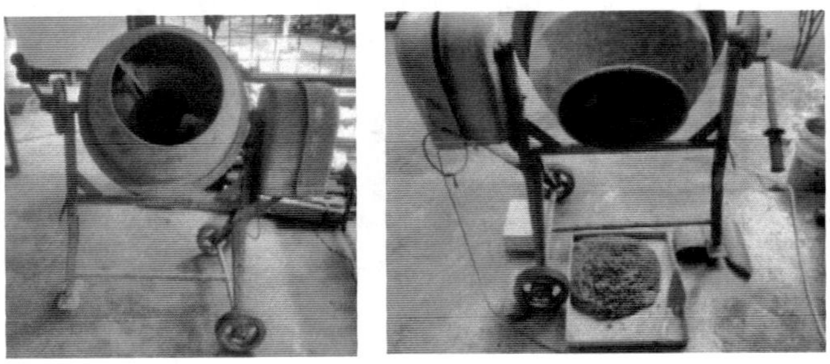

(a) (b)

Figure 4. (a) Mini mixer for stirring mortar and (b) dough pouring process into baking pan

- The water and NaOH were taken into the mixer.
- The mixer was turned on, so that all ingredients were mixed evenly for 5 minutes.
- The mixture was poured into a baking pan.
- The mortar mixture was added into the oil-coated mould. The mixture was put into 2 layers of mould, each layer was crushed evenly 32 times, and the top surface of the mortar was flattened (Figure 5)

- The test items were stored for 24 hours.

Figure 5. Mortar mould

2.2.4. Porosity Test on Mortar

The porosity is the ratio of pore volume (volume occupied by fluid) to the total volume of mortar (volume of the test object). The purpose of this test is to determine the percentage of the mortar pores to the volume of geopolymer mortar. The porosity affects the durability and compressive strength of the mortar. Reduced porosity leads to higher durability and compressive strength of the mortar. The porosity test stages are as follows:

- The mortar was removed from the mould after it is 1 days old.
- The mortar specimens were put into an oven with a temperature of 60 °C, as per the research in another study [11]. The test material in the oven was tested until the weight of the mortar became unchanged. The temperature of 60 °C was considered so that the mortar does not experience a shaking, resulting in increased mortar porosity.

Figure 6. Mortar covenant

- The test object was removed from the oven and waited until it is cold
- The weight of dry mortar (A) was weighed.
- The mortar was soaked until no bubbles were coming out of the mortar.
- The weight of mortar in water was considered as C.
- The mortar was lifted, and the surface of the mortar was wiped with a cloth to get the saturated surface dry (SSD) condition.
- The weight of the mortar in SSD condition was considered as B.

The mortar conditions are shown in Figure 7 to Figure 9.

(a) (b)

Figure 7. The condition of mortar (a) after unloading and (b) after the oven

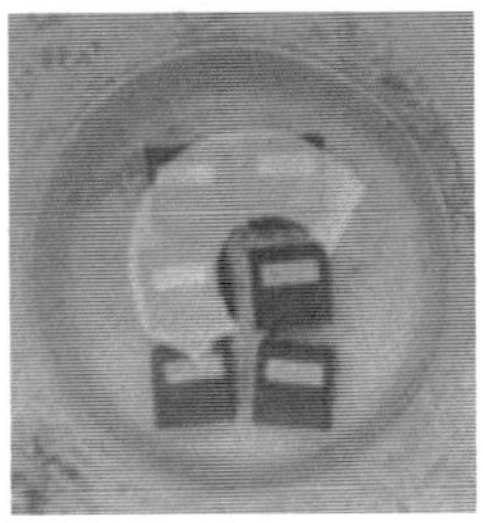

Figure 8. Soaking the mortar

(a) (b)

Figure 9. (a) Mortar in dry conditions and (b) mortar in SSD condition

Porosity of mortar can be calculated using the formula (1):

$$Porosity = \frac{B-A}{B-C} \: x \: 100\% \tag{1}$$

with:
A = the weight of dry mortar (kg)
B = the weight of mortar in SSD condition (kg)
C = the weight of mortar in water (kg)

3. Result and analysis

The results of porosity testing on the 7 variations, according to percentage of white soil in the mortar, are shown in Table 2.

Table 2. Mortar geopolymer porosity

Test object	Weight (kg)			B-A	B-C	Porosity (%)
	Dry Oven (A)	SSD (B)	In Water (C)			
1	0.25403	0.27327	0.12438	0.01923	0.14888	12.9
2	0.26500	0.28393	0.12697	0.01893	0.15696	12.1
3	0.27284	0.28947	0.12675	0.01662	0.16272	10.2
4	0.29838	0.31668	0.14053	0.01829	0.17614	10.4
5	0.28904	0.30617	0.13504	0.01712	0.17113	10.0
6	0.26350	0.28903	0.13504	0.02552	0.15398	16.6
7	0.26599	0.29222	0.13572	0.02623	0.15650	16.8

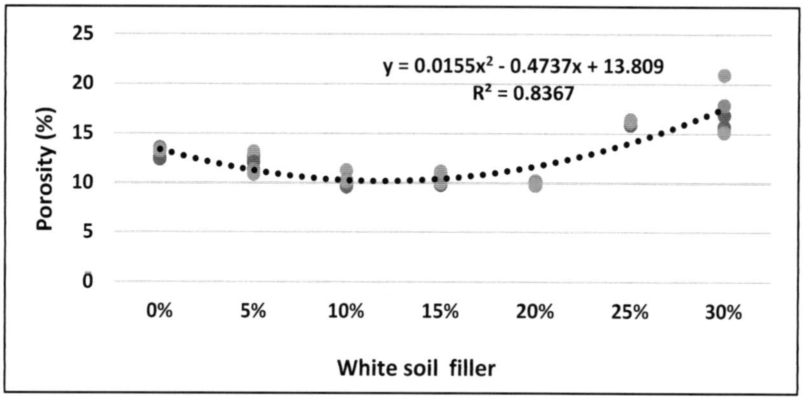

Figure 10. Regression of porosity value with white soil filler variations

Figure 10 shows the value of $R^2 = 0.8367$, which means that the white soil filler affects geopolymer porosity. The porosity of 83.67% is influenced by the white soil and the rest (100% - 83.67% = 16.33%) is influenced by other variables.

The equation to calculate optimum percentage of white soil to produce a minimum porosity value is as follows formula (2):

$$P = 0.0155x^2 - 0.4737x + 13.809 \quad (2)$$

with:

P = Mortar porosity (%)

x = White soil filler in mortar geopolymer (%)

P minimum occurs on $dP/d(x) = 0$

Then, $dP/d(x) = 0.031x - 0.4737$

From the above equation, the x and P values are obtained as follows:

$$x = \frac{0.4737}{0.031} = 15.28$$

By entering the value of x into the initial equation, it is obtained:

$$P = 0.0155x(15.28)^2 - 0.4737x(15.28) + 13.809$$
$$P = 10.19$$

From the result of the equation above, the value of $x = 15.28$. It means that the optimum value of white soil filler in geopolymer mortar is 15.28%, which will result in a minimum mortar porosity value of 10.19%.

4. Conclusion

Based on the results of the analysis, the following conclusions can be made in this study:

1. There is a reduction of porosity with the increment of white soil percentage of 5, 10, 15 and 20. The reduction are 0.9, 2.7, 2.5 and 2.9%, respectively. However, 25 and 30% of white soil have increased porosity by 3.7 and 3.8%. From the results of the analysis, it was concluded that the white soil is suitable to be used as a filler, with the percentage of filler being 5 to 20 of the binder. The white soil filler of 20% produced the smallest porosity.
2. The increase in white soil easily hardens the dough, making the compaction difficult. This is considered as the cause of higher porosity in the type of mortar with 25% filler and 30% white soil.

References

[1] Davidovits J 1988 Soft Mineralogy and Geopolymer. *Proceedings of The Geopolymer 88 International* Conference The Universitie de Technologie Compiegne France

[2] McCaffrey R 2002 Climate Change and Cement Industry Global Cement and Lime Magazine *(Environmental Special Issue)* p15-19

[3] Davidovits J 1994 Global Warming Impact on the Cement and Agregates Industriees. *World Resource Review*. 6(2) p263-278

[4] Ekaputri J J, Triwulan and Damayanti O 2007 Sifat Mekanik Beton Geopolimer Berbahan Dasar Fly Ash Paiton sebagai Material Alternatif *Jurnal Pondasi* V13(2) p124-133

[5] Hunggurami E, Simatupang P H and Lori A L 2015 Studi Kelayakan Penggunaan Tanah Putih Sebagai Pengganti Agregat Halus (Pasir) Terhadap Kualitas Beton *Jurnal Teknik Sipil* V4(1) p29-38

[6] Hunggurami E, Bunganaen W and Muskanan R Y 2014 Study Eksperimental Kuat Tekan dan Serapan Air Bata Ringan Cellular Lightweight Concrete dengan Tanah Putih sebagai Agregat *Jurnal Teknik Sipil* V3(2) p125-136

[7] As'at P, Anzila N A, Martyana D C and Hendra 2013 Kuat Tekan Beton Geopolimer dengan Bahan Utama Bubuk Lumpur Lapindo dan Kapur Konteks 7 Surakarta Material M129-36

[8] Satria J, Sugiarto A, Antoni and Hardjito D 2016 Karakteristik Beton Geopolimer Berdasarkan Variasi Waktu Pengambilan Fly Ash p1-8

[9] Parhulutan S and Fernando G 2018 Studi Kelayakan Pemanfaatan Campuran Fly Ash dan Tanah Putih Sebagai Semen Alternatif *Jurnal Karya Teknik Sipil* p8-10

[10] ASTM C33 2013 *Standard Specification for Concrete Aggregates. ASTM International. US.*

[11] Wallah S E 2014 Pengaruh Perawatan dan Umur Terhadap Kuat Tekan Beton Geopolimer Berbasis Abu Terbang *Jurnal Ilmiah Media Engineering* 4(1) p1-7

Acknowledgments

Special gratitude is expressed to the Faculty of Engineering, Diponegoro University, which has given permission to conduct research in the Diponegoro University laboratory and support research in the form of Competitive Grants TA 2019.

Determining predictor variables of HC, CO, and CO2 emissions using decision tree models

A Larasati[1], Y W Chen[2], A M Hajji[3], A F P Mahardika[1] and V E B Darmawan[1]

[1] Department of Industrial Engineering, Universitas Negeri Malang, Indonesia
[2] Department of Industrial Engineering and Management, Da-Yeh University, Taiwan
[3] Department of Civil Engineering, Universitas Negeri Malang, Indonesia

aisyah.larasati.ft@um.ac.id

Abstract. HC, CO, and CO_2 are three gasses emitted from heavy duty construction equipment, such as backhoe that contain carbon and hugely affect the environment. This research aims to determine predictor variables of HC, CO, and CO_2 emissions level using decision tree. The emissions level is categorized into three classes: low, medium, and high. The predictor variables decision tree are related to the backhoe operation and specification, including backhoe type, engine technology tier, RPM, MAP, horsepower, backhoe age, and intake temperature. This study runs 12 cycles for each model to classify HC, CO, and CO_2 level. The results indicate that each gas has a different order of the important level for the predictor variable. Decision tree model to classify HC emission level show the top three most important predictor variables are RPM, temperature, and MAP. On the other hand, the top three most important variables to classify CO are backhoe type, RPM, and temperature. The last decision tree model to classify CO_2 level show RPM, MAP, and backhoe type as the most important predictor variables.

1. Introduction

Mostly during construction operations, air pollutant emissions are produced from the exhaust of heavy equipment at the construction site [1]. Backhoe is one of heavy equipment used in the construction site. Air emissions produced by the equipment are vary, three of them are Hydrocarbons (HC), Carbon monoxide (CO), and Carbon dioxide (CO_2). CO_2 or carbon dioxide is one of gas that gives hugely impact on environment, whether it is resulted from bikes, car, or heavy equipment. Mostly, CO_2 is considered to be the major contributing component of the global warming [2]. Hydrocarbons (HC) are composed of unburned fuels as a result of lack in temperature. Normally, diesel engines emits low levels of hydrocarbons, because HC is commonly occurred at light loads [3]. CO or carbon monoxide are results from the incomplete combustion. CO is created when the process of oxidation does not occurred perfectly[3]. Even though CO produced during the droplets in diesel engine is big enough, it is minimal chances to occur in diesel engine, especially backhoe. Because of the combustion in diesel engine are lean combustion engines, nonetheless the concentration of HC and CO is minimal.

Mostly carbon dioxide are produced even the backhoe are in idle or non-idle conditions. The emissions value which have been increasing day after day making the company to work on reducing the pollutant emission from heavy equipment. Currently 35% reduction in usage time of backhoe leads to a reduction of approximately 15% in total emissions of backhoe on site construction [4]. Therefore estimating the emissions gas produced are giving impact on environment around the construction site. And by using predictive model are also can estimating the emissions produced by heavy equipment [5].

Currently, the emissions are estimated mostly from NONROAD model from the EPA, because it was difficult to estimate the emissions from backhoe [1]. That is why this study aims to use data mining with a classification model to estimate the class of the emissions released from backhoe. Decision tree model is one of a technique in data mining which is classified by following the path from the root of the tree to the leaf and it determines which label class it goes [6]. Decision tree is also considered easy to use and to understand. The tree are mostly concise. The large tree with a lot of nodes or branch is less understandable. The output of decision tree are just not a trees that gives nodes and class label that has been explained above, but also an accuracy rate that can be used to measure the accuracy of the tree model [7]. Decision tree can also extract some of the variance variables from the training data and then increases the accuracy of the lower level branches [8]. In brief, decision tree look alike shown in figure 1.

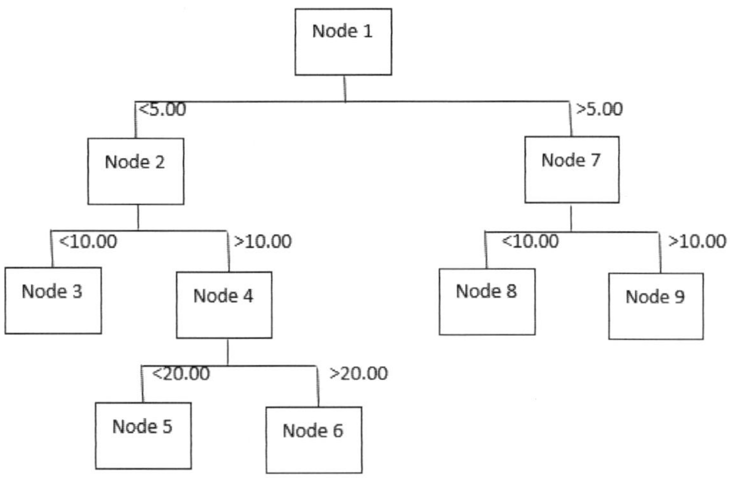

Figure 1. Decision tree

Based on Figure 1 above, every branch has nodes and it contains the classification class. It is why classification are playing important role in decision tree model. If the data isn't classified then the class label in the trees cannot be explained or probably there will not be any class label in the trees. Classification sometimes is not effective when dealing with numerical data, because it will be has very large number of ways to form the ranges of data and cover it. However, discretization of the numerical data can be a solution to the problem [9]. In this study, HC, CO, and CO_2 are classify into low, middle, high level. It means creating three class can be applied to classify the intention level of the gas emission. Because emissions of backhoe are gasses and it units are mass per time, creating three classes are suitable. This research focuses on the estimating emissions gas which is HC, CO, and CO_2 by classifying it classes into three using Decision Tree model.

2. Research Methodology

This study used secondary data about gas emissions produced by backhoe construction equipment such as CO, HC, and CO2. The secondary data was measured by a research team from North Carolina State University using portable emissions measurement system (PEMS) [10]. Each of gas produced by backhoe are measured in mass per time (g/s) because of 50% total emissions occurred during the operation time [11]. The dependent variables used are HC, CO, and CO_2 level, while the independent variables used are Normalized MAP (Manifold Absolute Pressure), RPM (Revolutions per Minute), Backhoe type, Backhoe model year (age), Temperature, Engine Technology Tier, and Horsepower. Decision Tree model is used to classify each emissions gas that uses a tree graph decisions, which concludes the value of dependent variables such as HC, CO, CO_2 by given the value of independent variables. The method used in this study briefly shown in Figure 2 where the process flows from data preparation to interpreting the output model.

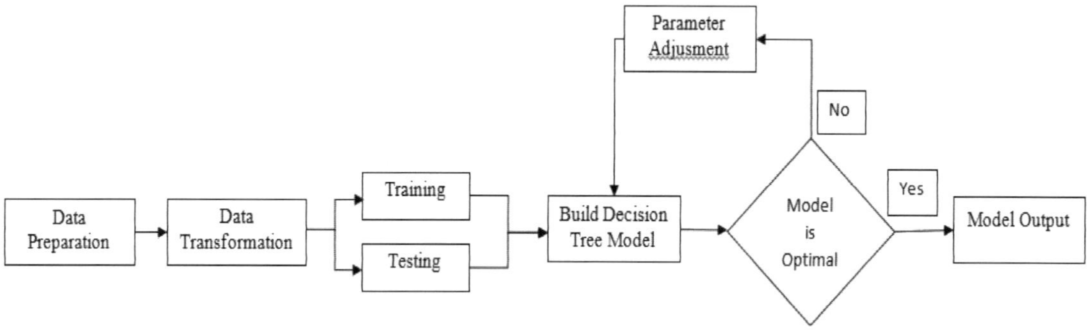

Figure 2. Flowchart of research method

Data preparation is performed on IBM SPSS Modeler by selecting data that are used in the model, and determining the dependent variable, which is HC, CO, and CO_2. Not only selecting variables and determining the target, preparation phase also handling missing value on each variables dataset. Missing value may lead to bias in the model. That is why it is important to not have missing values on the variables, especially on target variables [7]. This study also performs data normalization before finishing data preparation phase. Data of CO_2 is still measured by mass per time (g/s) but not for HC and CO that have been measured in g/hr. Thus, this study transforms CO_2 into mass per hour (g/hr). Other thing, since the range of HC and CO are much less than CO_2, this study do data normalization in order to make these three data set have the same range 0-1.

Data preparation are finished and the next step is binning the data. The dependent variables is classified by binning it into three classes based on equal count. But not all gasses can be classified equal. Because if it classified into 3, it will give 33.33% cases on each classes, and the range will be not evenly distributed. So every dependent variables is classified with different upper limit for every classes that can be shown in table 1.

Table 1. Upper limit value

Class	HC (gram/hour)	CO (gram/hour)	CO_2 (gram/hour)
1	6.0	13.5	2.0
2	12.0	29.5	4.0
3	182.0	2589.0	13.3

After that split the data into two datasets, training and testing. Each training and testing are contained of 50 points respectively. The next step is to put the model with decision tree. There are a lot Parameter used such as pruning value, minimum records per branch, and number of trials. Pruning value used to make a concise tree. By decreasing the value smaller means create more severe pruning, and accurate tree [12]. Set the minimum records branch also can be used to limit the number of branch splits of the tree, because it can grow exponentially large with the maximum value of it is 2^n [8]. Number of trials is a method that allows the model to boost it for improving the accuracy rate by forming the basis for building a subsequent tree that forced to focus on the error of the model before [12]. Basically, it is creating a lot of models to boost the accuracy rate. This model can be said optimal measured by the highest accuracy can get on testing datasets.

In this research, a total of twelve cycles had been running with a different parameter given, from cycle 1 until cycle 12. Every cycle has different pruning severity value, and minimum records per branch value from the smallest value until the highest value can be set. The smallest value can be given for pruning severity value is 0 while the highest is 100. And for minimum records per branch can be set for smallest is 2 while the highest given in this research is 64. Number of trials this research set is 50, so the model is created for 50 times to boost the accuracy rate. The model was stopped when it reached the lowest point of accuracy obtained in testing datasets. The last step is to see the output which cycle gave the highest accuracy it get.

3. Results and Discussion

By initializing different parameters with total of twelve cycles such as pruning value and minimum records of branch, the accuracy results can be seen in table 2.

Table 2. The Result of Accuracy of the model.

	Parameter	Accuracy Level		
		HC (%)	CO (%)	CO_2 (%)
Cycle 1	Pruning Value : 75 Min branch : 2	77.34	85.19	89.84
Cycle 2	Pruning Value : 35 Min branch : 10	77.10	85.03	89.79
Cycle 3	Pruning Value : 10 Min branch : 30	75.41	84.20	89.74
Cycle 4	Pruning Value : 100 Min branch : 10	74	82.90	89.12
Cycle 5	Pruning Value : 75 Min branch : 30	75.73	84.31	89.70
Cycle 6	Pruning Value : 100 Min branch : 30	72.75	82.35	89.25
Cycle 7	Pruning Value : 55 Min branch : 23	75.49	84.63	89.72
Cycle 8	Pruning Value : 45 Min branch : 10	77.25	85.25	89.69
Cycle 9	Pruning Value : 0 Min branch : 40	74.96	84.53	89.58
Cycle 10	Pruning Value : 25 Min branch : 30	75.43	84.38	89.70
Cycle 11	Pruning Value : 40 Min branch : 30	74.96	84.54	89.69
Cycle 12	Pruning Value : 75 Min branch : 64	74.16	83.67	89.27

In table 2 above, the accuracy rate given is taken from the output of the testing datasets. Cycle 1 is a simple parameter without changing any default value. The result of the accuracy is that HC gives 77.34%, CO gives 85.19%, and CO_2 gives 89.84%. Given pruning value is 75 and minimum branch is 2, CO_2 and HC simply result in the best accuracy, but CO does not perform at the same level. The highest accuracy of CO classification model is obtained in cycle 8, which gives accuracy of 85.25% with decreasing pruning value into 45 and increasing the minimum branch into 10. Based on the accuracy results, increasing the value of minimum records per branch eventually does not increase the accuracy, except for CO gas classification model. Normally, more complex and size of the branch in tree will give higher accuracy [13]. Because the range of data on each classes in CO data set is much higher than those ones in HC and CO_2 data sets, thus when the number of minimum value increases, the accuracy of CO classification model also increases.

Normalizing datasets also affects the accuracy level of the classification models. Moreover, data normalization ensures each independent variable in the classification model have a fair and equal numerical contribution. Thus, it is important that each class has an equal importance for the classification decision [14]. The existing of non-normal distributed data interferes and lead to bias in the algorithm, as a result it lowers the classification model performance, which is accuracy. Boosting the model may also increases the accuracy of the model. Because the given parameter is 50 for the number of trials, it means the classification model creates 50 times of model being run, which makes the model accuracy increased. The bigger the number of boosting cycle does not guarantee higher accuracy [13].

The other useful result from the classification model is the predictor importance level, which explain how important one variable relatively to the other variables. The predictor importance for each independent variable are shown in Table 3.

Table 3. Predictor Importance HC, CO, and CO_2

Independent Variable	Relative Importance Level		
	HC (%)	CO (%)	CO_2 (%)
RPM	66	33	55
Normal MAP	8	2	32
Backhoe type	-	40	9
Horsepower	-	1	3
Age (year)	8	-	1
Engine Tier	5	1	-
Temperature	13	23	-

Table 3 implies that most independent variables that gives impact on estimating emissions is RPM for three gasses (HC, CO, & CO_2). RPM simply gives significant correlation with the fuel of diesel engine. MAP or Manifold Absolute pressure also gives highly contributes especially on CO_2, because the ability and size of MAP on different type will affect the emissions produced [15].

The most important results obtained from decision tree is the rules that is used to classify an input. The rules of the best model to classify CO_2, CO, and HC are shown in Table 4, Table 5, and Table 6 respectively. Table 4 indicates that the most important predictor variables within the rules to classify CO_2 at low level and high level are similar. Meanwhile, the rules to classify medium level CO_2 is a bit different since it includes intake temperature as one of the most important predictor variable.

Table 4. Rules to Predict CO_2 Emissions

Rules	Number of Rules	Three most important predictor variable within the rule
Rules 1 (low level)	12	Backhoe type, MAP, RPM
Rules 2 (medium level)	38	MAP, RPM, intake temperature
Rules 3 (high level)	13	Backhoe type, MAP, RPM

Table 5 indicates that each rules contains relatively different predictor variable to classify different level of CO emissions. All rules uses backhoe type and RPM as two important variables. However, the third predictor variable is different for each level. The low level CO emissions rule includes MAP, the medium level rule includes intake temperature, and the high level rule includes horsepower.

Table 5. Rules to Predict CO Emissions

Rules	Number of Rules	Three most important predictor variable within the rule
Rules 1 (low level)	22 rules	Backhoe type, RPM, MAP
Rules 2 (medium level)	23 rules	Backhoe type, RPM, Intake temperature
Rules 3 (high level)	17 rules	Backhoe type, Horsepower, RPM

Table 6 indicates that the most important predictor variables within the rules to classify HC at medium level and high level are similar. Meanwhile, the rules to classify low level HC is a bit different since it includes horsepower as one of the most important predictor variable.

Table 6. Rules to Predict HC Emissions

Rules	Number of Rules	Three most important predictor variable within the rule
Rules 1 (low level)	37	Backhoe type, Horsepower, MAP
Rules 2 (medium level)	43	Backhoe type, MAP, RPM
Rules 3 (high level)	34	Backhoe type, MAP, RPM

Based on the number of rules shown in Table 4, Table 5, and Table 6, it shows that the decision tree to predict HC emissions contains the most rules, which are more than 30 rules for classifying each level. It implies that the decision tree model to predict the class of HC emissions is more complex that the other two models. Furthermore, the rules to classify the medium level of the three gas emissions (CO, CO_2, and HC) are the most complex rules. It explains that the variety combination of data occurs in the medium level gas emissions than in the high or low level.

4. Conclusion

Decision tree can easily model the emissions gas produced by backhoe based on its specification as independent variables. Classification model is used to determine the class of emissions into three classes (low, medium, high), even though not all of the gas emissions model has a good performance (accuracy) on it. CO_2 classification model has the best performance (highest accuracy) because of the range of classes is not as high as CO and HC data sets. Normalization on the data sets also greatly effects on classifying the classes of gas emissions. Normalization effectively increases the classification performance, so the accuracy can be maximized.

Based on the best accuracy model, it conclude that each gas has a different order of the important level for the predictor variables. Furthermore, each class in the decision tree model may not have the same three most important variables in its rule. Decision tree model to classify HC emission level show the order of the predictor variables from the most important to the least important are RPM, temperature, MAP, age and engine technology tier. The three most important variables used in the rule to determine the class of HC emissions are the same for rules 2 and 3, but not for rules 1. The decision tree model to classify CO emission show the order of the predictor variable based on its important level are backhoe type, RPM, temperature, MAP, horsepower and engine tier. The three most important variables used in the rule to determine the class of CO emissions are different for each rules. The last decision tree model to classify CO_2 level show the important rank of the predictor variables are RPM, MAP, backhoe type, horsepower, and engine tier. The three most important variables used in the rule to determine the class of CO_2 emissions are the same for rules 1 and 3, but not for rules 2. Moreover, the independent variable that has the biggest importance to estimate either HC, CO, or CO_2 is RPM.

References

[1] Wang B Z, et al 2018 Assessment and management of air emissions and environmental impacts from the construction industry *J. Environ Plan Manag* vol 0 no. 0, pp. 1–24.

[2] Lewis P and Rasdorf W 2017 Fuel Use and Pollutant Emissions Taxonomy for Heavy Duty Diesel Construction Equipment *J. Manag Eng.* vol 33 no. 2

[3] Res A and Keskin A 2014 *The pollutant emissions from diesel-engine vehicles and exhaust aftertreatment systems.*

[4] Jassim H S H 2017 *Predicting Energy Consumption and CO 2 Emissions of Excavators in Earthwork Operations* : An Artificial Neural Network Model.

[5] Hajji A M, et al 2017 *Combining Off-the-Job Productivity Regression Model with EPA's NONROAD Model in Estimating CO_2 Emissions from Bulldozer* Civ. Eng. Dimens vol. 19 no.

2, pp. 73–78.
[6] Shaun R, et al 2008 *Educational Data Mining 2008 The 1 st International Conference on Educational Data Mining.*
[7] Song Y and Lu Y 2015 *Decision tree methods : applications for classification and prediction* vol 27 no 2, pp. 130–135.
[8] Brydon M and Gemino A 2008 *Classification trees and decision-analytic feedforward control : a case study from the video game industry*, pp. 317–342.
[9] Shao J and Tziatzios A 2018 *Data & Knowledge Engineering Mining range associations for classification and characterization Data Knowl. Eng* vol 118 no September 2017, pp. 92–106.
[10] Lewis P, et al 2015 Engine variable impact analysis of fuel use and emissions for heavy-duty diesel maintenance equipment *Transp Res Rec* vol 2482 no. 1, pp. 8–15.
[11] Sepasgozar S M, et al 2019 Methods for monitoring construction off-road vehicle emissions : a critical review for identifying deficiencies and directions *Environ Sci. Pollut. Res* vol. 26 no 16, pp 15779–15794.
[12] Brown E C, et al 2003 *National Park vegetation mapping using multitemporal Landsat 7 data and a decision tree classifier* vol 85, pp. 316–327.
[13] Rawal B and Agarwal R 2019 *Improving Accuracy of Classi fi cation Based on C4 . 5 Decision Tree Algorithm Using Big Data Analytics* Springer Singapore.
[14] Singh D and Singh B 2019 Investigating the impact of data normalization on classification performance *Appl. Soft Comput J,* p. 105524.
[15] Park Y, et al 2016 *Control of the air system of a diesel engine using the intake oxygen concentration and the manifold absolute pressure with nitrogen oxide feedback.*

Acknowledgement
The authors would like to acknowledge funding support from Universitas Negeri Malang (UM), which has funded this research through PNBP - UM Research Grant 2019.

Financial economic cost on gravel road maintenance: study using HDM-4

F Suwarto [1] and S Fauziyah [1]

[1]Diponegoro University, Semarang, Indonesia

fardzanela@live.undip.ac.id

Abstract. This study demonstrates the economic benefit analysis from two alternative project on gravel road surface using HDM-4 to access the most appropriate project and maintenance standard both in term of pavement remaining life and also road user cost (RUC). The two alternative projects will be treated with different work standards. First Alterative namely "Without Project" will only be maintain the existing project with grading every six months, spot re-gravelling to replace 30% of material lost each year, and gravel resurfacing. Second Alternative namely "With Project" will be maintaining and upgrading the existing road pavement to bituminous pavement in year 2019 for section A001-01 and for section A001-02 in year 2020, by doing grading every six months and spot re-gravelling to replace 30% of material lost each year. In conclusion, from two proposed alternatives, although it requires more funds at earlier stages 'with project' alternative is the most appropriate project considering the number of maintenance followed, overall road performance, RUC, and the values of NVP and IRR.

1. Introduction

More than 57% of rural road in Indonesian districts is in non-stable condition. Currently, unsealed roads are still common in some areas in Indonesia. Based on road condition data from the ministry of public works, total unsealed road pavement in Indonesia in the end of year 2016 is 218.260 km [1]. Whereas rural road is very important, as the development of rural roads brings multiple socio-economic benefits to the rural areas that form a strong base of the National economy so that the existence of gravel road needs particular attention and preservation to be able to comprise with the traffic. Road networks are always significant for national productivity, local economic growth and individual social wellbeing [2].

The perseverance of the roadway is to carry on the road traffic safely and comfortably within the economic benefit capacity in term of both road user cost and road user agency, therefore several aspects have to be taken into consideration when designing and planning maintenance of a roadway. The examination can be analysed and assessed based on the feature of road geometric, pavement condition and traffic volume with different treatment of project and maintenance. The purpose of road maintenance itself is to reduce the degree of road deterioration on a continuous basis. Different treatment that carried out will have different maintenance effect, therefore it will lead to different performance as over period of time.

The difficulty of these interactions and certain numbers of data means that a computer modelling is necessary. The best-known example of a road investment model is the Highway development and management tool (HDM-4) [3]. The main area of analysis that can be undertaken using HDM-4 are Project analysis, Programme analysis and Strategy analysis [4]. For the strategic analysis purpose, the

HDM-4 will propose the appropriate maintenance approaches and develop the economic analysis of the project by derive optimal capital and recurrent maintenance and hence preserve the road network in a good condition on a sustainable basis [5]. The prioritization for pavement maintenance is done on the basis of Net Present Value by Agency Capital Cost (NPV/CAP) value [6].

HDM-4 is a comprehensive model for prioritizing pavement rehabilitation and maintenance, this model generates the total life – cycle condition and coasts, and provide economic decision for multiple alternative maintenance [7]. The application of HDM-4 has been extensively used for the management of road networks in many countries. [8] performed the HDM-4 to define road works program and investment priorities of 13 roads Bosnia and Herzegovina local road. The application also used in Mumbai India as a tool to plan pavement maintenance strategies on 60 km road length of urban roads to maximize the value of NPV and the desirable IRI [9]. The utilization of the model to generate fuel consumption also has been reported by [10] and [11] In this research HDM-4 will propose the appropriate maintenance approaches and develop the economic analysis of the project.

Hence the aim of this report is to analyse and estimate the economic benefits from gravel road segment by conducting two different strategic planning. The project analysis will consider the economic feasibility of two projects alternative. The objectives of this report are:

- Analyse the prediction of road deterioration with road roughness as the reference
- Analyse the road user cost (RUC) of two project alternatives
- Analyse economic comparison of two the project alternatives

2. Methodology

This report demonstrates the economic benefit analysis from two alternative project on gravel road surface using HDM-4 to access the most appropriate project and maintenance standard both in term of pavement remaining life and also road user cost (RUC). The operation of the program is defined on the three stages, the first stages is define road data requirement including road network and vehicle fleet, the second stage is define the maintenance and road improvement work in the Work Standard sector, then the last stages in the Project Analysis section is to define the candidate project and the treatment required in each project.

2.1. Case Study Data

The Case study data has been given to be analysed by HDM-4, both alternative project classified as a secondary road and located in sub-tropical/humid climate. The road network is divided into section 1 and section 2 with different characteristic in terms of road geometric and pavement structure as given in appendix 1. The road length is assumed as 20 km length for section one and 30 km length for section two and both section have 7m carriage width with no shoulder width. For the structure, both sections are assumed almost identically with the thickness of gravel 145mm for section one and 140 for section two then the road roughness is given as high as 8.5 IRI for section one and 9 IRI for section two.

Road traffic composition is assumed as low traffic class and the heaviest vehicle type will be 4+ axle arctic with the traffic growth maximum 4% per year as seen in table 1. In addition, the road sections is defined as State Road and vehicle fleet are within National Fleet.

Table 1. Traffic data

Vehicle Name	Annual Growth	Traffic Composition	
		Section One	Section Two
2-axle Truck	4%	32	27
3-axle Heavy Truck	4%	15	25
4+axle Artic	4%	15	15
Bus	4%	60	45

Table 1. Traffic data (cont.)

Vehicle Name	Annual Growth	Traffic Composition	
		Section One	Section Two
Motor Car	4%	115	95
Motorcycles	4%	35	35
Utilities van	4%	48	40
TOTAL		320	272

2.2. Road Work Standard

The two alternative projects will be treated with different work standards. First Alterative namely "Without Project" will only be maintain the existing project with grading every six months, spot re-gravelling to replace 30% of material lost each year, and gravel resurfacing. Second Alternative namely "With Project" will be maintaining and upgrading the existing road pavement to bituminous pavement in year 2019 for section A001-01 and for section A001-02 in year 2020, by doing grading every six months and spot re-gravelling to replace 30% of material lost each year.

For the purpose to carry out the project, upgrading and road maintenance standard have to be defined in the HDM-4 software under Work Standard sector, the associated works are given in table 2 as follow. For 'Without Project' alternative, the work assigned is Gravel Road Maintenance. Whereas for 'With Project', the road network was given three main works which are maintenance standard for gravel road prior to upgrading, improvement standard representing the upgrading works, and maintenance standard for paved road after upgrading.

Table 2. Road works standard for each project alternative

Project alternative	Section ID		Road Works Standards	Effective from year	Maintenance works/ Improvement type
Without Project Maintain Gravel Road	A001-01	M	Gravel Road Maintenance (GRAVEL)	2019	Upgrading (GRADE6) Spot re-gravelling (SPG100) Gravel resurfacing (RESURF)
	A001-02	M	Gravel Road Maintenance (GRAVEL)	2019	Grading (GRADE6) Spot re-gravelling (SPG100) Gravel resurfacing (RESURF)
With Project Upgrade Gravel Road	A001-01	M	Maintenance before upgrading (BEFORE)	2019	Grading (GRADE6) Spot re-gravelling (SPG100)
		I	Pave Section A001-01 in 2019 (PAVE01)	2019	Upgrading
		M	Crack sealing, patching and resealing paved road (SEAPAT)	2020	Crack sealing (SEAL) Patching (PATCH) Resealing (RSL7)
	A001-02	M	Maintenance before upgrading (BEFORE)	2019	Grading (GRADE6) Spot re-gravelling (SPG100)
		I	Pave Section A001-02 in 2020 (PAVE02)	2020	Upgrading
		M	Crack sealing, patching and resealing paved road (SEAPAT)	2021	Crack sealing (SEAL) Patching (PATCH) Resealing (RSL7)

IOP Conf. Series: Materials Science and Engineering **669** (2019) 012033 doi:10.1088/1757-899X/669/1/012033

The breakdown of road standard for maintenance works including the intervention level and costs are given in table 3. Intervention level is the limit point of the defect level where if the limit is exceeded then certain maintenance must be conducted. As seen from the table that the trigger of intervention level for maintenance before and after road upgrading is different. For the maintenance before road upgrading the trigger is the maximum year and the gravel thickness, while for the maintenance after road upgrading the trigger is the number of cracking and potholes.

Table 3. Intervention limit and maintenance cost

	Road works standard	Effective from year	Maintenance works	Code	Intervention	Max IRI	Costs	
							Economic	Financial
M	Gravel Road Maintenance	2019	Grading	GRADE6	2 times per year	-	720 per km	900 per km
			Spot re-gravelling	SPG100	GT < 100mm	-	12 per m3	12 per m3
			Gravel resurfacing	RESURF	GT < 50mm	-	12 per m3	14.4 per m3
M	Maintenance before upgrading	2019	Grading	GRADE6	2 per year	-	720 per km	900 per km
			Spot re-gravelling	SPG100	GT < 100mm	-	12 per m3	12 per m3
M	Crack sealing and patching paved road	2020 (section1) 2021 (section2)	Crack sealing	CRSEAL	5% Cracking	9	5 per m2	6 per m2
			Patching	PATCH	10 ≤ Potholes	9	12 per m2	14.4 per m2
			Resealing	RSL7	Interval of 7 years	12	4 per m2	5 per m2

The breakdown of road improvement is given in table 4. where B is before upgrading and A is after upgrading.

Table 4. Upgrading works detail

Section ID		A001-01	A001-02
Section Name		**Section One**	**Section Two**
Speed flow type and Accident class	B	Two lane road	Two lane road
	A	Two lane standard	Two lane standard
Traffic flow pattern	B	Free-Flow	Free-Flow
	A	Free-Flow	Free-flow
Road class	B	Secondary	Secondary
	A	Primary	Primary
Surface class	B	Unsealed	Unsealed
	A	Bituminous	Bituminous
Pavement type	B	Gravel	Gravel
	A	Surface Treatment on Granular Base	Surface Treatment on Granular Base
Carriageway width (m)	B	W1	W2
	A	7.4	7.4

Table 4. Upgrading works detail (cont.)

Section ID		A001-01	A001-02
Section Name		**Section One**	**Section Two**
Speed limit (km/h)	B	100	80
	A	120	100
Upgraded SN (excluding subgrade)		1.84	1.71
Economic Cost		306000 per km	255000 per km
Financial Cost		360000 per km	300000 per km

2.3. *Project Alternative*
When all candidate projects have been identified, HDM-4 can then be used to compare the life cycle performance and costs predicted under the existing regime of pavement management (i.e. the 'without-project' case) against the life cycle costs predicted for the periodic maintenance, road improvement or development alternative (i.e. the 'with project' case).

3. Result Analysis

3.1. *Project Maintenance*

3.1.1. *Without project*

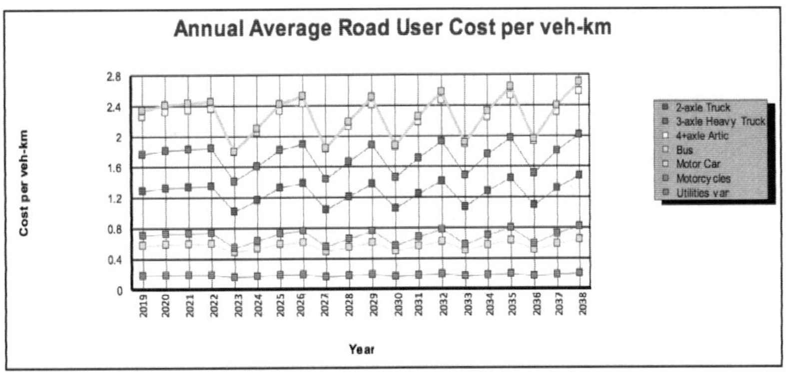

Figure 1. Annual average RUC section one without project

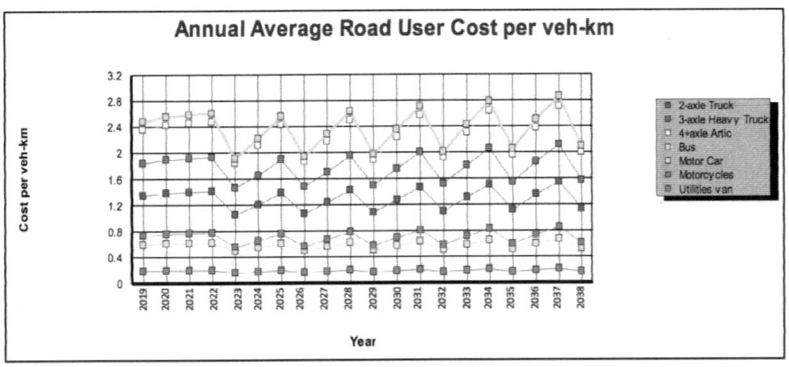

Figure 2. Annual average RUC section two without project

IOP Conf. Series: Materials Science and Engineering **669** (2019) 012033 doi:10.1088/1757-899X/669/1/012033

By not upgrading the roadway to a bituminous pavement, the Annual average RUC per veh-km will be increase gradually parallel with the increasing of road life. From the graph we can see that this road will need to carry on a routine maintenance every three years to comply with the deterioration of the road. After maintenance conducted, the RUC will be decreased significantly but then it will be rapidly increase again due to road deterioration by the increasing volume of vehicle.

3.1.2. With project

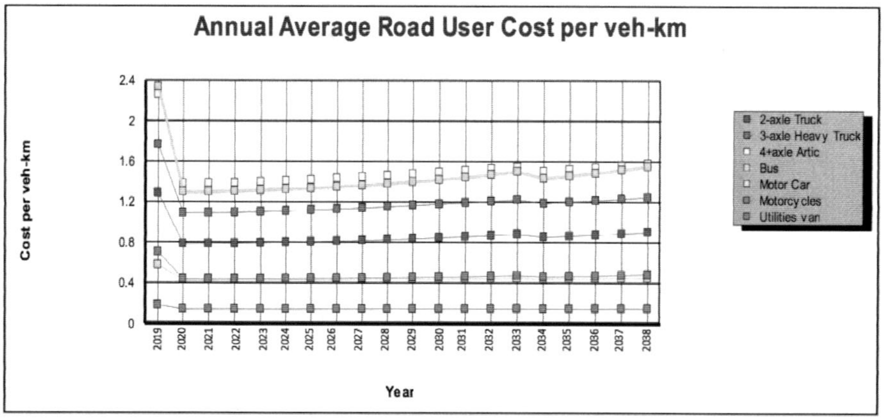

Figure 3. Annual average RUC section one with project

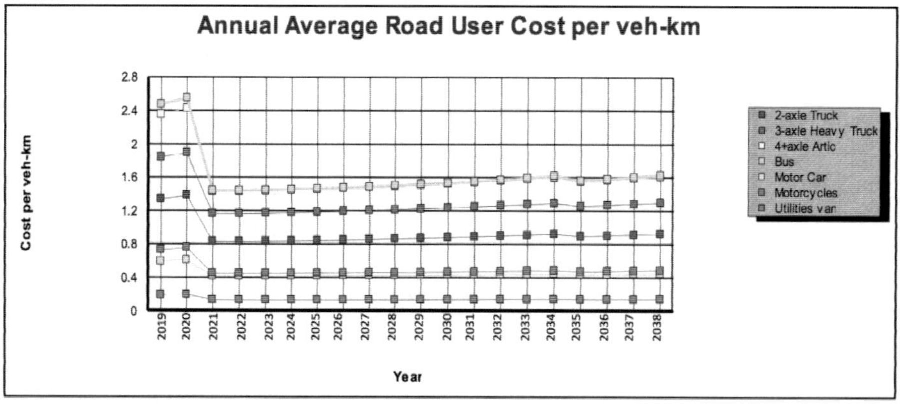

Figure 4. Annual average RUC section two with project

As seen from figure 3 and 4 both section one and section two shows the same behaviour, after the project was being implemented the Annual average RUC per veh-km is decreasing significantly. For heavy vehicle categorize the RUC decrease as much as 0.9 Dollar per km, while for light vehicle category the decreasing of RUC is not as much as the heavy vehicle category which is about 0.2 US Dollar/ km.

After the year of project was being conducted, the average RUC of the overall vehicle slowly start to increase, however the value of RUC until the end years of planning will not exceed the value of RUC before the project is being implemented. In addition, further routine road maintenance will be not required.

3.2. Summary of Economic Indicator

Table 5. Economic Indicator

Alternative	RAC	CAP	Increase in Agency Costs (C)	Decrease in User Costs (B)	Net Exogenous Benefits (E)	(NPV = B+E-C)	NPV/Cost Ratio (NPV/RAC)	NPV/Cost Ratio (NPV/CAP)	IRR
Without Project	1.822	0.893	0.000	0.000	0.000	0.000	0.000	0.000	0.000
With Project	14.272	14.161	12.450	20.053	0.000	7.603	0.533	0.537	17.6 (1)

With RAC = Present Value of Total Agency Costs
CAP = Present Value of Agency Capital Costs
NPV = Net Present Value
IRR = Internal Rate of Return

After define a series of treatment strategies, the optimum net present value (NPV) can be obtained. NPV is a method of appraising investment opportunities. The basic understanding of NPV is if the NPV is positive means that the project still has a value in the future. The economic summary indicates that With Project alternative giving a better investment opportunity at the end of analysis period in year 2038. At the beginning of analysis period, the Total Agency Cost (RAC) of With Project alternative is immensely high compare with without project alternative, consequently at the end of analysis period the Net Present Value which indicate the investment opportunities is higher than Without Project alternative. In addition, the Internal Rate of Return (IRR) of paved road also higher than the gravel road that is 17.6 compared with 0.00.

The internal rate of return (IRR) of a particular investment is the discount rate that, when applied to its future cash flows, will produce an NPV of precisely zero. In essence, it represents the yield from an investment opportunity. In accordance of the statement, in means that the value of NPV of with project will remain positive although subjects to discount rate at 17.6 %.

4. Conclusion

The treatment standard to which roads are maintained will affect the future performance of the road and also the total road cost. Consequently, it is essential to observe the variation in cost and overall road performance with different treatment maintenance standard to decide which treatment standard is economically optimal and appropriate.

By compare the road works between with project alternative and without project, we can see that there is an immense difference on the work carried on and the financial economic cost. On the With Project alternative both for section one and section two, it will be upgrading works in the beginning of the year and there will only two maintenances that will be carried out during 20 year of the analysis period. However, by conducted upgrading and only two maintenances, this project requires high economical and financial cost that is 7,332,800.0 US Dollar and 8,716,000.0 US Dollar for section one and 9,512,400.0 US Dollar and 11,328,000.0 US Dollar for section two but mostly the cost will used to fund road upgrading.

In contrast, considering Without Project Alternative, maintenance is carried on every year including grading beside spot regrevelling, and gravel resurfacing will need to be conduct as much as 6 times during 2012 - 2031. We can see from the table that if compared by With Project alternative, this project expenditure is considering as a very low-cost project, both in economic and financial cost specifically 1,708,284.0 US Dollar and 2,033,477.1 US Dollar for section one and 2,620,615.5 US Dollar and 3, 3,114,314.9 US Dollar for section two.

The alternatives project should be chosen based on the life cycle cost of maintenance and road user cost (RUC). Despite of the amount of cost that has to be incurred, although 'with project' alternative spend much more than 'without project' alternative however 'with project' alternative can sustain the road performance better than without project alternative. In terms of RUC, 'with project' alternative also give the best result with very low RUC if compared with 'without project'.

In conclusion, from two proposed alternatives, 'with project' alternative is the most appropriate project considering the number of maintenance will be carried out, overall road performance, RUC, and the values of NVP and IRR.

References

[1] Badan Pusat Statistik 2017 Panjang Jalan Menurut Jenis Permukaan. [internet] Jakarta: Badan Pusat Satistik; 2017 [cited 1 July 2019]. Available from: https://www.bps.go.id/linkTableDinamis/view/id/820

[2] BCA 2009 Groundwork for Growth: Building the Infrastructure that Australia Needs, Business Council of Australia

[3] Kerali H and Odoki J 2006 *Stannard E. Overview of HDM-4* Volume One The World Bank (Washington)

[4] Morosiuk G, et al 2006 Applications Guide. HDM-4 Highway Development and Management Series Manual Two

[5] Odoki J B et al 2013 *HDM-4 adaptation for strategic analysis of UK local roads* Inproceedings of the Institution of Civil Engineers-Transport Apr vol. 166 No 2, pp. 65-78 Thomas Telford Ltd

[6] Shah Y U, et al 2014 Evaluation of prioritization methods for effective pavement maintenance of urban roads *International Journal of Pavement Engineering* Mar 16;15(3):238-50.

[7] Khan M U, et al 2014 *Developing a new road deterioration model incorporating flooding.* InProceedings of the Institution of Civil Engineers-Transport Oct vol. 167 No 5, pp. 322-333 Thomas Telford Ltd.

[8] Ćutura B, et al 2016 Application of the HDM-4 Model on Local Road Network: Case Study of the Herzegovina-Neretva Canton in Bosnia and Herzegovina Transp Res Procedia 14:3021–30.

[9] Yogesh U S, et al 2016 *Adaptation of HDM-4 Tool for Strategic Analysis of Urban Roads Network.* Transp Res Procedia [Internet]. Elsevier B.V.; 2016;17(December 2014):71–80. Available from: http://dx.doi.org/10.1016/j.trpro.2016.11.062

[10] Jiao X and Bienvenu M 2015 *Field Measurement and Calibration of HDM-4 Fuel Consumption Model on Interstate Highway in* Florida. Int J Transp Sci Technol [Internet]. Tongji University and Tongji University Press; 2015;4(1):29–45. Available from: http://dx.doi.org/10.1260/2046-0430.4.1.29

[11] Perrotta F, et al *Verification of the HDM-4 fuel* consumption *model using a Big data approach: A UK case study* Transp Res Part D Transp Environ [Internet]. Elsevier; 2019;67:109–18. Available from: https://doi.org/10.1016/j.trd.2018.11.001

IOP Conf. Series: Materials Science and Engineering **669** (2019) 012034 doi:10.1088/1757-899X/669/1/012034

Experimental study of precast concrete beam-to-column connection under sustained loading

V A Noorhidana [1]

[1] Civil Engineering Department, Universitas Lampung, Jl. Sumantri Brojonegoro No.1 Gedong Meneng, Bandar Lampung, Indonesia, 35145

vera.agustriana@eng.unila.ac.id

Abstract. The long-term behaviour of an exterior precast concrete beam-column (PCBC) connection has been investigated in this study. The connection consists of a precast U-beam, precast column with corbel, and interlocking bars to connect the precast column and precast beam. The specimen was tested under sustained loading for 120days. The crack propagation, the development of strain and the deflection of the beam was observed throughout the test. The results show that the PCBC connection behaved in similar manner to correctly designed conventional reinforced concrete members under long-term loading. Flexural cracks occurred within the plastic hinge region of the beam. No cracks occurred in the joint core and column. The long-term behaviour of the PCBC connection was influenced by shrinkage and creep. The significant increase of deflection occurred in the first 15 days of testing, which is 37% of the total long-term deflection. The experimental long-term deflection has bigger value than the theoretical deflection evaluated using ACI 318-08.

1. Introduction

Precast concrete systems are a popular system in the recent years. These systems have some advantages in terms of quality, cost and time in comparison with the conventional system [1]. In addition, precast concrete systems satisfy the green construction rules, which have been developing to face the global warming issues.

Technical problems, such as the ease of implementation of the joints and the behavior of the connection are still a concern in precast systems. These problems could increase the construction time and costs. Because of that the design of the connection between precast elements are important. The complicated connection with high degree of precision should be avoided.

Many types of connections, in terms of beam to column connections in moment resisting frames, have been developed such as bolted, welded, cast-in-place (CIP) connection. Bolted connection could be the east method to install the precast elements on site, but require a high degree of precision in placing the steel plates or profiles. Welded connections need skilled labours in order to guarantee the quality of welding connection. Cast-in-place (CIP) connections are more monolithic connections and recommended for seismic resistant buildings. Unfortunately, this connection takes longer to construct due to the concrete has to gain its strength [2, 3].

A new precast concrete beam-column (PCBC) connection has been developed as presented in Noorhidana & Forth [4]. This joint connection consists of a precast beam with corbel and a precast column which are connected using interlocking bars and cast-in-place concrete.

PCBC connection behavior under static loading has been investigated and presented by Noorhidana & Forth [5]. The results showed that the specimen exhibited a flexural failure mode and behaved monolithically until failure. To understand the behavior of PCBC connection under other loading types, this paper will present the experimental investigation of the connection under sustained loading which represents the service load working on the structure. The result will be compared with those of the connection under short-time loading (static loading).

2. Experimental program

2.1. Test specimen

The beam-column specimen of this investigation consisted of one precast concrete beam and one precast concrete column joined using a cast-in-place (CIP) concrete, and represented an exterior beam-column joint of a moment resisting frame. The length of the column and the beam were determined based on the location of the contra-flexure points determined via a software analysis of a representative planar frame. Figure 1 and 2 show the reinforcement details of the PCBC specimens, while Figure 1(b) presents the precast concrete beam-column specimen as an isometric.

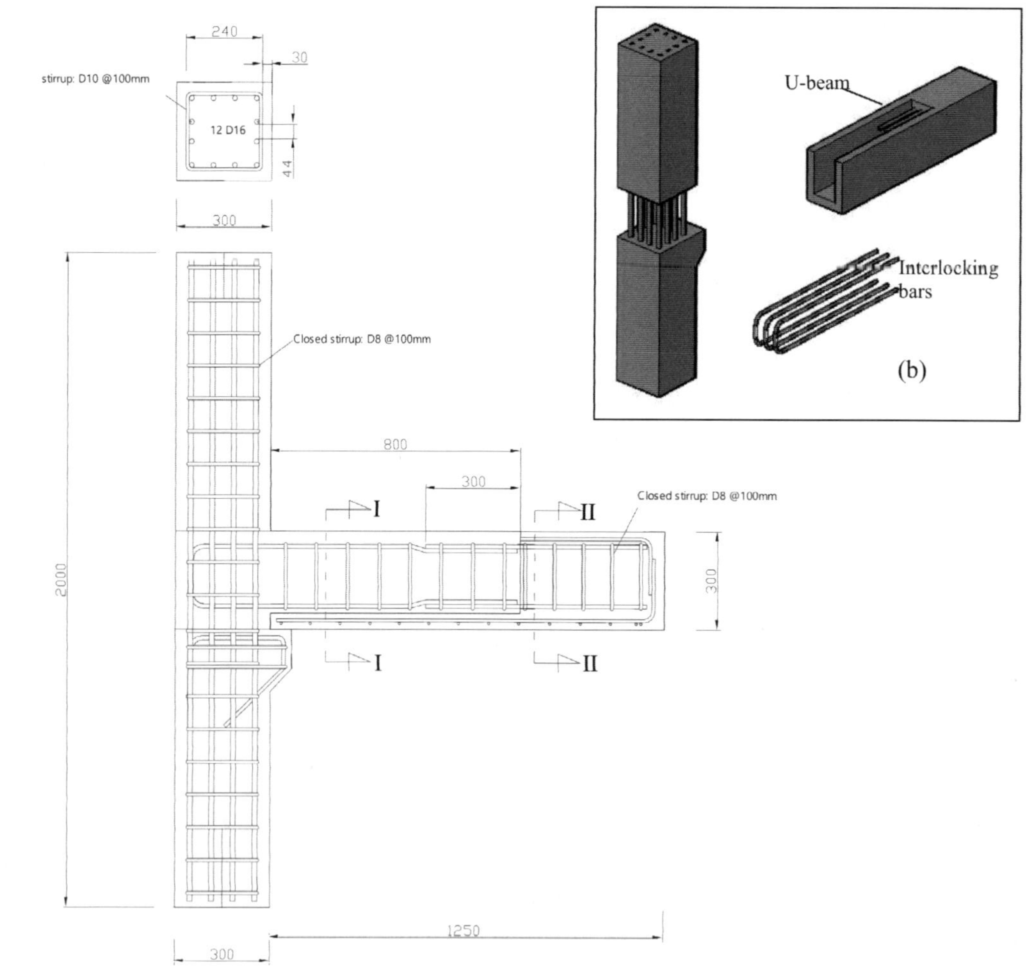

(a)

Figure 1. Reinforcement detail of PCBC specimen (a), Isometrics of precast concrete beam-column connections (b)

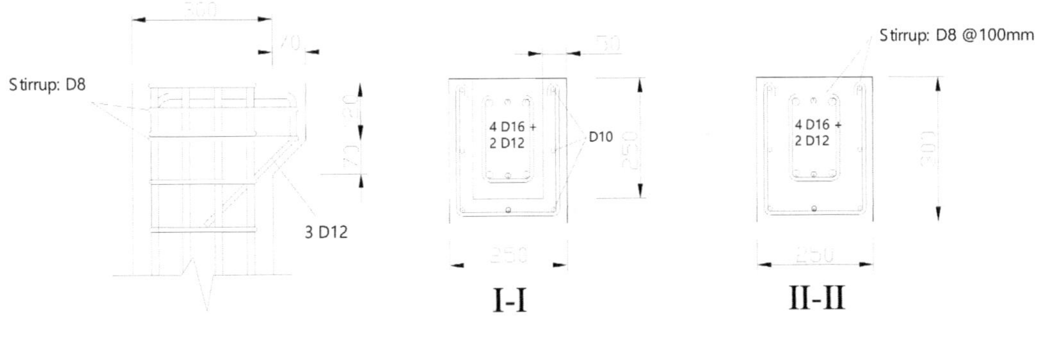

(a) (b)

Figure 2. Reinforcement detail of PCBC specimen, (a) Corbel reinforcement, (b) Beam cross sections.

Two precast concrete beam-column (PCBC) specimens were fabricated in this study, namely P1 and P5. P1 was loaded until failure. The testing was done in less than 1 hour (short-time loading). The maximum load of P1 then was used to determine the load level which was sustained for P5 testing. P5 was tested under long-tem loading. The specimen was loaded until 20kN (which is within a serviceability limit state), then that load was sustained until 120 days. The deflection of the beam tip and the crack was observed during testing.

2.2. Materials

There were 2 concrete mixes, i.e., one for the precast concrete beam and column (a design compressive strength of 30MPa) and one for the cast-in-place (CIP) connection (a design compressive strength of 45MPa). Table 1 shows the compressive strength of concrete; standard deviations were between 1.87MPa and 2.71MPa. The compressive strength was obtained from 100 mm x 100mm cubes, where the concrete cylinder strength fc' was taken as 80% of the cube strength fcu [6].

Four different diameters of steel bar were used for the PCBC specimens, i.e. 8 mm, 10 mm, 12 mm, and 16 mm. The yield strength of the steel reinforcing bars was 500MPa.

Table 1. Compressive strength of concrete

Specimens	Average cube strength f_{cu} (f_c')* (N/mm^2)	
	Precast concrete	CIP concrete
P1	38.07 (30.46)	59.12 (47.30)
P5	44.58 (35.66)	55.02 (44.02)

* Concrete cylinder strength was taken as 80% of the cube strength [6].

2.3. Test setup and instrumentation/loading procedure

The test set-up of the precast concrete beam-column (PCBC) Specimen P5 is explained as follows: as mentioned above, both precast column ends were restrained by steel plates which were bolted to the test rig, while the beam end was free. The vertical load was applied on the beam end in the negative direction (downward), which represented the gravity load. While, there was no axial load applied the precast column since this tends to enhance the joint shear strength; hence, this is a worst loading case scenario. The test-set up is presented in Figure 3.

One LVDT was located under the beam end to measure the deflection of the beam. Several DEMEC points were mounted on the surface of the beam of the specimen in order to measure the strain of the concrete surface. There are 5 positions horizontally. Each position consisted of 4 levels. The top and bottom levels were placed at the level of the longitudinal bars of the beam. The measurement of the strains was taken manually using a digital DEMEC strain gauge at specific days in order to evaluate the curvature development during the testing. The deflection of the beam tip and the crack was observed during testing.

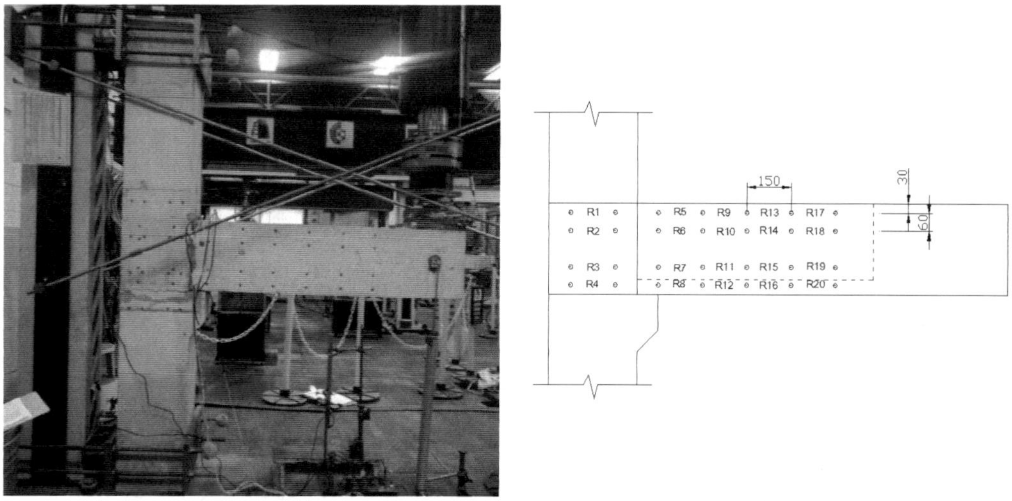

(a) (b)

Figure 3. (a) Test set-up of the PCBC connection, (b) Demec points' position

3. Results and discussion

3.1. Load-deflection curves

Specimen P1 was loaded under static loading until failure. The load-deflection relationship is presented in Figure 4(a). The maximum load was 59.80kN at the deflection of 23.5mm. After reaching the maximum load, the curve decreased and the specimen failed at a load of 50kN and the deflection of 46.5mm.

Specimen P5 was loaded by static loading until 20kN (about 30% of the maximum load of SpecimenP1), the load-deflection relationship is presented in Figure 3 (a). The graph shows that P5 curve is coincident with P1 curve. Then, the load of 20kN was sustained for 120 days. The deflection development was observed within this period.

Figure 4(b) presents the development of deflection of Specimen P5 under sustained loading of 20kN for 120 days. The deflection increased as the time increased. This is because of shrinkage, creep and other factors, such as loading configuration, the steel reinforcement, and the environment (i.e. temperature and humidity) during 120days of testing. The deflection at 0 day was 3.9mm. The deflection increased to 10.1mm after 120 day. So, the total long-term deflection was 6.2mm.

The significant increase occurred in the first 15 days of sustained loading, which was about 37.% of the total long-term deflection. The deflection increase became smaller with time which was showed by the curve become flat.

The experimental of long-term deflection obtained from Specimen P5 test was evaluated using ACI 318-08 [7]. The theoretical long-term deflection is presented in Figure 4(b) indicated by dash-line. As seen in the figure, the theoretical methods under-predict the experimental behaviour. As the theoretical deflections are based on a cantilever which is assumed to have full restraint at its support it is not surprising that these models under-predict the long-term experimental deflection.

IOP Conf. Series: Materials Science and Engineering **669** (2019) 012034 doi:10.1088/1757-899X/669/1/012034

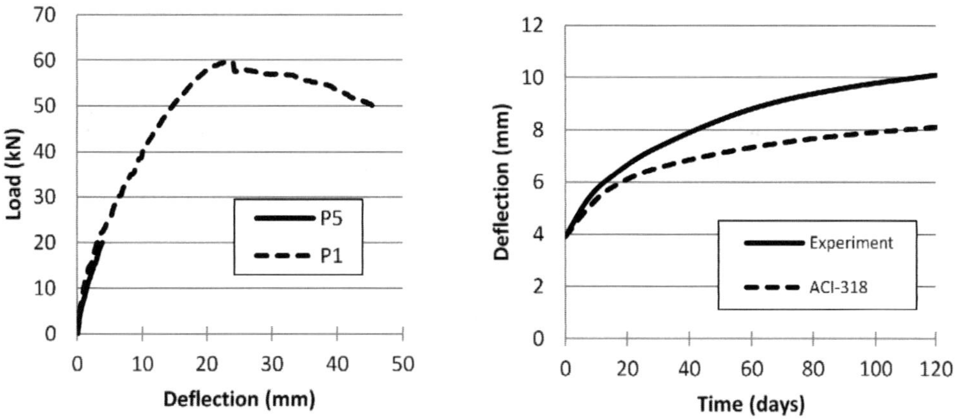

(a) (b)

Figure 4. (a) Load-deflection relationship, (b) Deflection-time relationship

(a) (b)

Figure 5. Crack-pattern of (a) Specimen P1, (b) Specimen P5.

3.2. Cracks Observation

The crack propagation of the PCBC connection under static loading until failure (Specimen P1) is presented in Figure 5(a). The first crack appeared on top side of the beam core, on the beam adjacent to the column, in line with the tip of the corbel. The cracks occurred both in the beam core and the wall of the precast U-beam. In general, all cracks were categorized as typical flexural cracks which were distributed along the cast-in-place (connection) region. Due to a high load level (about 55kN), the longitudinal bars of the U-beam started to slip which was indicated by a horizontal crack in the bottom of the U-beam which extended from the support until 200mm from the corbel. There was no significant cracks occurred in the joint core.

Under sustained loading at service limit state, the cracks pattern of PCBC connection (P5) was similar with those of P1, as can be seen Figure 5(b). The cracks occurred both in the beam core and the wall of the U-beam. The extension of cracks and new cracks throughout the sustained loading period were due to the effect of shrinkage and creep. No cracks occurred in the joint core because the applied load is relatively low.

3.3. Curvature developments of the U-beam

Figure 6 shows that the curvatures at Positions II-VI (of the U-beam) increased with time. The curvatures at III and V are bigger than those at II and IV (both at the beginning and during the long-term test), owing to the fact that vertical cracks occurred in regions III and V when the immediate load of 20kN applied. The cracks then lengthened during the sustained load applied. The curvatures at II and IV appear to be similar, with a dramatic increase at the age of 30 days; at this stage the cracks occurred on the top surface of the beam core.

The curvature at Position I (the joint core), as presented in Figure 6, decreased as the loading time increased. This behaviour is similar with the beam-column joint Specimen P1 under static loading, i.e. the joint experienced a compression strut mechanism.

The similarity of the curvatures in all zones (except Position I-joint core) suggests that beyond the initial cracking during loading and due to shrinkage restraint, the system appears to be quite homogeneous.

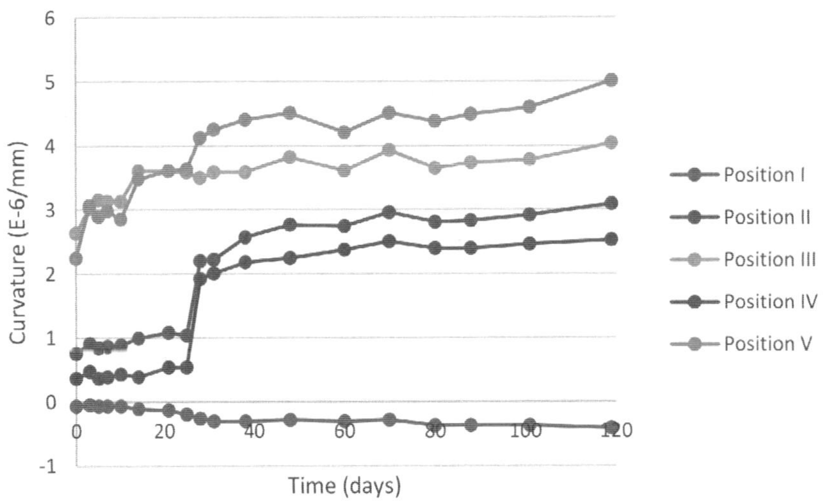

Figure 6. Curvature development with time of the U-beam of Specimen P5

4. Conclusion

- Specimen P1 (PCBC connection under short/static loading) performed good structural behaviour with flexural failure mode and behaved monolithically until failure.
- Specimen P5 (PCBC connection under sustained loading for 120 days) have similar behaviour with Specimen P1 in terms of load-deflection curves and crack pattern.
- The deflection and curvatures of the beam increased with time due to creep and shrinkage. The significant increase of deflection occurred in the first 15 days of testing, which is 37% of the total long-term deflection.
- The cracks are typical flexural cracks which were distributed along the cast-in-place (connection) region which is the plastic hinge region of the beam. No cracks occurred in the joint core and column.
- The experimental long-term deflection has bigger value than the theoretical deflection evaluated using ACI 318-08.

References

[1] Nurjaman H N, et al 2011 Sistem pracetak beton sebagai sistem konstruksi hijau: studi kasus perbandingan energi konstruksi dan dampak lingkungan di pembangunan rumah susun di

Batam. in *Seminar HAKI*. 2011. Jakarta, Indonesia.

[2] Bhatt P and D W Kirk 1985 Test on an improved beam column connection for precast concrete *ACI Journal*, p. 834-843.

[3] Ertas O S, et al 2006 Ductile connections in precast concrete moment resisting frames *PCI Journal*, p. 2-12.

[4] Noorhidana V A and Forth J P 2016 *Precast concrete beam-to-column connection using interlocking bars* An alternative in Concrete Plant International (CPI).

[5] Noorhidana V A and Forth J P 2016 An experimental study on precast concrete beam-to-column connection using interlocking bars. in *The 2nd International Conference on Concrete Sustainability* (Madrid, Spain).

[6] Kaung J S and H F Wong 2011 Effectiveness of Horizontal Stirrups in Joint Core for Exterior Beam-Column Joints with Nonseismic Design. Procedia Engineering 14: p. 3301-3307.

[7] ACI 318-08 2008 *Building Code Requirements for Structural Concrete (ACI 318-08) and Commentary* American Concrete Institute.

Acknowledgments

This research was sponsored by the Ministry of Research Technology and Higher Education, Republic of Indonesia, and the School of Civil Engineering, University of Leeds.

IOP Conf. Series: Materials Science and Engineering **669** (2019) 012035 doi:10.1088/1757-899X/669/1/012035

Development of surface ground motion and spectral acceleration based on modified shear wave propagation analysis

W Partono[1], F Kistiani[1], U C Sari[1], Haryadi[1], E Putro[1] and W R Ramadhan[1]

[1] Civil Engineering Department, Engineering Faculty, Diponegoro University, Semarang 50275, Indonesia

windu_bapake_dila@yahoo.com

Abstract. The shear wave propagation analysis (SPA) is a method for developing surface ground motion and spectral acceleration by conducting seismic wave propagation from bedrock to surface elevation. The bedrock elevation and soil profile from bedrock through surface elevation are important information needed for SPA. The shear wave velocity, damping ratio, soil density and shear modulus are the four parameters that are needed for developing SPA. This paper describes a modified SPA for developing surface ground motion and spectral acceleration by conducting two different soil deposit model, actual and modified bedrock elevation models. The difference in the two models depends on the bedrock elevation used for SPA. The actual bedrock elevation model is conducted based on the real bedrock elevation observed using microtremor test. However, the second model is performed based on the final depth of soil boring investigation. The study was conducted at the Engineering Faculty of Diponegoro University. Based on the microtremor observation, the bedrock elevation was predicted at 50 metres below the surface level. However, two soil boring investigations conducted in this study area were stopped at 30 metres below the surface level. The surface ground motion and spectral acceleration developed from both models are almost equal.

1. Introduction

The new American code ASCE/SEI 7-16 (Minimum Design Loads and Associated Criteria for Buildings and Other Structures) [1] introduced a different concept for developing surface spectral acceleration especially for site class SE (soft soil). The site specific analysis or shear wave propagation analysis (SPA) shall be performed as an alternate method for obtaining the site factor. The site factors are usually used for developing surface spectral acceleration. The surface spectral acceleration can be obtained by multiplying spectral acceleration at bedrock elevation with the site factor. The development of SPA needs the information of bedrock elevation and soil profiles from the bedrock through surface elevation. The bedrock elevation can be observed using soil boring investigation and will time consuming and expansive. Another method that can be used or is often used for developing or predicting bedrock elevation is microtremor investigation (array or single microtremor). The soil density (γ), shear wave velocity (VS), damping ratio (ξ) and shear modulus (G) for each soil layer above the bedrock elevation are other four parameters used for SPA. The development of SPA for obtaining surface spectral acceleration is difficult for most of the civil engineers in Indonesia, compared to the previous method used or introduced by SNI 1726:2012 [2] and ASCE/SEI 7-10 [3]. This paper describes a comparative study for the development of SPA by conducting two different approaches. The first approach (Model-A) is conducted by using predicted bedrock elevation observed from microtremor investigation. However, the second approach (Model-B) is performed using soil boring investigation data. Two methods are introduced in this study, because of the limitation of soil data above the bedrock elevation. The soil boring investigation is usually performed for the foundation design and construction purposes. The final depth of soil boring investigation is usually not equal to the depth of bedrock elevation. The study was performed at the Engineering Faculty of Diponegoro University, Semarang, Indonesia. Figure 1 shows the study area. Three soil boring investigations were performed with maximum 30 metres of depth. The soil boring investigations were stopped at 30 metres of depth, because the maximum standard penetration test (N-SPT) values observed in this area was greater than 60 at 20 - 22 metres below the surface level. The single and array microtremor investigations were also conducted in this area. The bedrock elevation was also observed and predicted at 50 metres below the surface level.

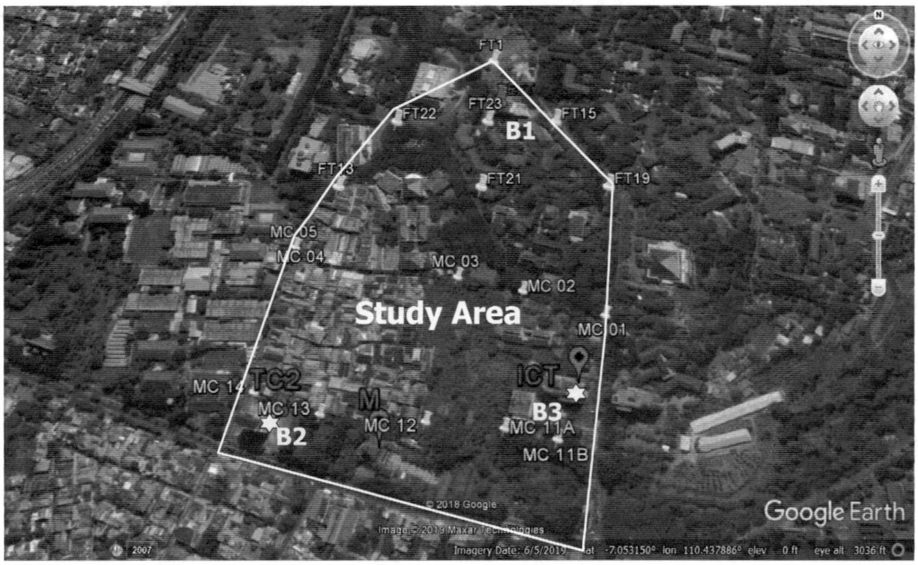

Figure 1. The study area

Figure 2 shows three example results of array microtremor investigation, N-SPT profile and shear wave velocity at the study area. This figure shows that the bedrock elevation can be predicted at 50 metres below the surface level. The hard soil data with N-SPT > 60 were observed at three boring positions B1, B2 and B3. Based on the Vs profile at two points, the rock elevation with Vs > 750 m/sec [2] was predicted at 30 metres below the surface level and the bedrock elevation with Vs > 1500 m/sec [2] was predicted at 50 metres below the surface level. Based on the data observed within the study area, the SPA was developed using two different models 50 metres (Model-A) and 30 metres (Model-B) soil profile models. Because of the limitations of soil or rock data below 30 metres, all data were arranged using the same data obtained from the soil boring investigation B1, B2 and B3.

2. Methodology

The development of surface spectral acceleration and surface ground motion is important for seismic resistance design and evaluation of buildings and other structures. The spectral acceleration can also be developed using SNI 1726:2012 or international code using amplification factor and spectral acceleration at the bedrock elevation. The developing of surface spectral acceleration and surface ground motion needs ground motion data, usually in terms of acceleration time histories. The acceleration time histories can be collected from national and international data base. The ground motion data used for

SPA can be collected based on the earthquake events close to the building position. The information of seismicity data, for all seismic sources close to the building position, will be collected for SPA [4, 5]. If there is no earthquake data, caused by all seismic sources close to the building position, the data can be collected from other earthquake events by conducting response spectral matching analysis. The modified ground motion, developed based on the spectral matching analysis, can be used for predicting surface spectral acceleration and surface ground motion caused by a specific earthquake, having specific magnitude and epicentre distance. Based on the new Indonesian Seismic Hazard Maps 2017 [6], two important seismic sources are located close to the study area (Lasem fault and Semarang fault sources), having a minimum distance of less than 10 km. Another information described in these seismic maps, is the maximum magnitude (6.5 Mw) that should be taken into account for these two seismic sources. Based on these two information (maximum magnitude and distance), the ground motion data used for SPA for the study area were collected from earthquake events having magnitude ± 6.5 Mw and maximum epicentre distance of 10 km.

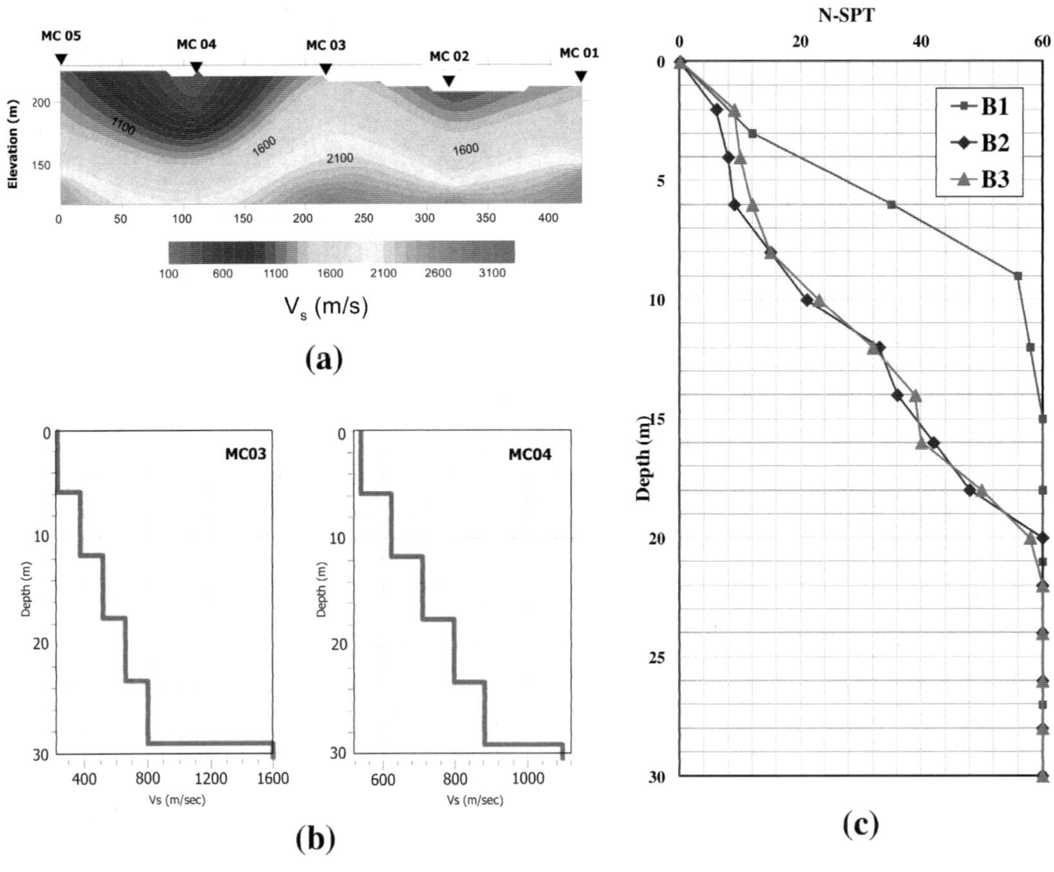

Figure 2. (a) Array Microtremor section MC01 – MC05, (b) shear wave velocity profile collected from different points and (c) N-SPT profile collected from boring no. B1, B2 and B3.

The surface spectral acceleration and surface ground motion for the study area were developed following two basic steps, such as response spectral matching analysis and SPA. The spectral matching analysis was performed for two directions of ground motion North-South (NS) and East-West (EW) directions. Figure 3 shows the spectral matching results for NS and EW directions of ground motion. The SPA is performed by conducting two different models. The soil profile from 0 through 30 metres depth is divided into 5 (five) layers with 6 m thickness for each layer. However, the thickness of soil or rock layer from 30 to 50 metres of depth is divided into two different layers. Figure 4 shows two different models used in this study. The Vs values used for each soil layer were developed using the average N-SPT data [7. 8, 9] from three different boring investigations and average Vs profiles obtained from 16 points microtremor investigation. The soil density used for each soil layer is calculated from the average of three boring investigations. Based on the soil boring investigations, all soil layers from 0-12 metres of depth is sandy silt, 12-24 metres of depth is hard soil and from 24-30 metres of depth is sandstone. Because of the limitation of soil or rock data from 30-50 metres of depth, the same sandstone parameter data were used at this layer.

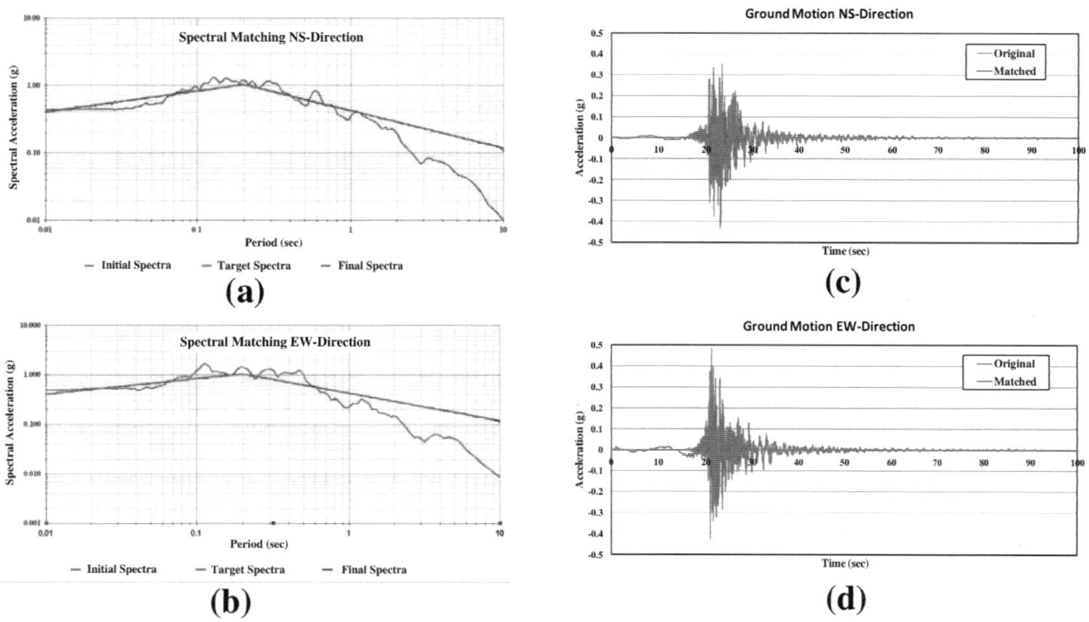

Figure 3. Spectral Matching for (a) NS and (b) EW directions, Original and Matched ground motion for (c) NS and (d) EW directions.

3. Results and Discussions

The development of surface ground motion and spectral acceleration was conducted using two different models of soil profile. Two matched ground motions at the bedrock elevation were used for SPA. The matched spectral acceleration was calculated at the bedrock elevation for both directions of ground motions. A comparative analysis was also performed by comparing the peak ground acceleration (PGA) profiles at each soil layers developed using the two soil models. Figure 5 shows two different spectral accelerations calculated at the surface elevation. The spectral acceleration calculated using Model-A is slightly greater compared to the second model (Model-B). The PGA profiles calculated using two different models also indicated the same results for both models. The PGA profiles calculated using Model 1 are greater than Model-B. The surface PGA difference calculated using both models are 0.1g.

The development of surface ground motion or acceleration time history is also performed using the same two models. Figure 6 shows the surface ground motion calculated using the two models. The ground motion calculated using Model-A is slightly greater compared to the same ground motion calculated using Model-B. Both NS and EW directions calculated using Model-A ere slightly greater than those of Model-B.

The development of surface spectral acceleration and surface ground motion can be simplified by reducing the maximum depth of soil boring investigation (Model B), the time requirement and expenditure, compared to the same requirements developed using Model-A. Based on the PGA profiles, it seems that the second model (Model-B) can be used as a substitute model for developing the surface ground motion.

IOP Conf. Series: Materials Science and Engineering **669** (2019) 012035 doi:10.1088/1757-899X/669/1/012035

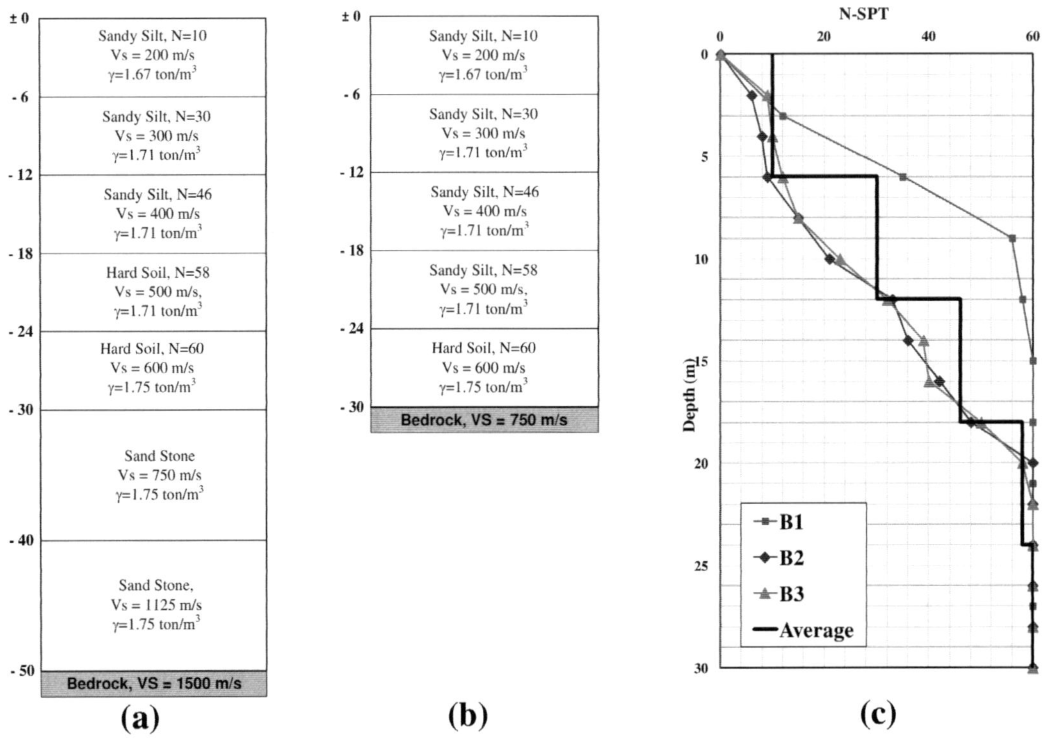

Figure 4. (a) Soil profile model 50 metre, (b) soil profile model 30 metre and (c) N-SPT profile collected from boring no. B1, B2 and B3 and N-SPT 30 metre.

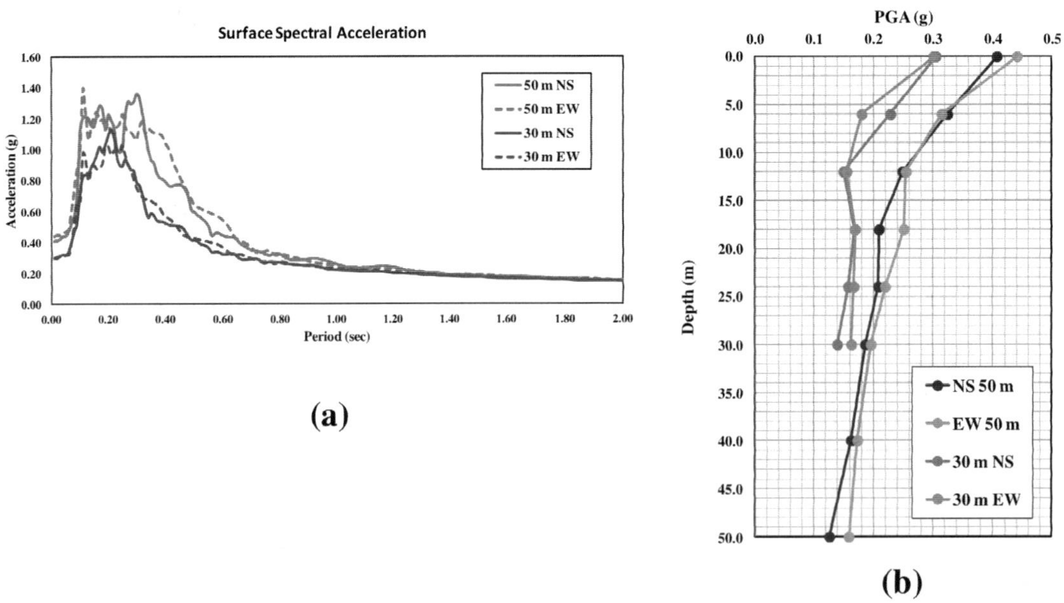

Figure 5. (a) NS and EW direction surface spectral acceleration for Model-A and Model-B and (b) PGA profile for Model-A and Model-B (b).

IOP Conf. Series: Materials Science and Engineering **669** (2019) 012035 doi:10.1088/1757-899X/669/1/012035

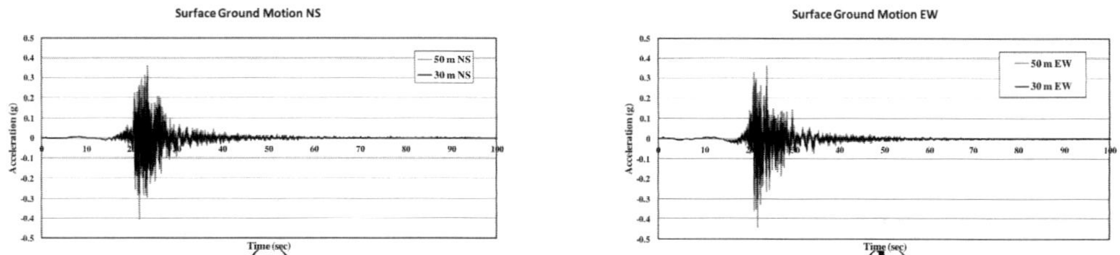

(a) (b)

Figure 6. (a) NS direction surface ground motion for Model-A and Model-B and (b) EW direction surface ground motion for Model-A and Model-B (b)

4. Conclusions

The development of surface ground motion and surface spectral acceleration was performed at the study area based on the shear wave propagation analysis (SPA). The development of surface spectral acceleration and surface ground motion is important for seismic resistance design and evaluation of buildings and other structures. The surface ground motion and spectral accelerations are developed using two different models (50 metre and 30 metre soil model). Three soil boring investigations and 16 points microtremor investigation are performed at the study area. A good correlation between N-SPT and Vs data are observed in the study area for soil data between 0 through 30 metre soil layer.

The surface ground motion and surface spectral accelerations are developed using two direction ground motions (NS and EW). The surface ground motion and surface spectral accelerations are also developed using two models, and the difference of surface PGA calculated using two models is approximately 0.1g. The surface ground motion and surface spectral accelerations calculated using 50 metre soil model is slightly greater compared to the 30 metre soil model. The 30 metre model can be used as a substitute model for developing surface ground motion and spectral acceleration.

References

[1] ASCE/SEI 7-16 2017 Minimum Design Loads and Associated Criteria for Buildings and Other Structures (Virginia: American Society of Civil Engineer) p 1

[2] SNI 1726:2012 2012 Tata Cara Perencanaan Ketahanan Gempa untuk Struktural Bangunan Gedung dan Non Gedung (Seismic Resistance Design Codes for Building and Other Structures), (Jakarta: Centre for Research and Development of Housing and Resettlement, Ministry of Public Works and Human Settlements) p 1

[3] ASCE/SEI 7-10 2010 *Minimum Design Loads for Buildings and Other Structures Design*. (Virginia : American Society of Civil Engineer) p 1

[4] Partono W, Wardani S P R, Irsyam M and Maarif S 2016 *Jurnal Teknologi* 78(8-5) 31

[5] Shiully A, Sahu R B and Mandal S 2017 *Journal of Geophysics and Engineering* 14(3) 466

[6] Pusat Studi Gempa Nasional 2017 *Peta Sumber Dan Bahaya Gempa Indonesia Tahun 2017* (*Indonesian Seismic Sources and Seismic Hazard Maps 2017*) (Jakarta : National Center for Earthquake Studies, Centre for Research and Development of Housing and Resettlement, Ministry of Public Works and Human Settlements) p 1

[7] Ohsaki Y and Iwasaki R 1973 *Soil and Foundations* 13 59.

[8] Ohta Y and Goto N 1978 Earthquake Engineering and Structural Dynamics 6 167

[9] Imai T and Tonouchi K 1982 *Proc. of Second European Symposium and Penetration Testing* (Amsterdam: The Netherlands) p 67

Acknowledgments

This research was financially supported by The Faculty of Engineering, Diponegoro University, Indonesia.

IOP Conf. Series: Materials Science and Engineering **669** (2019) 012036 doi:10.1088/1757-899X/669/1/012036

Efficiency analysis of Muskingum-Cunge method and kinematic wave method on the stream routing (Study case: upper Ciliwung watershed, Indonesia)

D Prawira [1], H Soeryantono [1], E Anggraheni [1] and D Sutjiningsih [1]

[1] Department of Civil Engineering, Universitas Indonesia, Depok, Indonesia

darmaprawira108@gmail.com, dwita@eng.ui.ac.id

Abstract. Flood routing is needed to determine the distribution of flood peak time and magnitude of peak flooding that occurs. Flood routing can be done on the river channel using the transform function principle. There are two methods of flood routing in reach that have significant differences based on general concepts and number of data requirements needed, namely Muskingum-Cunge method and Kinematic Wave method. This study aims to evaluate the efficiency of two flood routing methods, Muskingum-Cunge method and Kinematic Wave method using HEC-Geo HMS, the efficiency of each routing method is obtained from comparison of simulation's hydrograph peak discharge. results with the observation's hydrograph peak discharge. Results of the comparisons show that Kinematic Wave method give a large peak discharge which is closer to the observation than simulation results with Muskingum-Cunge method. However, difference of result from two methods are not too far from observation, although Kinematic Wave method uses more detailed data. Based on these results, it can be concluded that Muskingum-Cunge method is more efficient than Kinematic Wave method in conducting flood routing because it provides a hydrograph that is close enough to fewer data requirements.

1. Introduction

Flood routing is a procedure to determine time and magnitude of a flow at a point from a hydrograph that is known or assumed at one or more points upstream. Flood routing can be done on river channel using transform function [1]. The transform function is a principle that focuses on transformation function to get output based on input given.

If viewed from the number of data requirements needed and the general concepts, there are two transform function methods that have significant differences, namely the Muskingum-Cunge method and the Kinematic Wave method. Muskingum-Cunge method performs a lump-sum from storage value in a certain time range. This method focuses on the relationship between storage and discharge in a channel cross section. Meanwhile, Kinematic-wave method estimates unsteady flow equation by ignoring pressure [2]. As a result of the more detailed needs, so a statement arises that Kinematic Wave method gives more accurate results than Muskingum-Cunge method of conducting flood routing. However, many data requirements effect to a longer data search process and require greater costs.

Based on these comparisons, the question arises, namely how much flood routing efficiency uses the Muskingum-Cunge method compared to routing floods using the Kinematic Wave method. In this journal, the study aimed to evaluate the efficiency of flood routing using the Muskingum-Cunge method rather than routing the flood of the Kinematic Wave method in simulating existing data hydrographs. The efficiency of each routing method is obtained from comparison of simulation's hydrograph peak discharge. results with the observation's hydrograph peak discharge. Observation's data are obtained from recording water level data in Katulampa Weir. Katulampa Weir is an outlet of the Upper Ciliwung watershed, Indonesia.

2. Research Methodology

The study was conducted using HEC-HMS to make a flood-routing model through reach using Muskingum-Cunge method and Kinematic wave method. Each method has different data needs. The results of modelling in HEC-HMS are hydrographs. This study located at Ciliwung Watershed. Ciliwung river is the main river that affecting Jakarta. The section that is used as test location is last reach to Katulampa's Weir as we called Upper Ciliwung Watershed. This sub watershed has rapid infrastructure development that causes the hydrological regime change [3]. The observation point taken as existing hydrograph data recording point is outlet of Upper Ciliwung watershed, namely Katulampa's Weir. This study used secondary data to modelling the watershed. As for data needed like river maps, topographical maps, soil type maps, land use maps, rainfall data and river cross sections.

The first step is by delineating the Upper Ciliwung watershed using HEC-Geo HMS based on DEM maps and rivers maps. By using HEC-Geo HMS, data preparation is done in the form of slope, river length, and many more. Results of modelling on HEC-Geo HMS will be exported into HEC-HMS. In process of flood routing, rain data is needed based on rainfall-station data that affects the watershed. The rainfall data entered must be in accordance with the rain that occurred at the time of the observation. Therefore, a daily rainfall calculation is carried out. Rainfall data are entered during the modelling stage at HEC-HMS.

There are two simulations in this research namely the Muskingum-Cunge method as the first method and the Kinematic Wave method as the second method. The simulation hydrograph from two methods was compared with observational hydrograph data. Hydrograph comparison is done in terms of visuals (difference in peak discharge and duration of flooding, and prediction of hydrological model using Nash-Sutcliffe test). After the comparison, an evaluation of efficiency of flood routing can be carried out with both methods.

2.1. Flood routing method in reach

2.1.1. Muskingum-Cunge method

The Muskingum method is a hydrological routing method that solved by mass conservation laws and the relationship of discharge with storage [1]. This method models flood storage volumes in river channels with a combination of wedge storage and prism storage. Wedge storage is a storage change that occurs due to changes in inflow and outflow when a flood wave occurs in a river cross section. Whereas, prism storage is a constant volume of storage from a cross section along the river channel. During periods of increased flooding, wedge storage is positive and added to prism storage. It is better, during the decline in floods, wedge storage is negative and reduces the prism storage value.

Wedge storage is assumed as $KX\ (I - Q)$ and prism storage is assumed as KQ. So, it can be seen that the total storage equation of a river channel cross section for the Muskingum method is as follows:

$$S = KX + KX(I - Q) \tag{1}$$

The change in storage over time interval Δt are as follows:

$$S_j + 1 - S_j = K\{[XI_{j+1} + (1 - X)Q_j + 1] - [XI_j + (I - X)Q_j]\} \cdot \tag{2}$$

The change in storage can also be expressed:

$$S_{j+1} - S_j = \frac{(I_j + I_{j+1})}{2}\Delta t - \frac{(Q_j + Q_{j+1})}{2}\Delta t \tag{3}$$

Combining (2) and (3) and simplifying gives:

$$Q_{i+1}^{j+1} = C_1 Q_i^{j+1} + C_2 Q_i^j + C_3 Q_{i+1}^j . \quad (4)$$

where i and j are sequence of n. The values of C1, C2, and C3 are routing parameters which can be calculated with the following equation:

$$C_1 = \frac{\Delta t - 2KX}{2K(1-X)+\Delta t} . \quad (5)$$
$$C_2 = \frac{\Delta t + 2KX}{2K(1-X)+\Delta t}$$
$$C_3 = \frac{2K(1-X)-\Delta t}{2K(1-X)+\Delta t}$$

Note that $C_1 + C_2 + C_3 = 1$.

Cunge determines the value of K and X according to the Kinematic Wave equation there the value of K and X can be formulated as follows:

$$K = \frac{\Delta x}{c_k} = \frac{\Delta x}{dQ/dA} \quad (6)$$

and

$$X = \frac{1}{2}\left(1 - \frac{Q}{Bc_k S_o \Delta x}\right) \quad (7)$$

where Ck as celerity, and B as the channel width.

2.1.2. Kinematic wave method.

Kinematic Wave method model comes from the Saint-Venant equation. The Saint-Venant equation uses continuity equations and momentum equations. The general continuity equation states that inflow minus outflow equals the rate of change in storage [4] . The continuity equation is

$$\frac{\partial A}{\partial t} + \frac{\partial Q}{\partial x} = q \quad (8)$$

where q is the lateral inflow rate per unit channel length and A is the channel cross-sectional area. For overland flow model, the momentum equation can be expressed as follows:

$$S_f = S_0 - \frac{\partial y}{\partial x} - \frac{v}{g}\frac{\partial v}{\partial x} - \frac{1}{g}\frac{\partial v}{\partial t} \quad (9)$$

For shallow flow, bottom slope and the energy gradient are approximately equal and acceleration effects are negligible, so the momentum equation simplifies to:

$$So = Sf \quad (10)$$

where *So* is base slope of the channel and *Sf* is slope of friction.

2.2. Rainfall distribution

To obtain the rainfall data the average of watershed can be calculated by polygon thiessen method. Polygon thiessen method is calculated based on the equation as follows:

$$\bar{P} = \frac{A_1P_1 + A_2P_2 + A_3P_3 + \cdots + A_nP_n}{A} = \sum_1^n \frac{A_iP_i}{A} \quad (11)$$

where $\bar{P}$ is the regional average rainfall, n is number of observer station, *Pi* is rainfall of each observer station, and A_i is influence area from each observer station [5].

If there are no rain data obtained within 24 hours from results of the recording, the synthetic rainfall pattern is used to obtain rain data. This study uses the distribution of rain in West Java shown in Table 1.

Table 1. Rainfall distribution for West Java

Rain pattern (hourly)	Hourly Percentage (%)								Interval (hour/pattern)
	1	2	3	4	5	6	7	8	
3	68	24	8						1
4	26	61	10	3					1
5	11	54	28	6	1				1
6	12	54	24	6	3	1			1
7	50,5	25,5	12,6	6,5	3,4	1,2	0,3		1
8	12,3	50,2	4,4	7,7	21,5	2,4	1,2	0,3	1

2.3. *Model calibration*

Model calibration in this research visualizes the simulation hydrograph of the observed hydrograph. Visual comparison of hydrographs is based on three aspects, namely rising time, peak discharge and flood time. When the time rises, it is compared to the length of time when a new flood occurs and reach a peak discharge. At peak discharge, the difference of peak discharge will be compared between the simulation and the existing results. At time of the flood, the length of time needed will be compared to the flood from the new start until it approaches the base flow.

The Nash-Sutcliffe test is used for calibration. The efficiency coefficient of the Nash-Sutcliffe model is used to assess the predictive power of the hydrological model. The most important statistical indicator in determining the reliability of the model is R^2. R^2 shows the efficiency value of the hydrological model with its existing conditions. The R^2 indicator value is calculated based on equation (12) below :

$$R^2 = 1 - \frac{\sum (Q_0 - Q_m)^2}{\sum (Q_0 - \bar{Q}_0)} \quad (12)$$

where Q_0 is the simulation discharge (m^3/s), Q_m is the observational discharge (m^3/s), and $\bar{Q}_0$ is the average simulation discharge (m^3/s) [6]. The statistical indicator is to evaluate the performance of the model in terms of comparing the results of the model with the observed data. The optimal value for R^2 approaches one. The formulation of the R^2 equation is based on the Nash-Sutcliffe model efficiency indicator. Nash-Sutcliffe test has a range between 0 and 1. Nash-Sutcliffe test has several criteria shown in Table 2.

Table 2. Nash-Sutcliffe test criteria

Nash-Sutcliffe Efficiency (NSE)	Interpretation
NSE > 0,75	Good
0,36 < NSE < 0,75	Satisfied
NSE > 0,36	Not Satisfied

3. Result and Discussion

3.1. *Property identification of Upper Ciliwung watershed*

Model 1 uses the Muskingum-Cunge method, while model 2 uses the Kinematic Wave method. Upper Ciliwung watershed consists of seven sub-catchments and three reach. The seven sub-watersheds are W80, W90, W100, W110, W120, W130, and W140 sub-basins. Meanwhile, the three reach are R10, R30 and R40 (Figure 1).

The process of identifying the Ciliwung Watershed is carried out using HEC-Geo HMS. Results of identification are used as data in running HEC-HMS. The watershed properties sought is the area of each sub-watershed, reach length and slope, and Curve Number value for each sub-watershed. (Table 3 and Table 4).

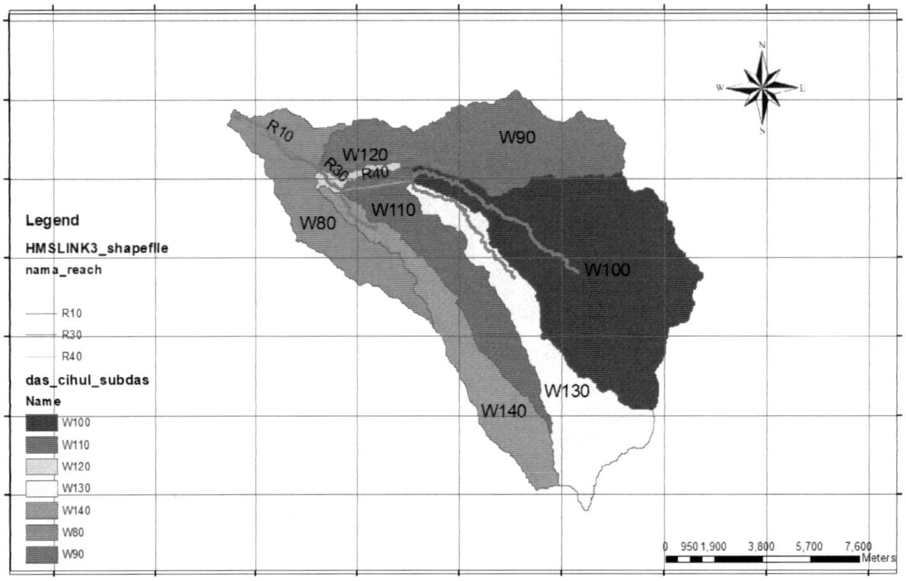

Figure 1. Sub-areas and reach of kelapa gading watershed

Table 3. Sub-areas properties data

Sub-area Name	A (Km^2)	*Curve Number*
W80	17.731	83
W90	26.036	76
W100	49.531	79
W110	16.683	80
W120	1.231	83
W130	22.271	78
W140	16.971	77

Table 4. Reach properties data

Reach Name	Length (m)	Slope (%)
R10	3933.847	0.02593
R30	1277.203	0.03837
R40	2694.670	0.03562

3.2. Observational hydrograph

Simulation results will be compared against observational data. Observation data that is compared is one day's discharge. Data selected is the discharge data on April 13th, 2017 at 07.00 to April 14th, 2017 at 07.00. This data was chosen because it has highest discharge in 2017. (Figure 2). Based on observation's hydrograph, it is known that peak discharge occurred in the 14th hour with the amount of discharge valued at 153,625 m^3/s. The time needed to reach peak of flood, which is for two hours and starts at the 12th hour and there is a flood for 12 hours.

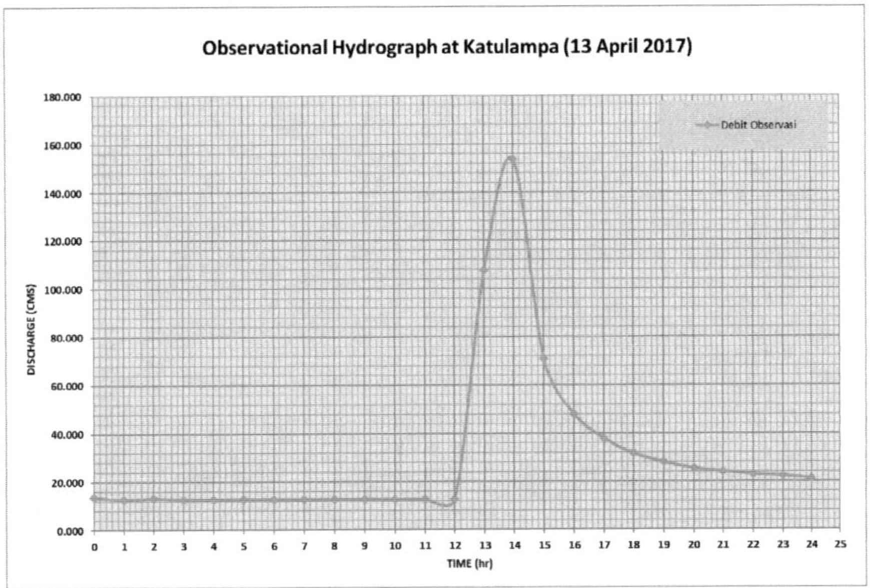

Figure 2. Observational hydrograph at Katulampa

3.3. Analysis of rain load

Rain load is calculated based on hourly rainfall data from several rain stations that affect Upper Ciliwung watershed. The rain load given in the simulation is regional rainfall data on the day according to time of observation. There are three rain stations that affect this watershed. The three rain stations are Gadog rain station, Cilember rain station, and Gunung Mas rain station. The extent of the influence of each rain station on the Upper Ciliwung watershed can be seen in Table 5. The selected rainfall data is rainfall data on April 13th, 2017 for the three rain stations. The amount of daily rainfall data selected and the results of calculations with the area of influence are shown in Table 5.

Table 5. Result of regional rainfall

Name of rainfall station	Influence percentage (%)	Rainfall Data (mm)	Result of influence area (mm)
Gadog	16.93	14.3	2.4
Cilember	32.06	41	13.1
Gunung Mas	51.01	26	13.3
	Daily Rainfall		28.8

After obtaining regional rainfall in the watershed, weighting is carried out on daily rainfall to obtain rainy days on that day. Hourly rain distribution is calculated based on the distribution of rain patterns in West Java shown in Table 1. Duration of rain distribution chosen based on observational hydrographs in Figure 2. Based on changes in the hydrographic discharge of the observation, it can be predicted that rain occurs for 3 hours. The distribution of rainfall for 3 hours on April 13th 2017 at 07:00 until April 14th 2017 at 07:00 is shown in Table 6.

Table 6. Result of hourly rain distribution

Date	Rain distribution of percentage	Rain distribution of Upper Ciliwung watershed
13Apr2017, 07:00	0%	0
13Apr2017, 08:00	0%	0
13Apr2017, 09:00	0%	0
13Apr2017, 10:00	0%	0
13Apr2017, 11:00	0%	0
13Apr2017, 12:00	0%	0
13Apr2017, 13:00	0%	0
13Apr2017, 14:00	0%	0
13Apr2017, 15:00	0%	0
13Apr2017, 16:00	0%	0
13Apr2017, 17:00	0%	0
13Apr2017, 18:00	68%	19.6
13Apr2017, 19:00	24%	6.92
13Apr2017, 20:00	8%	2.31
13Apr2017, 21:00	0%	0
13Apr2017, 22:00	0%	0
13Apr2017, 23:00	0%	0
14Apr2017, 00:00	0%	0
14Apr2017, 01:00	0%	0
14Apr2017, 02:00	0%	0
14Apr2017, 03:00	0%	0
14Apr2017, 04:00	0%	0
14Apr2017, 05:00	0%	0
14Apr2017, 06:00	0%	0
14Apr2017, 07:00	0%	0

3.4. Simulations of model hydrograph

Simulations were did on both models based on selected rain loads using HEC-HMS. Simulation results from both models are shown in Figure 3.

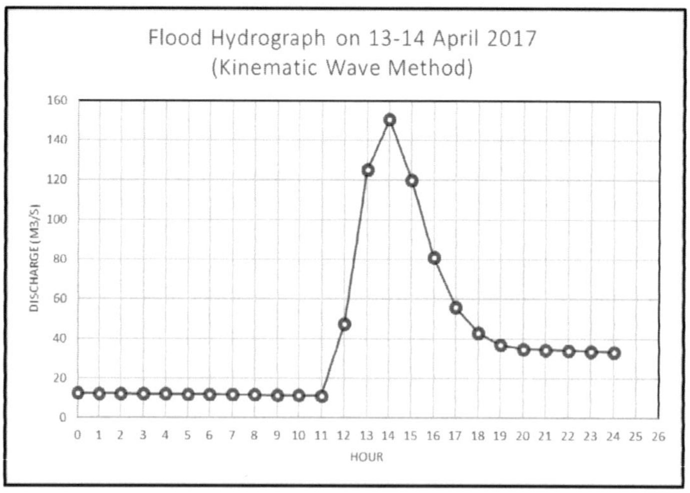

(a)

Figure 3. Flood hydrograph in outlet (a) Kinematic Wave method

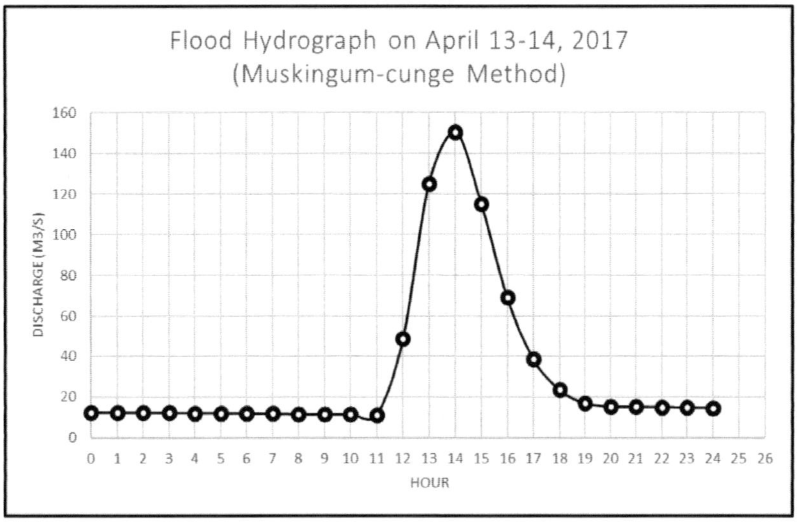

(b)

Figure 4. Flood hydrograph in outlet (b) Muskingum-Cunge method

From Figure 3, hydrograph of the simulation results using Kinematic Wave method has a simulated peak discharge of 150.6 m^3/s at the 14th hour. From the hydrograph of the simulation results, it can be seen that floods began to occur at the 11th hour and occurred for 13 hours. The time needed to reach peak discharge for three hours.

From Figure 3, hydrograph of the simulation results using Muskingum-Cunge method have simulated peak discharge of 150.5 m^3/s at the 14th hour. From simulation's hydrograph, it can be seen that floods began to occur at the 11th hour and occurred for 13 hours. The time needed to reach the peak discharge for three hours.

3.5. Comparisons of model hydrograph and observational hydrograph

Simulation's hydrograph of the two models were compared with the observational hydrograph in Katulampa's Weir. Comparison is done by comparing visually and conducting Nash-Sutcliffe test. Visual comparison of hydrograph is based on three aspects, namely rising time, peak discharge and flood time, while Nash-Sutcliffe test is compared based on hourly discharge.

3.5.1. Visual comparison

Simulation's hydrographs of both model and the observation hydrograph are juxtaposed in one graph in Figure 4 to find out the differences in the three forms of hydrograph. The comparison results are summarized in Table 7.

Based on Figure 4, it can be seen that from two hydrographs the simulation results have differences with the hydrograph of the observation. At the 1st to 11th hours, the two hydrographs of the simulation results have a large base flow which is quite similar to the hydrograph observation because the rain has not fallen on the watershed. After the 12th hour or after the rain falls into the watershed, the second form of hydrograph has begun to experience a difference because the size of the simulation discharge is greater than the observation discharge.

After reaching the peak discharge, the hydrograph of the simulation results using the Kinematic Wave method has a higher discharge than the observation hydrograph after reaching the peak discharge until the end of the running period. The discharge rate is above the base flow rate of the observation discharge. Whereas, the hydrograph discharge from the simulation results with the Muskingum-Cunge method initially had a higher discharge than the observation hydrograph, but at the 17th hour the amount of the discharge from the simulation results was smaller than the observation. From the differences in the discharge values of the two simulations with respect to observations on the recession side, it can be seen that the hydrograph of the simulation results using the Kinematic Wave method is in an overestimate condition and means that the results have entered within the range of observational data. Meanwhile, the hydrograph of the simulation results using the Muskingum-Cunge method is underestimated and means that the results have not reached the range according to the observation data.

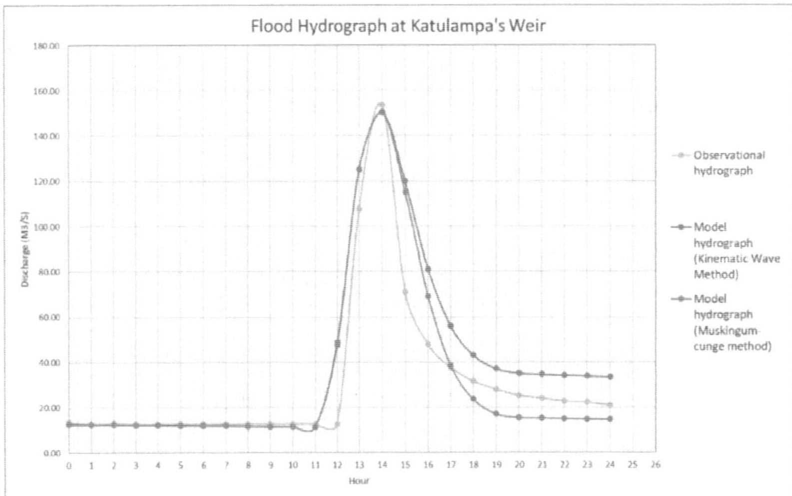

Figure 5. Comparisons of model hydrograph and observational hydrograph

Table 7. Recapitulation of observational discharge and simulation discharge

Point of	Peak discharge (m³/s)			Time to peak (hour)			Flood time (hour)		
observation	Qobservation	$Q_{K\text{-}W}$	$Q_{M\text{-}C}$	Qobservation	$Q_{K\text{-}W}$	$Q_{M\text{-}C}$	Qobservation	$Q_{K\text{-}W}$	$Q_{M\text{-}C}$
Katulampa Weir	153.63	150.6	150.5	3.0	2.0	2.0	12.0	13.0	13.0

Based on Table 7 shows that the discharge simulation results are close enough to the observation discharge. From the two hydrographs of the simulation results, show the peak length of time and the same time of flooding between the two. The two hydrographs of the simulation results have a peak time that is one hour faster than the observation hydrograph. However, the duration of the flood is longer than one hour compared to the hydrograph of observation. Longer flood times are caused by the time needed to reach the peak faster. The large difference in peak discharge from the simulation results with observational discharge is not much different. The percentage of the large difference in peak discharge from the simulation results with the peak discharge observation of 1.97% for the simulation using the Kinematic Wave method and 2.03% for the simulation using the Muskingum-Cunge method.

Based on the value of the percentage difference, the peak discharge value from the watershed model simulation with Kinematic Wave method is closer than the other method. The value of the peak discharge from Kinematic Wave method simulation according to the initial statement that this method gives closer results than the results of Muskingum-Cunge method simulation. This is because Kinematic Wave method requires detailed cross-sectional data that is sized according to the field. Channel cross-section data is needed in calculating flow velocity and discharge so that the more detailed channel cross-section data, the more accurate flood routing is carried out. However, Muskingum-Cunge method yields results that are not much different and are still close enough to the observation discharge.

Therefore, just routing the flood using Muskingum-Cunge method is enough to provide results that are close enough with less data. Based on the results of visual comparison, the Muskingum-Cunge method is considered more efficient than the Kinematic Wave method. This statement is based on the results of the simulation that are close enough to observations with fewer data requirements.

3.5.2. Nash-sutcliffe Test

Based on the results of the Nash-Sutcliffe test shown in Table 8, the R^2 value for the Kinematic Wave model is 0.773 and R^2 for the Muskingum-Cunge model is 0.833.

Table 8. Result of Nash-Sutcliffe test

Nash-Sutcliffe Test	R^2
Muskingum-Cunge Method	0,833
Kinematic Wave Method	0,773

Based on Table 2, it can be said that the two simulation results are in a good range so the level of accuracy of the simulation results on the observation discharge is quite good. From the two R^2 values, it is known that the R^2 value for the simulation results of the Kinematic Wave model is smaller than the R^2 value for the Muskingum-Cunge model simulation results. The Nash-Sutcliffe efficiency value shows the difference in the value of each debit data per hour from the discharge of the simulation results and observation discharge. The results of the Nash-Sutcliffe test showed that flood routing with the Muskingum-Cunge method was closer than the Kinematic Wave method when reviewing the hourly discharge. However, these two methods already have results that are close enough to do flood routing because they are categorized as good in the NSE test.

Based on the results of the Nash-Sutcliffe test, it is known that the Muskingum-Cunge method is more efficiently used in routing hourly debits with less data requirements than the Kinematic Wave method because it provides closer predictive power.

4. Conclusions

Based on comparison methods of simulation hydrograph, the result of two methods are quite close to observation. So, the method with less data show a more efficient method. Therefore, it can be stated that Muskingum-Cunge method (a method with less data requirements) is more efficient than Kinematic Wave method (a method that needs more detailed data) in simulating existing data recording hydrograph.

References

[1] Chow V T, et al 1988 *Applied Hydrology.* United Of States: McGraw-Hill, Inc.

[2] US Army Corps of Engineers : Institute of Water Resources 2000 HEC-HMS Technical Reference Manual

[3] Sutjiningsih D, et al 2018 Water-sensitivity assessment of regional spatial plan based on the relation between watershed imperviousness and aquatic ecosystem health. *IOP Conference Series: Earth and Environmental Science 140 012061.*

[4] Bedient P B, et al 2013 *Hydrology and Floodplain Analysis Fifth Edition.* Great Britain: Clay Ltd, St Ives plc.

[5] Limantara L M 2010 *Hidrologi Praktis* (Bandung: Lubuk Agung)

[6] Deviana A D 2011 Kajian Pemodelan Spasial Banjir Untuk Mendukung Kebijakan Sempadan Sungai dan Tata Ruang Wilayah. *FTSL ITB.*

[7] Soemarto C D 1987 *Hidrologi Teknik.* Surabaya: Usaha Nasional.

[8] US Army Corps of Engineers Institute of Water Resources 2013 HEC-Geo HMS User's Manual 10.1.

[9] US Army Corps of Engineers Institute of Water Resources. 2013 HEC-HMS User's Manual.

[10] Yustika R D, et al 2014 *Penggunaan Model Hidrologi di Sub-DAS Upper Ciliwung* Informatika

Pertanian, 197-204.

Acknowledgement
The authors express their appreciation as the study was conducted using the research funding for indexed international publication "Hibah Penugasan Publikasi Internasional Terindeks 9 (PIT 9) with contract number NKB-0053/UN2.R3.1/HKP.05.00/2019" fiscal year 2019 granted by Universitas Indonesia.

Performance of SCS unit hydrograph and kinematic wave methods application on rural and urbanized watershed

D Sharaswati [1], H Soeryantono [1], E Anggraheni [1] and D Sutjiningsih [1]

[1] Civil Engineering Department, Faculty of Engineering, Universitas Indonesia, Depok, 16452, Indonesia

dwita@eng.ac.ui.id

Abstract. Flood routing in the catchment area can be simulated by transform method. Based on the complexity of data requirement, there are two significant differences of the transform method in watershed routing, namely the SCS Unit Hydrograph and Kinematic Wave. The purpose of this study is to analyze the effectiveness of those two methods by comparing two hydrographs using HEC-Geo HMS on Urban and Rural Watersheds. Comparison has been simulated by calculating the ratio of peak hydrograph discharge to each calculation condition and the level of hydrograph compatibility. Another possibility to analyze the effectiveness of two methods by plotting accumulative discharge values from the two hydrographs produced by the SCS Unit Hydrograph method using the Kinematic Wave method. The results of the hydrograph comparison indicate that Kinematic wave method which requires more detailed data more suitable for Urban Watershed than SCS Unit Hydrograph. However, for the Rural watershed, SCS Unit Hydrograph shows satisfactory result.

1. Introduction

A procedure for determining or predicting the time and magnitude of flood flow at a point on a river based on data that is known or assumed in the upstream part is called Flood Routing [1].Analysis of flood routing can be reviewed at two locations, namely in the river channel section and in the catchment area of the watershed. Flood Routing in the channel is an analysis of water flow on the connecting channel of each Sub-DAS or reach in the watershed [2].Whereas flood routing in the catchment area is an analysis of water flow at each channel level in the Sub-watershed (catchment area).

Flood routing conducted in the catchment area can use the transform method component in the additional ArcGIS Geo HMS software tools and HEC HMS software. Based on the level of data needed, there are two significant differences in the Transform Method, namely the SCS Unit Hydrograph and Kinematic Wave. SCS Unit Hydrograph method is a flood routing method in the Lump Sum watershed. While the Kinematic-wave method estimates, the need for more complex data such as taking into account every dimension of the channel found in the watershed [3].

The use of these two methods has advantages and disadvantages of each in conducting flood searches. Kinematic Wave Method requires more data than the SCS Unit Hydrograph method. With more data requirements in the kinematic wave method, there is an assumption that the level of accuracy will be greater than the SCS Unit Hydrograph Unit method of conducting flood routing. However, with many data requirements, data search time will be longer and can require greater costs, and vice versa with fewer data requirements, costs will be lower. This causes more choices of faster and cheaper methods, the SCS Unit Hydrograph. Based on the characteristics of urban (urban) and rural (village) watersheds, each trait has a level of complexity. Urban properties are more suitable to use kinematic wave methods because there will be more components that must be more detailed than rural (village) properties. Whereas rural property can be more representative using a lump sum method such as the SCS Unit Hydrograph.

Based on the consideration of the use of the method, there are questions whether the consequences are if the nature of the Urban Watershed uses a lump sum method such as the SCS Unit Hydrograph and vice versa as well as how efficient the use of both methods are tested on urban and rural properties. The efficiency of the use of the methods are seen by comparing the results of the flow hydrograph due to the use of methods on flood routing in the catchment area (Transform method) which are tested on both the nature of the watershed, Urban and Rural. So, based on the results of the comparison, it can be seen the level of suitability of the method used in the Transform Method component on the nature of the watershed reviewed.

2. Research methodology and material

The study began by determining the location of the study based on the characteristics of a sub-watershed in the Upper Ciliwung watershed, namely sub-watersheds dominated by rural, urban and existing conditions (rural-urban). The characteristics of a sub-watershed are determined based on the impervious cover level of each sub-watershed. In this study, determining the characteristics of the sub-watershed using impervious cover charts adapted by CWP in 1998 [4]. After determining the location of the study, the properties of the sub-watershed were carried out using the help of HEC-Geo HMS software on Arc-GIS. The results of modelling in this application will be input into the HEC-HMS application.

In the flood routing process, rainfall data that affect the location of the research have needed. Regional rainfall at the study site was influenced by three rain stations including Gunung Mas Station, Cilember Station, and Gadog Station, which were conducted using the Thiessen polygon method. Then Analysis Frequency Data as design rainfall or return period using rainfall data 2008-2017 [5]. The results of the frequency data analysis were tested for compatibility using the Chi square test and the Smirnov test [6]. The flow hydrograph has generated with the help of HEC-HMS software by inputting the data properties and design rainfall of the research location. Then the running process has carried out using the SCS Hydrograph and Kinematic Wave method in each study location to produce flow hydrograph.

In achieving the research objectives, it is to review the efficiency of the use of the SCS Unit Hydrograph and Kinematic Wave method on the nature of Urban and Rural Watersheds. The efficiency of using the method will be seen based on the comparison of the results of the hydrograph of the four calculation conditions. Comparison is done from two ways, namely calculating the ratio of peak hydrograph discharge to each calculation condition and the level of hydrograph compatibility. The hydrograph suitability level has seen by plotting accumulative discharge values from the two hydrographs produced by the SCS Unit Hydrograph method using the Kinematic Wave method.

2.1. Rural and urban classifications in each Sub-DAS in the Ciliwung Hulu watershed

Based on the impervious value of the Upper Ciliwung Sub-watersheds, the characteristics of watershed can classified using Impervious cover model chart that was adapted from CWP in 1998[4]. The nature of Urban Watershed can be represented in the Crises cover values exceeding 25% and Rural Properties can be represented by Cisukabirus Sub watershed, which has impervious cover values below 10%. In addition, the study has a location that has both urban and rural characteristics by taking sub-watersheds that have impervious cover values between 10% - 25% (Figure 1 and Table 1).

Table 1. Impervious Cover for each Cilwiung Hulu Sub-watershed

Sub Watershed	Impervious Cover (%)
Ciliwung Hulu Sub Watershed	20.80
Cisarua Sub Watershed	15.73
Cisuren Sub Watershed	11.45
Cisukabirus Sub Watershed	7.67
Ciesek Sub Watershed	18.25
Ciseseupan Sub Watershed	34.36
Cibalok Sub Watershed	23.65

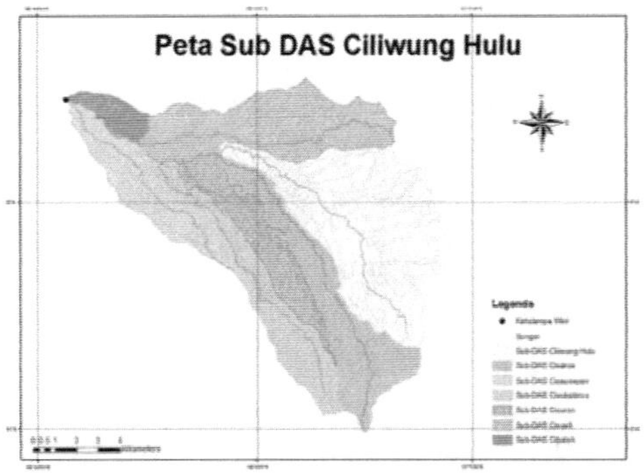

Figure 1. Map Ciliwung Hulu Watershed

2.2. Rural and urban classifications in each Sub-DAS in the upper Ciliwung watershed Determination of CN value in the Ciliwung Hulu watershed

Based on the 2014 Technical Guidelines for National Soil Classification [7], it can be seen that both types of soil are included in the type of andosol where Hapludands is a type of andosol soil Umbrik and Udivitrands are types of andosol vitric soil. Based on HSG, the two types of soil can be classified sequentially into soil types B and C and presented in Table 2.

Table 2. Classification of Soil Type Types According to HSG

Sub Watershed	HSG
Ciliwung Hulu Sub Watershed	B & C
Cisarua Sub Watershed	B & C
Cisuren Sub Watershed	B & C
Cisukabirus Sub Watershed	B & C
Ciesek Sub Watershed	B
Ciseseupan Sub Watershed	B
Cibalok Sub Watershed	B

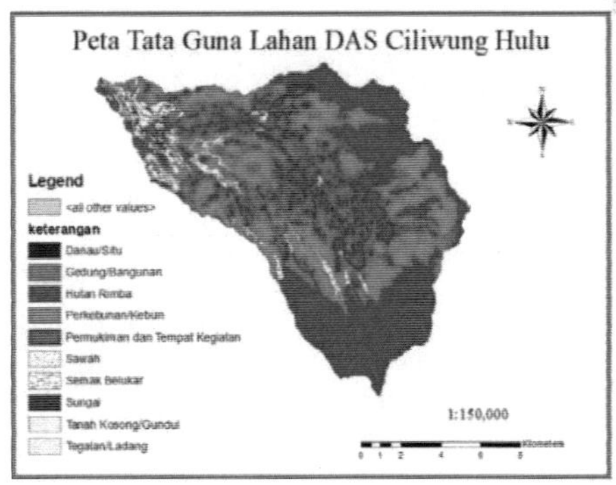

Figure 2. Map land use of upper Ciliwung watershed

Referring to the value of the CN by the USDA [1] and Classification of soil types according to the type of HSG, then the resulting value of the CN composite in their respective Regional Sub-Ciliwung Upper shown in Table 3.

IOP Conf. Series: Materials Science and Engineering **669** (2019) 012037 doi:10.1088/1757-899X/669/1/012037

Table 3. The composite CN value in the territory of each of the Upper Ciliwung sub-watersheds

Sub Watershed	CN composite
Ciliwung Hulu Sub Watershed	73
Cisarua Sub Watershed	81
Cisuren Sub Watershed	71
Cisukabirus Sub Watershed	77
Ciesek Sub Watershed	69
Ciseseupan Sub Watershed	77
Cibalok Sub Watershed	79

2.3. Hydrological parameters in 3 Sub-watershed-study sites

Delineation of the Research Sites directly uses tools at HEC-Geo HMS to estimate hydrological parameters to provide sub basin properties [8]. The processed products in HEC-Geo HMS are general data on the completeness of a watershed such as river length, river slope, longest river flow, CN value, loss rate of various types of land and land use of a watershed and can estimate the time lag in Sub-watershed. Property of the research sites presented in Table 4.

Table 4. Property of the Cisukabirus Sub-watershed, Ciseseupan Sub-Watershed and Cisarua Sub-Watershed

Properties Sub Watershed	Cisukabirus Sub Watershed	Ciseseupan Sub Watershed	Cisarua Sub Watershed		
			1	2	3
Area Sub Watershed (km2)	16.743	12.448	4.988	2.665	16.273
Slope Sub Watershed	0.41	0.095	0.17	0.15	0.46
CN Sub Watershed	76.696	77.23	82.76	82.27	77.39
River Length (m)	3099.721	2752.23	6217.856	308.210	8813.135
River Slope	0.040	0.027	0.04194	0.067402	0.078215
TimeLag (min)	81.68	135.04	53.4	37.4	57.4

2.4. Design rainfall

This study uses rainfall data from 3 observation stations which are considered to represent 3 sub-watersheds (Cisukabirus, Ciseseupan and Cisarua watersheds), namely Gadog Station, Cilember Station, and Gunung Mas Station. Analysis of the frequency of rainfall data has done to get the planned rainfall. Determination of regional rainfall has determined by the Thiessen Polygon method [2], which takes into account the broad percentage of influence of each station on the total watershed area.

The design rainfall with a return period of 5, 10 and 25 years shown in Table 6. However, in this research the results of the hydrograph compared to the rain load are only using a 10-year return period for all methods in the flow simulation because it only calculates the suitability of the hydrograph at a certain time. Calculations using frequency analysis with the length of the 10-years rainfall data (2008-2017) then performed the Chi-square test and Smirnov test [6].

Table 5. The design rainfall for each of the research sites

Return Period (year)	Xt (mm/hour)		
	Cisukabirus Sub Watershed	Ciseseupan Sub Watershed	Cisarua Sub Watershed
5	122.526	139.592	124.854
10	131.957	155.035	134.391
25	142.817	172.516	144.031

3. Results and discussion

3.1. Simulation flow hydrograph

The results of the simulation flow that occurred at the research sites (Cisukabirus Sub-watershed, Ciseseupan Sub-watershed, Cisarua Sub-watershed) using the SCS Unit method Hydrograph and Kinematic wave can be seen from the peak discharge (Qp), Time of peak (Tp), and rain duration values summarized in Table 6, 7 and 8. The results of the hydrograph for each method are given in Figure 3, Figure 4 and Figure 5.

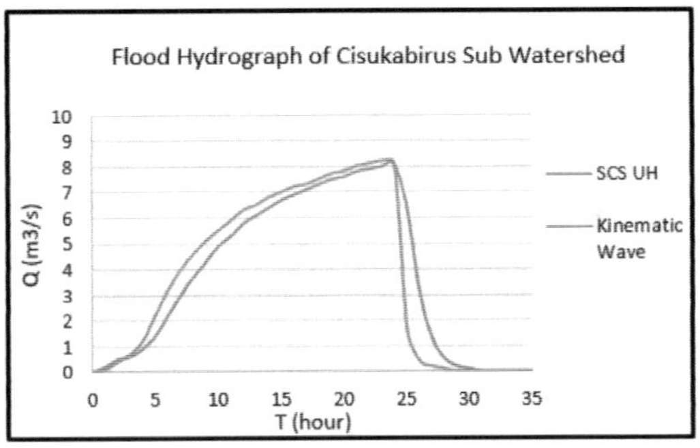

Figure 3. Flood Hydrograph of Cisukabirus Sub –watershed

Table 6. Results of simulation flow in the Cisukabirus sub-watershed

Parameter	SCS Unit Hydrograph	Kinematic wave
Q Peak (Qp) (m3/s)	8.1	8.2
Time of Peak (Tp)	00:00	23:00
Duration (hour)	31	30

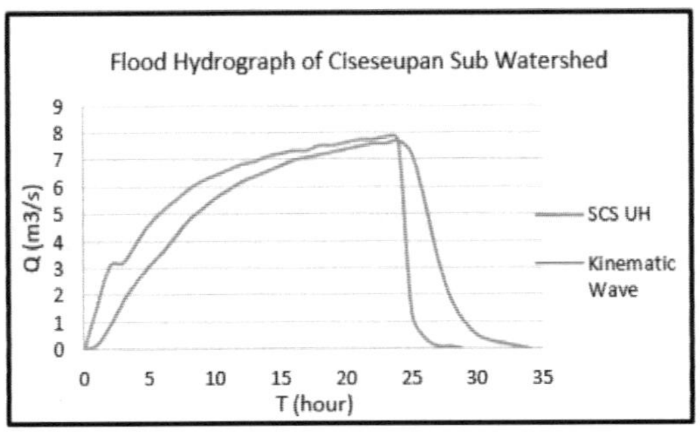

Figure 4. Flood Hydrograph of Ciseseupan Sub –watershed

Table 7. Results of simulation flow in the Ciseseupan sub-watershed

Parameter	SCS Unit Hydrograph	Kinematic wave
Q Peak (Qp) (m3/s)	7.7	7.8
Time of Peak (Tp)	00:00	23:00
Duration (hour)	34	29

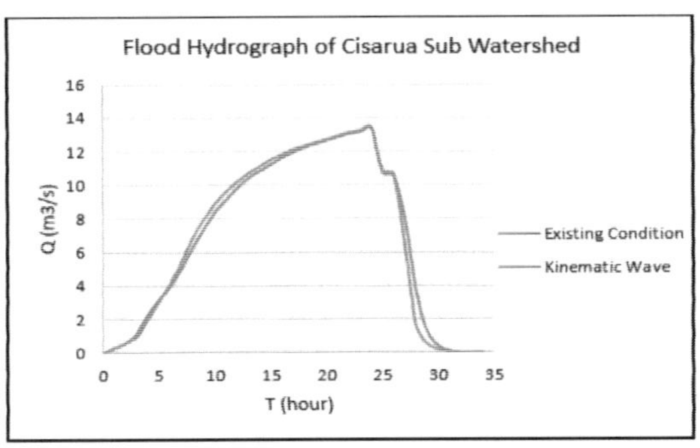

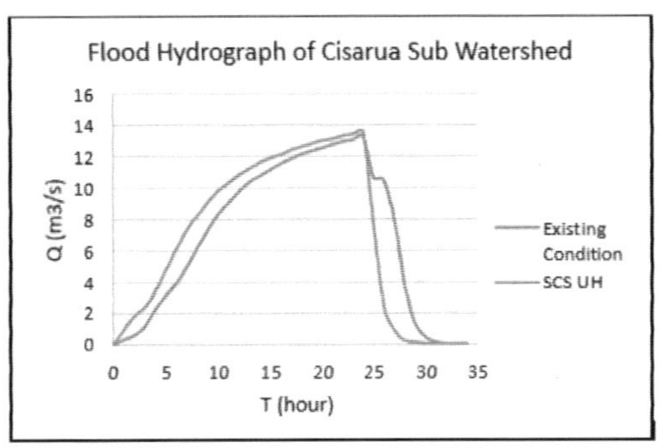

(a) **(b)**

Figure 5. Flood Hydrograph of Cisarua 1 Sub –watershed (a) Flood Hydrograph of Cisarua 2 Sub –watershed (b)

Table 8. Results of simulation flow in the Ciseseupan sub-watershed

Parameter	Existing condition	SCS unit hydrograph	Kinematic wave
Q Peak (Qp) (m3/s)	13.3	13.4	13.2
Time of Peak (Tp)	00:00	00:00	23:00
Duration (hour)	32	32	33

3.2. Comparison results of flow simulation hydrographs

Based on the comparison between the use of the SCS UH method and the kinematic wave in the Cisukabirus sub-watershed, it appears that the smallest percentage is when using the SCS UH method (Table 9). This shows that the Cisukabirus sub-watershed, which is assumed rural, is entirely more suitable using the SCS UH method on the Transform Method. In addition, viewed from the results of the graph of the accumulative debit relationship between the SCS UH method and the Kinematic wave, it can be seen that the direction of the slope line is not biased towards the Kinematic wave method (Figure.7a). With the two balanced results, it can be said that the Cisukabirus sub-watershed does not have a significant effect when using the SCS UH and Kinematic Wave methods.

Table 9. Percentage Suitability of SCS UH and Kinematic wave methods

Research Sub Watershed	Transform Method	Percentase (%)	Suitability Method
Cisukabirus Sub	SCS UH	1.2	SCS UH
Watershed	KW	4.8	
Ciseseupan Sub	SCS UH	1.3	Kinematic
Watershed	KW	1.2	wave
Cisarua Sub	SCS UH	0.75	Neutral
Watershed	KW	0.75	

Based on the comparison between the use of the SCS UH method and kinematic wave in the Ciseseupan Sub-watershed, it appears that the smallest percentage value is when using the kinematic wave method (Table 9). In addition, viewed from the results of the graph (Figure 7b), the relationship of accumulative debit of the SCS UH method with the kinematic wave shows that the line has more slope towards the kinematic wave method. This shows that the Ciseseupan Sub-watershed, which is assumed urban, is entirely more suitable using the kinematic wave method in the transform method model.

From the results of the two comparisons, it can be said that Urban Sub-watersheds are very influential if the method of flood routing in the catchment area uses a method that is lump sum due to the significant difference in results.

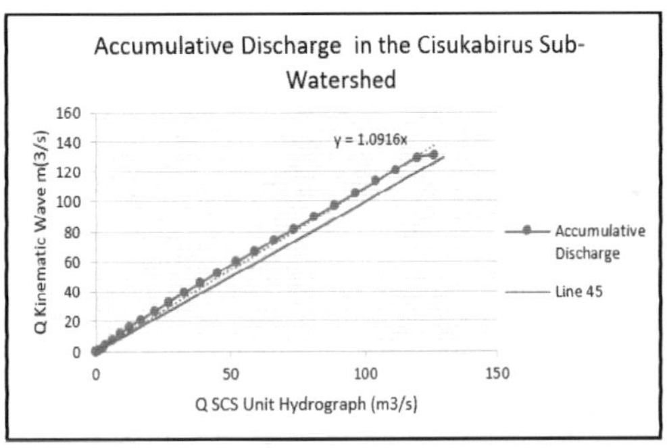

(a)

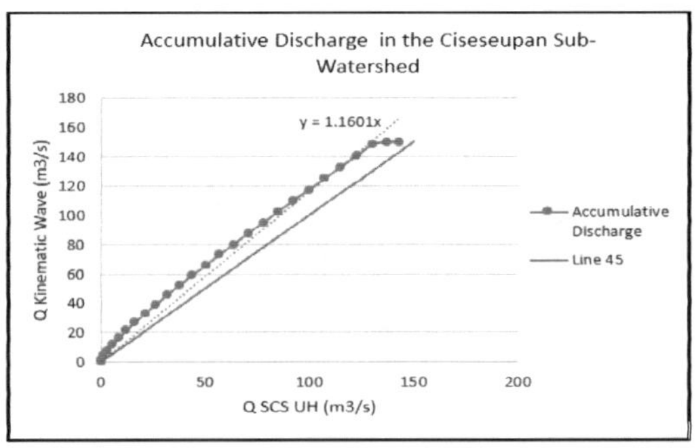

(b)

Figure 6. Accumulative discharge in the Cisukabirus Sub-watershed (a) Accumulative discharge in the Ciseseupan Sub-watershed (b)

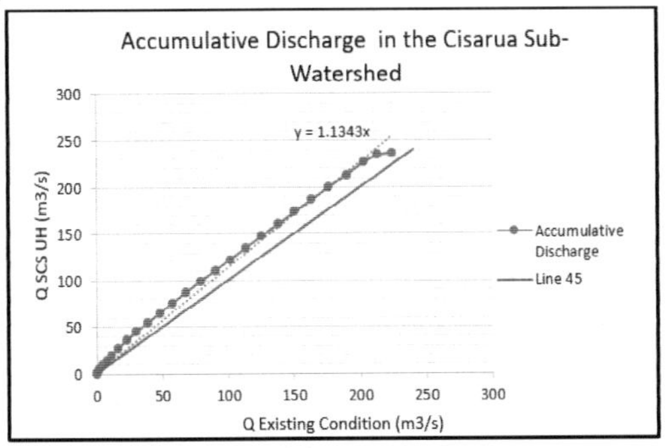

(a)

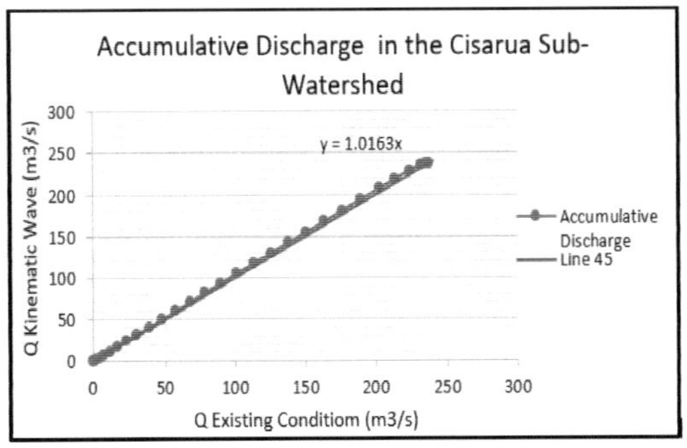

(b)

Figure 7. Accumulative discharge in the Cisarua Sub-watershed 1 (a) Accumulative discharge in the Cisarua Sub-watershed 2 (b)

Based on the comparison between the use of the SCS UH method and kinematic wave in the Cisarua sub-watershed, it can be seen that the percentage of existing conditions with the use of the kinematic wave and SCS UH methods did not differ (Table 9). Moreover, in terms of the results of the accumulative discharge graph SCS UH method with the condition of the existing seen that the direction of the line slightly inclined towards the SCS UH method (Figure 7a). As for the use of the kinematic method wave direction graph lines are neutral (Figure 7b).

4. Conclusion

Refers to figure 7a, Cisukabirus sub- watershed with Rural sub-watersheds will be more efficient when using the SCS unit hydrograph method because of the suitability of the results of the percentage ratio of peak discharge and accumulative discharge direction of the two methods that indicate compatibility with the SCS unit hydrograph method. In addition, with the SCS UH method, the use of methods is easier, faster and cheaper without reducing the level of accuracy of the results of the hydrological analysis of a watershed.

Refers to figure 7b, Ciseseupan sub- watershed with urban sub-watersheds can use both methods (SCS UH and kinematic wave) but it will be more efficient to use the Kinematic Wave method to weigh the level of accuracy of the results obtained because the already complex nature of the watershed requires more detailed data analysis. So as to get the results of a more optimal hydrological analysis even though it requires more time and money when analysing the hydrology of a watershed.

Refers to the figure 8 a and b, Cisarua sub-watershed with existing conditions (both urban and rural) will have an effect if using the SCS UH method on the transform method component. So if the sub-watershed, which is dominated by rural characteristics, it is more efficient to directly use the SCS UH because the results of the flow hydrograph are not very different when using the kinematic wave method. So as to maximize the time and costs needed to analyze the hydrology of a watershed.

References

[1] Chow V D 1988 *Applied Hydrology* New York: McGraw Hill, Inc.
[2] Sosrodarsono I S 2003 *Hidrologi untuk Pengairan* Jakarta: PT. Pradnya Paramita.
[3] Feldman A D 2000 *Hydrologic Modelling System HEC - HMS Technical Reference Manual.* Washington, DC: U.S Army Corps of Engineers.
[4] Schueler T R 2009 *Is Impervious Cover Still Important?* 2.
[5] Haan C 1997 *Statistical Methods in Hydrology* Iowa State University Press.
[6] Triatmodjo B 2008 *Hidrologi Terapan* Yogyakarta: Beta Offset.
[7] Technical Guidelines for National Soil Classification 2014.
[8] HEC- Geo HMS User's Manual 10.1 2013 Washington DC: US Army Corps of Engineers; Institute of Water Resources.

Acknowledgement
The authors express their appreciation as the study was conducted using the research funding for indexed international publication "Hibah Penugasan Publikasi Internasional Terindeks 9 (PIT 9)" fiscal year 2019 granted by Universitas Indonesia

Design of drainage system on industrial area Kampung Cina, Dobo City, Aru Islands Regency, based on Eco drainage using with HEC-RAS 4.0

B Rizaldi [1], G Idfi[1], E Suwarno[1], Mujiyono[1]

[1] Department of Civil Engineering, State University of Malang, Indonesia

gilang.idfi.ft@um.ac.id

Abstract. This study aim to plan a drainage system in the Industrial area of Kampung Cina Dobo City that can overcome the problems of inundation occurred. The inundation is caused by the drainage channel that has not been properly arranged and is still in the form of a natural drainage channel, most of which still use the road drainage network whose condition is not functioning properly due to sediment and garbage which results in many drainage channels being cut off. Data processing is carried out with two analyzes including hydrological analysis and hydraulic analysis. The hydraulics analysis includes channel section planning and drainage system modeling with HEC-RAS 4.0 software. In conventional drainage discharge flowing in the channel before the end of drain amounted to 1,136 m^3/sec. Once added a retention pond with a capacity of 10.358 m^3 with 5 scenario dimensions spillway. The results of modeling obtained by the reduction of the maximum flood flow is 1,07 m^s / sec with dimensions of aperture is 8 m x 0,8 m and lower the water level in the upstream channel by 31 cm and 44 cm in the downstream channel.

1. Introduction

The existing condition of the drainage system in Dobo City drainage channel is still not well defined, it is still a natural drainage channel. Most use a network of drainage path is not functioning properly because it was covered by sediment, trash, and the number of drainage channels cut off, especially in the area of Industrial Estate Kampung China Dobo City, therefore some areas or settlements in the Industrial Area Kampung China frequent inundation in case of heavy rains, and coupled with relatively gentle topography, so that the flow rate is relatively slow water runoff, as well as the conditions of inundation will increase when high tide the sea can result in rainfall runoff water can not be directly discharged into the sea.

Based on information compiled from the information media in the city of Maluku, describing "Dobo has ocean and land surface are almost equivalent. That means Dobo land is flat, so if there is a big wave when the tide will surely be creeping up on land once flooded residential areas. The high sea levels as well as one of the problems which results in inundation because the drainage channel is no longer able to accommodate the volume of water from the overflowing tide and rainwater [8]. Forresolve problems that occurred in the city, especially in the Industrial Area Kampung China Dobo City, should be a good drainage system planning and an environmental insightful.

Unavailability of a drainage system that is able to overcome the problems puddle in the industrial area of Kampung China City Dobo make planning and rebuilding a new drainage channel is needed. Especially with the existing drainage conditions during which it was unable to avoid the influence of tides increasingly urgent for the construction of a drainage system that is able to overcome the problem of backflow or back water caused by tidal sea water. The drainage system in the wake granted based on a sustainable drainage system or Eco Drainage to meet existing demands on zero delta Q policy and utilization of the water before it is released into the ocean in order to more beneficial for society [1].

2. Methodology

The method used in this project is analysis covering the primary data and secondary data. The primary data used in this planning includes the measurement of the topography of the area and survey the population about inundation occurred. While secondary data used in this planning includes daily rainfall data, land use data, the data watershed area, sea level height data pairs, the data geometric regions.

In the data processing is done by two analysis includes analysis of hydrology and hydraulics analysis [5]. Hydrological analysis was conducted on the analysis of the probability distribution, the probability distribution of the test, the calculation of rainfall plan, analysis of the concentration time, rainfall intensity analysis, and the latter is the calculation of design flood to divide the area into several sub-watershed planning (sub-catchment) [4]. In the analysis of a cross-channel hydraulics include planning and drainage network modeling with HEC-RAS software 4.0 to know the profile of water level does occur backflow (back water) or not, and to know the dimensions of the planned able to accommodate the flood flow or not [3].

Furthermore, the planning of eco drainage system by adding a retention pond to the planned drainage channel [2]. Planning is done to determine the extent and volume of the planned retention pond. Following that is to do modeling for disposing of water through the channel into an overflow bin with the building, will be obtained in the form of technical data storage capacity and reduction of the discharge channel [6]. Based on modeling results of the planning phase was completed followed by making inferences and suggestions,

3. Analysis and Discussion

Flood discharge plan is an important part and the basis for determining the capacity and dimensions of buildings water [9]. In determining the amount of design flood are done by using rational method whereby flow used at peak flow flood with rain return period of 25 years.

Table 1. Amount of plan discharge for each channel

No.	Channel	Q (m3 / s)	No.	Channel	Q (m3 / s)	No.	Channel	Q (m3 / s)
1	S1.Ka	0080	25	Box S9	0085	49	S19.Ka	0087
2	S1.Ki	0075	26	S10. Ka	0055	50	S19.Ki	0047
3	Box S1	0080	27	S10. Ki	0059	51	Box S19	0628
4	S2.Ka	0066	28	Box S10	0151	52	S20.Ka	0395
5	S2.Ki	0057	29	S11.Ka	0057	53	S20.Ki	0093
6	Box S2	0066	30	Box S11	0155	54	Box S16	0624
7	S3.Ka	0061	31	S11.Ki.1	0016	55	S21.Ka	0882
8	S3.Ki	0055	32	S11.Ki.2	0043	56	S21.Ki	0191
9	S4. Ka	0166	33	S12.Ka	0006	57	Box S15	1,059
10	S4. Ki	0191	34	S12.Ki	0010	58	S22.Ka	1,136
11	S5.Ka	0052	35	S13.Ka	0541	59	S22.Ki	0267
12	S5.Ki	0044	36	S13.Ki	0167	60	S23.Ka.1	0657
13	S6.Ka	0,098	37	S14.Ka	0509	61	S24.Ka	0078
14	S6.Ki	0095	38	S14.Ki.1	0025	62	Box S24	0736
15	S7.Ka 1	0046	39	S14.Ki.2	0034	63	S24.Ki	0043
16	Box S4	0212	40	S15.Ka	0176	64	S23.Ka.2	1,131

No.	Channel	Q (m3 / s)	No.	Channel	Q (m3 / s)	No.	Channel	Q (m3 / s)
17	S7.Ka.2	0449	41	S15.Ki	0172	65	S23.Ki	0326
18	S7.Ki	0052	42	S16.Ka	0229	66	S25.Ka	0544
19	Box S7	0449	43	S16.Ki	0192	67	Box S22	1,659
20	S8.Ka	0069	44	S17.Ka	0025	68	S25.Ki	0053
21	S8.Ki	0056	45	S17.Ki	0021	69	S26	1,540
22	Box S8	0069	46	S18.Ka	0054	70	outlet 1	1,926
23	S9. Ka	0041	47	S18.Ki	0055	71	outlet 2	1,592
24	S9. Ki	0036	48	Box S18	0221			

Hydraulics analysis performed to plan the dimensions of the channel capable of design flood. Hydraulics analysis includes the calculation of channel capacity, flow rate, and the amount of flow that is able to be streamed by the channel.

Table 2. Calculation of hydraulics analysis

Channel name	Width m	Height m	Q Hydraulics m3/s	Channel name	Width m	Height m	Q Hydraulics m3/s
Box S1	0.4	0.3	0117	S13.Ka	0.8	0.7	0740
S2.Ka	0.4	0.3	0117	S13.Ki	0.5	0.4	0225
S2.Ki	0.3	0.3	0072	S14.Ka	0.8	0.8	0740
Box S2	0.4	0.3	0117	S14.Ki.1	0.4	0.3	0117
S3.Ka	0.4	0.3	0117	S14.Ki.2	0.4	0.3	0117
S3.Ki	0.4	0.3	0117	S15.Ka	0.6	0.5	0382
S4. Ka	0.5	0.4	0225	S15.Ki	0.5	0.4	0225
S4. Ki	0.5	0.4	0225	S16.Ka	0.6	0.5	0382
S5.Ka	0.4	0.3	0117	S16.Ki	0.6	0.5	0382
S5.Ki	0.4	0.3	0117	S17.Ka	0.4	0.3	0117
S6.Ka	0.4	0.3	0117	S17.Ki	0.4	0.3	0117
S6.Ki	0.4	0.3	0117	S18.Ka	0.4	0.3	0117
S7.Ka 1	0.4	0.3	0117	S18.Ki	0.4	0.3	0117
Box S4	0.6	0.5	0305	Box S18	0.6	0.5	0382
S7.Ka.2	0.6	0.5	0458	S19.Ka	0.4	0.3	0117
S7.Ki	0.4	0.3	0117	S19.Ki	0.4	0.3	0117
Box S7	0.8	0.7	0740	Box S19	0.8	0.7	0740
S8.Ka	0.4	0.3	0117	S20.Ka	0.8	0.7	0458
S8.Ki	0.4	0.3	0117	S20.Ki	0.4	0.3	0117
Box S8	0.4	0.3	0117	Box S16	0.8	0.7	0863
S9. Ka	0.4	0.3	0117	S21.Ka	1	0.9	0987
S9. Ki	0.4	0.3	0117	S21.Ki	0.6	0.5	0382
Box S9	0.4	0.3	0117	Box S15	1	0.9	1,431
S10. Ka	0.4	0.3	0117	S22.Ka	1	0.9	1,431
S10. Ki	0.4	0.3	0117	S22.Ki	0.6	0.5	0382
Box S10	0.5	0.4	0225	S23.Ka.1	0.8	0.7	0740
S11.Ka	0.5	0.4	0225	S24.Ka	0.5	0.4	0225
Box S11	0.5	0.4	0225	Box S24	1	0.9	0987
S11.Ki.1	0.4	0.3	0117	S24.Ki	0.6	0.5	0382
S11.Ki.2	0.4	0.3	0117	S23.Ka.2	1	0.98	1,431
S12.Ka	0.4	0.3	0117	S23.Ki	0.6	0.5	0382
S12.Ki	0.4	0.3	0117	S25.Ka	0.8	0.7	0863

Channel name	Width	Height	Q Hydraulics	Channel name	Width	Height	Q Hydraulics
	m	m	m3/s		m	m	m3/s
				Box S22	1.2	1.1	2,424
				S25.Ki	0.4	0.3	0117
				S26	1	0.9	1,610
				outlet 1	1.4	1.3	3,761
				outlet 2	1.4	1.3	3,761

The dimensions used for calculating the hydraulic discharge are able to accommodate the planned flood discharge, it is then added with a free board of 10 cm on each channel. Material used in the channel using precast concrete (U-ditch).

Table 3. Channel dimension

Channel name	Length Channels	dimensions Channel			Channel name	Length Channels	Dimensions Channel		
	m	b		h		m	b		h
S1.Ka	117.9	400	X	400	S13.Ki	53.7	500	X	500
S1.Ki	118.3	400	X	400	S14.Ka	95	800	X	800
Box S1	8	400	X	400	S14.Ki.1	38	400	X	400
S2.Ka	65.9	400	X	400	S14.Ki.2	49.5	400	X	400
S2.Ki	66.7	400	X	400	S15.Ka	73.8	600	X	600
Box S2	8	400	X	400	S15.Ki	77.0	500	X	500
S3.Ka	61.2	400	X	400	S16.Ka	36.8	600	X	600
S3.Ki	61.9	400	X	400	S16.Ki	38.3	600	X	600
S4. Ka	36.9	500	X	500	S17.Ka	28.3	400	X	400
S4. Ki	37.8	500	X	500	S17.Ki	25.1	400	X	400
S5.Ka	48.	400	X	400	S18.Ka	84.2	400	X	400
S5.Ki	48.7	400	X	400	S18.Ki	81.2	400	X	400
S6.Ka	54.3	400	X	400	Box S18	5	600	X	600
S6.Ki	48.5	400	X	400	S19.Ka	71.2	400	X	400
S7.Ka 1	40.5	400	X	400	S19.Ki	69.7	400	X	400
Box S4	8	600	X	600	Box S19	5	800	X	800
S7.Ka.2	29.2	600	X	600	S20.Ka	49	800	X	800
S7.Ki	77.7	400	X	400	S20.Ki	48	400	X	400
Box S7	8	800	X	800	Box S16	5	800	X	800
S8.Ka	78.9	400	X	400	S21.Ka	54.9	1000	X	1000
S8.Ki	79.3	400	X	400	S21.Ki	58.8	600	X	600
Box S8	8	400	X	400	Box S15	5	1000	X	1000
S9. Ka	37.2	400	X	400	S22.Ka	74.6	1000	X	1000
S9. Ki	37.2	400	X	400	S22.Ki	81.4	600	X	600
Box S9	8	400	X	400	S23.Ka.1	33.1	800	X	800
S10. Ka	71.1	400	X	400	S24.Ka	63.2	500	X	500
S10. Ki	72.6	400	X	400	Box S24	5	800	X	800
Box S10	8	500	X	500	S24.Ki	66.6	600	X	600
S11.Ka	116	500	X	500	S23.Ka.2	113.8	1000	X	1000
Box S11	6.	500	X	500	S23.Ki	70.3	600	X	600
S11.Ki.1	50.5	400	X	400	S25.Ka	37.8	800	X	800
S11.Ki.2	59	400	X	400	Box S22	5	1200	X	1200
S12.Ka	63	400	X	400	S25.Ki	32.8	400	X	400
S12.Ki	59	400	X	400	S26	71.4	1000	X	1000

Channel name	Length Channels	dimensions Channel			Channel name	Length Channels	Dimensions Channel		
	m	b		h		m	b		h
S13.Ka	53.3	800	X	800	outlet 1	20	1400	X	1400
					outlet 2	20	1400	X	1400

Eco drainage modeling in this drainage planning use to eliminate or reduce the flood flow through a channel that does not occur in the event of advance puddle tide[7]. Eco drainage type that will be used in drainage planning in Industrial Area Kampung Cina Dobo City is the type of pitcher capacity is an interim (retarding Basin). With the division of two zones, where zone A using Eco drainage retention pond type (retarding Basin) and while zone B can not use the retention pond type because of the limited land can be used for pond reservoirs are relatively small and tekendala with public facilities such as churches, schools , health centers and others.

Retarding basin planned depth of 5.00 meters which can accommodate the flood flow at 10358.3 m^3 is equipped with inlet channels and outlet channels. In this plan that will be reviewed is the amount of inflow into retarding basin that is affected by the scenario dimension of building openings spillway channel (S22.Ka). Retarding basin also comes with disposal channels (outlet) with dimensions of 1.4 meter x 1.2 meter and planned with 2 doors with dimensions of 0.8 metersx 0.5 meters.

Scenario dimension opening side spillway and comparing the dimensions of the opening scenario building side spillway with inflow value into retarding basin and water level on the downstream channel S22.Ka. Modeling performed with 5 scenario opening dimension building side spillway.

- **Scenario 1 with an opening spillway b = 2 mh = 0.8 m**
  In modeling the scenario 1 can reduce the water level in the upstream channel of = 0.14 m and 0.29 m in the downstream channel. As well as data obtained with flow passing through the channel is Q US = 1.14 m^3 / sec and a flow that comes out or overflow into the retarding basin is Q Leaving = 0.60 m^3 / sec, the obtained values of flood flow remaining in the channel at Q DS = 0.524 m^3 / sec. It can be concluded based on the value of the output obtained from scenario 1, which can overwhelm the flood flow into the retarding basin of 0.60 m^3 / sec, so that the remaining flow on channel is 0.52 m^3 / sec.
- **Scenario 2 with openings spillway b = 4 m, h = 0.6 m**
  In the second scenario modeling can reduce the water level in the upstream channel of = 0.14 m and 0.25 m in the downstream channel. As well as data obtained with flow passing through the channel is Q US = 1.14 m^3 / sec and a flow that comes out or overflow into the retarding basin is Q Leaving = 0.54 m^3 / sec, the obtained values of flood flow remaining in the channel of QDS = 0.60 m^3 / sec, It can be concluded based on the value of output obtained from the second scenario that can bestow floow flow into retarding basin are 0.54 m^3 / sec, so that the remaining flow on channel is 0.60 m^3 / sec.
- **Scenario 3 with openings spillway b = 6 m, h = 0.8 m**
  In the third scenario modeling can lower the water level in the upstream channel is = 0.26 m and 0.41 m in the downstream channel. As well as data obtained with flow passing through the channel is Q US = 1.14 m^3 / sec and a flow that comes out or overflow into the retarding basin is Q Leaving = 0.99 m^3 / sec, the obtained values of flood flow remaining in the channel is Q DS = 0.15 m^3 / sec. It can be concluded based on the value of the output obtained from third scenario can bestow flood flow into the retarding basin is 0.99 m3 / sec, thus remaining flow on the channel is 0.15 m^3 / sec.
- **Scenario 4 with openings spillway b = 8 m, h = 0.8m**
  In the fourth scenario modeling can reduce the water level in the upstream channel is = 0.31 m and 0.44 m in the downstream channel. As well as data obtained with flow passing through the channel is Q US = 1.14 m^3 / sec and a flow that comes out or overflow into the retarding basin

is Q Leaving = 1.07 m^3 / sec, the obtained values of flood flow remaining in the channel id Q DS = 0.06 m^3 / sec. It can be concluded based on the output value obtained from the four scenarios can bestow flood flow into the retarding basin is 1.07 m^3 / sec, so that the remaining flow on the channel is 0.06 m^3 / sec

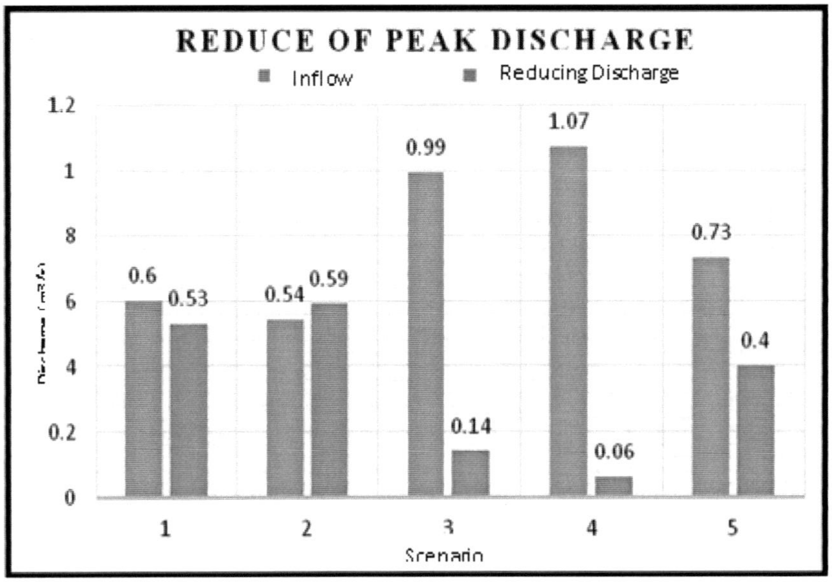

Figure 1. Graphic of comparation with inflow and reducing discharge

Here is the water level in the channel S22.Ka with scenario variations in the show at 5.17 picture is the scenario 1, 2, 3, 4 and 5:

= without a retarding basin = Scenario 3
= Scenario 1 = Scenario 4
= Scenario 2 = Scenario

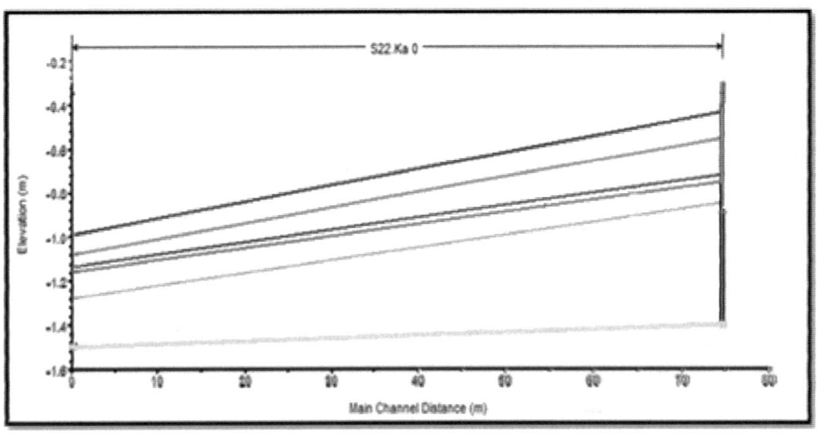

Figure 2. Water surface elevation condition

- **Scenario 5 with spillway openings b = 10 m, h = 0.6 m**
  In the fifth scenario modeling can reduce the water level in the upstream channel is = 0.17 m and 0.32 m in the downstream channel. As well as data obtained with flow passing through the channel is Q US = 1.14 m^3 / sec and a flow that comes out or overflow into the retarding basin

of Q Leaving = 0.73 m^3 / sec, the obtained values of flood flow remaining in the channel is Q DS = 0.41 m^3 / sec. It can be concluded based on the value of output obtained from 5 scenarios that can bestow flood flow into the retarding basin is 0.73 m3 / sec, so that the remaining flow on the channel is 0.41 m^3 / sec.

4. Results and conclusion

Based on the results of data analysis planning and modeling results in software HEC-RAS 4.0 can be concluded that:

- Based on the calculation results obtained design flood obtained flow on the downstream drainage channel of outlet 1 is 1,926 m^3 / sec and Outlet 2 amounted to 1,592 m^3 / sec.
- Channel section dimension planned has the smallest dimensions 0.4 mx 0.4 m and the dimensions of the largest cross-section of the channel is 1.4 mx 1.4 m. has met and is able to accommodate design flood with 25 years return period rainfall.
- In conventional drainage without retarding basin obtained of water level in the upstream channel S22.Ka as high as 0.97 m from the bottom of the channel, in the upstream and downstream as high as 0.51 m from the base of the channel.
- Planned retarding basin with depth 5 meters can accommodate the flood flow at 10358.3 m^3
- In this planning area A can only use this type of Eco Drainage retarding basin because area B does not have enough land for the retarding basin and the location also bordering public facilities.
- After the addition of eco drainage retarding basin and be equipped with spillway building with different dimensions, retarding basin can reduce flood flow through the channel which was originally 1.13 m^3 / sec can reduce flood flow to 1.07 m3 / sec in the scenario 4 with dimensions of side spillway b = 8 m and h = 0.8 m.

References

[1] G. Idfi 2017 2017 *Perbandingan Model Aliran Banjir Unsteady Flow dan Steady Flow pada Sungai Ngotok Ring Kanal.* Jurnal Bangunan, 22(2), 31-40.
[2] Istiarto 2014 *HEC-RAS Steady or Unsteady Flow Analysis* (istiarto.staff.ugm.ac.id)
[3] Anggrahini 1997 *Hidrolika Saluran Terbuka* (CV. Citra Media, Surabaya)
[4] V. T. Chow 1992 *Open Channel Flow., (*Erlangga, Jakarta)
[5] Sri Harto Br 1993 *Analysis of Hydrology. (*PT Gramedia Utama, Jakarta)
[6] Lensley, Ray K, Franzini, Joseph B. 1991 *Teknik Sumber Daya Air Jilid II*, (CV. Citra Media, Surabaya)
[7] C.D. Soemarto 1987 *Hydrology Engineering,* (Erlangga, Jakarta.)
[8] E. Suhartanto 2008 *The Manual Book of HEC-HMS and Application at Water Resource Management*, (CV Citra, Malang)
[9] B. Triatmojo 2008 *Hidrologi Terapan* (Beta Offset, Yogyakarta)

Analysis of flood inundation in North Sunter on the North Sunter Polder system performance

S Ignatius[1], H Soeryantono[1], E Anggraheni[1] and D Sutjiningsih[1]

[1]Departement of Civil Engineering, Universitas Indonesia, Depok, Indonesia

santoign5@gmail.com

Abstract. More than 40% area in Jakarta is below mean sea level which it causes inundation to occur every year in Jakarta. In order to solve this problem, the government developed Jakarta flood control management. Polder system is one of drainage system for areas that cannot drain water by gravity. There is one polder system located at North Jakarta named North Sunter Polder System. However, the inundation still occurs every year in that area. Based on the problem, the aim of this study is to analyse the cause of flood inundation in the area, it can be known by simulation using the HEC-HMS and HEC-RAS applications. Based on the simulation results it is known that the cause of flood inundation in the North Sunter Polder system is due to the less optimum pump operation in the polder system. So that to overcome the flood inundation can be done by increasing pump capacity and changing the pump operating elevation.

1. Introduction

Jakarta as the capital city and one of the most populous cities in Indonesia is often flooded. It is happened due the plains of Jakarta which were mostly below mean sea level and have fairly flat contours, so many plains in Jakarta cannot drain water by gravity. To resolve the flood inundation problem in Jakarta, a drainage system plan was made. The main idea in the planning of the drainage system was originally planned in the Dutch colonial era, but it was actualized in a study conducted by NEDECO in 1973 [1].

Based on the masterplan of the drainage system in Jakarta from NEDECO, it can be seen that to resolve the inundation problem in Jakarta is by making a polder system. The polder system is a drainage system made for areas that cannot drain water by gravity [2]. From the NEDECO masterplan, it can be seen that some areas in Jakarta that cannot drain water by gravity are made into polder systems. One of them is the North Sunter Polder System located in the northern part of Jakarta. North Sunter is one of the economically strategic areas in Jakarta. This is due to the large number of industrial companies and labor in the region [3]. Although this economically strategic area is formed in a polder system, in fact the inundation still occurs in this area.

In 2019 there were inundation in various places in the North Sunter area. On January 30, 2019 based on data from the BPBD DKI Jakarta, floods occurred at Gaya Motor Street and Swasembada Raya Street. Because inundation that often occur in the North Sunter region that are economically strategic, it is necessary to do research to find out the causes of inundation in the North Sunter Polder System. In this case the research was based on the last flood that occurred in the area which was on January 30, 2019. Based on the existing problems the authors assumed several possibilities that caused flooding in the North Sunter Polder System, namely pumps that are not able to work, compressed inlets, non-integrated channels, lack of main channel capacity, and lack of micro-drainage capacity. For the assumption of the causes of flooding in the North Sunter Reservoir Polder system, it can be investigated using simulations using the HEC-HMS and HEC-RAS application to determine the causes of inundation in the North Sunter Polder system.

2. Material and method

2.1. Study area

North Sunter located in Tanjung Priok district, North Jakarta. The North Sunter Polder System has a service area of 1,159.04 ha. North Sunter Polder System is bordered by Java Sea in the north, North Sunter reservoir street in the south, Laksamana Yos Sudarso Street in the east, and West Sunter reservoir street in the west. The map of catchment boundary of North Sunter Polder system is presented in *Figure 1* [4].

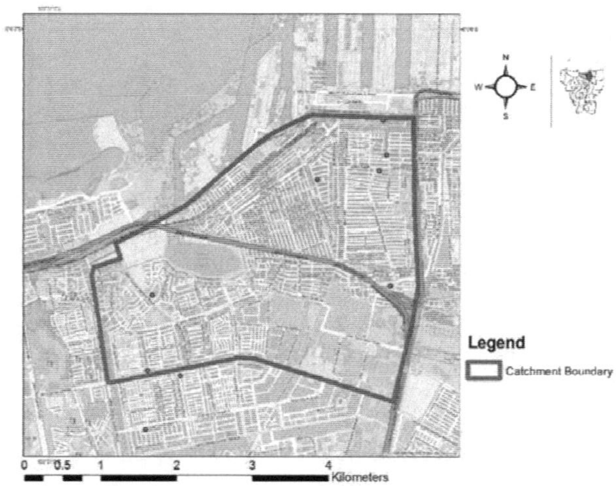

Figure 1. Catchment boundary of North Sunter Polder System

The components of North Sunter Polder System consist of main channel, reservoir, and pump. Based on the design planning of North Sunter Polder System, the catchment area is divided into 9 sub-catchments area. Eight of nine sub-catchment area has one main channel. Every sub-catchment flow through the main channel to the North Sunter Reservoir and pumped directly to the sea. The sub-catchment and the components of North Sunter Polder System is presented in *Figure 2.* The North Sunter Reservoir has an area of 30.31 ha with an average depth of 5 m. There are five pumps in North Sunter Polder System with two different capacities. Three pumps have a capacity of 2,475 m3/s, while the other two have a capacity of 5 m3/s and all the pumps have different operating elevations. The operating elevations of the pumps in North Sunter Polder System are tabulated in *Table 1.*

Table 1. Operating elevation of pumps

Pump	Capacity(m3/s)	Operating Elevation (m)	
		On	Off
1	2.475	-1	-1.1
2	2.475	-0.5	-1
3	2.475	0	-0.5
4	5	0.1	0
5	5	0.1	0

IOP Conf. Series: Materials Science and Engineering **669** (2019) 012039 doi:10.1088/1757-899X/669/1/012039

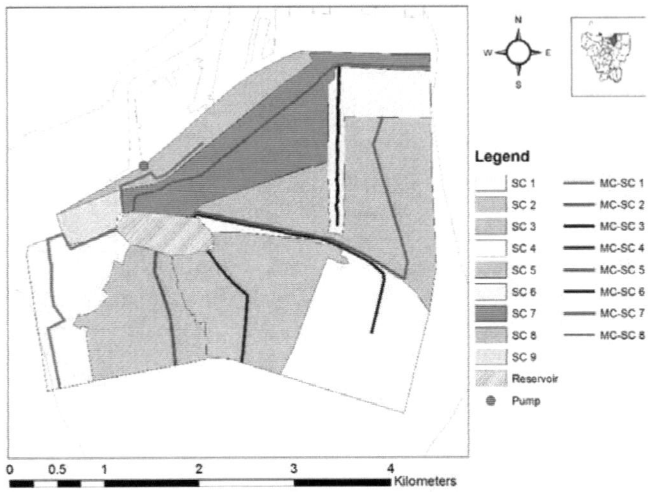

Figure 2. The sub-catchments and components of North Sunter Polder System

The data used in this study are mostly obtained from the design planning of North Sunter Polder System itself. Some data like channel cross-section are obtained from field survey. The measurement of cross-section from every main channel are based on the station from the design planning. A program with geographic information system is needed to achieve the geographical data like area and length. In this study a program called ArcMap is used to find the geographical data from the study location, that is area of sub-catchments and length of the main channel. The sub-catchment properties data from ArcMap and field survey are tabulated in *Table 3.*

2.2. Daily rainfall

The rainfall calculation is done based on rain gauges that affect the study area. In this case, Thiessen Polygon method used to determine the rain gauges that affect North Sunter area. The result from Thiessen Polygon method is that rainfall in North Sunter area affected by two rain gauges, Tanjung Priok and Kemayoran. Tanjung Priok rain gauge affect the rainfall in North Sunter around 91.66% of the area, meanwhile Kemayoran rain gauge is around 8.34%. The rainfall used in this study is daily rainfall during inundation, which was on January 30, 2019. The daily rainfall from two rain gauges is multiplied by the percentage affect of each rain gauge to determine the regional rainfall. The regional rainfall in North Sunter on January 30, 2019 is tabulated in *Table 2.*

Table 2. Regional rainfall in North Sunter on January 30, 2019

Rain Gauge	Rainfall on January 30, 2019 (mm)	Percentage (%)	Regional Rainfall (mm)
Tanjung Priok	127.6	91.66	124.2
Kemayoran	86.6	8.34	

2.3. Rainfall distribution

Based on regional rainfall that has been obtained from the affect of two rain gauges, then the rainfall is distributed into hourly rainfall using Wanny rain distribution. In Wanny rain distribution, it must be known how long the rain occurred in North Sunter on January 30, 2019. Based on rainfall history data that occurred in North Sunter on January 30, 2019, the rain starts from 01.00 to 04.00, so the rainfall pattern that is used in this study from Wanny rain distribution is 3 hours rain pattern. The hourly-rainfall distribution is tabulated in *Table 5*

Table 3. The sub-catchment properties data

Sub Catchment	Area (ha)	Main Channel					
		Station (m)	Shape	TW (m)	BW (m)	y (m)	S
SC 1	100.28	0 - 572	Rectangular	10.3	10.3	2.5	0.00018
		572 - 1946	Trapezoidal	7.7	4.8	1.64	0.0002
SC2	115.32	0 - 499	Trapezoidal	5.28	4.8	3.24	0.0039
		499 - 1455	Trapezoidal	3.94	2.48	2.8	0.0011
SC 3	160.12	0 - 1053	Rectangular	4.75	4.75	2.88	0.0004
SC 4	170.96	0 - 1480	Rectangular	6	6	4.2	0.00013
		1480 - 2120	Rectangular	6.28	6.28	2.38	0.00143
		2120 - 2780	Trapezoidal	4.57	3.4	2.36	0.0002
SC 5	198.66	0 - 1560	Rectangular	6	6	3.7	0.000384
		1560 - 2220	Rectangular	4.7	4.7	2.2	0.00034
		2220 - 3870	Rectangular	5.1	5.1	2.95	0.00061
SC 6	67.85	0 - 183	Trapezoidal	17.86	17.43	4	0.005
		183 - 1585	Trapezoidal	12.86	12	3.5	0.00071
SC 7	157.57	0 - 982	Rectangular	12	12	3.6	0.00081
		982 - 1477	Trapezoidal	9.66	7.4	2.9	0.00061
		1477 - 2634	Trapezoidal	12.66	11.83	3.6	0.00242
		2634 - 3232	Rectangular	11	11	3.2	0.00033
		3232 - 3593	Trapezoidal	24.6	22	5.2	0.00069
SC 8	61.3	0 - 165	Rectangular	2.8	2.8	1.2	0.00303
		165 - 290	Rectangular	2.8	2.8	1.2	0.00184
		290 - 644	Rectangular	2	2	1.07	0.00209
		644 - 1047	Rectangular	1.7	1.7	0.89	0.00124
SC 9	22.9			-			

2.4. *Flood hydrograph*

Estimation of the flood hydrograph is simulate using HEC-HMS application. The input data as parameter of the simulation are sub-catchment properties, weighted CN, the pumps and reservoir data, and the hourly rainfall distribution [5]. In this study, the simulation used SCS-CN for loss method and kinematic wave for transform method. SCS-CN used based on the available data, but kinematic wave used because of this concept suitable with urban area [6], like North Sunter. The estimation of weighted CN based on 2017 land use map from Badan Informasi Geospasial which is presented in *Figure 3* and the weighted CN of each sub-catchment are tabulated in *Table 4*. The CN number for each sub-catchment was based on runoff curve number table for selected land uses [7]. The outputs of simulation using HEC-HMS are the flood hydrograph for each sub-catchment in North Sunter on January 30, 2019. This flood hydrographs will be used as input in simulation with HEC-RAS application.

Table 4. Weighted CN of each sub-catchment in North Sunter

Sub-Catchment	SC 1	SC 2	SC 3	SC 4	SC 5	SC 6	SC 7	SC 8	SC 9
Weighted CN	83	84	85	81	84	85	84	85	70

Table 5. Hourly rainfall distribution

Date	Time	% Wanny Rainfall Distribution	Hourly Rainfall (mm)
30-Jan-19	0:00	0%	0
30-Jan-19	1:00	68%	84.32
30-Jan-19	2:00	24%	29.76
30-Jan-19	3:00	8%	9.92
30-Jan-19	4:00	0%	0
30-Jan-19	5:00	0%	0
30-Jan-19	6:00	0%	0
30-Jan-19	7:00	0%	0
30-Jan-19	8:00	0%	0
30-Jan-19	9:00	0%	0
30-Jan-19	10:00	0%	0
30-Jan-19	11:00	0%	0
30-Jan-19	12:00	0%	0
30-Jan-19	13:00	0%	0
30-Jan-19	14:00	0%	0
30-Jan-19	15:00	0%	0
30-Jan-19	16:00	0%	0
30-Jan-19	17:00	0%	0
30-Jan-19	18.00	0%	0
30-Jan-19	19:00	0%	0
30-Jan-19	20:00	0%	0
30-Jan-19	21:00	0%	0
30-Jan-19	22:00	0%	0
30-Jan-19	23:00	0%	0
31-Jan-19	0:00	0%	0

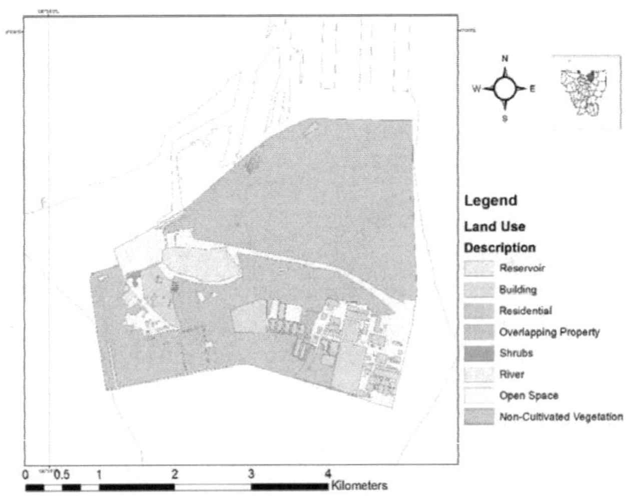

Figure 3. Land use map of North Sunter area

2.5. Determine the causes of inundation using HEC-RAS

HEC-RAS application is one-dimensional flow model which is capable of performing calculation of steady flow, unsteady flow, sediment, and water quality analysis for a network of natural or constructed channels [8]. The analysis that used in this study is unsteady flow analysis. The input data as the parameter of the simulation are network and geometry of channel, data of pumps and reservoir, and flow hydrograph from the simulation with HEC-HMS. In this case unsteady flow simulation used as a simulation run.

The outputs of HEC-RAS simulation are maximum water level of each main channel cross-section during the simulation time which is from January 30, 2019 00:00 until January 31, 2019 00:00 based on the flow hydrograph from HEC-HMS with the existing main channel. The result will show which main channel cross-section has inundation. Then it compared to the inundation area data in North Sunter on January 30, 2019.

After comparing the result, there will be another simulation run, but there will be some changes to the pump's capacity and operating elevation to find out whether the flood inundation is caused by the pumps or channel. In this case, the maximum changes of pump capacity based on the maximum pump capacity in Indonesia which is 10 m^3/s. If after doing some changes in pump capacity and operating elevation, there is no flood inundation, then the cause of inundation is caused by a pump that is not optimum.

3. Results and discussion

Flow hydrographs from HEC-HMS simulation results for each sub-catchment in North Sunter Polder System is presented in *Figure 4*. Meanwhile, the result of HEC-RAS simulation after the flow hydrograph was made into an input in HEC-RAS is presented in *Figure 5-7*.

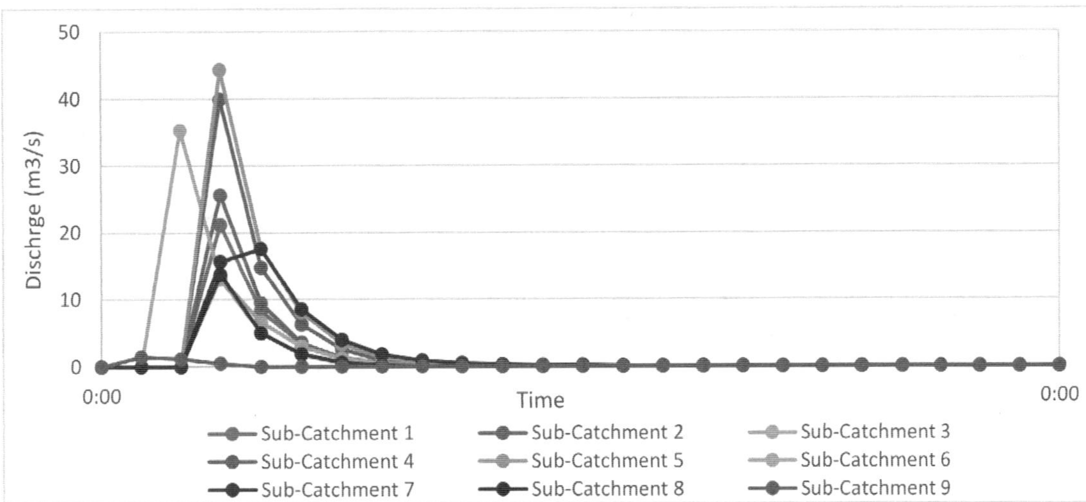

Figure 4. Flow hydrographs of each sub-catchments

Based on the HEC-RAS simulation results, it can be seen that there is flood inundation on the main channel in Sub-Catchment 4 Sta.2780 and Sub-Catchment 5 Sta.3870. It can be seen that the water level exceeds the existing bank station. Flood inundation on the main channel in Sub-Catchment 4 occurs at the same location as flood inundation data on 30 January 2019, which is on Gaya Motor Street. Then the flood inundation in Sub-Catchment 5 also occurred at the same location, which is on Swasembada Raya Street. Therefore, the results of HEC-RAS simulation are in accordance with the conditions in the North Sunter Polder System on January 30, 2019.

After knowing the location of the inundation based on HEC-RAS simulation results, some changes will be made to the pump operation. The purpose of this changes is to reducing flood inundation at the study location, so that the cause of inundation could be known in the North Sunter Polder System.

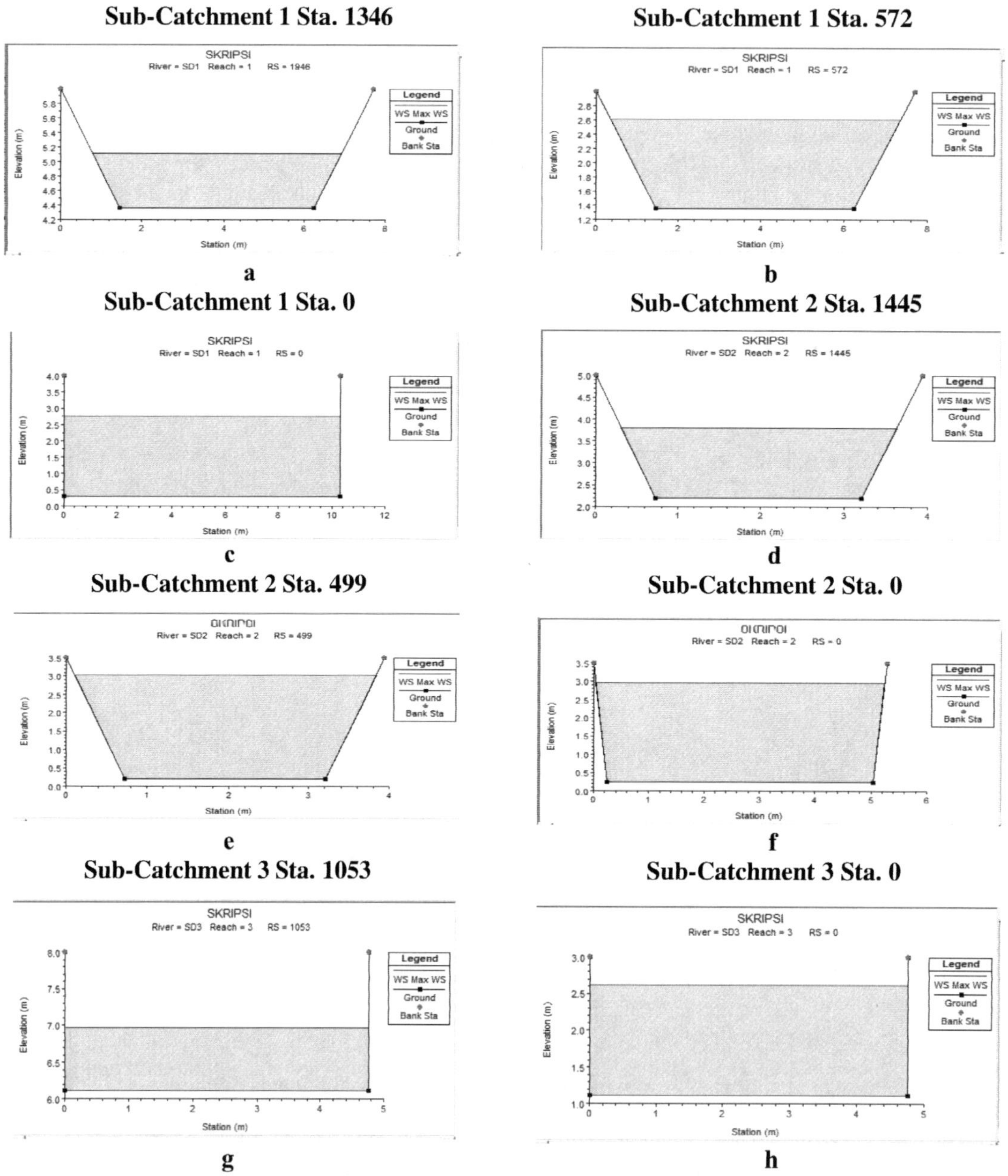

Figure 5. Inundation in Sub-Catchment1 Sta. 1346 – Sub-Catchment 3 Sta. 0 (a-h)

Sub-Catchment 4 Sta. 2780

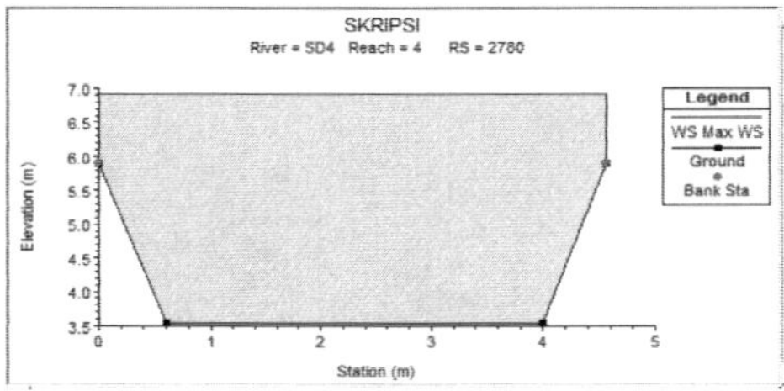

i

Sub-Catchment 4 Sta. 2120

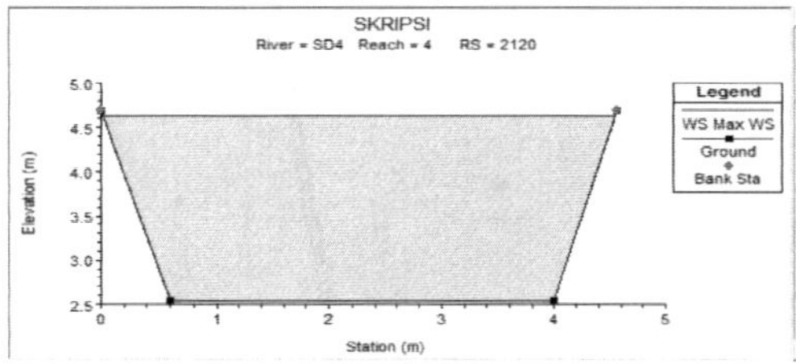

j

Sub-Catchment 4 Sta. 1480

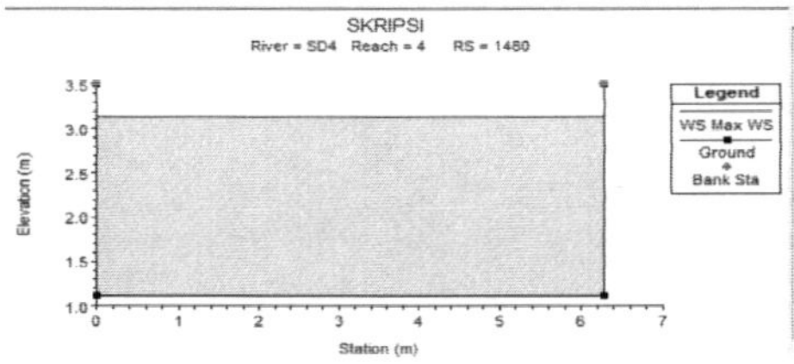

k

Sub-Catchment 4 Sta. 0

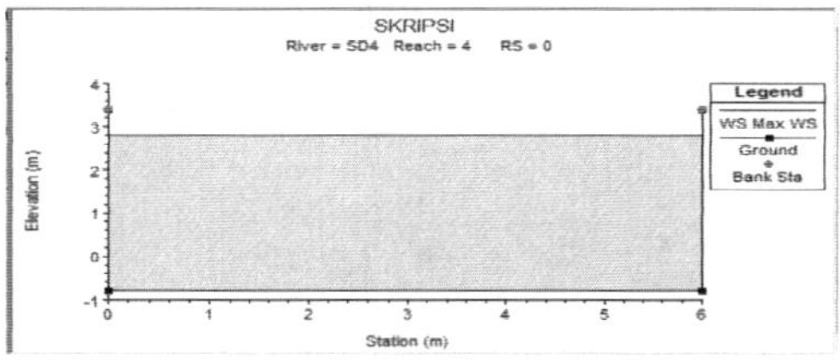

l

Sub-Catchment 5 Sta. 3870

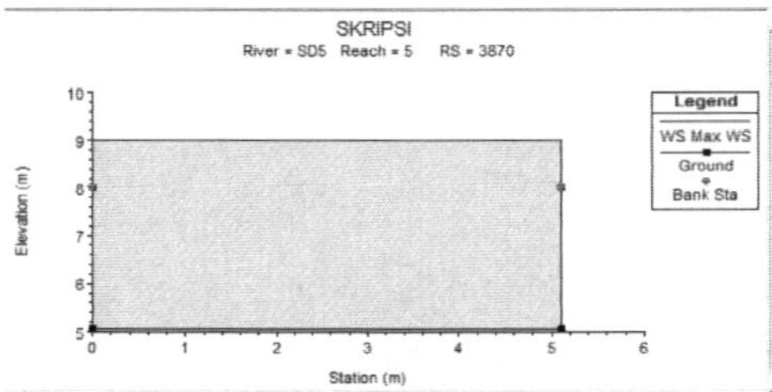

m

Sub-Catchment 5 Sta. 2220

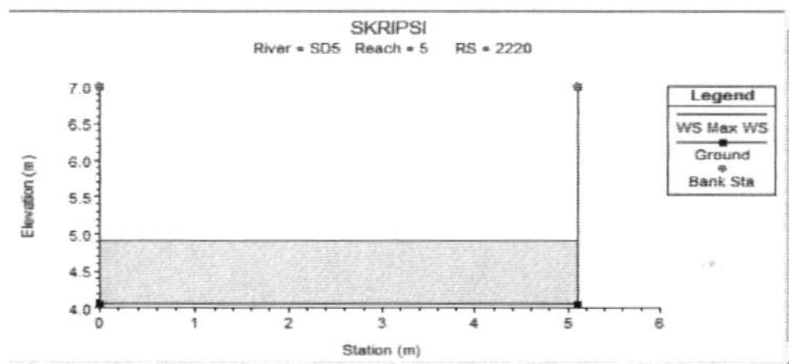

n

Sub-Catchment 5 Sta. 1560

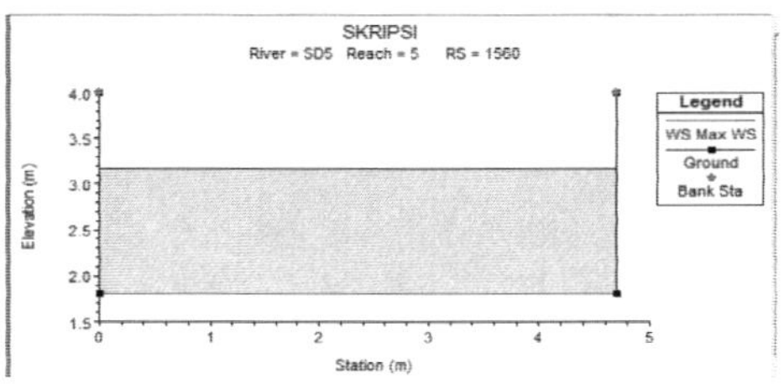

o

Sub-Catchment 5 Sta. 0

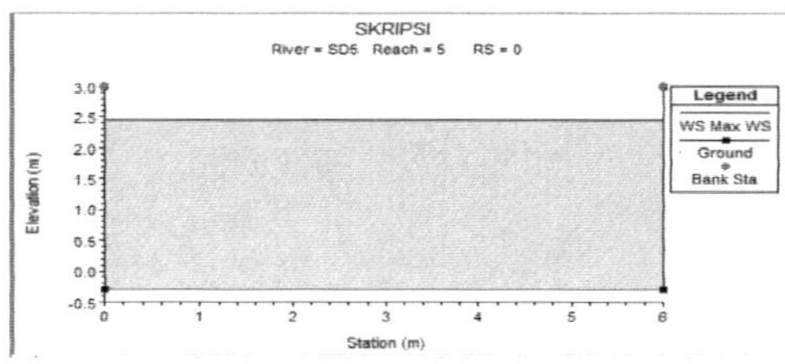

p

Sub-Catchment 6 Sta. 1585

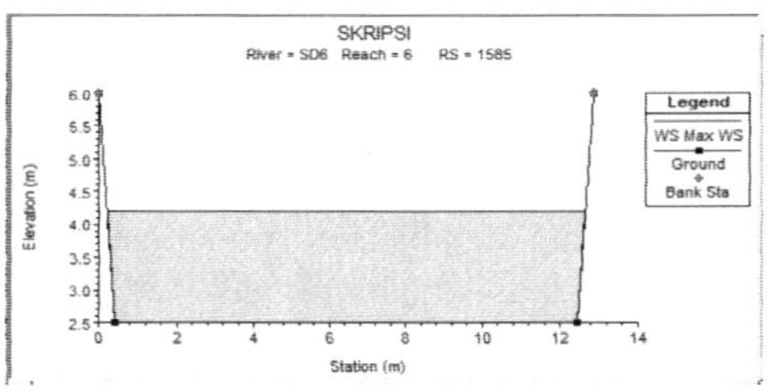

q

Sub-Catchment 6 Sta. 183

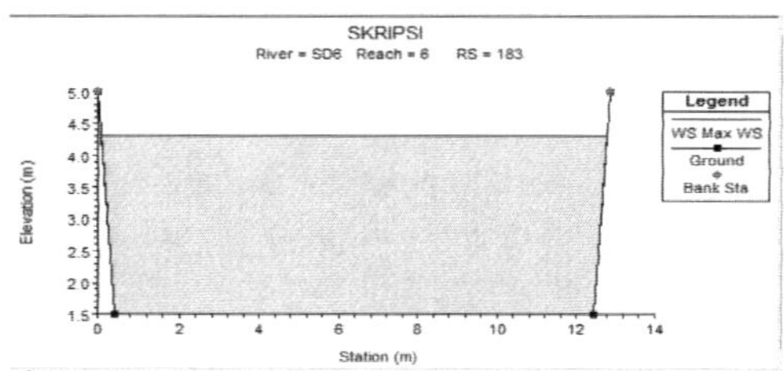

r

Figure 6. Inundation in Sub-Catchment 4 Sta. 2780 – Sub-Catchment 6 Sta. 183 (i-r)

IOP Conf. Series: Materials Science and Engineering **669** (2019) 012039 doi:10.1088/1757-899X/669/1/012039

Sub-Catchment 6 Sta. 0

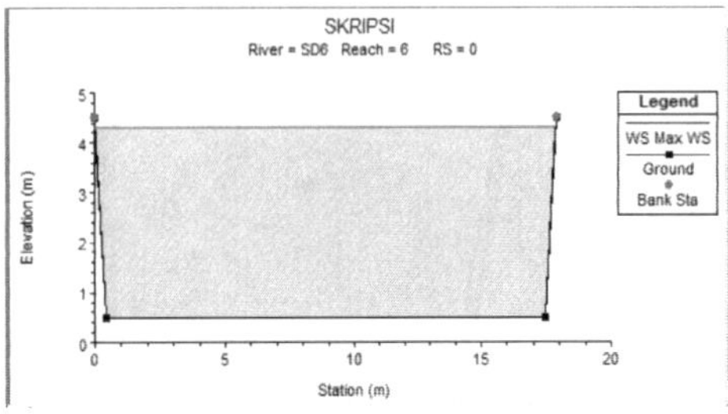

s

Sub-Catchment 7 Sta. 3593

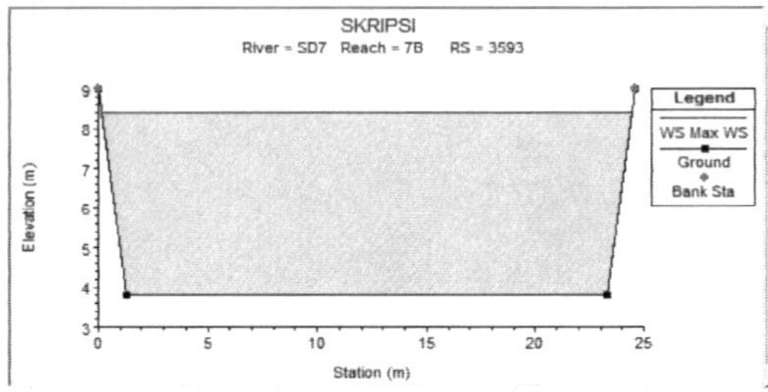

t

Sub-Catchment 7 Sta. 3232

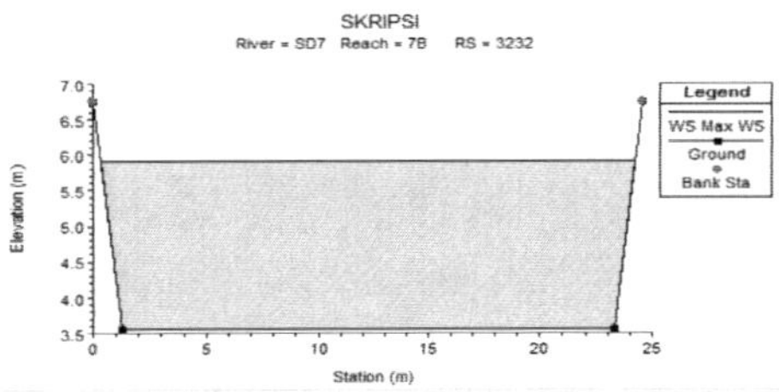

u

Sub-Catchment 7 Sta. 2634

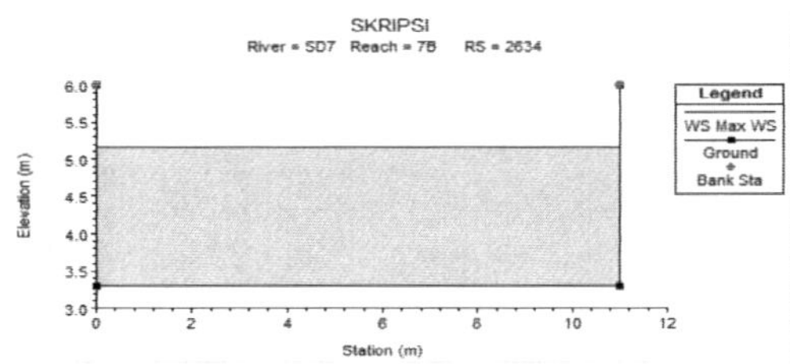

v

Sub-Catchment 7 Sta. 1477

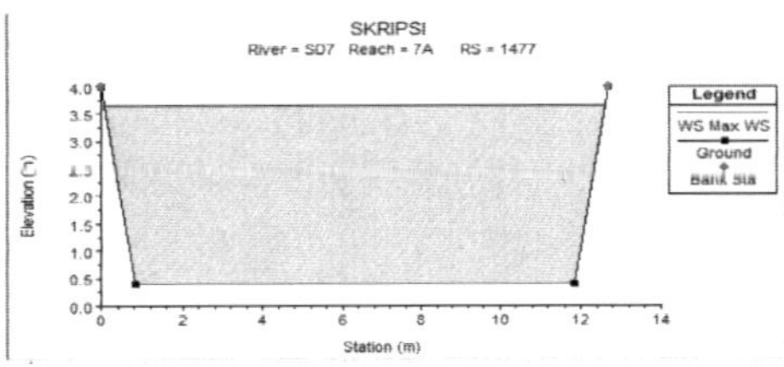

w

Sub-Catchment 7 Sta. 982

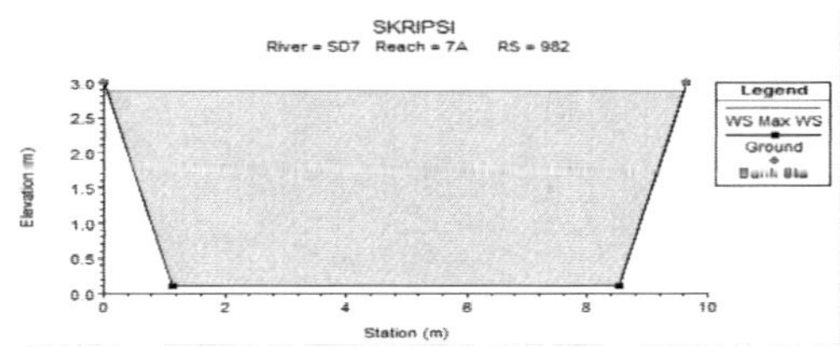

x

Sub-Catchment 7 Sta. 1477

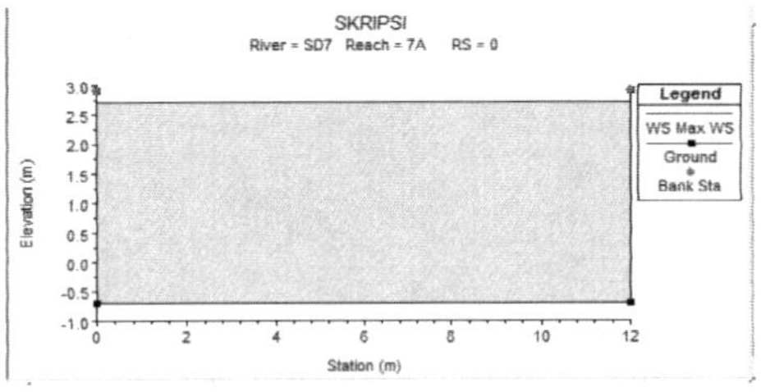

y

Figure 7. Inundation in Sub-Catchment 6 Sta. 0 – Sub-Catchment 7 Sta. 1477 (s-y)

The changes in pump operation are carried out by trial and error. There are many changes to pump operations have been carried out by HEC-RAS simulation, but there is only one simulation will be explained. The selection of this simulation is based on best simulation results. In this case the changes in pump's operation was made by adding one of the pump's capacity and lower the operating elevation. Adding pump capacity means the volume of water that pumped every second will increase and lowered the pump operating elevation means the pump operation will start earlier. The changes in pump operation is tabulated in *Table 6*. The results of new HEC-RAS simulation with changes in pump operation presented in *Figure 8*.

Table 6. Changes in pump operation

Pump	Capacity (m3/s)	Operating Elevation (m)	
		On	Off
1	5	-1	-1.3
2	2.475	-0.5	-1.1
3	2.475	0	-0.7
4	5	0	-0.5
5	5	0	-0.5

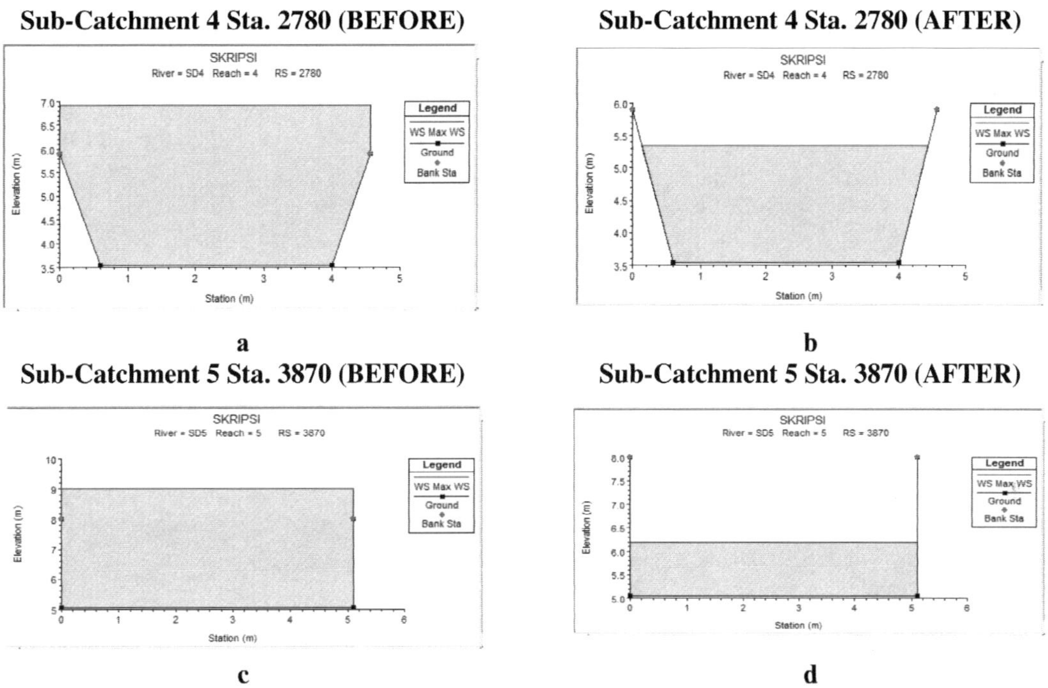

Figure 8. The results of new HEC-RAS simulation after changes in pump operation

Based on the results of the new HEC-RAS simulation in the condition of additional pump capacity and changes in pump operating elevation it can be seen that this condition makes no flood inundation on the main channels in Sub-Catchment 4 Sta. 2780 and Sub-Catchment 5 Sta. 3870. From the results of the simulation, the water level elevation in Sub-Catchment 4 decreased by ± 1, 7 m and the water level elevation in Sub-Catchment 5 decreased by ± 2.5 m.

From the simulation results it can be concluded that by adding pump capacity and changes in pump operating elevation as shown in *Table 6*, it is effective in resolving flood inundation problem in the North Sunter Polder System. From these results it can also be seen that the cause of flooding in the North Sunter Polder System is because the existing pump cannot work optimally. This happens due to lack of pump capacity and the length of time the pumping starts so that there is flood inundation.

4. Conclusions

Based on the HEC-RAS simulation result, there is flood inundation on main channel in Sub-Catchment 4 Sta. 2780 and Sub-Catchment 5 Sta. 3870 and the results of HEC-RAS simulation are in accordance with the conditions in North Sunter Polder System on January 30, 2019. Based on the simulation results of HEC-RAS, the causes of floods in the Sunter Reservoir North Polder System occur due to the lack of optimum pump operation. So, flood inundation in the North Sunter Polder System can be overcome by increasing the capacity of one pump and lower the pump operating elevation.

Furthermore, it is necessary to do more research and study, starting from field surveys to conducting simulations with applications to determine the causes of flood inundation in the North Sunter Polder System. Then, measurement of the channel cross-section should be carried out at a closer interval to determine the actual channel conditions, because the possibility of flood inundation can be caused by main channel conditions which not detected by the HEC-RAS simulation. And the last, the process of measuring the channel cross-section must be carried out in detail by calculating the sediment conditions that exist in the main channel to obtain the simulation results that are close to the existing conditions.

References

[1] NEDECO, *Detailed engineering design report nos. DKI 42 warakas; Nos. DKI 44 papanggo.*
[2] Yudi R K and Nugroho A M 2017 Perencanaan Sistem Polder Wilayah Semarang Timur, *Jurnal karya Teknik Sipil* 265-275
[3] BPS Kota Jakarta Utara, Kecamatan Tanjung Priok dalam angka 2018
[4] Dinas Tata Air Provinsi DKI Jakarta, Dinas tata air dalam angka 2015
[5] USACE 2013 *Hydrologic modeling system HEC-HMS User's Manual Version 4.0*
[6] Bedient P B, Huber C W and Vieux B E 2013 *Hydrologic and Floodplain Analysis*
[7] Chow V T, Maidment D R and Mays L W 1988 *Applied hydrology*
[8] USACE 2010 *Hydraulic reference manual v.4.1.*

Acknowledgement

The authors express their appreciation as the study was conducted using the research funding for indexed international publication "Hibah Penugasan Publikasi Internasional Terindeks 9 (PIT 9)" fiscal year 2019 number NKB-0053/UN2.R3.1/HKP.05.00/2019 granted by Universitas Indonesia.

IOP Conf. Series: Materials Science and Engineering **669** (2019) 012040 doi:10.1088/1757-899X/669/1/012040

Green Cimahi watershed for balancing water supply and flood control

A B Safarina[1], I Karnisah[2], A S Permana[3] and D Sufianto[4]

[1]Civil Engineering Department, Universitas Jenderal Achmad Yani, Jalan Terusan Jenderal Sudirman PO BOX 148, Cimahi, 40511, Indonesia
[2]Civil Engineering Department, Politeknik Negeri Bandung, Jalan Kampus Polban Ciwaruga, Bandung, 40559, Indonesia
[3]Electrical Engineering Department, Universitas Jenderal Achmad Yani, Jalan Terusan Jenderal Sudirman PO BOX 148, Cimahi, 40511, Indonesia
[4]Government Administration Department, Universitas Jenderal Achmad Yani, Jalan Terusan Jenderal Sudirman PO BOX 148, Cimahi, 40511, Indonesia

ariani.budi@lecture.unjani.ac.id

Abstract. Lack and damage of clean water and also flooding occurred in Cimahi river basin. Water sources polluted by industry and human activities. The purpose of this study was to design the water balance model of the Cimahi watershed as a Green Watershed. The study locations were the Cimahi river, Cilember river, Cibeureum river and Cisangkan river, which flow through the city of Cimahi. The method used is the analysis of the rainfall runoff model with the hydrograph of convolution method in the four watersheds, land use analysis, runoff coefficient analysis, dilution model analysis, groundwater analysis and demographic analysis. The results of this study is Design Green Cimahi Watershed that re-functioning local drainage to maintain the hydrological storage capacity of all watersheds in the city of Cimahi, normalizing Cilember river to reach the width of the river by 5.6 m from the previous width of 1.5 m, and the river capacity which was 4.8 m^3 / s to 42.37 m^3 / s so that enough to accommodate Q_{50} flood discharge, enforcing the rule that every industry must process its waste through a waste processing plant, building Pasirkaliki retention ponds that can function for groundwater conservation, flood control and clean water needs.

1. Introduction

Water as a natural resource is very important and vital for survival. In various countries, water is mostly used for domestic, industrial, recreational, agricultural and cultural purposes. Population growth, agriculture, urbanization and industrialization have caused a reduction in the amount of surface water [1]. Water is distributed in nature as surface and ground water in different forms and sources which are oceans, seas, rivers, streams, lakes, ponds, wells, boreholes and springs. Rivers are among the oldest water bodies in the world. Water is a resource that has many uses and also supports all forms of life and affects our health, lifestyle, and economic well- being [2]. Water resources on this earth consisting of surface water, rainwater and groundwater can be represented as a hydrological model in the form of watersheds [3]. Lakes that are a type of surface water are easily accessible water resources, and accommodate more than 95% liquid surface fresh water on the Earth's surface, for domestic, agricultural, and industrial water supply [4].

The drinking water problem is the pattern of distribution of rainfall and water quality as in Zhejiang Province, which is located on the east coast of China. Maintaining the quality of reservoir water is very important for water security and socio-economic development. The number of reservoirs have experienced increased pressure and degradation, increased waste disposal with sustainable economic development and urban construction, severely damaging the reservoir environment and affecting reservoir function [5]. Common problems with the distribution of clean water in urban areas are poor water quality, continuous service, inadequate disinfection and old infrastructure that is not repaired. To overcome this, in the city of Bandung a community-based program was conducted for the Cikapundung riverbank community that was representative of the community with socio-economic and geographical vulnerability [6].

Several provinces on the Mindanao Island of the Philippines have been flooded due to extreme rain during the Lingling and Jangmi storms which resulted in the overflowing of rivers and lakes. To reduce the impact of the disaster, a hydrological model of the Tago River basin was created and a runoff rainfall simulation was carried out using HEC HMS. The basin model, which is a physical representation of the watershed, was developed using the 10-m Synthetic Aperture Digital Elevation Model (SAR DEM) and river network in the delineation of the watershed, using information from the land cover map [7]. Mitigation of the Karnali flood disaster in the western part of Nepal was carried out by increasing the lag time from 2-3 hours to 7-8 hours which was informed through a community-based flood early warning system. The method used is a time series model that comes from a database that can be interpreted physically. The challenge is to communicate probability predictions to the farthest communities [8].

Cimahi River that flows along the City of Cimahi, West Bandung district and Bandung Regency is the largest river in the city of Cimahi besides other rivers, namely the Cibeureum river, Cisangkan river and Cilember river. This river network has contributed a lot to the water resources of Cimahi City, including the Sukawana Reservoir. The significant problem of water resources in the city of Cimahi is flooding and lack of clean water. Floods in the city of Cimahi often occur both in the northern areas such as Citeureup, Cipageran, and also Central Cigugur, Cibeureum and Melong. The depth of the flood can reach one meter so it is prone to accidents and turns off traffic flow for several hours. The cause of this overflow is that the local drainage is not functioning and the river border is narrowed so that the river is no longer able to accommodate the discharge that occurs [9-10].

The lack of clean water in the city of Cimahi occurs in a long dry season where there is very little or no rainfall for more than six months. Drought occurred in the Utama village, Cibeber village, Cipageran village, Cibabat village, Pasirkaliki village, Cimahi village and Leuwigajah village. In this condition, the availability of clean water cannot depend on rainfall, but the ground water supply in the city of Cimahi is not sufficient to meet the community's clean water needs. The industry in the city of Cimahi which has taken significant groundwater extraction at various depths has disrupted the balance of the water system in the city of Cimahi. Besides that, the uncontrolled disposal of industrial waste also made the water quality of Cimahi city highly polluted [11].

The purpose of this study was to design the water balance model of the Cimahi watershed as a Green Watershed. The study locations were the Cimahi river, Cilember river, Cibeureum river and Cisangkan river, which flow through the city of Cimahi.

2. Methods of the study

This research consists of four stages of activity, namely measuring field data in the form of river hydrology and hydraulics, analyzing situation maps intersecting regional areas and river areas, analyzing existing Cimahi river watershed systems consisting of water availability and water requirements and the design of Cimahi water systems in the form of Green Cimahi watershed.

2.1. Hydrology analysis method

The parameters needed in hydrology analysis in this study are the availability of water, namely the discharge plan for each river both for drinking water needs and design flood discharge. In addition, groundwater availability will also be analyzed. Discharge per mm effective rainfall each river is calculated using the Nakayasu modification method that has been calibrated based on the characteristics of the watershed [12].

$$Q_p = \frac{0.99AR_o}{3{,}7(0{,}1T_p + T_{0,3})} \quad (1)$$

Peak discharge (Q_p) is the function of watershed area (A), watershed characteristic coefficient (C = 0.99), unit rainfall (R_o), time lag (Tp) and time required to discharge reduction up to 30% peak discharge ($T_{0.3}$). Design rainfall is calculated using Gumbel frequency analysis while effective rainfall is calculated using a runoff coefficient derived from the land use map.

The water needs analyzed are only clean water needs. The need for clean water is calculated based on the population of the city of Cimahi and the standard of urban clean water needs of each person, namely 144 liters /person/ day. Clean water needs are analyzed spatially and temporally so that the water distribution in all villages can be known.

2.2. Hydraulic analysis method

The parameters of the river hydraulic data measured are river cross-section, river slope, flow velocity, flow depth and riverbed elevation [13]. This parameter is used to determine the maximum river hydraulic capacity calculated using the following Manning equation,

$$Q = \frac{1}{n} A\, R^{2/3} S^{1/2} \quad (2)$$

where Q is discharge, A is wet cross-sectional area, R is hydraulic radius, S is the slope of the energy line and n is roughness coefficient from Manning.

2.3. Method of mapping intersections of river areas and regional areas

In analyzing water drainage both on land and in rivers, it is very necessary to make intersection maps of river areas and regional areas. River boundaries show patterns of drainage in accordance with topography and regional boundaries indicating administrative and regional management.

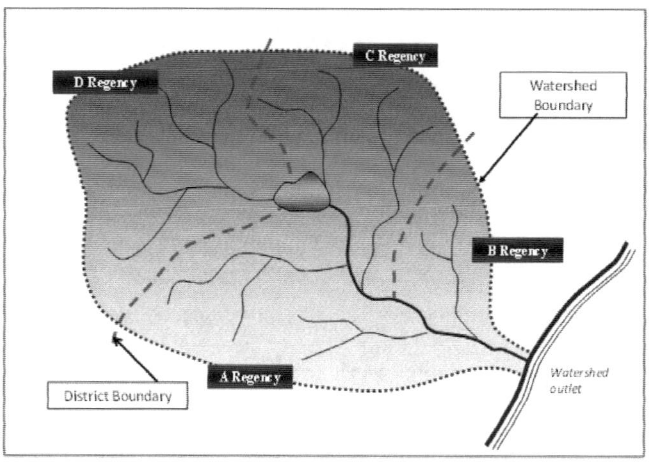

Figure 1. Intersection map of river areas and regional areas

2.4. Method of analyzing existing cimahi river watershed systems

To get the condition of existing city water system in Cimahi, the existing conditions of water availability and water requirements are calculated. In addition, disaster conditions were also reviewed, namely the problem of lack of clean water and flooding. The availability of surface water is viewed from discharge of four major rivers in the city of Cimahi, namely the Cimahi River, Cibeureum River, Cisangkan River and Cilember River. Water availability is also viewed from groundwater discharge in the Cimahi area. If the groundwater discharge cannot serve water needs during the dry season, it will be planned for a dual function reservoir, namely flood control during the rainy season and water supplier during the dry season. The flood analysis of Cimahi city which was reviewed was the heaviest flood area, namely the Melong and Central Cigugur villages originating from the overflow of the Cilember river.

2.5. Green Cimahi watershed design method

The criteria for the design of Green Cimahi watershed water system are the achievement of a water balance management between surface water, land cover and ground water, without floods, drought and water pollution so as to be able to serve the water needs of the people of Cimahi. The design output in the form of the Green Cimahi watershed map is equipped with a design note for temporal and spatial water availability and water needs.

3. Results and discussion

3.1. Study area

The city of Cimahi is geographically located between 107^{0}30 '30"- 107^{0}34 '30" BT and 6^{0}50 '00 " - 6^{0}56 '00" South Latitude. The total area of Cimahi City is 40.2 km2 (4,020 Ha). The city of Cimahi is included in the West Java Province and includes 3 Subdistricts consisting of 15 villages, namely, North Cimahi Subdistrict consisting of 4 villages, Central Cimahi Subdistrict consisting of 6 villages and South Cimahi Subdistricts consisting of 5 villages.

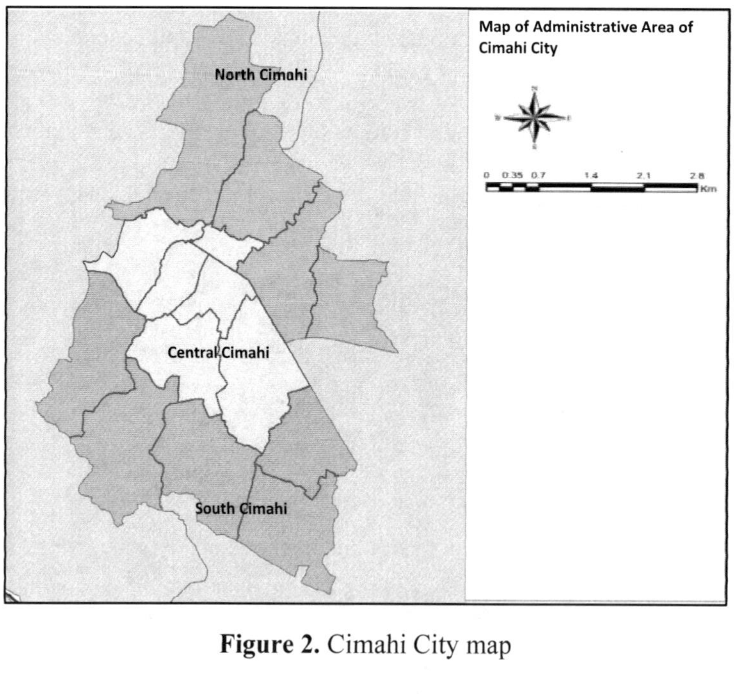

Figure 2. Cimahi City map

The city of Cimahi is located in the upstream of Citarum River as a part of Bandung Basin and one of the Citarum River valleys. The river that passes through the City of Cimahi is the Cimahi River, Cibeureum river, Cibaligo (or Cilember) river, and Cisangkan River, meanwhile, the springs in Cimahi are Cikuda springs and Cisintok springs. Intersection map of those rivers and Cimahi Subdistricts is shown in Figure 3 below,

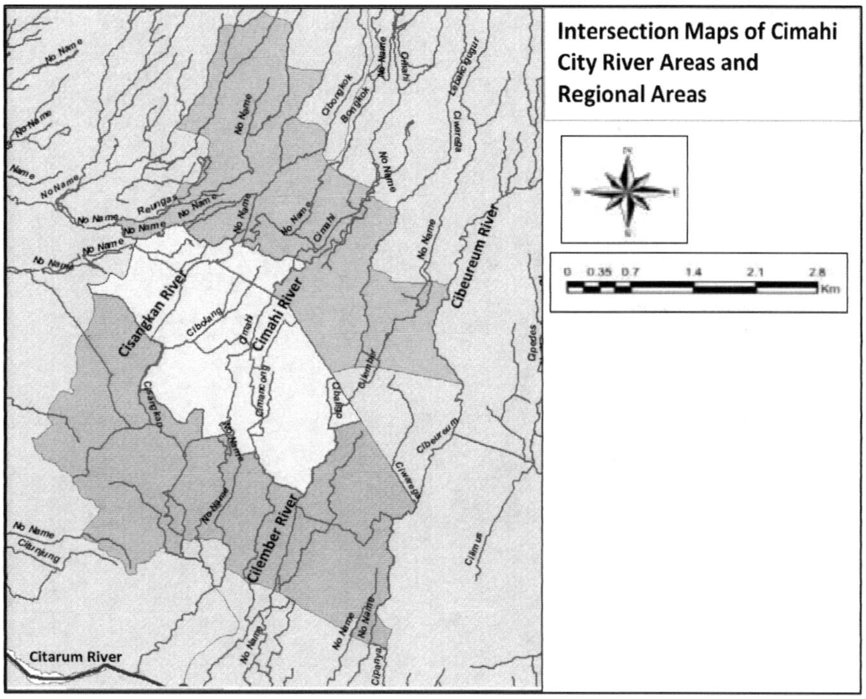

Figure 3. Intersection maps of Cimahi City river areas and regional areas

3.2. Water need and water availability existing

Clean water needs are calculated based on the population of Cimahi city in 2017 with the standard of clean water needs is 144 liters / second / person. Clean water needs are calculated monthly and distributed to each village as shown in table 1 below,

Table 1. Cimahi City water need existing

Subdistrict	Villages	Area (km2)	Population (people)	Clean Water Need (m^3/s)
North Cimahi	Cipageran	5.94	73,870.00	0.12
	Citeureup	1.27	15,794.00	0.03
	Cibabat	2.87	35,691.00	0.06
	Pasir Kaliki	3.23	40,168.00	0.07
Central Cimahi	Padasuka	0.84	14,596.00	0.02
	Setiamanah	1.98	34,404.00	0.06
	Cimahi	2.25	39,096.00	0.07
	Karang Mekar	1.37	22,763.00	0.04
	Baros	2.25	39,096.00	0.07
	Central Cigugur	1.37	23,805.00	0.04
South Cimahi	Cibeber	3.33	51,612.00	0.09
	Cibeureum	3.93	60,911.00	0.10
	Leuwigajah	2.75	42,622.00	0.07
	Utama	3.13	48,512.00	0.08
	Melong	3.8	58,896.00	0.10

Water availability is calculated from the monthly discharge of the Cimahi River, Cibeureum River, Cisangkan River, and Cilember River at the location as shown in the map in Figure 4. For the discharge of clean water design, a 5-year return period is used, while the flood discharge design is used for a 50-year return period.

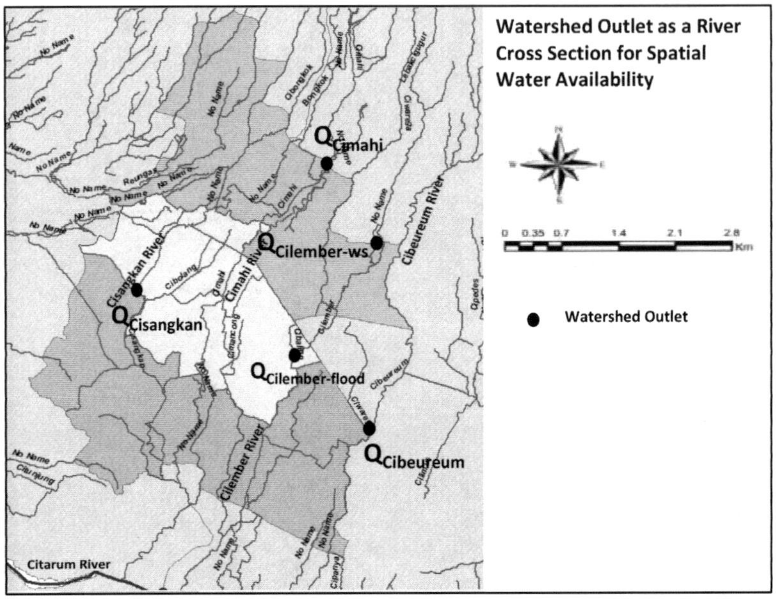

Figure.4 Watershed outlet for river water availability

The amount of water availability each month for the four rivers above at the outlet is shown in table 2

Table 2. Monthly water availability

River Name	Notation	January	February	March	April	Mei	June	July	August	September	October	November	December
Cimahi	Qav (m^3/s)	24.77	26.13	25.39	30.83	33.33	16.67	14.93	10.59	16.43	31.31	32.78	28.68
Cibeureum	Qav (m^3/s)	30.55	32.23	31.32	38.03	41.11	20.57	18.42	13.06	20.27	38.62	40.43	35.38
Cisangkan	Qav (m^3/s)	12.51	13.20	12.83	15.57	16.83	8.42	7.54	5.35	8.30	15.81	16.56	14.49
Cilember	Qav (m^3/s)	17.30	18.25	17.74	21.54	23.28	11.65	10.43	7.40	11.48	21.87	22.90	20.04

In table 2 it can be seen that the smallest amount of water in one year is in August and the highest one is in November.

3.3. Flood analysis of cilember watershed

Floods due to the overflowing of the Cilember river caused a large amount of inundation in the villages of Melong and Central Cigugur. This level of flooding is a heavy category because it can occur in light rainfall even without rain. Flooding can occur if rainfall falls in the upstream area. The cause of the Cilember river flood is due to the narrowing of the very large river border so that the existing border is only 30% of the actual border. Due to the narrowing of this border, the capacity of the Cilember river starts in the Central Cigugur to the downstream decreases up to 70% as shown in table 3 below,

IOP Conf. Series: Materials Science and Engineering **669** (2019) 012040 doi:10.1088/1757-899X/669/1/012040

Table 3. Overflow of Cilember river downstream

Disccharge	January	February	March	April	Mei	June	July	August	September	October	November	December
Q_{flood} (m^3/s)	25.93	27.73	26.09	36.24	39.02	21.23	20.80	14.44	23.45	41.20	35.56	31.49
$Q_{cap\ existing}$ (m^3/s)	4.80	4.80	4.80	4.80	4.80	4.80	4.80	4.80	4.80	4.80	4.80	4.80
$Q_{overlow}$ (m^3/s)	21.13	22.93	21.29	31.44	34.22	16.43	16.00	9.64	18.65	36.40	30.76	26.69
$Q_{overlow}$ (%)	81.49	82.69	81.60	86.76	87.70	77.39	76.93	66.76	79.53	88.35	86.50	84.76

as shown in table 3, overflow discharge every month is average 80%, it means only 20% of runoff flows can be accommodated by Cilember river.

3.4. Groundwater and water quality analysis

The geographical position of Cimahi City is very strategic and has the potential as a center for service activities, especially industry, education, tourism and trade, causing an increase in the extraction of groundwater resources. Based on data on water potential in Cimahi City, it is known that potential water sources in Cimahi City consist of ground water (springs, shallow groundwater and deep groundwater) and surface water (lakes/ponds and Cimahi River). Cimahi City is in the Bandung-Soreang Groundwater Basin with unconfined aquifer potential of 795 million m^3 / year and confined aquifer potential of 117 million m^3 / year [14].

Groundwater extraction is generally carried out mostly by using boreholes for industrial businesses, other commercial businesses or communities, where some groundwater extraction has exceeded the allowable capacity. A number of areas in the Bandung basin that have been included in the red zone include industrial estates in Cimahi, namely Leuwigajah village, South Cimahi subdistrict.

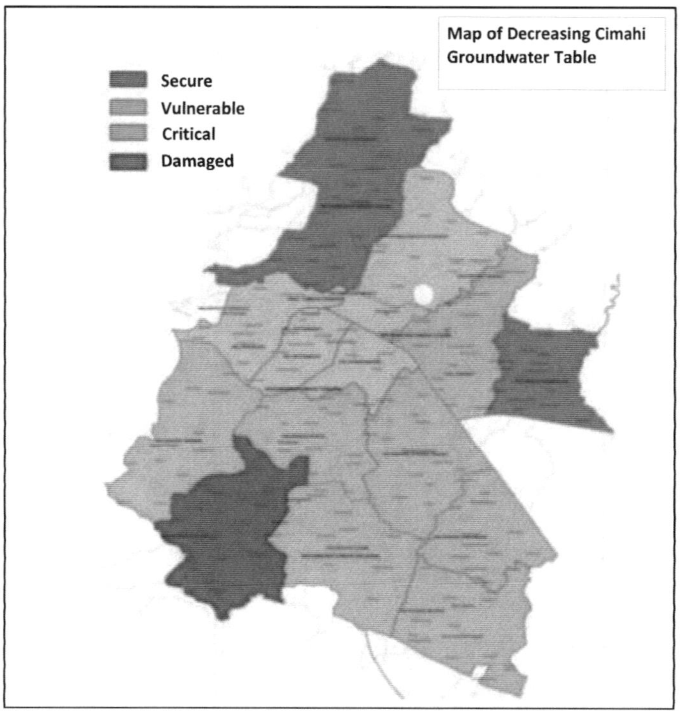

Figure.5 Map of decreasing cimahi groundwater table

An industrial plant in Cimahi city is 90% located in the Utama village and taking and removing water from the Cilember river. This is the dominant cause of the narrowing of the border of the Cilember river which has caused flooding. Waste from the factory waste without waste treatment is also a cause of pollution of river water and ground water in the city of Cimahi. In the monitoring period in September 2017, almost all of the rivers in the city of Cimahi had the status of the quality of heavy polluted water with the heaviest pollutant index of 100.96 in the Cilember river. The quality of ground water in Cimahi city is only 53% which still meets clean water quality standards [15].

3.5. Green Cimahi watershed design with river flow spatial distribution

There are 3 Green Cimahi Watershed design criteria, first the balance of water quantity between availability and needs, the second watershed balance between the land cover and ground water, third quality of surface water and ground water that meets the quality standards of clean water quality.

The balance of the water quantity in this design will be used the concept of the spatial and temporal distribution of four rivers in Cimahi, namely the Cimahi River, Cibeureum River, Cisangkan River and Cilember River with the outlet as shown in Figure 4. Temporal distribution is the amount of water at each river outlet reviewed in the month distribution. The spatial distribution of water distribution means that water is distributed in space, that is the distribution of each river. The spatial water balance quantity distribution of each river is shown in the following table 4,

Table 4. The spatial water balance quantity distribution of each river

River Name	Villages	Area (km2)	Population (people)	Clean Water Need (m^3/s)
Cimahi	Citeureup	1.27	15,794.00	0.03
	Cibabat	2.87	35,691.00	0.06
	Central Cigugur	1.37	23,805.00	0.04
	Karang Mekar	1.37	22,763.00	0.04
	Baros	2.25	39,096.00	0.07
	Utama	3.13	48,512.00	0.08
	Total Load			**0.31**
Cibeureum	Melong	3.8	58,896.00	0.10
	Cibeureum	3.93	60,911.00	0.10
	Total Load			**0.20**
Cisangkan	Setiamanah	1.98	34,404.00	0.06
	Padasuka	0.84	14,596.00	0.02
	Cibeber	3.33	51,612.00	0.09
	Leuwigajah	2.75	42,622.00	0.07
	Cipageran	5.94	73,870.00	0.12
	Cimahi	2.25	39,096.00	0.07
	Total Load			**0.43**
Cilember	Pasirkaliki	3.23	40,168.00	0.07
	Total Load			**0.07**

Based on the table above, the heaviest load of water needs is Cisangkan River. This is in accordance with the topographic position and for the efficiency of the clean water pipeline because the Cisangkan River is located in the western part of the city of Cimahi according to the location of the villages served. Although the Cibeureum River has a large discharge, its topography is at the lower elevation and its location at the easternmost border of the city of Cimahi borders the city of Bandung. The Cilember River is given the burden of one village taken from the upstream part of the outlet because in the lower part of the Cilember river there has been damage, namely narrowing of the river border and heavy pollution of water quality due to industrial waste.

The discharge at each river outlet in this design is distributed temporally ie every month. Discharge per month is calculated based on the distribution of effective rainfall each month. Table of distribution of monthly water availability and monthly water needs of each river outlet are shown in table 5.

Table 5. Distribution of monthly water availability and monthly water needs

River Name	January	February	March	April	Mei	June	July	August	September	October	November	December
Cimahi												
$Q_{av}(m^3/s)$	24.77	26.13	25.39	30.83	33.33	16.67	14.93	10.59	16.43	31.31	32.78	28.68
$Q_{need}(m^3/s)$	0.31	0.31	0.31	0.31	0.31	0.31	0.31	0.31	0.31	0.31	0.31	0.31
% of Water Used	1.25	1.18	1.22	1.00	0.93	1.86	2.07	2.92	1.88	0.99	0.94	1.08
Cibeureum												
$Q_{av}(m^3/s)$	30.55	32.23	31.32	38.03	41.11	20.57	18.42	13.06	20.27	38.62	40.43	35.38
$Q_{need}(m^3/s)$	0.20	0.20	0.20	0.20	0.20	0.20	0.20	0.20	0.20	0.20	0.20	0.20
% of Water Used	0.65	0.62	0.64	0.53	0.49	0.97	1.08	1.53	0.99	0.52	0.49	0.56
Cisangkan												
$Q_{av}(m^3/s)$	12.51	13.20	12.83	15.57	16.83	8.42	7.54	5.35	8.30	15.81	16.56	14.49
$Q_{need}(m^3/s)$	0.43	0.43	0.43	0.43	0.43	0.43	0.43	0.43	0.43	0.43	0.43	0.43
% of Water Used	3.41	3.24	3.33	2.74	2.54	5.07	5.66	7.98	5.14	2.70	2.58	2.95
Cilember												
$Q_{av}(m^3/s)$	17.30	18.25	17.74	21.54	23.28	11.65	10.43	7.40	11.48	21.87	22.90	20.04
$Q_{need}(m^3/s)$	0.07	0.07	0.07	0.07	0.07	0.07	0.07	0.07	0.07	0.07	0.07	0.07
% of Water Used	0.39	0.37	0.38	0.31	0.29	0.57	0.64	0.90	0.58	0.31	0.29	0.33

Based on table 5, the surface water of the four large Cimahi rivers if distributed spatially, the percentage of use is below 8%. Thus the available surface water is still very much, even though it is used for industrial, agricultural, public facilities such as offices, hospitals, school buildings etc. Calculation of water availability is analyzed based on very low wet year probabilities, namely a 5 year return period, so this it's safe enough for statistical calculations.

Analysis of the temporal distribution for the number of runoffs per month can answer the drought problems that have occurred in Cimahi over the past few years. The drought that occurs shows that runoff does not flow into the river. This is because the falling rainfall cannot reach the river, but flows as a flood on the land. As a result, excess rainfall is not in accordance with the calculation of river hydrology in a river network stream in the watershed. The inhibition of runoff flow into the river is due to the non-functioning of local drainage in almost all areas of the city of Cimahi which are experiencing drought. So the solution to the water drought in the city of Cimahi is the re-functioning of all the local drainage networks in the city of Cimahi.

The mitigation of the Cilember river flood due to the narrowing of the river border can be overcome by normalizing the downstream Cilember river to restore the Cilember river storage capacity. Calculation of river capacity for normalization is shown in the following table 6,

Table 6. The Dimension of river cilember cross section for normalization

Geometry Variable	Before Normalization	After Normalization By Government	Evaluation Design by This Research
Slope of channel wall (z)	0.643	0.743	0.743
Channel bed width (B),m	1.500	4.800	5.600
Depth max (Ymax), m	0.900	1.462	1.500
Top Width (T), m	2.660	6.400	7.830
Roughness Coef (n)	0.025	0.025	0.025
Energy Slope	0.014	0.014	0.014
Qcapacity (m^3/s)	4.800	23.550	42.374

The monthly design flood discharge for the Cilember river before, after normalization and after evaluation design is shown in table 7 below,

Table 7 Monthly Cilember capacity discharge before and after normalization

Discharge	January	February	March	April	Mei	June	July	August	September	October	November	December
Qflood (m^3/s)	25.93	27.73	26.09	36.24	39.02	21.23	20.80	14.44	23.45	41.20	35.56	31.49
Qcap existing (m^3/s)	4.80	4.80	4.80	4.80	4.80	4.80	4.80	4.80	4.80	4.80	4.80	4.80
Qcap design by Government (m^3/s)	23.55	23.55	23.55	23.55	23.55	23.55	23.55	23.55	23.55	23.55	23.55	23.55
Qcap Evaluation design (m^3/s)	42.37	42.37	42.37	42.37	42.37	42.37	42.37	42.37	42.37	42.37	42.37	42.37

River water pollution is overcome by enforcing the rule that every industry must process its waste through a waste processing plant so that liquid waste discharged into the river complies with waste water quality standards.

Cimahi municipal groundwater conservation can be done by building retention ponds that can function double, namely for groundwater conservation, flood control and clean water needs. Retention ponds can be built in areas with a relatively low elevation, namely in Pasirkaliki village at an elevation of 755 m MSL.

4. Conclusion

Green watershed can be designed by first identifying problems that occur in a watershed. The main criteria for green watershed are hydrological storage capacity, the balance between availability and water requirements, maintenance of land cover and groundwater conservation. Green Cimahi Watershed can be built by re-functioning local drainage to maintain the hydrological storage capacity of all watersheds in the city of Cimahi. The problem of narrowing the border of the Cilember river which caused flooding in Melong and Central Cigugur villages can be improved by normalizing the river to reach the width of the river by 5.6 m from the previous width of 1.5 m, and the river capacity which was 4.8 m^3 / s to 42.37 m^3 / s so that enough to accommodate Q_{50} flood discharge with a maximum value of 41.2 m^3 / s and occurs in November. River water pollution is overcome by enforcing the rule that every industry must process its waste through a waste processing plant so that liquid waste discharged into the river complies with waste water quality standards. Cimahi municipal groundwater conservation can be done by building retention ponds that can function double, namely for groundwater conservation, flood control and clean water needs. Retention ponds can be built in areas with a relatively low elevation, namely in the Pasirkaliki village at an elevation of 755 m msl.

References

[1] Amos T K 2014 F-Hydra Index for Evaluating Groundwater Vulnerability in Data Scarce Regions. *Journal of Environment and Ecology* 5(1): 1-14.

[2] Igwe P U, Chukwudi C C, Ifenatuorah F C, Fagbejs I F and OKK C A. 2017. A Review of Environmental Effects of Surface Water Pollution *International Journal of Advanced Engineering Research and Science* 4(2): 128-137.

[3] Safarina A B, Salim H T, Hadihardaja I K and Syahril M B K 2011 Clusterization of SyntheticUnit Hydrograph Methods Based on Watershed Characteristics. *International Journal of Civil and Environmental Engineering* 11(6): 76-85

[4] Lee H, Durand M, Jung H C, Alsdorf D, Shum C K and Sheng Y 2010 Characterization of Surface Water Storage Changes in Arctic Lakes Using Simulated SWOT Measurements *International Journal of Remote Sensing* 31(14): 3931-3953

[5] Gu Q, Deng J, Wang K, Lin Y, Li J, Gan M, Ma L and Hong Y 2014 Identification and Assessment of Potential Water Quality Impact Factors for Drinking-Water Reservoirs *International Journal of Environmental Research and Public Health* 11:6069-6084

[6] Muntalif B S, Nastiti A, Roosmini D, Sudradjat A, Meijerink S V and Smits A J M 2017 Household Water Supply Strategies in Urban Bandung, Indonesia: Findings and Implications

for Future Water Access Reporting *Journal Engineering Tecgnology and Science* 49(6): 811-832

[7] Santillan J R, Amora A M, Santillan M M, Marqueso J T, Cutamora L C, ServianoJ L and Makinano M 2016 Assessing the Impacts of Flooding Caused By Extreme Rainfall Events Through Combined Geospatial And Numerical Modeling Approach The International Archives The Photogrammetry Remote Sensing and Spatial Information Sciences, Vol XLI-B8, 2016 XXIII ISPRS Congress, 12-9 July 2016, Prague, Czech Republic

[8] Smith P J, Brown S and Dugar S 2017 *Community-based Early Warning Systems for Flood Risk Mitigation* in Nepal Natural Hazards and Earth System Sciences. 17: 423-437.

[9] Safarina A B and Ramli 2015 Analisis Keterpaduan Jaringan Sungai dan Sistem Drainase Sebagai Solusi Banjir Kota Cimahi *SNIJA Proceeding* 2015 :169-174

[10] Safarina A B, Ramli and Hashuro M S S 2017 Two Alert Flood Early Warning System Method Based on Rainfall Runoff Model *International Journal of Civil Engineering and Technology* 8(12); 1032-1044

[11] Cimahi Government 2019 Buku II Laporan Utama Dokumen Informasi Kinerja Lingkungan Hidup Daerah Cimahi 2019

[12] Safarina A B 2012 Modified Nakayasu Synthetic Unit Hydrograph Method for Meso Scale Ungauge Watersheds *International Journal of Engineering Research and Applications* 2(4): 649-654

[13] Safarina A B, Karnisah I and Permana A S 2019 Study of The Effect of River Basin Morphology Change on Threshold Parameters in Cimahi Flood Early Warning System IOP *Conf.Series: Earth and Environmental Science* 273 (2019) 012013: 1-8

[14] Cimahi Government 2018 Executive Summary Dokumen Informasi Kinerja Lingkungan Hidup Daerah Cimahi 2018

[15] Cimahi Government 2018 Laporan Utama Dokumen Informasi Kinerja Lingkungan Hidup Daerah Cimahi 2018

Acknowledgments
This research is fully supported by Ristekdikti Competitive Research, 7/E/KPT/2019, and UNJANI Competitive Research SKEP/133/UNJANI/V/2019. The authors fully acknowledged Ministry of Research Technology Higher Education (Ristekdikti) and Jenderal Achmad Yani University for the approved fund which makes this important research viable and effective.

A comparative study of deterministic approach for assessing liquefaction potential

H Farichah[1]

[1]Department of Civil Engineering, Faculty Engineering, Universitas Muhammadiyah Surabaya, Jl. Raya Sutorejo No. 59, Surabaya, East Java, Indonesia 60113

himatulfarichah@ft.um-surabaya.ac.id

Abstract. SPT-based methods have been developed to evaluate liquefaction potential due to seismic loading. This method was evolved over time modified and updated by some researchers. This paper performed a comparative analysis of deterministic approach for assessing liquefaction potential. The selected case study site is in Taiwan which subjected to the 1999 Chi-Chi Earthquake. Furthermore, the oldest and most well-known method proposed by Seed et al. [3] was selected to identify the liquefaction potential. Moreover, a method from Idriss and Boulanger. [6] in which stated as the most suitable method for a certain case in Indonesia, was also conducted. The recent updated method of Cetin et al. [12] which has considered some seismic parameter from the recent seismic events was also utilized in quantifying liquefaction potential. Despite the fact that liquefaction potential is expressed by the factor of safety in deterministic method, some liquefaction parameters including r_d (stress reduction coefficient), CSR (Cyclic stress ratio), $(N_1)_{60CS}$ (fines-corrected $N_{1,60}$ value), and FS (factor of safety) were quantified by these methods and then compared. The results show that r_d and CSR calculated by Cetin et al. [12] is lower than that of Seed et al. [3] and Idriss and Boulanger [6]. The FS from Idriss and Boulanger [6] is lower as compared to the two methods; Seed et al. [3] and Cetin et al. [12].

1. Introduction

Soil liquefaction is one of the secondary hazardous of earthquake. Soil liquefaction, phenomenon induced by shaking of earthquake, is the increment of pore water pressure causes the soil lose its shear strength significantly. Moreover, it can lead to ground failures which cause structural damage and loss of life. Thus, soil liquefaction is a topic of considerable interest to civil engineers particularly geotechnical engineers. Liquefaction potential of soil can be assessed by utilizing two general types of approaches: (1) use of laboratory testing of undisturbed samples, and (2) use of empirical relation based on in-situ tests, such as cone penetration test (CPT), standard penetration test (SPT), shear wave velocity (Vs), and Becker penetration test [1]. However, the use of laboratory testing is sufficiently difficult and costly to be performed. Whereas, empirical approach based on in-situ tests is more common in practice and quite feasible to be performed. Among these in-situ tests, the oldest and still the most frequently used to evaluate the liquefaction potential of soil is SPT. SPT-based methods to assess soil liquefaction potential will be focused in this study.

Numerous SPT-based methods to evaluate liquefaction potential have been developed by some researchers. The SPT-based method for evaluating soil liquefaction potential which has been used over the last four decades was originally proposed by Seed and Idriss [2]. Actually, this method was evolved over time modified and updated by some researchers. A modification proposed by Seed et al. [3] is One of the most commonly accepted and used in practice which was confirmed by the discussion in NCEER Working Group [4, 5]. Seed et al. [3] employed 126 case history data points were screened for data quality by the original investigators, and so represented an excellent starting point. Although widely used in practice, this method is somewhat dated, and does not make use of an increasing body of field case history data from seismic events that have occurred since 1984.

After two decades, Idriss and Boulanger [6] proposed SPT-based method procedures to investigate liquefaction potential. In substance, Idriss and Boulanger [6] is the updated version of Idriss and Boulanger [7]. Idriss and Boulanger [7] primarily used cases summarized in the databases compiled by Seed et al [8] and cetin et al. [9,10] were screened for data quality. Those databases were then updated by Idriss and Boulanger [6] to derive updated triggering correlation. The updated database includes considerably more case histories and that the updated database continues to support the previously derived triggering correlation. The total number of case histories in the updated database is 230, of which 115 cases had surface evidence of liquefaction, 112 cases had no surface evidence of liquefaction, and 3 cases were at the margin between liquefaction and no liquefaction. Unsurprisingly, a study from mase [11] performed reliability study of SPT-based liquefaction analysis stated that Idriss and Boulanger [6] is the most suitable method with the smallest error weighted factor, although the case study is specified for liquefaction case due to earthquake of magnitude M_w=8.4 in Bengkulu City, Indonesia.

After a decade, Cetin et al. [12] cast new regression model for seismic soil liquefaction triggering based on the data improvements. Some improvements were made by compiling the databases from Cetin et al. [9], Idriss and Boulanger [13-14]. Some databases of Cetin et al. [9] were excluded due to some reasons. Meanwhile, Idriss and Boulanger [13] SPT database has 230 case histories, and many are those have been screened and compiled by Cetin et al. [9]. Idriss and Boulanger [13] has added a new group data consist of 33 case histories. More recently, Boulanger and Idriss [14] compiled 24 additional cases. Thus, there are 57 new cases which were then re-evaluated by Cetin et al. [15-16] with screening criteria of Cetin et al. [9]. The screening results showed that 13 of the new Idriss and Boulanger [13-14] case histories are satisfied the Cetin et al. [9] screening criteria, and were added to the database utilized in Cetin et al. [12]. According to this updated data bases, Cetin et al. [12] developed an equation for liquefaction assessment in which only the coefficients are different from the Cetin et al. [9].

Assessment of liquefaction potential can be addressed by using deterministic and probabilistic approach. In a deterministic approach, the occurance of soil liquefaction is defined by the factor of safety (FS). The FS is defined as the ratio of cyclic resistance ratio (CRR) over cyclic stress ratio (CSR). Moreover, if the FS is smaller than or equal to 1 then the soil liquefaction is predicted to occur. Differently, no soil liquefaction is concluded to occur if FS>1. The purpose of this study is to make a comparative analysis of liquefaction potential expressed as r_d, CSR, $(N_1)_{60CS}$, and factor of safety (FS) according to several methods proposed by Seed et al. [3], Idriss and Boulanger [6], and Cetin et al. [12].

2. Liquefaction assessment methods

2.1. *Seed et al. [3]*

Evaluation of liquefaction potential basically take account two parameters: (1) the level of cyclic loading on the soil caused by the earthquake, expressed as CSR; and (2) the resistance of the soil to liquefaction, expressed as a CRR. However, the differences between one and the other methods are the considered parameters to compute them. According to Seed et al. [3], the CSR and CRR were depended on magnitude (M_w), peak horizontal ground surface acceleration (a_{max}), r_d, effective overburden, (σ'_v), fines content (FC), and SPT blow count, N (as shown in Figure 1).

2.2. *Idriss and boulanger [6]*

Idriss and Boulanger [6] proposed a liquefaction potential analysis by considering some parameters, including the:

- r_d and K_σ (correction for cyclic stress ratios)

- Earthquake magnitude affects the MSF (magnitude scaling factor) and r_d
- N-SPT correction factors, such C_R (Correction for sampler configuration details), C_N (Overburden correction factor for penetration resistances), C_S (correction for sampler configuration), C_B (correction for borehole diameter), C_E (correction for hammer energy ratio – ER)
- Effective overburden, (σ'v) affects the C_N
- Fines content affects the CRR and has poorly defined effects on all the other relationships

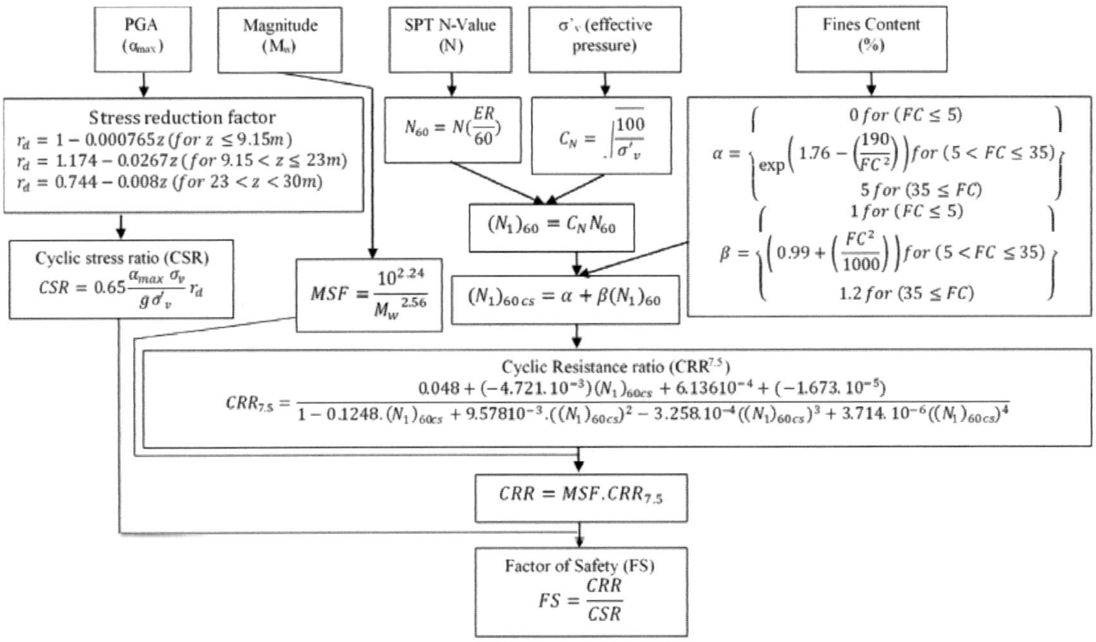

Figure 1. Flowchart of Seed's method

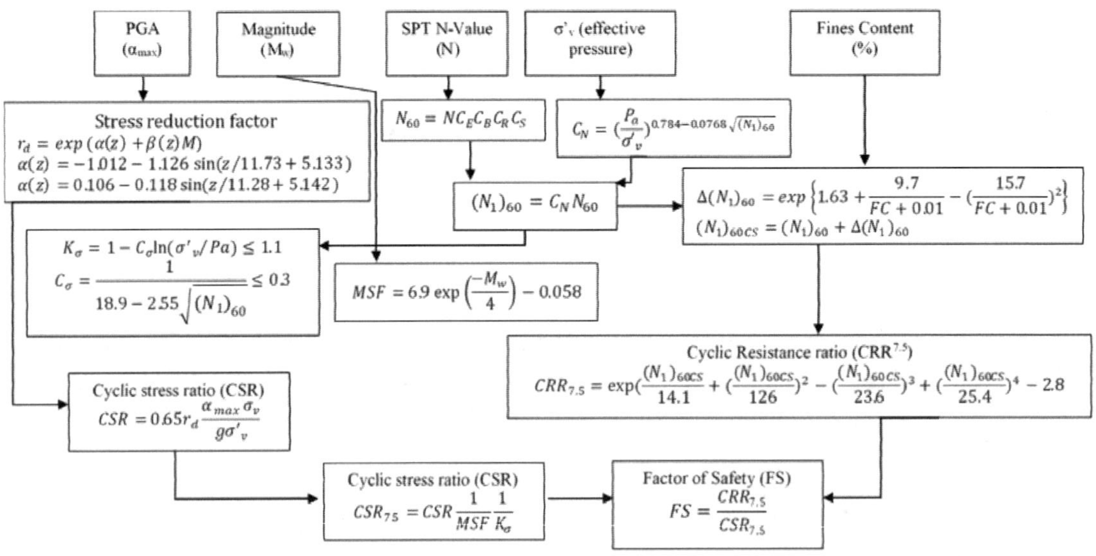

Figure 2. Flowchart of Idriss and Boulanger's method

2.3. *Cetin et al. [12]*

A recent liquefaction triggering curves was presented by Cetin et al. [12] by updating the data bases. This method is the slightly different as compare to Seed et al [3] and Idriss and Boulanger [6] since the FS is calculated from PL (Probability of triggering of liquefaction) which is calculated first. As compared with some studies from over a decade ago, this new SPT-based triggering curves proposed by Cetin et al [12] have shifted to slightly higher CSR-levels for a given $(N_1)_{60CS}$ greater than 15 blows/ft, but the correlation curves remain necessarily unaffected at $(N_1)_{60CS}$ values less than 15 blows/ft. The procedures to use Cetin et al. [12] are clearly documented in Cetin et al. [17].

- rd-values estimated by using the predictive rd relationship developed by Cetin and Seed [1]. The consistent use of rd relationship of Cetin and Seed [1] is presented on an illustrative seismic soil liquefaction triggering assessment problem in Cetin et al. [17]. rd requires the estimation of a representative shear wave velocity for upper 12 m of the soil site.

$$r_d = \frac{\left[1+\frac{-23.013-2.949a_{max}+0.999.M_w+0.0525.V_{s,12m}}{16.258+0.201e^{0.341(-d+0.785.V_{s,12m}+7.586)}}\right]}{\left[1+\frac{-23.013-2.949a_{max}+0.999.M_w+0.0525.V_{s,12m}}{16.258+0.201e^{0.341(0.785.V_{s,12m}+7.586)}}\right]} \quad (1)$$

- $N_{1,60}$ (standard penetration test blow count value corrected for overburden, energy, equipment and procedural factors) values are the product of the averaged N values for a stratum corrected for effective normal stress (C_N), hammer energy (C_E), equipment rod length (C_R), equipment sampler (C_S), borehole diameter (C_B), and procedural effects to fully standardized $N_{1,60}$ values as given below:

$$N_{1,60} = NC_NC_EC_RC_SC_B \quad (2)$$

- The resulting (new) recommended seismic soil liquefaction triggering relationships are presented in the following equation and the new model coefficients are θ_1= 0.00167, θ_2= 27.352, θ_3= 3.958, θ_4= 0.089, θ_5= 16.084, θ_6= 11.771, θ_7= 0.392, σ_c = 2.95.

$$P_L(N_{1,60}, CSR_{\sigma'_{v},\alpha=0,M_w}, M_w, \sigma'_v, FC) = \Phi\left(-\frac{(N_{1,60}(1+\theta_1 FC)-\theta_6\, ln\left(CSR_{\sigma'_{v},\alpha=0,M_w}\right)-\theta_2\, ln(M_w)-\theta_3\, ln\left(\frac{\sigma'_v}{P_a}\right)+\theta_4 FC+\theta_5}{\sigma_c}\right) \quad (3)$$

$$CRR(N_{1,60}, M_w, FC, P_L) = exp\left(-\frac{(N_{1,60}(1+\theta_1 FC)-\theta_2\, ln(M_w)-\theta_3\, ln\left(\frac{\sigma'_v}{P_a}\right)+\theta_4 FC+\theta_5+\sigma_c\Phi^{-1}(P_L)}{\theta_6}\right) \quad (4)$$

$$FS = \frac{CRR(P_L=50\%)}{CRR(P_L)} = exp[-0.251.\Phi^{-1}(P_L)] \quad (5)$$

$$CSR = 0.65r_d\frac{a_{max}\,\sigma_v}{g\sigma'_v} \quad (6)$$

$$CSR_{7.5} = CSR\frac{1}{K_\sigma}\frac{1}{K_{Mw}}\frac{1}{K_\alpha} \leq 0.6 \quad (7)$$

$$K_\sigma = \left(\frac{\sigma'_v}{Pa}\right)^{-\theta_3/\theta_6}; 0.8 \leq K_\alpha \leq 1.6\,,\ K_{MW} = \left(\frac{M_W}{7.5}\right)^{-\theta_2/\theta_6}; 5.5 \leq M_W \leq 8.4 \text{ and } K_\alpha = 1 \quad (8)$$

3. Research method

A sample site selected for this analysis is located at the south western plain of Taiwan. The simplified geological profile of the site consists of five layers as listed in Table 1 [18]. The ground water table is 2.0 m below the ground surface. For liquefaction analysis, only two sandy soil layers, which are located at G.L.-2.0 to -6.0m (Layer 2) and G.L.-8.0~-12.0m (Layer 4), are considered to be liquefiable. Hence, the analysis presented below is focused on these two layers. A scenario earthquake of magnitude M_w=7.6 Chi-Chi earthquake was used in the calculations. Analyses were carried out for a_{max} levels at 1g. Thereafter, the liquefaction potential was analyzed by conducting three methods; Seed et al. [3], Idriss and Boulanger [6], and Cetin et al. [12].

Table 1. Simplified geological profile of the site [18]

Depth (m)	N-SPT	Soil Types	ɣ (KN/m³)	Fines (%)
1	12	SM	19.2	5
2	12		19.2	5
3	12	SM	19.3	5
4	12		19.3	5
5	12		19.3	5
6	12		19.3	5
7	5	ML	19.4	57
8	5		19.4	57
9	15	SM	19.3	5
10	15		19.3	5
11	15		19.3	5
12	15		19.3	5

4. Results and discussion

A comparative study of these methods for identifying liquefaction potential can be focused on some parameters. Since in general, basically the assessment procedures of liquefaction potential from each method are the same. However, the differences are the considered parameter and the equation to compute them. Four parameters will be compared in to simplify this comparative study.

The first parameter will be discussed is r_d, in which all methods are considering this parameter to calculate the CSR. r_d was computed by using method proposed by Seed et al. [3], Idriss and Boulanger [6], and Cetin et al. [12]. r_d were then plotted with respect to the depth as shown in Figure 3. Obviously, it shows the value of r_d decreases as depth increasing. Moreover, the r_d from Cetin et al. [12] is smaller as comparing to that of Seed et al. [3] and Idriss and Boulanger [6]. It had been suspected that the differences in the r_d, K_α, and C_N relationships may have played a significant role, but it was found that they had a relatively minor effect on the resulting liquefaction triggering correlations [13].

Figure 4. shows the comparison of CSR computed by Seed et al. [3], Idriss and Boulanger [6], and Cetin et al. [12]. The line CSR versus Depth of Seed et al. [3], and Idriss and Boulanger [6] are coincident at the depth of 1.0 to 8.0 m and slightly different at the depth of 8.0 to 12.0 m. Moreover, the line CSR versus Depth from Cetin et al. [12] coincides with the line from Seed et al. [3], and Idriss and Boulanger [6] at the depth of 1.0 to 3.0 m. However, after depth of 3 m, the CSR of Cetin et al. [12] is smaller than that of the other methods.

The comparison of $(N_1)_{60CS}$ calculated by three methods can be seen in Figure 5. The line CSR vs Depth from Seed et al. [3], and Cetin et al. [12] is coincident at the depth of 9.0 to 12.0 m. Moreover, the line CSR vs Depth from Idriss and Boulanger [4], and Cetin et al. [12] is coincident at the depth of 6.0 to 8.0 m. In general, the $(N_1)_{60CS}$ of Seed et al. [3] is slightly greater from that of the other methods.

Figure 6. shows the comparison of FS form these three methods. Some certain depths, 1.0 to 2.0 m is definitely not liquefiable since the water table is 2.0 m below the ground surface and the depth of 6.0 to 8.0 m is not sandy soil. According to Figure 6, all the soil layers are liquefiable except those certain depths. Surprisingly, the FS from Seed et al. [3] resembles to the FS from and Cetin et al. [12]. Meanwhile, the FS from Idriss and Boulanger [6] is lower as compared to the two methods; Seed et al. [3], and Cetin et al. [12].

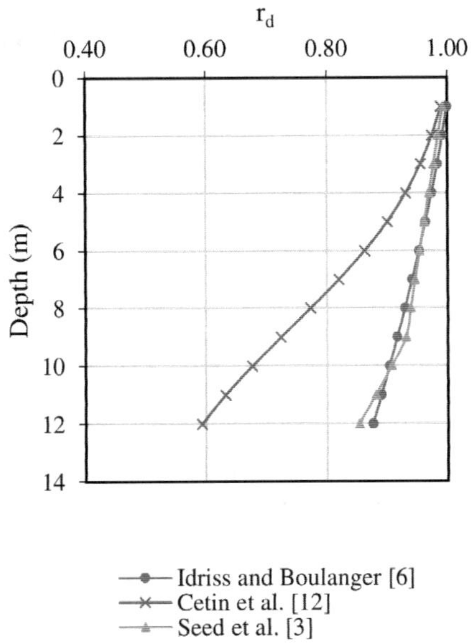

Figure 3. Plots of r_d

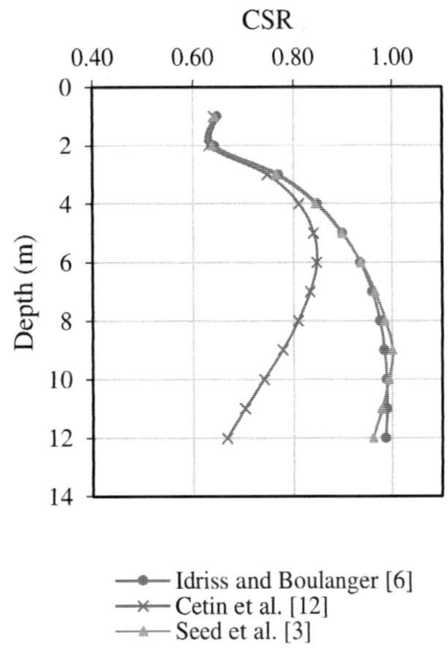

Figure 4. Plots of CSR

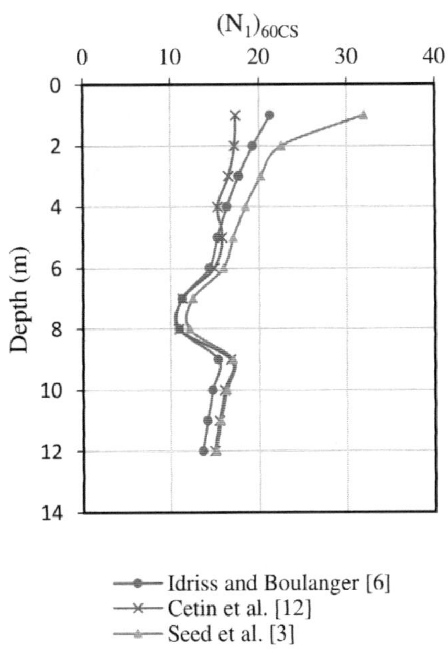

Figure 5. Plots of $(N_1)_{60CS}$

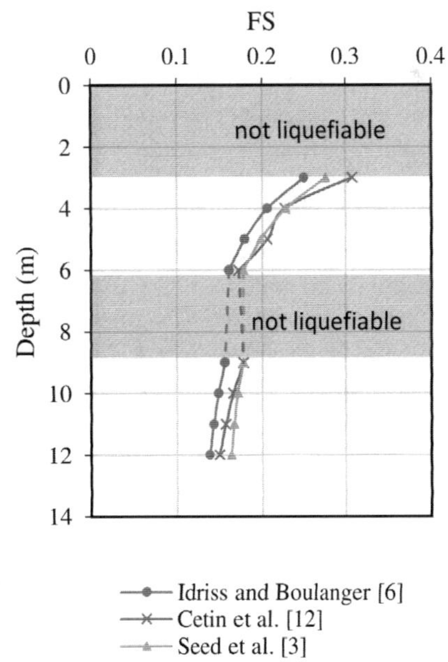

Figure 6. Plots of FS

5. Conclusions

In summary, liquefaction is considered as one of critical issues in geotechnical engineering. Some SPT-based methods have been developed to evaluate liquefaction potential. This paper presents a comparison analysis of liquefaction potential expressed as factor of safety (FS) according to several methods proposed by Seed et al. [3], Idriss and Boulanger [6], and Cetin et al. [12]. Some comparisons of liquefaction parameters have been performed, including r_d, CSR, $(N_1)_{60CS}$, and FS. r_d and CSR calculated by Cetin et al. [12] is lower than that of Seed et al. [3] and Idriss and Boulanger [6]. The $(N_1)_{60CS}$ of Seed et al. [3] is slightly greater from that of the other methods. Surprisingly, the FS from Idriss and Boulanger [6] is lower as compared to the other two methods; Seed et al. [3], and Cetin et al. [12]. It is necessary to consider more databases for comparative study

References

[1] Cetin K O and Seed R B 2004 *Nonlinear shear mass participation factor r_d for cyclic shear stress ratio evaluation* vol. 24, pp. 103–113

[2] Seed H B and Idriss I M 1971 Simplified procedure for evaluating soil liquefaction potential, *Journal of the Soil Mechanics and Foundations Division* ASCE 97 (SM9, Proc. Paper 8371) 1249-1273.

[3] Seed H B, Tokimatsu K, Harder Jr L F and Chung R M 1985 Influence of SPT procedures in soil liquefaction resistance evaluations. *Journal of Geotechnical Engineering* 111(12)

[4] Youd T L and Idriss I M 1997 National Center for Earthquake Engineering Research, (NCEER) proceeding of the NCEER workshop on evaluation of liquefaction resistance of soils, Technical Rep No NCEER-97-0022.

[5] Youd T L, et al 2001 Liquefaction resistance of soils; summary report from the 1996 NCEER and 1998 NCEER/NSF workshop on evaluation of liquefaction resistance of soils *J. Geotech Geoenviron. Eng.* 127(10), 817-833

[6] Idriss I M and Boulanger R W 2008 *Soil Liquefaction during Earthquakes*, Monograph MNO-12, Earthquake Engineering Research Institute, Oakland, CA

[7] Idriss I M and Boulanger R W 2004 Semi-empirical procedures for evaluating liquefaction potential during earthquakes . In: Doolin D, editor. *Proceedings of 11th international conference on soil dynamics and earthquake engineering and 3rd International conference on earthquake geotechnical engineering, 1*. Stallion Press;vol. 1, pp 32–56

[8] Seed H B, Tokimatsu K, Harder L F and Chung R M 1984 The influence of SPT procedures in soil liquefaction resistance evaluations (Earthquake Engineering Research Center Report No. UCB/EERC-84/15). Berkeley, CA: Dept. of Civil and Environmental Engineering, University of California at Berkeley

[9] Cetin K O et al. 2004*Standard Penetration Test-Based Probabilistic and Deterministic Assessment of Seismic Soil Liquefaction Potential* no. December, pp. 1314–1340

[10] Cetin K O 2000 *Reliability-based assessment of seismic soil liquefaction initiation hazard* [Dissertation submitted in partial fulfillment of the requirement for the degree of doctor of philosophy]. University of California at Berkeley

[11] Mase L Z 2018 *Studi Kehandalan Metode Analisis Likuifaksi Menggunakan SPT Akibat Gempa 8 , 6 M w , 12 September 2007 di Area Pesisir Kota Bengkulu* vol. 25, no. 1, pp 53–60

[12] Cetin K O, Seed R B, Kayen R E, Moss R E S and Bilge H T 2018 SPT-based probabilistic and deterministic assessment of seismic soil liquefaction triggering hazard *Soil Dyn. Earthq. Eng.*, vol. 115, no. July 2017, pp 698–709

[13] Idriss I M and Boulanger R W 2010 SPT-based liquefaction triggering rocedures (Report UCD/CGM-10/02) Davis, CA: Center for Geotechnical Modeling, Department of Civil and Environmental Engineering, University of California; p 136

[14] Boulanger R W and Idriss I M 2014 *CPT and SPT based liquefaction triggering procedures* (Report No. UCD/CGM-14/01). Davis, CA: Center for Geotechnical Modeling, Department of Civil and Environmental Engineering, University of California; p 134

[15] Cetin K O, Seed R B, Kayen R E, Moss R E S, Bilge H T, Ilgac M, et al. 2016 *Summary of SPT based field case history data of the updated 2016 Database* (Report no: METU/ GTENG 08/16-01) METU Soil Mechanics and Foundation Engineering Research Center

[16] Cetin K O, Seed R B, Kayen R E, Moss R E S, Bilge H T, Ilgac M, et al. 2018 *Dataset on SPT*

based seismic soil liquefaction 20:544–8. Elsevier

[17] Cetin K O, Seed R B, Kayen R E, Moss R E S, Bilge H T, Ilgac M, et al. 2018 *The use of the SPT based seismic soil liquefaction triggering evaluation methodology in engineering hazard assessments. Methods*

[18] Hwang J H, Chen C H and Juang H 2005 Liquefaction hazard analysis: a fully probabilistic method *Earthquake engineering and soil dynamics*

Risk study on supply chain management in construction (Case study: Building projects in Indonesia)

A Nurdiana[1] and M N Sholeh[1]

[1]Vocational School, Diponegoro University

asri@live.undip.ac.id

Abstract. The concept of supply chain in construction is very influential on the improvement of projects performance. In outlining the problems in the supply chain, it need to be identified the risks therein related to the smoothness of the supply chain process. Risk is an event / uncertainty condition, when it happens may lead to positive or negative impact [1]. The purpose of this study are to determine the identification of risks on supply chain in construction projects, analyze the risks, and analyze the response of risks. Primary data obtained through a questionnaire and interview methods. Data processing method that used in this research were Promethee method for risk analysis, and descriptive statistical methods at risk response. From this research, it is known that the identification of risks at supply chain in construction projects can be seen from the potential failure of achievement of Key Performance Indicators in Supply Chain Management (SCM). From risk analysis with Promethee method, it is known that the highest risk in SCM is the change of price from sub-contractors. Based on the analysis for risk response, it is known that the contractor has developed mitigation measures to risk response.

1. Introduction

The construction industry has become the center of attention of the government in the last period. Recorded in 2019, the Ministry of PUPR received a ceiling of Rp. 102.01 trillion for infrastructure development, including allocations for the Water Resources (SDA) sector of Rp. 38.9 trillion, Bina Marga of Rp. 39.2 trillion, Cipta Karya of Rp. 13.4 trillion, and Housing Provision of Rp. 7.8 trillion. From data of BPS, it also showed that the value of construction projects has increased from year to year. It indicates that the position of the infrastructure is already regarded as a strategic action in the development of a country.

Various aspects in the construction project is currently receiving more attention. The goal is to unravel the problems that can cause poor performance of the project. One aspect that often studied is the Supply Chain Management (SCM).

The concept of supply chain in the construction is very influential on performance improvement at projects. By the appropriate design patterns of supply chain, it is expected that each supply chain actors can make a major contribution to the efficiency and productivity of its own activities [2]. Basically supply chain is a set of supplier and customer that are connected, each customer are the supplier for further downstream organization [3].

From the research on Supply Chain Performance Analysis at Building Construction Projects, it can be concluded that the pattern of supply chain network design is influenced by: method contract that used, scope of business by owner, and procurement strategy by the contractor. Recommendations are offered in improving supply chain performance by implementing information systems and better coordination between the parties that involved in the production process, by establishment the long-term cooperation relationship between the contractor, subcontractors, and suppliers [4].

From the research on Supply Chain Structure Analysis Bridge Construction at Work, the results conducted on contractors, subcontractors and suppliers in the construction of bridges with some of the key aspects like relationships with subcontractors and suppliers such as, the type of procurement, contract type, payment systems, and the interaction between the two companies, shows the relationship between the cooperative relationship and the effectiveness of the implementation of construction work [5]. This can be used as a reference to identify risks in construction projects.

Supply chain performance affects the availability of materials, tools, and labor in construction projects, that are factors to achieve good project performance. In outlining the problems in the supply chain, it needs to be identified the risks therein that related to the process in it. Risk is an event / uncertainty condition, when it happens may lead to positive or negative impact in terms of the scope of work, cost, quality and time [1]. This identification of risks can be analyzed to determine the impact on the performance of the project, that are the cost, quality and time.

The purpose of this study is to determine the identification of risks on supply chain in construction projects, analyze risks on supply chain in construction projects, and analyze the response of risks at supply chain in construction projects.

2. Literature review

In the process of writing of this article, several literature studies were carried out on the research topics related to supply chain and risk management in construction projects.

2.1. State of the art of research about supply chain

Supply Chain is an item of network activity or cooperative procurement of goods or services that work together and intertwined with each other to create and distribute goods or services. Meanwhile, supply chain structure is the arrangement of an item or networking activity procurement of goods or services that work together and intertwined with each other to create and distribute goods or services [5].

The concept of supply chain in the construction is very influential on performance improvement projects. By the appropriate design patterns of supply chain, it is expected that each supply chain actors can make a major contribution to the efficiency and productivity of its own activities [2].

The characters of Supply Chain Management are as follows: Inventory Management Approach, Cost Efficiensies, Time Horizon, Amount of Mutual Information Sharing and Monitoring, Amount of Coordination of Multiple Levels of The Channel, Joint Planning, Compatibility of Corporate Philosophies, Breadth of Supplier Base, Channel Leadership, Sharing of Risk and Reward, Speed of Operations [6].

There are two patterns of supply chain in construction projects, particularly in building projects. The pattern is a general pattern and specific patterns [7]. The general pattern is a pattern of relationships that occur in stages in accordance with the hierarchy in a contractual relationship pattern. The specific pattern is a pattern of relationships that have different properties to the general pattern. The general pattern and specific patterns are described as Figure 1.

2.2. State of the art of research about risk management

Risk is an uncertain event, and this event will allow a positive or negative impact to the goals and objectives of the project. Risks in construction projects must be managed properly so the result of the project will be good and safe for stakeholders, and will generate profits for the interested parties in project [1]. Group process on risk management according to PMBOK are as follows: Plan Risk Management, Risk Identification, Qualitative Risk Analysis, Quantitative Risk Analysis, Risk Action Plan, Monitoring and Control Risks [8].

In construction projects, research on risk management has been widely performed. An effective risk management is to correctly identify the important risks and allocate these risks in the contract. The risk identification is performed by classifying the risks that exist [9].

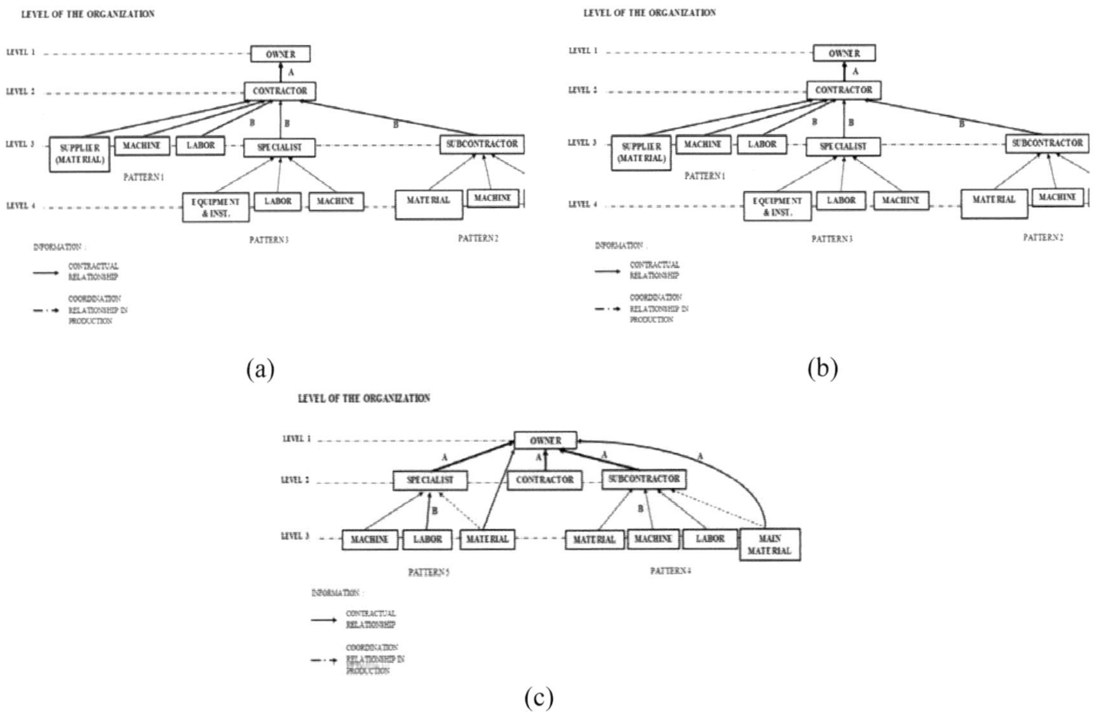

Figure 1. (a) General Pattern of Supply Chain in Construction, (b) Specific pattern of supply chain in construction - a direct relationship between the owner and specialist/subcontractors, (c) Specific pattern of supply chain in construction - a direct relationship between the owner and specialist/subcontractors in the procurement of certain materials

The risk categories on highway projects from owner perception are in seven categories: Political risk, such as changes in policy, tax increases, tariff setting is not appropriate, the rate increase is not appropriate, and changes in the structure of governance; Construction Risk, such as improper design, land acquisition, project delays, conditions on the ground, and the failure of the construction; Operation and Maintenance Risk; Legal Risk and Contracts; Risk of Revenue Generation; Financial Risk; Risk of Force Majeure [10]. Categories of risk in construction projects is the external risks, the risks of economic and financial, technical and contractual risks, and risk management [11]. While the risk categories of highway projects, according to the perception of various stakeholders that are contractor, owner, consultant of planning, supervisory consultants, and the society are construction risk; political, legal, and contractual risk; and economic risk [12].

2.3. *State of the art of research about promethee*

Promethee is one of the models in the system of Multi Criteria Decision Making (MCDM). Promethee is able to analyze a problem that has complex criteria.

Promethee (Preference Ranking Organization Method for Enrichment Evaluation) is one method of determining the priority that using the value of dominance criterion in relation outrangking [13]. In Promethee, there are six (6) criteria: insensitive, indifference, linear, criterion, linear with indifferent and gaussian.

After deciding the specified criteria, these criteria must be given a value (weights) so that these criteria can be operationalized. Scoring for each criterion is solved by weighting. If each criteria has the value or weight, then the analysis can be implemented using Promethee. The most ideal alternative can be seen in accordance with the ranking from this analysis [14].

3. Research methods

This research object is Supply Chain Management in construction, particularly in building projects. In this case, researchers analyzed risks in the Supply Chain Management, then be analyzed using Promethee methods. Stages in this study can be described as Figure 2:

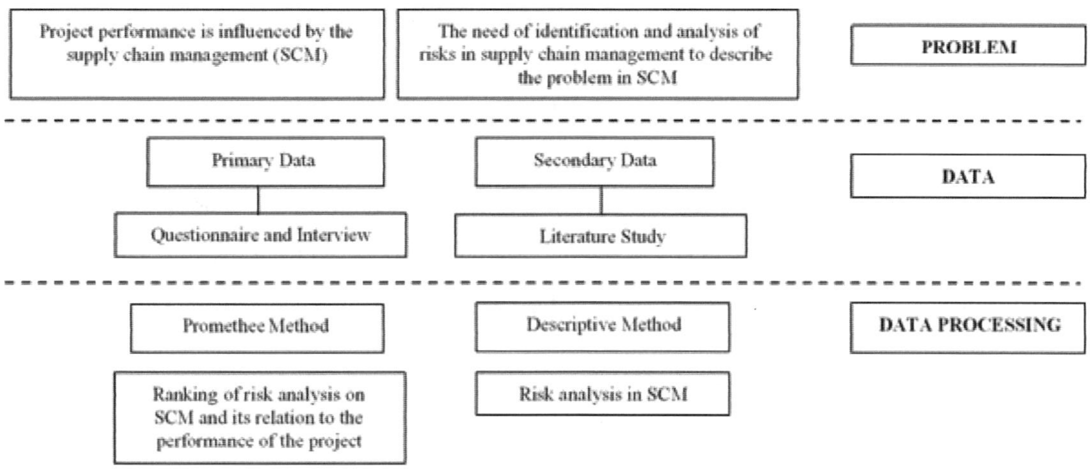

Figure 2. Flowchart of research

In the first stage (risk identification), researchers conducted an analysis of the implementation of Key Performance Index in the SCM on the terms of a previous study. Risks were identified if the Key Performance Index in SCM not achieved, which is obtained through the interview method. The Key Performance Index in SCM at construction were a perfect order fulfillment, order fulfillment cycle time, upside supply chain flexibility, upside supply chain adaptability, downside supply chain adaptability, overall value at risk, supply chain management cost, cost of goods sold, cash to cash cycle time, supply chain return on fixed assets, return on working capital, the utilization of information technology, and the behaviour of suppliers [15].

In the second phase (risk analysis), risks was measured by the method Promethee to get the risks ranking. This method was chosen to review the ranking of risk that is influenced by the factors of cost, quality, and project time. The data for probability and impact of risks that was to assess the risk ranking, obtained through interviews and questionnaires. In this questionnaire also included a response strategy for each risk.

Respondents in this study were the contractor on the building project. The respondents were the decision maker on the project. Data retrieval to respondents was conducted through questionnaires and interviews.

4. Data processing and discussion

Based on interviews and questionnaires in the first phase, the identification of risks based on the Key Performance Index on SCM are as Table 1:

Table 1. Identification of risk at SCM in construction

No.	KPI	Identification of Risk	Exist or not? Yes	No
Technical Risk (Supply Chain with Subcontractors)				
1	Perfect order fulfillment	Works of sub-contractor was not implemented in full (sub-contractor does not finish the job)	√	
2	Perfect order fulfillment	Works of sub contractors defective or not in accordance with specifications	√	
3	Order fulfillment cycle time	Works of sub-contractor does not start on time in accordance to SPK or letter of agreement between the contractor and sub-contractor	√	
4	Upside supply chain flexibility	Sub-contractors are not flexible to the number of days required to meet the design changes.	√	
5	Upside supply chain adaptability	Sub-contractors are not adaptive to additional work		√
6	Downside supply chain adaptability	Sub-contractors are not adaptable to job reductions		√
7	Overall value at risk	Imprecision in the selection of sub-contractors	√	
8	Return on working capital	Inadequacy of existing workers in sub contractors		√
9	Utilization of Information Technology	The lack of use of technology by sub contractors	√	
Economic and Financial Risks (Supply Chain with Subcontractors)				
10	Cost of goods sold	There were changes in Price from sub-contractors that are different with the previous offer price	√	
11	Supply chain management cost	Payment of the owner has been synchronized with the needs on the project	√	
Technical Risk (Supply Chain with the Supplier)				
12	Perfect order fulfillment	The material sent is not full	√	
13	Perfect order fulfillment	The material sent is not in accordance with specifications	√	
14	Order fulfillment cycle time	Late delivery of materials	√	
15	Upside supply chain flexibility	Supplier inflexible with the number of days that needed to meet the changing demand		√
16	Upside supply chain adaptability	Supplier is not adaptive to additional work		√
17	Downside supply chain adaptability	Supplier is not adaptive to the reduction of jobs		√
18	Overall value at risk	Inaccuracy in supplier selection		√
19	Return on working capital	Incompetence supplier		√
20	Utilization of Information Technology	The lack of use of technology by suppliers	√	
Economic and Financial Risks (Supply Chain with the Supplier)				
21	Cost of goods sold	There were changes in the price of materials from suppliers	√	

Risk identification in the supply chain in construction projects derived from the possibility of wasn't achieved the KPI in SCM, where the KPI on construction projects there are 13 kinds [15]. Risk identification in the supply chain is obtained through interviews with the respondents' (contractor). From interviews, it is obtained the identification of risks in the supply chain in construction. There are 21 risk, where the eight among these risks are not occur in the project, according to some respondents.

Furthermore, from the risks that were identified, there were the value of probability and risk impact. The impact of risk grouped into three, there are the impact on cost, quality, and time.

Table 2. Risk analysis at SCM using Promethee method

No	Risk	Value of Risk			Risk Response
		Cost	Quality	Time	
R1	Works of sub-contractor was not implemented in full (sub-contractor does not finish the job)	15,75	12,25	10,50	Mitigation
R2	Works of sub contractors defective or not in accordance with specifications	6,00	12,00	9,00	Mitigation
R3	Works of sub-contractor does not start on time in accordance to SPK or letter of agreement between the contractor and sub-contractor	10,50	14,00	7,00	Mitigation
R4	Sub-contractors are not flexible to the number of days required to meet the design changes.	12,00	16,00	8,00	Mitigation
R5	Imprecision in the selection of sub-contractors	10,00	14,00	8,00	Mitigation
R6	The lack of use of technology by sub contractors	10,50	7,50	7,50	Mitigation
R7	There were changes in Price from sub-contractors that are different with the previous offer Price	20,25	11,25	15,75	Mitigation
R8	Payment of the owner has been synchronized with the needs on the project	25,00	22,50	15,00	Mitigation
R9	The material sent is not full	25,00	22,50	17,50	Mitigation
R10	The material sent is not in accordance with specifications	12,00	10,50	10,50	Mitigation
R11	Late delivery of materials	16,00	16,00	12,00	Mitigation
R12	The lack of use of technology by suppliers	15,75	8,75	8,75	Mitigation
R13	There were changes in the price of materials from suppliers	20,25	20,25	4,50	Mitigation
	Average	15,31	14,42	10,31	

From the value of the probability and impact of these risks, then it was searched the risk value from the multiplication of these two variables. Risk values are obtained by processing the data using Promethee methods. The results of data processing with Promethee method are presented in Table 2.

Table 2 shows the value of risk that are divided into the impact categories to the cost, quality and time. The high-impact risk in project cost is risks R8 and R9; high-impact risk in project quality is risks R8 and R9; while the high-impact risk in project duration is the risk R9.

Of all the risks that have been identified, these risks 38% impacted to the project cost. Then, 36% risks impacted to quality of the project, and 26% risks impacted to duration of the project.

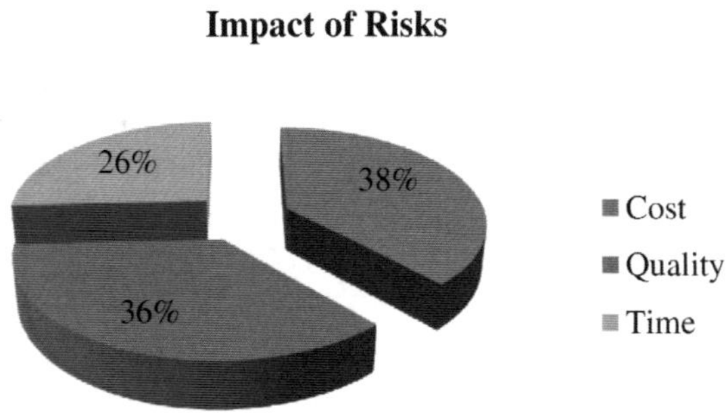

Figure 3. Diagram of impact of risks

From the above results, then the risks were processed using Promethee methods to determine risk values and the ranking of risk in the supply chain at construction. The results of processing using Promethee methode are as follows (Table 3).

Table 3. Risk ranking of SCM in construction

Rank	Code	Risk	Score
1	R7	There were changes in Price from sub-contractors that are different with the previous offer Price	7,667
2	R9	The material sent is not full	6,667
3	R12	The lack of use of technology by suppliers	3,000
4	R8	Payment of the owner has been synchronized with the needs on the project	3,000
5	R1	Works of sub-contractor was not implemented in full (sub-contractor does not finish the job)	2,000
6	R10	The material sent is not in accordance with specifications	1,667
7	R11	Late delivery of materials	1,667
8	R2	Works of sub contractors defective or not in accordance with specifications	-2,667
9	R6	The lack of use of technology by sub contractors	-4,000
10	R5	Imprecision in the selection of sub-contractors	-4,333
11	R13	There were changes in the price of materials from suppliers	-4,333
12	R4	Sub-contractors are not flexible to the number of days required to meet the design changes.	-4,333
13	R3	Works of sub-contractor does not start on time in accordance to SPK or letter of agreement between the contractor and sub-contractor	-6,000

From the data processing using Promethee method, it is known that the risk in the first rank in the supply chain activity is risk R7, and risk at the lowest ranking is the risk R3.

The advantages of using Promethee method to processing data is that it can be learned the ranking of risks according to three variables impact of risk, that are cost, quality, and time. Whereas from the other methods of risk analysis, it only can be known the impact of risk in general.

5. Conclusions and recommendations

The conclusion that can be drawn from this research is the identification of risks in the supply chain is derived from the possibility of weren't achieved the KPI in the supply chain. The identification of risks in the supply chain on a building project there are 13 kinds.

The highest risk that is impacted to the project cost is the payment of the owner that is difficult to synchronized with the needs on the project, and the risk of shipping materials that are not full or not appropriate. The highest risk that is impacted to the project quality is the payment from the owner that is difficult to synchronized with the needs on the project, and the risk of shipping materials that are not full or is not appropriate. While the highest risk that is impacted to the duration of the project is the delivery of materials that are not full. Based on the risk analysis using Promethee method, it is known that the highest risk in the supply chain at building projects is the risk of the price changes from sub-contractors that are different from the previous offer price.

Based on the analysis of the risk response of supply chain at building projects, it is known that there are risk mitigation on each risks and the measures to against the possible risks.

References

[1] Wang, Qing S, et al. 2004 *Risk Management Framework for Construction Projects in Developing Countries* Construction Management and Economics 22, page 237-252

[2] Nurwega, Maddeppungeng and Suryani 2014 Analisis Pola Dan Kinerja Supply Chain Pada Proyek Konstruksi Bangunan Perumahan *Jurnal Konstruksi* Vol 5 No 2 Agustus 2014

[3] Hanfield R B and Nichols E L 1999 *Introduction to Supply Chain Management, Prentice-Hall* Upper Saddle River, New Jersey

[4] Yuliatin and Rohaesih 2013 Analisis Kinerja Supply Chain Pada Proyek Konstruksi Bangunan Gedung *Tugas Akhir Jurusan Teknik Sipil Universitas Sultan Ageng Tirtayasa*

[5] Maulani Suraji and Istijono 2014 Analisis Struktur Rantai Pasok Kontruksi Pada Pekerjaan Jembatan *Jurnal Rekayasa Sipil* Oktober 2014

[6] Martha C, Cooper, Lisa M and Ellram 1993 *Characteristics of Supply Chain Management and the Implications for Purchasing and Logistics Strategy*

[7] Wirahadikusumah and Susilawati 2006 Pola Supply Chain pada Proyek Konstruksi Bangunan Gedung. *Jurnal Teknik Sipil* Vol 13 Tahun 2006

[8] Project Management Institute 2013. A Guide to the Project Management Body of Knowledge Third Edition PMBOK, Pennsylvania: Project Management Institute, Inc

[9] Andi 2006 The Importance and Allocation of Risks in Indonesian Construction Projects *Construction Management and Economics* 24 p 69-80

[10] Abednego, Martinus P, and Ogunlana S O 2006 Good Project Gevernance for Proper Risk Allocation in Public-Private Partnership in Indonesia *International Journal of Project Management* 24 p 622-634

[11] Wiguna, Artama I P, and Scott S 2006. Relating Risk to Project Performance in Indonesian Building Contracts *Construction Management and Economic* 24 p 1125-1135

[12] Nurdiana, Wibowo and Hatmoko 2015 Sensitivity Analysis Of Risk From Stakeholders' Perception (Case Study : Semarang-Solo Highway Project Section I (Tembalang-Gedawang)). *Proceedia Engineering* 125 (2015) 12 – 17

[13] Brans J P, Mareschal B and Vincke P 1986. How to Select and Rank Projects: The PROMETHEE Method *European Journal of Operations Research*, Vol. 24, 228-238

[14] Wayangkau and Gianditha H 2014 Analisis Manajemen Risiko Pada Proyek Pembangunan Bendungan (Studi Kasus : Bendungan Titab Di Bali, Bendungan Jatibarang Di Kabupaten Semarang Dan Bendungan Diponegoro Di Semarang) *Tesis Fakultas Teknik Universitas Diponegoro.*

[15] Sholeh M N 2016 Perancangan Indikator Pengukuran Kinerja Rantai Pasok Konstruksi Dengan Pendekatan Model SCOR *Tesis Fakultas Teknik Universitas Diponegoro*

IOP Conf. Series: Materials Science and Engineering **669** (2019) 012043 doi:10.1088/1757-899X/669/1/012043

Experimental analysis on the performance of concrete beam due to the effect of knitted bamboo reinforcement ratio

D Setyowulan[1], D Nuralinah[1], E Arifi[1] and F D Prasetyo[1]

[1]Civil Engineering Department, Universitas Brawijaya, No. 169 MT Haryono Street, Malang, East Java, Indonesia

desy_wulan@ub.ac.id

Abstract. Bamboo is an abundant material in Indonesia which has been previously studied as a construction material and reinforcement. The performance of concrete beam with knitted bamboo reinforcement are studied, especially on the flexural strength and crack behaviour. Three concrete beams with the dimension of 15 x 25 x 160 cm and three slices of bamboo in each knit are investigated. The outer skin surface of this knit is soaked NaOH with sikadur and sand as coating material. The results found that the flexural strength of concrete beam for the first and second model was 3.56 MPa and 6.15 MPa, respectively. Pure bending crack occurred on all concrete beam models.

1. Introduction

Bamboo is an abundant material in Indonesia, as a tropical climate country. The selection of bamboo as a building material already studied because of its advantages, such as high tensile strength and relatively low price than steel [1]. Previous research on the using of bamboo as a construction material and reinforcement has been studied [2]. From this study, it has been suggested that the use of bamboo strips as reinforcement in column should not be for the purpose of imparting compressive strength but rather than to induce elasticity in the cross section, which in turn guide against sudden failure.

Furthermore, the application of bamboo as a knit reinforcement already applied at the concrete wall in resisting lateral load [3] and embedded into concrete beam which is compared to the plain bamboo reinforcement [4]. The result shows that it contributes on the increasing of friction force between reinforcement and concrete as compared with plain bamboo reinforcement.

2. Experimental methods

The present experimental study examined the behavior of concrete beam with variation of knitted bamboo reinforcement subjected to three-point bending load [5]. Three concrete beams were investigated with the details on structural model and test set-up in the following section.

2.1. Structural modelling

The dimension was 15 x 25 x 160 cm, represent the width x height x length of the RC beam with concrete covers of 30 mm. The target compressive strength was 20 MPa. The knitted bamboo reinforcements with the dimension of 4 mm x 4 mm were applied in each beam as shown in Figure 2. Each knit contained three slices of bamboo with outer skin surface which were soaked with NaOH, with the pattern in Figure 1 [4]. The coating material consist of sikadur and sand were used in order to reduce the shrinkage and to provide the skin friction of bamboo, respectively.

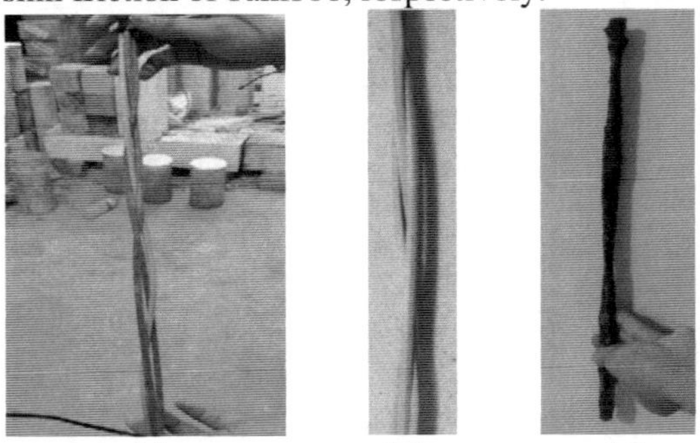

(a) (b) (c)

Figure 1. Knit pattern, (a) modeling of knit pattern which was refer to (b) [4], (c) knit pattern with sikadur and sand coating [4]

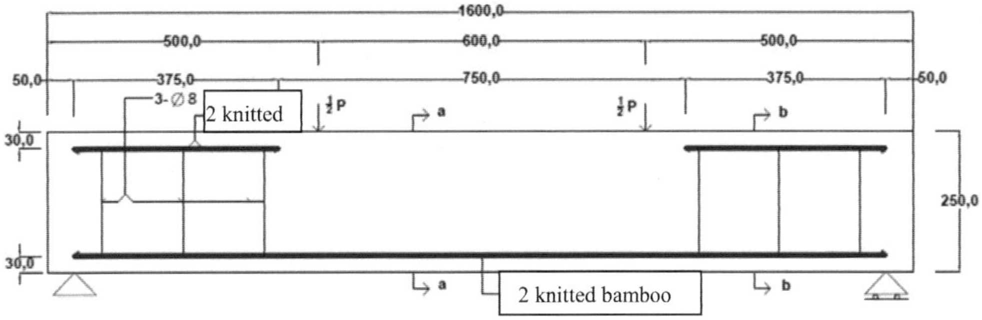

(a) First model

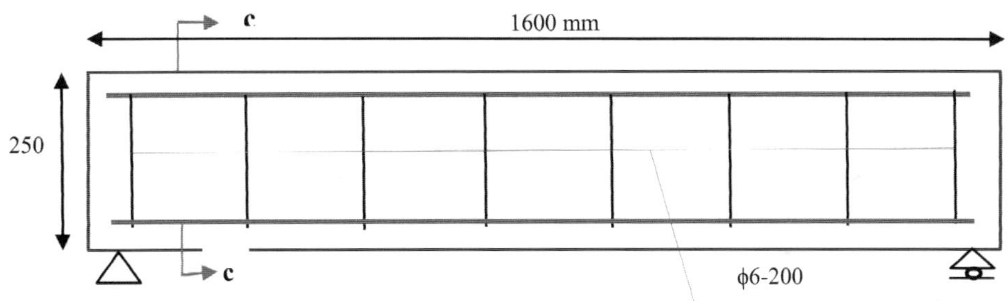

(b) Second model

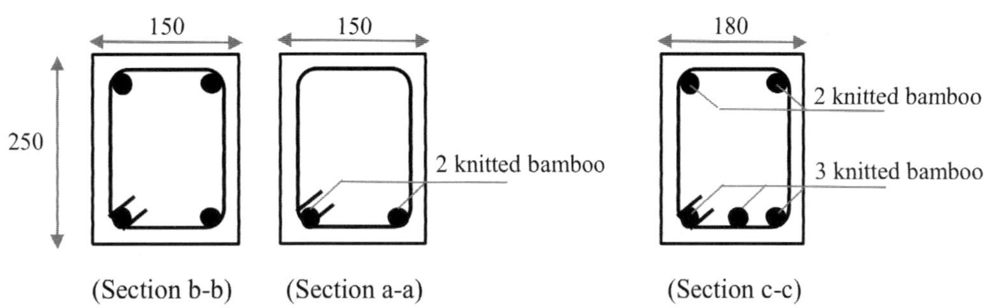

(c) Cross section of RC beam in the first and second model

Figure 2. Two different variation of knitted bamboo reinforcement (unit: mm)

2.2. Pull-out test of knit bamboo reinforcement

The tensile strength of knit bamboo reinforcement was experimentally tested previously with pull-out test method [4]. Based on this paper, it can be suggested that the first variation of knit bars applied in this research had the greatest average number of the tensile stress. Furthermore, this type of knit model could improve the bonding stress between bamboo and concrete due to its improvement to the friction force between bamboo reinforcement and concrete. In addition, prior research determined the tensile strength of this material was 216 MPa [6], which was slightly lower than the tensile strength of undeformed steel bar.

2.3. Compressive strength test of cylinder concrete

The total of six-cylinder concrete test specimens for the first model was examined previously by compressive strength machine, with the averaged result of 29.34 MPa, as shown in Table 1. From this result, it can be noted that the third mixing has the smallest results which is possible due to inconsistency compaction process. While the averaged compressive strength for the second model was 27.5 MPa. According to these results, it can be seen that the compressive strength for each model exceed the target strength of 20 MPa.

Table 1. The results of compressive test specimens for the first model

Specimens	Compressive Strenghth (MPa)	Averaged Compressive Strength (MPa)
1A	31.31	29.34
1B	30.18	
2A	28.87	
2B	32.22	
3A	28.87	
3B	24.57	

2.4. The flexural strength test for RC beam

The flexural strength for every beam was tested by using three-point bending load method depicted in Figure 3. In order to examine the maximum deflection of this structural model, LVDT was located under the mid-part of beam. In addition, crack was observed throughout loading produced by hydraulic jack and transferred using spreader beam. The flexural strength was calculated by equation (1).

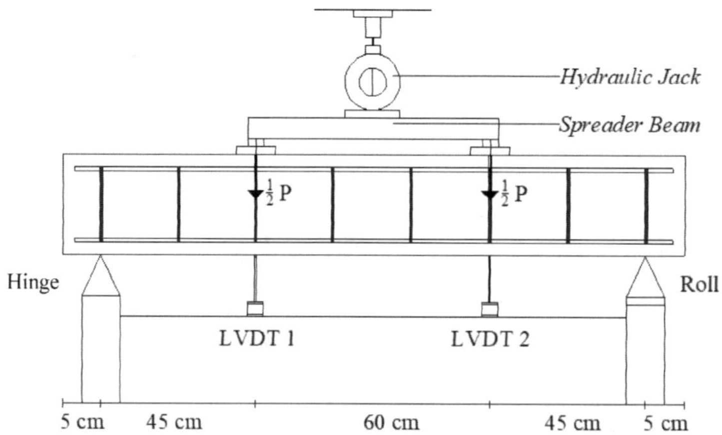

Figure 3. Bending test set-up in laboratory [6]

$$\sigma_{max} = \frac{PL}{bh^2} \tag{1}$$

where

P = the maximum load (N)
L = length of the beam (mm)
b = width of the beam (mm)
h = height of the beam (mm)

3. Results and discussions

3.1. The maximum load

The maximum load of reinforced concrete beam specimen was calculated theoretically with the moment capacity of the tested beam. In addition, the result would be compared to the experimental data. The theoretical calculation for the first model is presented below, shown in Figure 4. According to this analysis, it could be seen that the maximum theoretical load about 19.64 kN. While the experimental data are shown in Figure 5. It is depicted that the averaged maximum load is 22.23 kN, which is 13.18% larger than theoretical calculation. In addition, the averaged result is 46.09 kN which is twice larger than the previous model. This condition is due to difference on knit bamboo reinforcement modelling and width of beam.

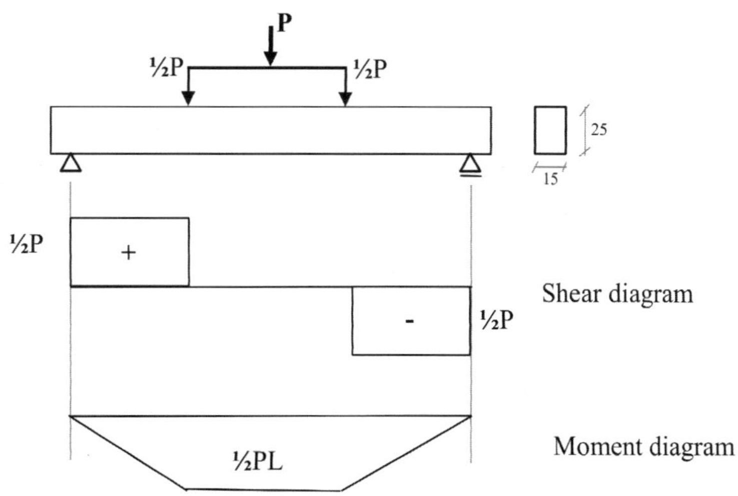

Figure 4. Structural model

Dimension of the RC beam:

Length of the beam (L) = 150 cm

Width of the beam (b) = 15 cm

Height of the beam (h) = 25 cm

C = T

0,85.f'c.b.a = As.fy

$$a = \frac{As.fy}{0{,}85.f'c.b}$$

$$= \frac{96 \ x \ 216{,}076}{0{,}85 \ x \ 29{,}34 \ x \ 150} = 5.55 \text{ mm}$$

Mn = As.fy.(d-a/2)

= 96 x 216,076 x (220 - 5,546/2) = 450.6 kgm

Mn = ½ P.L1

P = 2002.67 kg = 19.64 kN

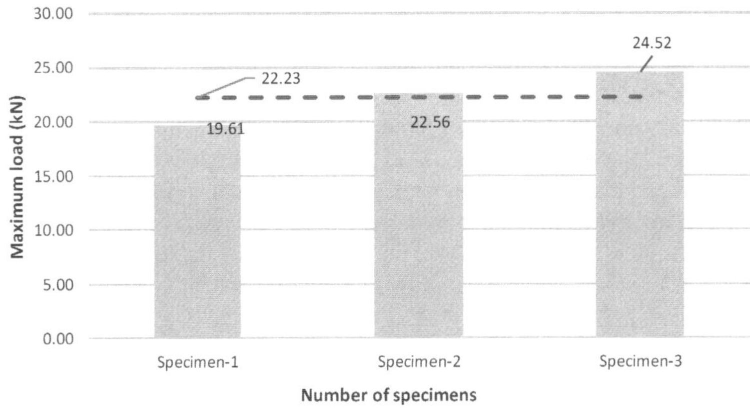

Figure 5. The maximum experimental load for the first model

3.2. Load vs deflection curve

Deflection of the tested beams were observed during the bending test analysis, depicted in Figure 7. According to the experimental results, it could be determined that the maximum deflection for the first model occurs in the third beam model with the value of 4.23 mm. The average data will be equal to 4.00 mm. Moreover, the experimental deflection result for the second model is 2.22 mm, which is smaller than the first model. It indicates that knit bamboo reinforcement capable to restrain the tensile force due to external load.

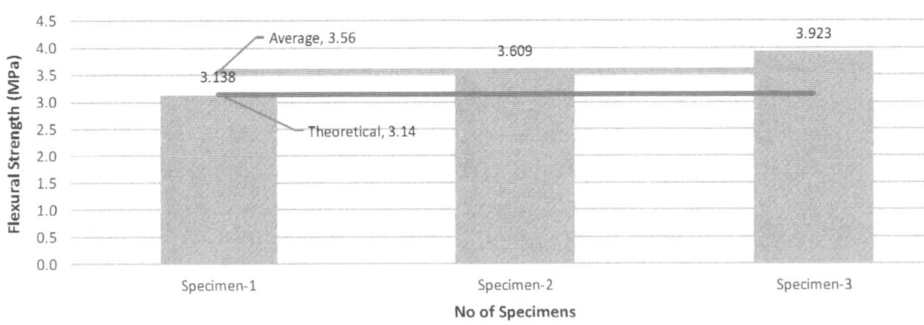

Figure 6. The flexural strength of RC beam for the first model

3.3. Crack propagation

Each specimen has a similar crack behaviour, which propagates firstly around middle of the beam as a pure bending area. Then, collapse occur immediately after increasing load of 100 kg with the maximum crack width of 6 mm for the first and second specimen, and 2 mm for the third specimen. Experimental results of the crack propagation can be seen in Figures 8 and 9 for the first and second model, respectively.

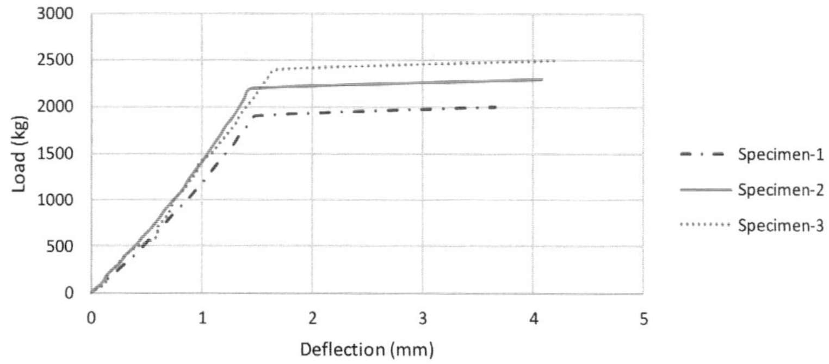

Figure 7. Load vs deflection curve for the first model

3.4. Crack propagation

Each specimen has a similar crack behaviour, which propagates firstly around middle of the beam as a pure bending area. Then, collapse occur immediately after increasing load of 100 kg with the maximum crack width of 6 mm for the first and second specimen, and 2 mm for the third specimen. Experimental results of the crack propagation can be seen in Figures 8 and 9 for the first and second model, respectively.

(a) Specimen-1

(b) Specimen-2

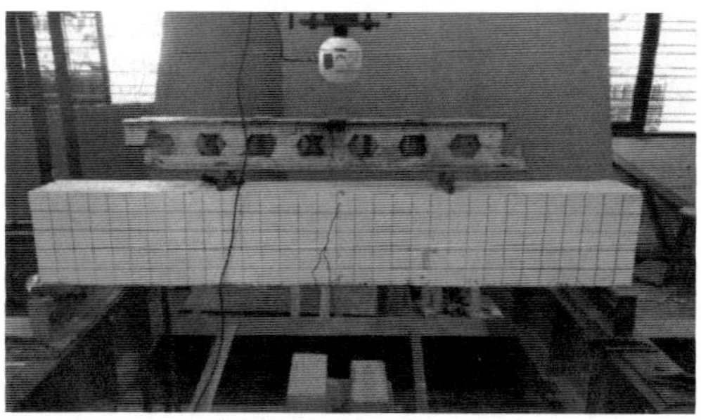

(c) Specimen-3

Figure 8. Crack propagation of the concrete beam for the first model

(a) Specimen-1

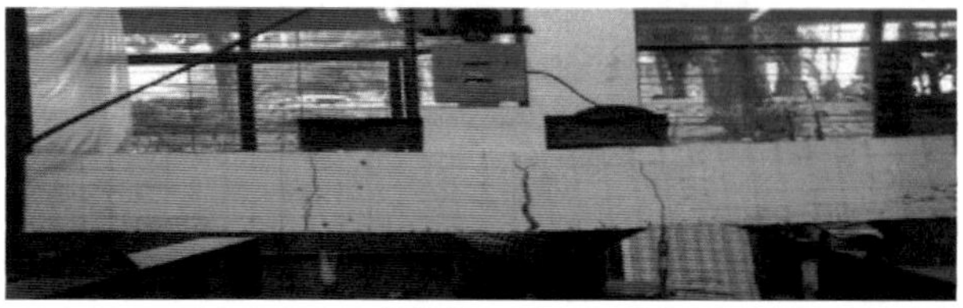

(b) Specimen-2

Figure 9. Crack propagation of the concrete beam for the second model

4. Conclusions

In this work, the knit bamboo was used as the reinforcement of concrete beam. Based on the experimental observation, the following can be deduced:

- The average flexural strength of the first model and second model were 3.56 MPa and 6.15 MPa, which are larger than theoretical calculation.
- The crack behaviour for the first and second model could be categorized as pure bending crack. The concrete beam collapsed directly after increasing load on the first crack.

References

[1] Morisco, 1999. Rekayasa Bambu. Nafiri, Offset, Yogyakarta.

[2] Singh, P. K., Jodhani, A., Sing., A. P., 2016. Bamboo as Construction Material and Bamboo Reinforcement. International Journal of Civil and Structural Engineering Research. Vol 4, Issue 1, pp. 312-323.

[3] Aditama, V., Dewi, S. M., Zacoeb, A., 2018. Bamboo Knit Reinforcement Concrete Plates as Walls to Resist Lateral Load. International Journal of Novel Research in Civil Structural and Earth Science. Vol. 5, Issue 2, pp. 8-13.

[4] Nuralinah, D., 2019. The pull-out test on knit bamboo reinforcement embedded into concrete beam. MATEC Web of Conferences 258, 01007 (2019).

[5] SNI 4431:2011. Cara uji kuat lentur beton normal dengan dua titik pembebanan. Badan Standardisasi Nasional, Jakarta.

[6] Prasetyo, F. D., Nuralinah, D., Arifi, E., and Setyowulan, D. 2018. The effect of Recycled Coarse Aggregate (RCA) Variation on the Flexural Strength of Knitted Bamboo Reinforced Concrete Beam. Jurnal Rekayasa Sipil.

Water quality indices for rainwater quality assessment in Bandung urban region

N Y Hasan[1], Driejana[2], A Sulaeman[3] and H D Ariesyady[2]

[1]Doctoral Program of Environmental Engineering, Faculty of Civil and Environmental Engineering Bandung Institute of Technology
[2]Faculty of Civil and Environmental Engineering, Bandung Institute of Technology
[3]Faculty of Mathematics and Natural Science, Bandung Institute of Technology

niayhasan@gmail.com

Abstract. Water quality indices (WQI) using STORET method has been used to assess rainwater quality in the Bandung urban regions, based on monitoring data at three locations (Coblong, Sumur Bandung, and Buah Batu). Rainwater samples were analyzed by detecting pH, SO_4^{2-}, NO_3^-, Cl^-, and heavy metals (As, Cd, Cr, Pb, and Zn) compared with Indonesian Government Regulation No. 82 Year 2001. Rainwater quality showed polluted based on the parameters of the water quality indices analysis using the Class I and Class II criteria. Rainwater quality in Bandung urban region were lightly polluted (Coblong) and moderately polluted (Sumur Bandung and Buah Batu). Rainwater can potentially become a water sources alternative for domestic use and urban farming in Bandung urban region, but it requires further treatment for better quality.

1. Introduction

The water quality indices is a single number to state the quality of water sources, the determination of water quality indices is carried out through 4 (four) steps: (1) parameter selection, (2) parameter transformation which transforms unit differences and dimensions to the general scale, (3) parameter weighting, and (4) aggregation of sub-indices to produce the final indices value [1].

The research on the water quality indices in Izombe of Niger Delta Region in monitoring the level of rainwater pollution and borehole water as the primary water source in the region compared to the drinking water quality standards from World Health Organization (WHO) showed that the parameters of temperature and color in rainwater exceeded WHO standards for drinking water. The results of analysis showed that the pH of rainwater samples was acidic with a pH range of 5.1-6.4 (average 5.8). This indicated that the rainwater was more acidic than borehole water [2].

The rainwater analysis in Africa to identify rainwater parameters as compared to water quality standards from WHO showed that rainwater in the Metropolis City of Uyo was not safe for drinking, where the total value of Pollution Index (PI) had indicated presence of Ni, Cd, Pb, and Fe which caused rainwater have a risk when consumed [3].

The use of rainwater in Indonesia has been regulated according to the Regulation of the Minister of Environment No. 12 Year 2009 about Rainwater Utilization were activities to collecting, using, and adsorbing rainwater into soil as groundwater conservation. Rainwater as raw water in Indonesia must meet water quality standard requirements based on Government Regulation No.82 Year 2001 about

Management of Water Quality and Water Pollution Control. The most widely used water quality indices method in Indonesia is the STORET (Storage and Retrieval) method.

Analysis of the water quality indices on raw water sources, especially rainwater, has shown that the water quality indices analysis method can produce recommendations regarding the feasibility of using rainwater as raw water for drinking water and domestic needs.

The Bandung urban region has high rainfall every year between 2,000-3,000 mm/year, but rainwater utilization must consider the potential contamination from air pollution due to human activities and industrialization both local and long-distance pollutant sources. Therefore, the water quality indices analysis can be used as a method for rainwater quality analysis in the Bandung urban region. This paper aims to explain the results of rainwater quality indices analysis in the Bandung urban region using the STORET method to determine the feasibility of rainwater before being utilized.

2. Material and Method

2.1. Study Region

Weekly rainwater samples were collected in bulk polyethylene bottle during rainfall from different locations at 3 (three) locations in the urban regions of Bandung: (1) Coblong region (06°55'11,07"S 107°36'04,54" E), (2) Sumur Bandung region (06°55'11,07"S 107°36'39,34"E), and (3) Buah Batu region (06°53'26,69"S 107°36'43,73"U), as explained in Figure 1. Sampling was carried out in 2016 and 2017, samples were collected and analyzed in the laboratory.

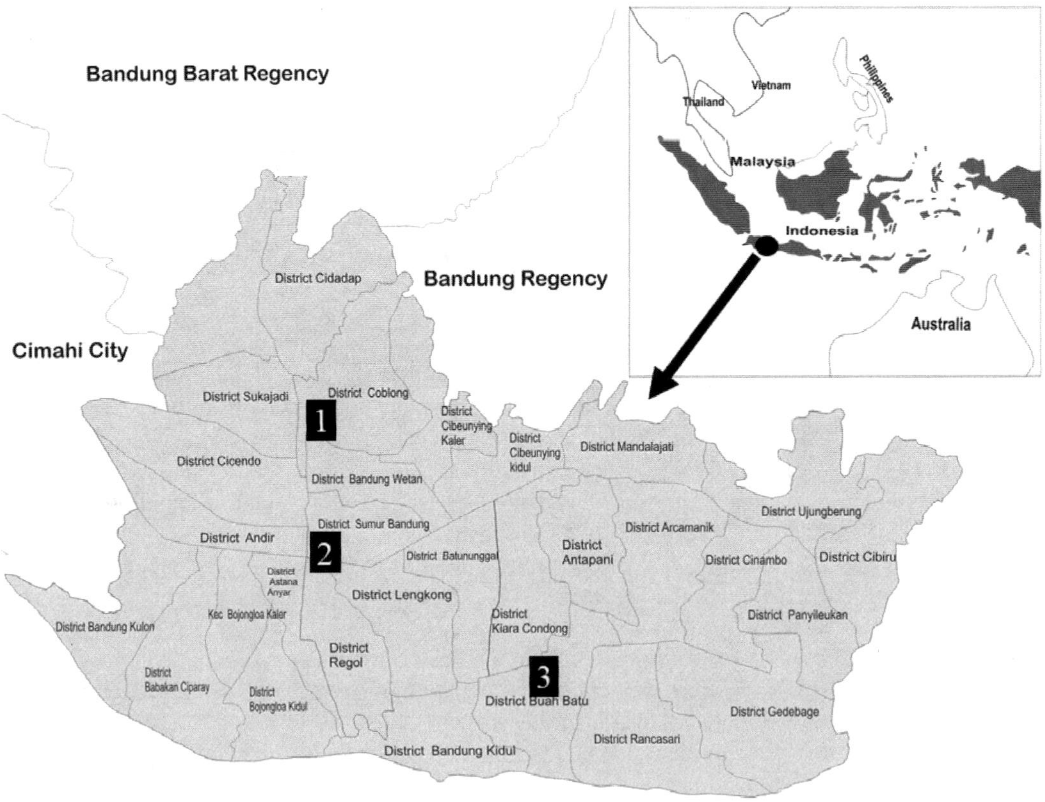

Figure 1. Sampling location in Bandung urban region

IOP Conf. Series: Materials Science and Engineering **669** (2019) 012044 doi:10.1088/1757-899X/669/1/012044

2.2. Rainwater analysis

The pH of rainwater varied between 3.0-7.5 pH units, as measured using a pH meter with a measurement accuracy of ± 0.01 pH units at a measurement temperature of 25°C. The pH meter has been calibrated with standard solution of pH 4, pH 6.86, and pH 10.

Ion chromatography (IC) was used to measure rainwater concentration of sulfate ions, nitrate and chloride, while heavy metals (As, Cd, Cr, Pb, and Zn) using a multi-element method with inductively couple plasma (ICP) [4]. The detection limits for heavy metal analysis were As (< 0.002 mg/L), Cd (< 0.001 mg/L), Cr (< 0.001 mg/L), Pb (< 0.001 mg/L), and Zn (< 0.002 mg/L).

2.3. Data Analysis

STORET refers overall to "STORage and RETrieval" was an electronic data system for water quality monitoring data developed by EPA. The STORET method is one commonly used method for determining the status of water quality in Indonesia. This method identifies which parameters meet or exceed water quality standards. The STORET method compares the water quality data with the adjusted water quality standard to determine water quality status.

The parameters analyzed to obtain indices values were pH, SO_4^{2-}, NO_3^-, Cl^-, As, Cd, Cr, Pb, and Zn. The parameter values were then compared with water quality criteria according to Government Regulation No. 82 Year 2001, as described in Table 1.

Table 1. Criteria of water quality based on class

Parameter	Unit	Class	
		I	II
pH	-	6 – 9	6 – 9
SO_4^{2-}	mg/L	400	(-)
NO_3^-	mg/L	10	(-)
Cl^-	mg/L	600	(-)
As	mg/L	0,05	1,00
Cd	mg/L	0,01	0,01
Cr	mg/L	0,05	0,05
Pb	mg/L	0,03	0,03
Zn	mg/L	0,05	0,05

The water quality classification used in this study was Class I and Class II criteria. Class I, is the raw water designated for drinking water, and/or other designation which requires water quality that is the same as the utility. Class II, is the water designated for water recreation facilities/ infrastructure, cultivation of freshwater fish, livestock, water for irrigating plantations, and/or other designations that require water quality that is the same as those uses. Thus, it is expected that rainwater can potentially be used as an alternative for raw water for drinking water and other domestic needs such as irrigating plants in urban regions (urban farming).

The determination of the status of water quality is through the value system from USEPA by classifying water quality in four classes, as described in Table 2. The classification of water quality base on STORET index were good condition, lightly polluted, moderately polluted, and heavy polluted.

Table 2. Classification of water status base on the STORET indices

Class	Criteria	Score	Quality status
A	Very good condition	0	Good condition
B	Good condition	$-1 \leq x \leq -10$	Lightly polluted
C	Moderately good condition	$-11 \leq x \leq -30$	Moderately polluted
D	Bad condition	≤ -31	Heavy polluted

3. Result and Discussion

3.1. Rainwater concentration

The concentration of rainwater parameter based on the result of data collection shows in Table 3. The results showed that the pH of rainwater in the Bandung region wass 3.13-7.06. The concentration of sulfate ions in rainwater shows a range of 8.86-17.82 mg/L, nitrate with 0.56-7.02 mg/L and chloride with 0.12-2.74 mg/L. The range of heavy metal concentrations detected in the Bandung urban region were As (< 0.002-0.046), Cd (< 0.001-0.011 mg/L), Cr (< 0.001-0.011), Pb (< 0.001-0.593) and Zn (< 0.002-0.665).

Table 3. Characteristic of rainwater in Bandung urban region

Parameter	Unit	Coblong			Sumur Bandung			Buah Batu		
		Min.	Max.	Mean	Min.	Max.	Mean	Min.	Max.	Mean
pH	-	3.13	7.06	5.40	4.10	6.70	5.50	4.30	6.86	5.52
SO_4^{2-}	mg/L	0.86	6.50	3.15	1.54	9.24	4.01	1.00	17.82	3.92
NO_3^-	mg/L	0.69	4.19	1.66	0.69	3.90	1.75	0.56	7.02	1.68
Cl^-	mg/L	0.12	2.74	0.42	0.14	1.39	0.43	0.12	1.44	0.41
As	mg/L	< 0.002	0.034	0.01	< 0.002	0.046	0.011	< 0.002	0.032	0.010
Cd	mg/L	< 0.001	0.011	0.010	< 0.001	0.014	0.001	< 0.001	0.007	0.001
Cr	mg/L	< 0.001	0.011	0.001	< 0.001	0.004	0.001	< 0.001	0.002	<0.001
Pb	mg/L	< 0.001	0.153	0.016	< 0.001	0.593	0.040	< 0.001	0.110	0.016
Zn	mg/L	< 0.002	0.015	0.002	< 0.002	0.326	0.016	< 0.002	0.665	0.022

3.2. Rainwater quality indices

Rainwater quality indices in Bandung urban region base on STORET method showed different quality status indices between different location, because of small spatial difference in this region. The rainwater quality indices result shows in Table 4. The rainwater quality indices at Coblong region based on the STORET calculation is categorized as lightly polluted with a score of -10. Parameters that did not meet Class I and Class II criteria were minimum pH value (3.13), average pH value (5.4), and maximum concentration of Pb (0.153 mg/L). Other metal parameters, Cd in Coblong region showed that the maximum concentration did not meet the quality standards Class II (0.011 mg/L).

Rainwater quality indices in Sumur Bandung region was categorized as moderately polluted with a score of -20. Parameters that did not meet Class I and Class II criteria were minimum pH values (4.10), and average pH value (5.50). On the other hand, the parameters of heavy metals that did not meet the quality standards Class I and Class II criteria were maximum concentration of Cd (0.014 mg/L), maximum and average concentration of Pb (0.593 mg/L and 0.040 mg/L), and maximum concentration of Zn (0.326 mg/L).

Rainwater quality indices in Buah Batu region was categorized as moderately polluted with a score of -12. Parameters that did not meet the Class I and Class II criteria were minimum pH values (4.3) and average pH value (5.52). The parameters of heavy metals that did not meet the quality standards Class I and Class II criteria include maximum concentration of Pb (0.110 mg/L), maximum and average concentration of Zn (0.665 and 0.022 mg/L).

Table 4. The result of the STORET index for rainwater in the Bandung region

Sampling region	Class I			Class II		
	Score index	Class	Quality Status	Score index	Grade	Quality Status
Coblong	-10	B	Good condition	-10	B	Good condition
Sumur Bandung	-20	C	Moderately polluted	-20	C	Moderately polluted
Buah Batu	-12	C	Lightly polluted	-12	C	Lightly polluted

The heavy metal of rainwater especially Pb and Zn in Sumur Bandung region was influence of local emissions of motorized vehicle activities in Bandung urban centers, while Zn emissions in Buah Batu region was possible from local industries emission. Coblong region was an urban area with a densely populated and dominated by domestic activities. The differences of rainwater in 3 (three) region in Bandung urban area were impacted of different landuse.

The dominant anthropogenic sources of heavy metals in rainwater represent coal combustion, automobile exhaust, and industrial emissions [5,6,7,8]. Heavy metals from rainwater accumulate in the biosphere and may cause adverse human health and environmental effects [9,10].

Exposure to extreme water pH values on human health can irritate the eyes, skin, mucous membranes, and gastrointestinal tract in sensitive people. A pH value of less than 4 can cause red eyes and irritation, and if the pH is less than 2.5, it will cause irreversible and extensive damage to the epithelial tissue [11].

EPA explained that the solubility of oxides of nitrogen and sulfur dioxide in acid rain is related to health problems, especially irritation of the eyes and lung diseases such as asthma and bronchitis, NO_x is a significant contributor to the formation of ozone in the atmosphere which also has disruptive health effects on resources such as contaminating food and water.

Rainwater utilization, such as rainwater harvesting systems, must pay attention to regional characteristics such as topography, season conditions, and air pollutants that can be sources of rainwater contaminants and their effects on health [12,13]. The rainwater can then be used for drinking water, agricultural irrigation, urban greening, and other needs that require good water quality [14].

4. Conclusion

Determining rainwater quality in the Bandung urban region was important before rainwater was using for water resources of drinking water or other uses such as agriculture in urban regions (urban farming). Utilization of rainwater, it must pay attention to the quality of rainwater in each urban region despite the results of the analysis of rainwater quality showing the presence of heavy metals and do not meet the quality standards for Class I and Class II criteria. This is due to prevent the impact of rainwater utilization on human health. Therefore, with the detection of heavy metals such as Pb in rainwater, the possibility of Pb accumulation must be considered, especially in plants if consumed directly.

References

[1] Abbasi T and Abbasi S.A. 2012 Water Quality Indices *Elsevier United Kingdom*

[2] Ubouh, E.A., Akhionbare, S.M.O. 2013 Effectiveness of Water Quality Index in assessing water resources characteristic in Izombie, Oguta Local Government Region of Imo State, Nigeria, *International Journal of Advanced Biological Research* **3(1)**, 31 – 35

[3] Udousoro, I.I., dan Unanaowo, A.E. 2015 Rainwater Quality Assessment in Uyo Metropolis using Water Quality Index, *Nigerian Journal of Chemical Research* **20**, 1 – 9.

[4] American Public Health Association (APHA). 1998 Standard Method for the Examination of Water and Wastewater 22nd Edition, *American Public Health Association*.

[5] Kaya, G. and Tuncel, G.1997 Trace element and major ion composition of wet and dry deposition in Ankara, Turkey, *Atmosphere Environment* **31**, 3985–3998.

[6] Hu, G. P. and Balasubramanian, R. 2003 Wet deposition of trace metals in Singapore, *Water Air Soil Pollution* **144**, 285 – 300.

[7] Cheng, M. C., You, C. F., Lin, F. J., Huang, K. F., and Chung, C. H. 2011 Sources of Cu, Zn, Cd and Pb in rainwater at a subtropical islet offshore northern Taiwan, *Atmospheric Environment* 45, 1919–1928,2011.

[8] Chon, K., Kim, Y., Bae, D.H., 4, and Cho, J. 2015 Confirming anthropogenic influences on the major organic and inorganic constituents of rainwater in an urban area, *Drinking Water Engineering Science* **8**, 35 – 48.

[9] Barrie, L. A., Lindberg, S. E., Chan, W. H., Ross, H. B., Arimoto, R., and Church, T. M. 1987 On the concentration of trace metals in precipitation, *Atmosphere Environment* **21**, 1133 – 1135..

[10] Báez, A. P., Belmont, R. D., García, R. M., Padilla, H. G., and Torres, M. C. B. 2007 Chemical composition of rainwater collected at a southwest site of Mexico City, Mexico, *Atmosphere Resources* **86**, 61–75.

[11] Guidelines for drinking-water quality, 2nd ed. Vol. 2. Health criteria and other supporting information. *World Health Organization*, Geneva, 2003.

[12] Sazakli E., Alexopoulos A., dan Leotsinidis M. 2007 Rainwater harvesting, quality assessment and utilization Kefalonia Island, Greece, *Water Research* **41**, 2039 – 2047.

[13] Lee, M., Kim, M., Kim, Y., Han, M. 2017 Consideration of rainwater quality parameters for drinking purposes: A case study in rural Vietnam. *Journal of Environmental Management* **200**, 400– 406.

[14] Laisheng, L., Linghua, L., Leixiang, W., Jiapeng, W., dan Weijie, H 2016 Problems and countermeasures on the safety of rainwater harvesting for drinking in China. *MATEC Web of Conferences* **6 (14006)**, 1 – 6.

[15] Anake, W. U., Benson, N. U., Akinsiku, A.A., Eromosele, C. O., Adeniyi, I. O. 2014 Assessment of trace metal in drinking water and groundwater sources in Ota, Nigeria **4(5)**, 1 – 4

[16] Mudgal, V., Madaan, N., Mudgal, A., Singh, R.B., dan Mishra, S. 2010 Effect of Toxic Metals on Human Health, *Nutraceuticals Journal* **3**, 94 – 99.

Acknowledgements

This paper is part of Riset KK dan Inovasi ITB 2017 work entitled "*Prediksi Dampak Kesehatan Lingkungan Deposisi Basah Logam Berat di Wilayah Perkotaan Bandung*" funded by Institut Teknologi Bandung.

IOP Conf. Series: Materials Science and Engineering **669** (2019) 012045 doi:10.1088/1757-899X/669/1/012045

The experimental compressive strength of normal and self-compacting concrete with recycled coarse aggregate

D Nuralinah[1], E Arifi[1] and D Setyowulan[1]

[1] Civil Engineering Department, Brawijaya University, Jl. MT. Haryono 167 Malang, East Java, Indonesia

devi@ub.ac.id

Abstract. Self-Compacting Concrete (SCC) is an innovation in the technology of making concrete that does not require a compaction process to fill the formwork. In this study, recycled coarse aggregate (RCA) was used as constituent material in both normal and SCC concrete. The aggregate was obtained from ready mix concrete waste with the quality of K300-K350, which was crushed into a size of 5-10 mm (25%) and 10-20 mm (75%). Concrete was cast into concrete cylinders, size of 15x30 cm then treated and tested at 28 days. In SCC concrete, superplasticizer is used, but fly ash was not used in the concrete mixture. The compressive test results show that normal concrete has a higher compressive strength than self-compacted concrete. This is due to the high amount of coarse aggregate that used in normal concrete. The results are almost corresponding with modulus of elasticity.

1. Introduction

The need for concrete material is increasing along with the many construction developments. The utilization of resources, such as gravel becomes larger. Massive natural stone mining causes environmental damage. With the increase in the need for concrete, making the constituent materials from nature experience over-exploitation and also the use of unbalanced natural resources, causing the material constituent of concrete to rise in price.

This material was chosen as an alternative of natural aggregate due to a large amount of waste from building destruction that was dumped carelessly. Partial disposal of waste are used as waste material and some is disposed of in open fields. Improper use of waste will cause damage to the environment. Used construction waste that has high-quality concrete can be managed into new aggregates of different qualities and can be used as recycled coarse aggregate (RCA).

In addition to environmental problems, one of the problems in concrete production is the lack of skilled workers in this field. As a solution, in the early 1980s, Japan began to develop SCC (self-compacting concrete) concrete. This concrete was widely used in the early 1990s, SCC is concrete that has high flowability and can fill concrete mold spaces without compaction or only requires a small amount of vibration to compact them. In the implementation of construction, the construction of concrete SCC can be applied with a smaller number of workers, so that the costs incurred also become more efficient.

The main difference between SCC and conventional concrete is the composition of the concrete mixture, where the portion of the filler is quite large, about 40% of the total volume of the concrete mixture. This filling material is fine-grained sand with maximum grain size is less than 0.125 mm.

Large portions of this filler cause concrete mixtures tend to behave as pasta [1]

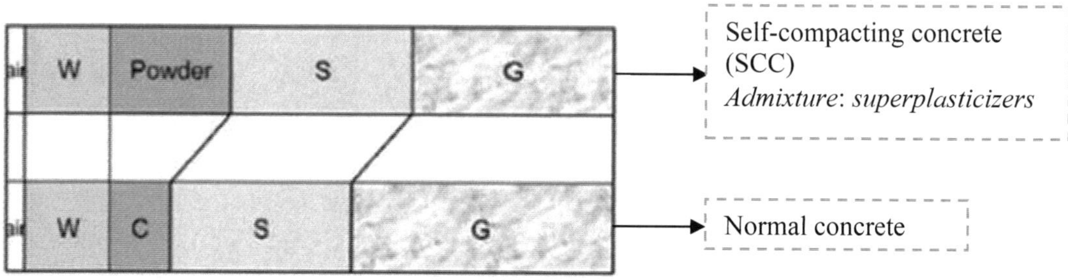

Figure 1. Comparison of SCC and normal concrete mixture proportions (Okamura and Ouchi, 2003)

From Figure 1, it can be seen that the difference in the proportion of the mixture between SCC concrete and normal concrete is the amount of coarse aggregate (G) and the amount of powder (P) to cement (C). Powder is cement which is combined with other fine particles (grain size <0.125 mm) such as fly ash and silicafume. In addition, admixture is used in SCC such as superplasticizers. There is no significant difference in the amount of water (W) and sand (S) [1].

The SCC concrete compressive strength test results showed that recycled coarse aggregate could reduce the compressive strength by 12.26% when natural stone aggregates were used (from 50.74 MPa to 44.52 MPa). However, when combined with round natural aggregates, the compressive strength increased by 62.78% (from 27.35 MPa to 44.52 MPa) [2].

In this study, the use of recycled coarse aggregate was used not only in normal concrete but also in self compacting concrete. So it is necessary to test the compressive strength to determine the characteristic of material, such as the modulus of elasticity.

2. Experimental methods

Cylindrical specimens, whose size is 15cm x 30 cm, were made from a mixture of normal concrete and self-compacting concrete, with 100% recycled aggregate composition. The aggregate was obtained from ready mix concrete waste with the quality of K300-K350, which was crushed into a gradation of 5-10 mm (25%) and 10-20 mm (75%). The use of gradations based on the results of the sieve analysis aims to produce homogeneous and aggregate interlocking in concrete. Before the manufacture of cylinder test material, aggregate analysis which included analysis of water content, analysis of specific gravity in SSD conditions, analysis of content weight and the aggregate absorption were carried out. The mix proportion of both normal and self-compacting concrete is explained in Table 1.

Table 1. Mix proportion of concrete

Type of concrete	Cement (kg)	Water (kg/lt)	Fine aggregate (kg)	Coarse aggregate (kg)	Admixture/ Superplasticizer (%)
Normal concrete	410	240.64	781.26	868.10	-
Self-compacting concrete	410	205	926.75	758.25	1

In the self-compacting concrete, testing of the filling ability of the mixture with a slump flow was conducted after mix design. Slump test was carried out in normal concrete. Concretes were cast in the cylinder and after the formwork was removed, the specimens were cured and tested after 28 days. Research procedure of testing of specimen can be seen in Figure 2.

Table 2. Cylinder specimens

No	Coarse aggregate	Type of concrete and specimen	Designed compressive strength	Admixture	Sample amount
1	RCA 100%	Normal (BRCA)	20 MPa	-	6
2	RCA 100%	SCC (SBR)	20 MPa	Superplasticizer (Viscocrete 3115)	3

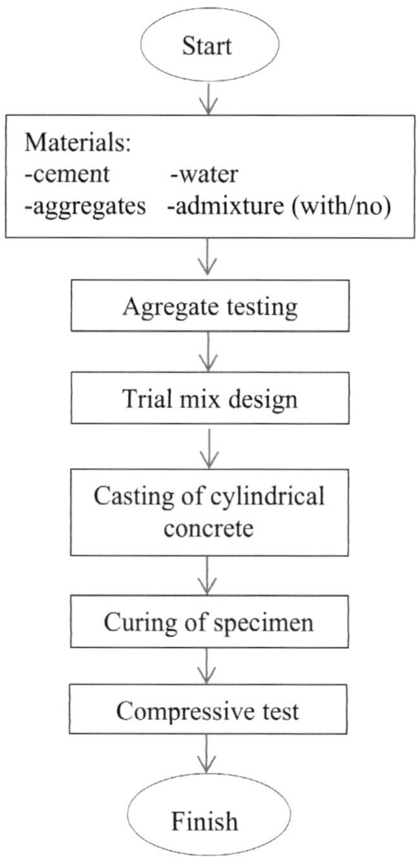

Figure 2. Research of procedures

2.1. Method of analysis

From the results of testing the compressive strength, elastic modulus can be analyzed based on changes in longitudinal displacement at each loading stage of 10 kN using an extensometer. The mechanical properties of concrete with recycled coarse aggregates will be compared to the use of normal concrete and self-compacted concrete.

In calculation of modulus of elasticity, it is compared with SNI standard and ACI standard for validation purpose. The equations of the modulus elasticity are as follows:

$$E_c = \frac{S_2 - S_1}{\varepsilon_2 - 0{,}00005} \quad (1)$$

$$S_2 = 0{,}4 \times f'_c$$

IOP Conf. Series: Materials Science and Engineering **669** (2019) 012045 doi:10.1088/1757-899X/669/1/012045

S_1 = strength when ε_1 = 0,00005 (obtained from regression curve of experimental strength-strain)

$E_c = 0{,}043 \text{ x } W_c^{\ 1{,}5} \text{ x } f'_c{}^{\ 0{,}5}$ (SNI-T-15-1991) (2)

$E = 4730 \text{ x } \sqrt{f'c}$ (ACI 318-89) (3)

3. Results and discussions

From aggregate testing, the results of water content, content weight, specific gravity and absorption of recycled coarse and fine aggregates were obtained, as seen in Table 3. Besides that, the results of testing the physical properties of fresh concrete and hard concrete can be obtained.

3.1. Results of aggregate examination test

Table 3. Aggregate examination test

Examination	Recycled coarse Aggregate	Fine aggregate
Water content (%)	5.09	4.37
Weight of content in rodded (gr/cc)	2500.78	2436.33
Weight of content in sholeved (gr/cc)	2230.08	2075.39
Density	2.44	2.48
Absorpsion (%)	2.73	6.79

3.2. Test results for the physical properties of fresh concrete

3.2.1. Slump height or slump flow test.

In normal concrete, a slump test was carried out to determine homogeneity and workability of mortar in fresh concrete with certain viscosity expressed by a slump value. In this research, the designed slump value was between 7 cm-15 cm. While, slump flow for testing of self-compacting concrete must have a good filling ability, which was designed between 60-85 cm. The slump height or slump flow test results can be seen in Table 4.

Table 4. Slump height or slump flow results

Specimens	Mixing variation	Slump height or flow	Average
BRCA-1a	Normal concrete-RCA	14 (slump height)	14.33 (slump height)
BRCA-1b	Normal concrete-RCA	14 (slump height)	
BRCA-1c	Normal concrete-RCA	15 (slump height)	
SBR-1	SCC-RCA	63 (slump flow)	61 (slump flow)
SBR-2	SCC-RCA	58 (slump flow)	
SBR-3	SCC-RCA	62 (slump flow)	

3.3. Test results for physical properties of hard concrete

3.3.1. Compressive test result.

Results of cylindrical compressive test can be seen in Figure 3.

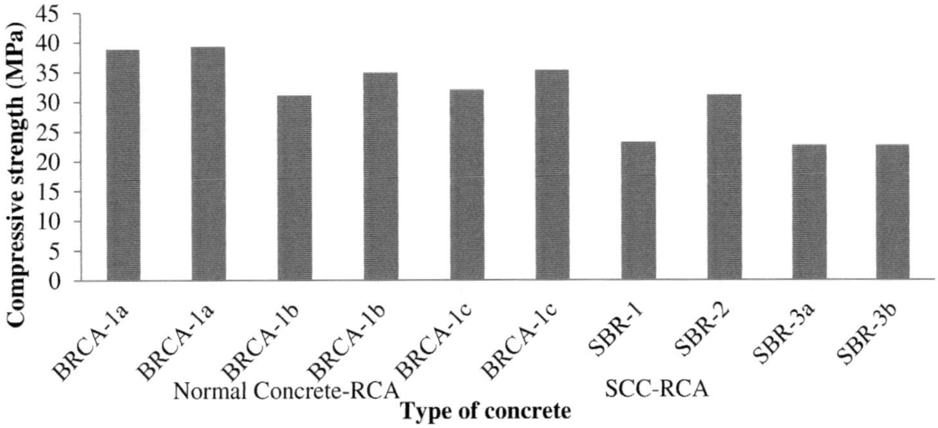

Figure 3. Compressive test result of concrete cylinder at 28-day

Figure 3 shows that the compressive test of normal concretes is higher than that of self-compacting concrete. Regarding the mix proportion in Table 3, it can be explained that the compressive test of normal concrete is significantly affected by the high amount of coarse aggregate.

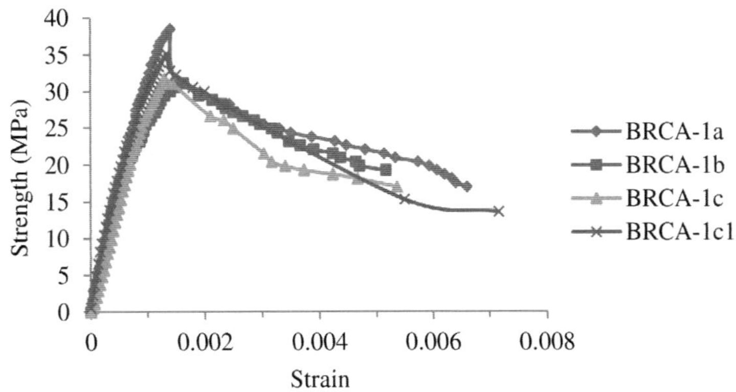

Figure 4. Relationship between strength and strain of normal concrete with recycled coarse aggregate

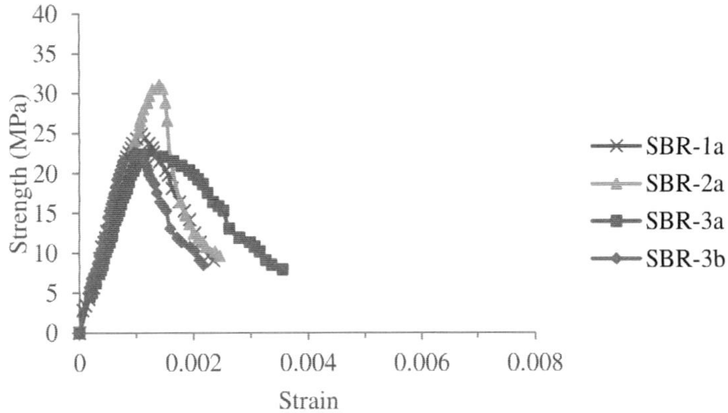

Figure 5. Relationship between strength and strain of self-compacting concrete (SCC) with recycled coarse aggregate

In Figure 4, normal concrete with recycled coarse aggregate (specimen BRCA-1a) results the highest compressive stress of 38.46 MPa with a deformation value of 0.0015 mm. While, in self compacting concrete with recycled coarse aggregate (SBR-2a specimen), the maximum compressive stress value is 31.11 MPa and deformation is 0.0014 mm, as shown in Figure 3.3.

Therefore, the compressive test of normal concrete is higher than the compressive test of self-compacting concrete, corresponding with the strain value. It might be caused there is no cement substitute using, such as fly ash that used in the SCC concrete, and high amount of coarse aggregate that used in the normal concrete.

According to the concrete compressive test results, calculated elastic modulus of each specimen can be obtained and compared each other with SNI and ACI standard specifications, as seen in Table 5.

Table 5. Modulus of elasticity of specimens

No	Specimens	Calculated elastic modulus (MPa)	Averaged elastic modulus (MPa)	Elastic modulus of SNI code (MPa)	Elastic modulus of ACI code (MPa)
1	BRCA-1a	36793	35789.25	29316.32	25488.01
2	BRCA-1b	35152			
3	BRCA-1c	29495			
4	BRCA-1c1	41717			
5	SBR-1	26280	25174.250	25455.78577	23709.36
6	SBR-2	24457			
7	SBR-3a	21715			
8	SBR-3b	28245			

Table 5 shows that there is correlation between result of compressive strength and elastic modulus of concrete. Among the specimens, calculated elastic modulus of self compacting concrete specimens almost coincide with validation result based on SNI and ACI.

4. Conlusion

From research that explained above, it can be concluded that:

- The compressive strength of normal concrete is higher than that of self-compacting concrete.
- Modulus of elasticity of both normal and self-compacting concrete corresponds with compressive test results. Averaged elastic modulus based on calculation agree with validation result based on SNI and ACI standard specification code.

References

[1] Okamura, H. and Ouchi, M. 2003. Self-Compacting Concrete, Journal of Advanced Technology, Vol.1, No.5-15, April 2003

[2] Widiantoro, Kurnia (2011). "Pengaruh Penggunaan Material Daur Ulang pada Beton Memadat Sendiri terhadap Kuat Tekan, Modulus Elastisitas, dan Kuat Lentur. Fakultas Teknik. Universitas Sebelas Maret. Surakarta.

[3] Sari, Rika Amenetya (2017). "Pengaruh Komposisi Fly Ash Terhadap Kuat Lentur Beton Porous dengan Variasi Komposisi Agregat Kasar Daur Ulang (RCA)". Fakultas Teknik. Universitas Brawijaya. Malang.

[4] SK SNI T- 15 – 1990 – 03 : Tata Cara Pembuatan Rencana Campuran Beton Normal

[5] Mulyanto, T. 2015. Analisis Sifat Mekanis Beton SCC Mutu Tinggi Dengan Pemanfaatan Teknologi High Volume Fly Ash Concrete. Universitas Muhammadiyah: Surakarta

[6] EFNARC (2005). The European Guidelines for Self-Compacting Concrete: Spesification, Production and Use, UK, May, 2005

IOP Conf. Series: Materials Science and Engineering **669** (2019) 012046 doi:10.1088/1757-899X/669/1/012046

The effect of sea water soak on asphalt porous characteristics with and without waste of rubber tire waste as an asphalt addictive

O S T Mawarni[1] and M Fauziah [1]

[1]Civil Engineering Study Program, Faculty of Civil Engineering and Planning, Universitas Islam Indonesia

oktivasiwi@yahoo.com

Abstract. The use of rubber tire scraded in the developement of porous asphalt technology as a part of flexible pavement aims to minimize the adverse effects of transportation facilities. The use of rubber tire scraded as an asphalt modifier in the pavement can provide several benefits, including reducing reflective cracking on overlays, reducing maintenance cost, increaseing resistance to new pavement cracking and rutting, increasing skid resistance an pavement age. This paper presents laboratory investigation on the effect of seawater immersion on porous asphalt performance using rubber tire scraded. Method use in laboratory investigation is based on AAPA 2004 and Bina Marga 2010. The results showed the permeability all types of mixture indicator "moderate drainage". The results of testing the Marshall characteristics showed a decrease in the value of stability, flow, and MQ in the porous asphalt mixture with Pen 60/70 asphalt and rubber asphalt as the duration of sea water immersion increased. The porous asphalt mixture using Pen 60/70 asphalt and rubber asphalt has Index of Retained Strength values which tend to decrease with increasing sea water immersion duration. Indirect Tensile Stength test results due to seawater submersion indicate a decrease in ITS value in porous asphalt mixtures with Pen 60/70 asphalt and rubber asphalt. The Cantabro test results showed an increase along with the increasing duration of seawater immersion.

1. Introduction

Transpotation is an importan part of a country's infrastructures. Hence, road infrastructure requires special attention in terms of safety and comfort. Some coastal areas in Indonesia often experience tidal flodd that cause the sea water to inundate the land, causing losses and disturbing community activities. Roads located in coastal areas have problems with seawater inundation. Seawater is a solition that has a coarsional nature and can cause damage to pavement. Porous asphalt is a mixture that is designed to have a higher porosity than other types of pavement. Porous asphalt is planned to have a large cavity so that water can permeate freely, so that it can provide a level of savety for road users, especially when it rains. Porous asphalt can also reduce glare from the road surface during the day or night. In addition to using porous asphalt as a solution to reduce road pavement damage, the use of added material rubber tire powder in pavement can provide the advantage of reducing reflective cracking on overlays, reducing maintenance costs, increasing resistance to new pavement cracking and rutting, increase skid resistance and pavement life.

Based on the describtion above, there is desire to continue the research that has been carried out by Harmadana [1], which uses porous asphalt mixture and added ingredients of rubber tire powder with a level of 3.5%, 4.5%, 5.5%, and 6.5%. from the result of these studies the use of rubber tire powder as an added material can increase stability so that it can meet exixting requirements[1]. The proportion of the addition of rubber tire powder at a value of 5.5% produce the best effect. Therefore, the authors plan to conduct research using porous asphalt mixture with ingredients plus rubber tire powder by (0% and 5.5%) and soaking in the form of seawater. This research was conducted to see the effect of seawater on porous asphalt by using added rubber tire powder materials with the aim of analyzing the extent of the influence of seawater on porous asphalt durability and the effect of adding rubber tire powder as added ingredients.

2. Research methodology

The method used in this study is an experimental reseach method that is a research method that is carried out by conducting experiments to obtain data and then the data is processed to obtain a comparison with the specification and requirements used the spesification and prodecures carried out refer to the Australian Asphalt Pavement Association, the Indonesian Nasional Standard, and the General Spesifications of Bina Marga 2010. The aggregates used came from Clereng, Kulonprogo, DIY. Asphalt used is Pertamina Pen 60/70 asphalt. The rubber tire powder used was obtained from Gulon, Salam, Magelang, Central Java. The seawater used is obtained from the North Coast of Kaliwage, Semarang.

The test begins with an examination of the physical properties of the material, then the proceed with the making of speciment to determine the optimum asphalt content (OAC). The next step is making test specimens using Pen 60/70 asphalt and asphalt mixed with rubber tire powder and assembled in seawater for testing Marshall Standard, IRS, ITS, Asphalt Flow down, Permeability and Cantabro test.

3. Results and discussion

Marshall, IRS, ITS, Asphalt Flow Down, and Cantabro test were conducted at the Highway Laboratory, Faculty of Civil Engineering and Palnning, Universitas Islam Indonesia. Permeability testing is carried out at the Transportation Laboratory, Faculty of Engineering, Universitas Gajah Mada. The test carried out refer to the AAPA 2004, SNI, and Bina Marga 2010.

3.1. Physical Properties of Material and Seawater

Testing the physical properties of material includes testing the characteristics of coarse aggregates and fine aggregates, filler testing, asphalt testing and seawater testing. The test results can be seen in Table 1, Table 2, Table 3, Table 4 and Table 5 below.

Table 1. Examination Result of AC 60/70

NO	Test Type		Specification	Result
1	Spesific Gravity		≥ 0,1	1,0751
2	Penetration	(0,1 mm)	60 - 70	65,6
3	Ductility	(cm)	≥ 100	157
4	Flash Point	(°C)	≥ 232	280
5	Fire Point	(°C)	≥ 232	281
6	Solubility TCE	(%)	≥ 99	100
7	Softening point	(°C)	≥ 48	49,5

Table 2. Examination Result of Rubber Tires Asphalt 5,5%

NO	Test Type	Specification	Result
1	Spesific Gravity	≥ 0,1	1,0277
2	Penetration (0,1 mm)	Min. 50	51,3
3	Ductility (cm)	≥ 50	63
4	Flash Point (°C)	≥ 232	259
5	Fire Point (°C)	≥ 232	274
6	Softening Point (°C)	≥ 54	55,5

Table 3. Examination Result of Rough Aggregate

NO	Test Type	Specification	Result
1	Spesific Gravity	> 2,5	2,6447
2	Water Absorbtion on Aggregate (%)	< 3,0	1,54648
3	Adhesion Test (%)	> 95	99,5
4	Abrasion Test (%)	< 40	26,6

Table 4. Examination Result of Smooth Aggregate

NO	Test Type	Specification	Result
1	Spesific Gravity	> 2,5	2,6271
2	Water Absorbtion on Aggregate (%)	< 3,0	2,35
3	*Sand Equivalent* (%)	> 50	71,2409

Table 5. Examination Result of Sea Water

No	Parameter	Unit	Result	Test Method
			A.003	
1	pH	-	6,82	SNI 06-6989.11-2004
2	Chloride (Cl)	mg/L	12.240	SNI 6989.19:2009
3	Sulfate (SO)	mg/L	20,9	SNI 6989.20:2009

3.2. Optimum Asphalt Content (OAC)

Determination of optimum asphalt content in porous asphalt mixture is calculated based on planning criteria from the Australian method, by means of a 35% cantabro loss value set to obtain the minimum bitumen content, then the 18% VITM value is set to obtain the maximum asphalt content, then both values plotted on the asphalt flow down graph. The optimum bitumen level is the sum of the average asphalt rough value with the asphalt flow down value as follows.

3.2.1. Cantabro Value

Cantabro Loss testing is carried out to determine the weight loss in the mixture carried out an abrasion test with a Loss Angeles machine and also as a parameter to determine OAL as required by AAPA. Test specimens which are left idle for 48 hours at room temperature and at least 6 hours before testing the temperature must be kept at room temperature. Before the test object is inserted into the Los Angeles drum machine. The image of the Cantabro test results can be seen in Figure 1 below.

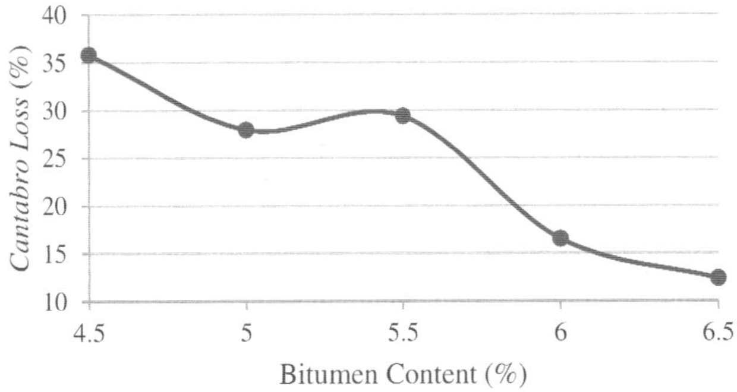

Figure 1. Relation between Asphalt Level and Cantabro Loss Value

From Figure 1 it can be seen that the cantabro value decreases with increasing proportion of bitumen content. The decrease in weight loss means that it is resistant to impact. This is caused by the increasing percent of bitumen content because with increasing asphalt content the binding strength of the mixture between aggregate and asphalt is getting better so that it increases the ability of porous asphalt mixture to withstand impact.

3.2.2. VITM Value

VITM is the percentage of air cavity in the mixture against the total volume of aggregate and asphalt mixture. Based on the results of the test, it is obtained a graph of the VITM value at various asphalt levels as shown in Figure 2 below.

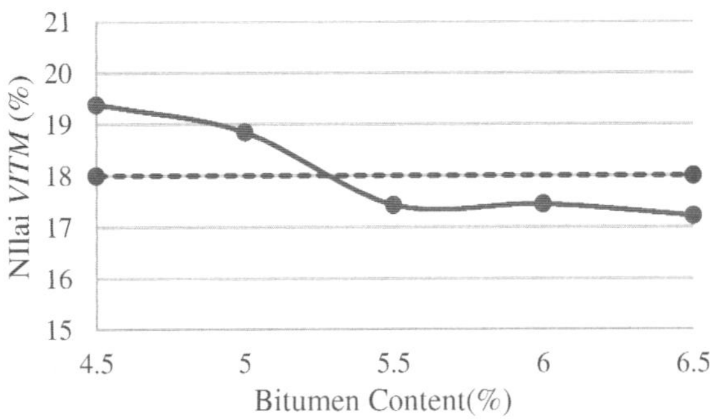

Figure 2. Relation between Asphalt Level with VITM Value

The results of testing the VITM value in Figure 2 show that the value of VITM gets smaller with increasing asphalt content, because the air cavity is filled with asphalt and increases the volume of the air cavity and the mixture becomes more dense.

3.2.3. Asphalt Flow Down Value

Asphalt Flow Down testing is carried out to determine the maximum asphalt content that can be homogeneously mixed with aggregates without the separation of asphalt separation and also as a parameter to determine KAO as required by AAPA (2004). The picture of the Asphalt Flow Down test results can be seen in Figure 3 below.

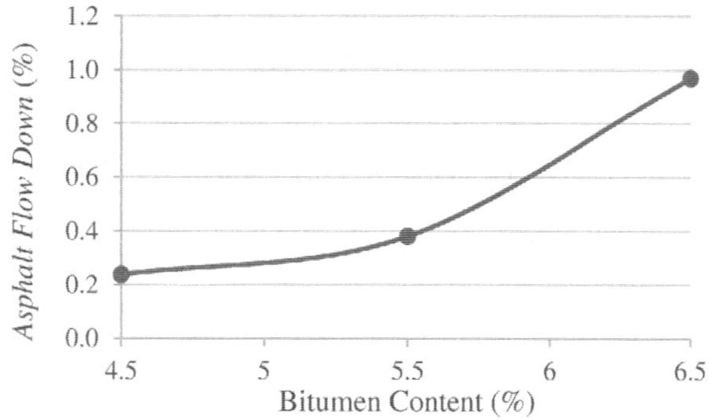

Figure 3. Relation Betwen Asphalt Level with Asphalt Flow Down Value

Based on Figure 3. it can be seen that the value of Asphalt Flow Down increases along with the addition of asphalt levels, however the increase in the value of Asphalt Flow Down is not significant. The increase in the results of the Asphalt Flow Down test shows that the separation between asphalt and aggregate increases with increasing asphalt content, this is due to the film enveloping the aggregate thicker with increasing percentage of bitumen so that the asphalt attached to the outside will melt and separate from the mixture.

The value of the test results of Cantabro, Marshall and Asphalt Flow Down that meet the requirements of the above parameters, obtained optimum asphalt content from porous asphalt mixtures using Pen 60/70 asphalt, which is 5.18%.

3.3. *Overview of Characteristic of Permeability*

Permeability testing graph can be seen in Figure 4. below.

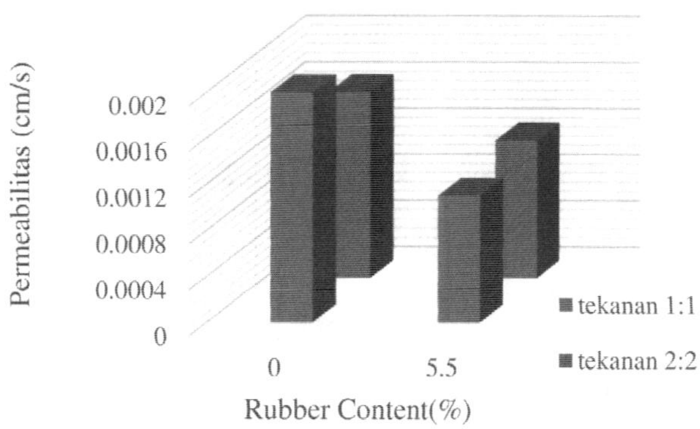

Figure 4. Relation of Rubber Asphalt Bonds to the Permeability Coefficient

Based on Figure 4. it can be seen that the two porous asphalt mixes that use Pen 60/70 asphalt and rubber asphalt are of moderate drainage. This is because the addition of rubber tires fills the cavity between the bitumen and the aggregate of small cavities in the mixture (VITM) causes the water to become difficult to seep into the porous asphalt mixture. Thus, the use of rubber tires is quite capable of producing a watertight mixture.

The research conducted by Juliansyah [2] also mentions that the addition of rubber tire waste as an additive to overcome the decline in the performance of superpave mixtures due to rainwater immersion, by using a mixture of added rubber tires becomes more impermeable so it is more difficult to get away

IOP Conf. Series: Materials Science and Engineering **669** (2019) 012046 doi:10.1088/1757-899X/669/1/012046

with water[2]. The difference with the research conducted by Juliansyah [2] is that the type of mixture used is included in a tight gradation, so that by adding rubber tires it produces a mixture of poor drainage.

In contrast to the research conducted by Bahtiar [3] which examined the porous asphalt mixture using oil sludge substrates, the permeability value increased with increasing oil sludge. Permeability value with oil sludge substitution has a higher permeability value than the research conducted by researchers using added rubber tire material, which means that the asphalt mixture with oil sludge has a better permeability value[3]. The highest permeability of porous asphalt mixture with oil sludge is 4.33 × 10-2 cm / second while the porous asphalt mixture with rubber tires is 1.19 × 10-3 cm / second. The equation from Bahtiar's research (2018) is that the type of mixture used is included in the category of open gradations that easily pass water.

3.4. Overview of Marshall Characteristics in OAC

Based on the test results of Marshall, Cantabro, and Asphalt Flow Down, the value of OAL is 5.18%. Samples were made using OAL which had been obtained and then immersed using sea water with variations in immersion duration of 48 hours and 96 hours. For comparison, some samples are made without sea water immersion. The results obtained can be seen in the following discussion.

3.4.1. VITM

The results of laboratory tests on porous asphalt mixes soaked in sea water can be seen in Figure 5. below.

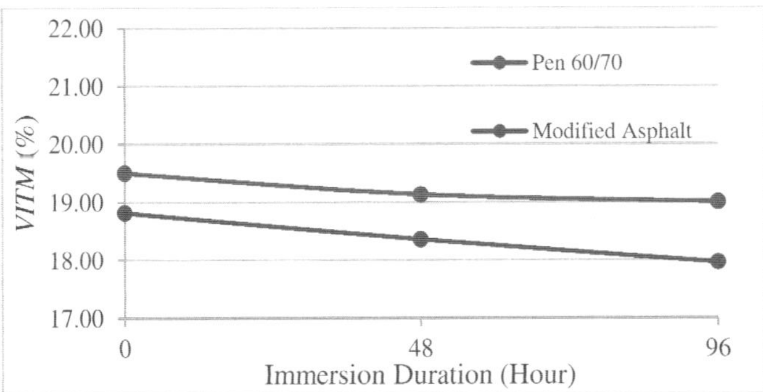

Figure 5. Relation between Immersion Duration with VITM Value

In Figure 5. it can be seen that the porous asphalt mixture with rubber asphalt binding material has a smaller VITM value compared to the 60/70 Pen bond mixture. This condition is caused by the air pores that have been filled with bitumen and other fine minerals. The small VITM value causes the mixture to be more waterproof, so it can reduce the bitumen oxidation process and increase the asphalt durability.

Based on graph 5 it can be seen that the more immersion time, the value of VITM decreases, this is because the minerals contained in seawater can fill the cavity to be smaller. . From the test results, the value of VITM with Pen 60/70 asphalt is 19.50%, 19.13%, and 19% while rubber asphalt is 18.81%, 18.36% and 17.96%. The test results meet the requirements of AAPA which is 18%, but in mixtures using added rubber tire material with a duration of 96 hours bathing falls below the minimum requirements.

The same can be seen in the study of Bahtiar [3] which examined the porous asphalt mixture using oil sludge substitution, the increasing oil sludge substitution the value of VITM decreased. This is because the asphalt that has been substituted with oil sludge has a more watery nature compared to Pen 60/70 asphalt and has a high mineral content, causing the minerals contained to fill the cavity to be smaller. The difference from Bahtiar's research is the type of substitution used.

3.4.2. Stability

The results of the laboratory test on porous asphalt mixes soaked in sea water can be seen in Figure 6. below.

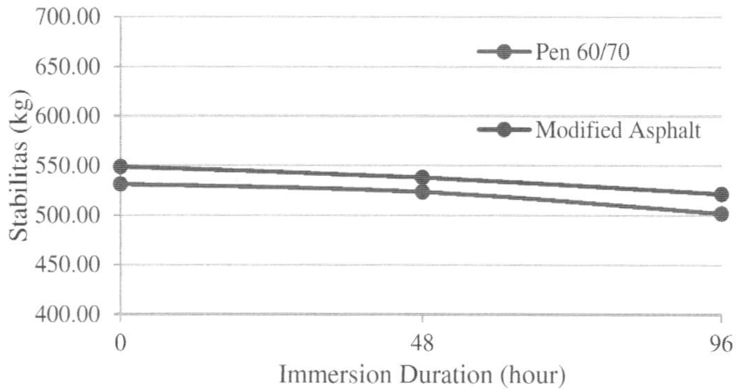

Figure 6. Relation between Immersion Duration with Stability Value

In Figure 6. it can be seen that the value of the stability of the mixture decreases with the duration of immersion that is getting longer. This shows that there is an effect of decreasing the bond between aggregate and asphalt due to sea water entering the mixed cavity containing various kinds of chemical compounds. Based on the graph above, it can be seen that the use of rubber asphalt in the mixture has a better performance in resisting the load under conditions of seawater tendam.

The stability value of porous asphalt mixture using material added to rubber tires larger than the mixture using Pen 60/70 asphalt. This is caused by the stiffer nature of rubber asphalt Pen 60/70, indicated by the smaller penetration value of rubber asphalt from Pen 60/70 asphalt.

In a study conducted by Juliansyah [2] who examined superpave mixtures using added rubber tire materials, also stated that the addition of rubber tires in the mixture caused the value of stability to increase. The difference from Juliansyah's research (2017) is that the porous asphalt mixture with added rubber tire material has a lower value of stability. This is because the porous asphalt mixture has a bitumen cavity (VITM) which is larger than the superpave mixture.

In a study conducted by Yusuf [4] which examined the performance of a mixture of Stone Matrix Asphalt with Pen 60/70 and Starbit E-55 due to the long seawater immersion, it can be seen that the stability value decreased with increasing sea water immersion duration. This happens because sea water can reduce the ability of asphalt to bind to the aggregate [4]. The stability value of the mixture of porous asphalt using material added by rubber tires soaked in seawater has decreased in accordance with the research conducted by the researcher

3.4.3. Flow

Flow is a state of change in the shape of a mixture that occurs due to a load to the collapse limit expressed in units of length (mm). The value of the flow to sea water bath can be seen in Figure 7. In Figure 7. it can be seen that the porous asphalt mixture with Pen 60/70 asphalt and rubber asphalt has decreased due to seawater immersion. This shows that under conditions of submerged seawater, porous asphalt using rubber asphalt binding material has good ability to withstand vertical deformation. Mixtures that use rubber asphalt have better resistance to vertical deformation than mixtures that use Pen 60/70 asphalt, this is because the rubber asphalt is more rigid.

The same thing can be seen from the study of Yusuf [4] who examined the performance of the mixture of Stone Matrix Asphalt with Pen 60/70 and Starbit E-55 due to sea water immersion, flow value decreased with increasing sea water immersion duration. the value of flow of porous asphalt mixtures with added rubber tire material meets the requirements of AAPA.

In a study conducted by Juliansyah [2] the addition of rubber tire waste as an additive to overcome the decrease in the performance of superpave mixtures due to rainwater immersion, the flow value decreased slightly as the duration of rainwater bathing increased. The addition of rubber tires can also affect the flexibility of the asphalt mixture, the more the rubber content increases the flow value will decrease.

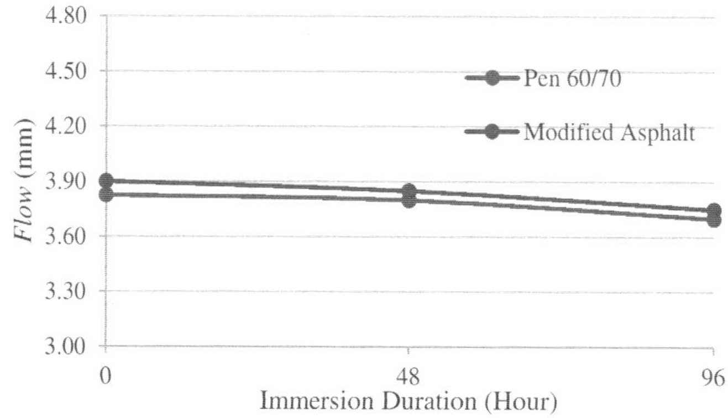

Figure 7. Relation between Immersion Duration with Flow Value

3.4.4. Marshall Quotient (MQ)

The Marshall Quotient value is the result of the stability between flow and flow. Marshall Quotient value for seawater immersion can be seen in Figure 8. below.

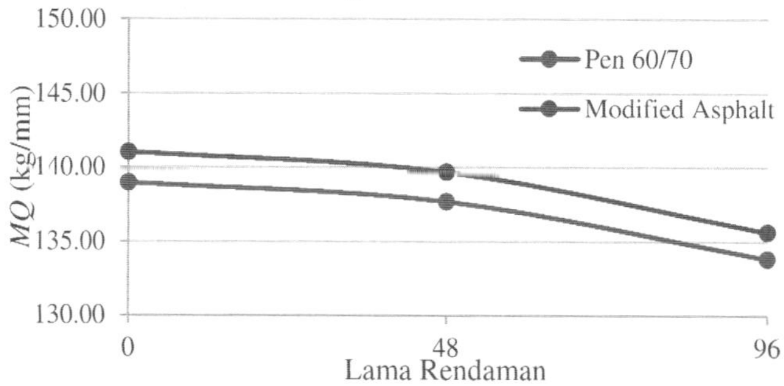

Figure 8. Relation between Sea Water Immersion Duration with MQ Value

In Figure 8. it can be seen that the MQ value has decreased, obtained from the comparison of the value of stability and the value of flow decreasing due to seawater immersion. The decrease in the MQ value in the porous asphalt mixture shows that the longer the mixture is soaked in seawater, the more flexible the mixture in accepting the load will be. The porous asphalt mixture using rubber asphalt has a higher MQ value than the mixture using Pen 60/70 asphalt, this is because rubber asphalt has a more rigid nature than Pen 60/70 asphalt.

The same can be seen in the study of Juliansyah [2] that the addition of rubber tire waste as an additive to overcome the decrease in the performance of superpave mixtures due to rainwater immersion, ie Marshall Quotient values will decrease with increasing duration of rainwater bathing. The addition of rubber tires also affects the decrease in the MQ value because of the effect of decreasing the bond between aggregate and asphalt after rainwater bathing.

This research is slightly different from Yusuf [4] who examined the mixture of Stone Matrix Asphalt with Pen 60/70 binding material which experienced an increase in MQ value before being held up to 96 hours of holding, then decreased after being immersed for 96 hours. In contrast to the mixture of Stone Matrix Asphalt that uses Starbit E-55 asphalt, the results have similarities to what the researchers did, namely the MQ value has decreased with increasing duration of sea water immersion.

3.5. Overview of Retained Strength (IRS) Index

The index of retained strength was obtained from the immersion process, to determine the strength and agility of the mixture after experiencing a 24-hour immersion process at a temperature of 60oC for soaking for 0.5 hours at 60oC. The IRS value graph can be seen in Figure 9. below.

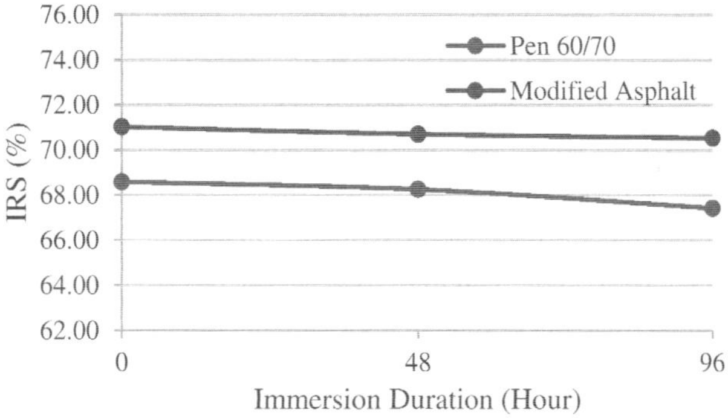

Figure 9. Relation between Immersion Duration and IRS Value

Based on Figure 9. it can be seen that the IRS value decreases with increasing duration of seawater immersion, this is influenced by a decrease in the value of stability caused by the amount of water entering the mixture causing the mixture to become weak in accepting the load. The graph above shows that the mixture using rubber asphalt has a higher IRS value than the mixture using Pen 60/70 asphalt, this is because the rubber asphalt is stiffer than Pen 60/70 asphalt.

In a previous study conducted by Juliansyah [2] the addition of rubber tire waste as an additive to overcome the decline in the performance of the superpave mixture due to rainwater immersion, as the duration of rainwater increased, the IRS value increased. These results occur because of the chemical influence of the superpave mixture soaked for 96 which increases the value of stability. In addition, in the same study showed that the IRS value using rubber asphalt was higher than the mixture using Pen 60/70 asphalt due to changes in the characteristics of the mixture due to the effect of adding rubber tires.

In a study conducted by Yusuf [4] who examined the mixture of Stone Matrix Asphalt using Pen 60/70 binding material and Starbit E-55 asphalt has different results from research conducted by researchers where IRS values increased with increasing immersion duration.

3.6. Overview of Indirect Tensile Strength (ITS) Characteristic

Indirect Tensile Strength is the ability of the pavement layer to withstand the tensile strength caused by the vehicle load. Based on the test results obtained the value of the relationship between the length of immersion and the Indirect Tensile Strength value can be seen in Figure 10 below.

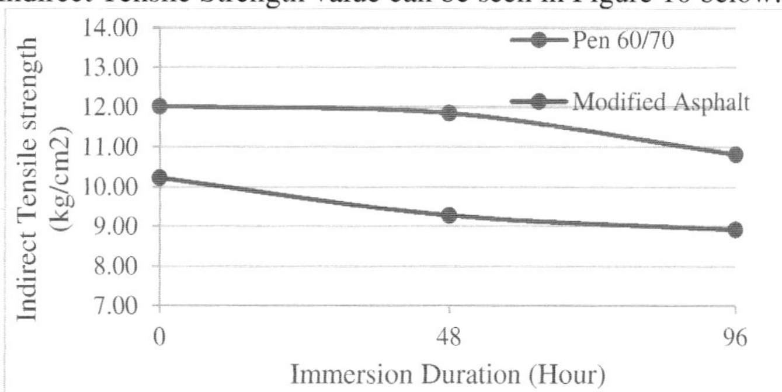

Figure 10. Relation between Immersion Duration and ITS Value

Based on the graph in Figure 10, the value of ITS tends to decrease with increasing immersion duration in sea water with a mixture of Porus made from asphalt Pen 60/70 and rubber asphalt. This is because the longer the duration of immersion causes the asphalt to soften and the bond between the asphalt and aggregate becomes reduced which causes the strain strength of the mixture to be reduced so that the mixture is easily cracked. The mixture using rubber asphalt has a lower ITS value than the one using rubber asphalt.

In a study conducted by Julainsyah [2] the addition of rubber tire waste as an additive to overcome the decrease in the performance of the superpave mixture due to rainwater immersion, also stated that as the soaking time increases the ITS value decreases and with the addition of rubber tires to make the mixture stable. The difference from the research of Juliansyah [2] with research conducted by researchers is on the type of mixture used and the type of soaking water used.

3.7. Overview of Cantabro Characteristics

Cantabro Test is a test conducted to determine the resistance of the test object to wear by using the Loss angeles machine. This test aims to determine the weight loss of the test object after an abrasion test. The image of the Cantabro test results can be seen in Figure 11 as follows.

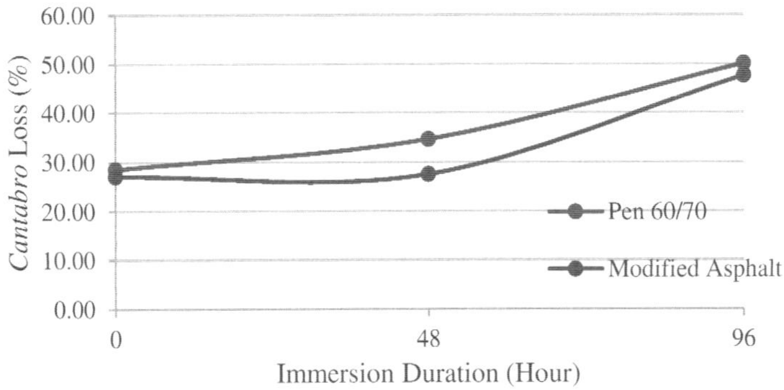

Figure 11. Relation between Immersion Duration and Cantabro Value

Based on the graph in Figure 11. it can be seen that the Cantabro value tends to increase with increasing immersion duration in seawater with a porous mixture of Pen 60/70 asphalt and Rubber asphalt. This means that the increasing duration of seawater immersion makes the mixture less resistant to impact. The mixture with rubber asphalt binding material is more resistant to impact or abrasion compared to the mixture with Pen 60/70 binding material.

The same can be seen from the research of Julainsyah [2] the addition of rubber tire waste as an additive to overcome the decrease in the performance of superpave mixtures due to rainwater immersion, namely the value of cantbro loss increases with increasing rainwater immersion. The addition of rubber tires also has an effect on the value of cantabro loss, the increasing the rubber tire level the lower the cantabro loss value. This shows that the addition of rubber tires can withstand heavy losses due to collisions.

In a study conducted by Bahtiar [3] which examined porous asphalt mixtures using oil sludge substitution, namely the value of cantabro decreased with increasing levels of oil sludge substitution, the decrease in cantabro test results showed that the resistance of the mixture to impact tended to increase, this was due the loss of weight in the mixture is getting smaller with the percentage of addition of oil sludge substitution.

4. Conclusion and suggestion

4.1. Conclusion

Based on research and analysis of porous Asphalt mixture performance data using ingredients added by rubber tire waste to the effect of seawater immersion, the following conclusions are obtained.

- The results of testing the physical properties of Pen 60/70 asphalt and rubber asphalt show insignificant density values, while the results of testing ductality, concentration, solubility in TCE, flash point, and burn point show significant results.

- Rubber tires can be used as an ingredient to add Pen 60/70 asphalt. This can be seen from the tests of Marshall, Cantabro and AFD whose results meet the requirements of AAPA (2004). Porus Asphalt mixture which has been added to rubber tires has changed the Marshall characteristics, namely the value of stability, flow, and MQ which has increased. The percent weight loss (cantabro) decreases with the addition of rubber tires.
- The permeability value of porous asphalt mixtures using 60/70 pen asphalt and asphalt using 5.5% rubber tire material indicates the results of the "moderate drainage" indicator. The mixture using rubber asphalt has a higher tightness than the mixture using Pen 60/70 asphalt.
- Effect of seawater immersion on Marshall characteristics, namely a decrease in the value of stability, flow and MQ in porous asphalt mixtures with Pen 60/70 asphalt and rubber asphalt as the duration of seawater immersion increases. The longer the porous asphalt mixture soaked in sea water causes a decrease in the mix ability to accept the load.
- The porous asphalt mixture using Pen 60/70 asphalt and rubber asphalt has an Indirect of Retained Strength value which tends to decrease with increasing sea water immersion duration.
- Indirect Tensile Strength test results due to seawater immersion indicate a decrease in ITS value in porous asphalt mixtures with Pen 60/70 asphalt and rubber asphalt. This is because the longer the duration of soaking causes the bitumen to soften and the bond between the asphalt and aggregate decreases so that it is easy to crack.
- The Cantabro test results show an increase along with the addition of sea water immersion duration. The mixture with rubber asphalt binding material is more resistant to impact or abrasion compared to Pen 60/70 asphalt bonding material, this is due to rubber asphalt which has strength and elasticity which can reduce impact on impact.

References

[1] Harmadhana, S. 2016. Study of Laboratory Characteristics of Porous Asphalt by Using Crumb Rubber as Additive Material. *Tugas Akhir*. Universitas Sumatera Utara. Medan.

[2] Juliansyah, R. 2017. Addition of Rubber Tire Waste as an Additive to Overcome the Decrease of Superpave Mixed Performance due to Rainwater Immersion. *Tugas Akhir*. Universitas Iaslam Indonesia. Yogyakarta.

[3] Bahtiar, M. 2017. Effect of Oil Sludge as 60/70 Bitumen Penetration Substitution on the characteristics of Porus Asphalt Mixes. *Tugas Akhir*. Universitas Islam Indonesia. Yogyakarta.

[4] Yusuf, N.A.N. 2017. Performance of Mixture of Asphalt Stone Matrix with Pertamina Asphalt Pen 60/70 and Starbit E-55 Due to Sea Water Immersion. *Tugas Akhir*. Universitas Islam Indonesia. Yogyakarta.

IOP Conf. Series: Materials Science and Engineering **669** (2019) 012047 doi:10.1088/1757-899X/669/1/012047

The gaps of passive strategy option in Indonesia's green building regulation

Sahid[1], Y Sumiyati[2] and R Purisari[1]

[1,1]Department of Architecture, Pembangunan Jaya University
[2]Faculty of Law, Bandung Islamic University

sahid@upj.ac.id

Abstract. Implementing building regulations that could encourage stakeholders to apply the green building concept is one of the effort that could be done to realize a green city. Implementation of green building passive strategies, through the elaboration of building design that fit to the condition of the surrounding environment, are believed to be a solution to achieve optimal green building performance. Currently, the government has set a regulation as representation of green building policy (at least three green building regulations, namely Green Building Decree of Ministry of Public Works and Public Housing Regulation, Governor DKI Regulation of Green Building and Bandung's Major Regulation of Green Building) and experts have also set up a rating system to support the green building policy. This study aims to determine the accommodation of passive strategies in regulations, as a policy representation of green building, by matching the availability of strategic options on those three regulations. It seems that efforts are still needed to complement and affirm the need for the application of the passive strategy of green buildings in order to obtain maximum benefits from the implementation of the concept of green building.

1. Introduction

Building as one of the city components plays an important role in realizing the vision of a city. Green city can be realized if the building inside also has the same vision and spirit, namely green buildings. To expand the implementation and to align views on the realization of green buildings, some cities set policies in the form of regulations. The regulation manages the role of stakeholders related to buildings and mentions the criteria that need to be fulfilled by stakeholders so that the building can be called a green building.

Principally, green building can accommodate two types of strategies, namely passive and active strategies. Active strategy is generally understood as a strategy to obtain the expected environmental conditions (especially thermal and visual conditions) by relying on the use of efficient electrical and mechanical equipment, while passive strategies are understood as a strategy to realize expected environmental conditions by relying on the design of buildings to make use of it at once anticipate environmental conditions. Although a building with certain conditions can only rely on passive strategies, but related to human dependence on electronic equipment, lighting requirements and air conditioning for 24 hours, the use of active strategies in certain conditions and time frames is inevitable.

Yeang [1] said that by considering the need for environmental conditioning to support human activities, these two strategies can be conditioned to work together on a building even though it is recommended to optimally utilize passive strategies before active work strategies. Attempts to rely entirely on passive strategies are impossible to implement if humans expect a permanent condition of the inner environment at all times because of outdoor conditions that change over time. On the other hand, the efficiency of mechanical and electrical equipment that is getting better cannot be used as an excuse to ignore the application of passive strategies or take advantage of passive strategies only to fulfill the formalities of compliance with regulations for example. Efforts to prioritize passive strategies as optimal as possible before the active strategy works should be a priority in implementing green buildings.

2. Methodology

To find out the availability green building policies to accommodate the green building's passive strategy, the researchers conducted a literature study to trace the opinions of some experts on passive strategies through books, journals, and proceedings. Furthermore, an analysis of the green building policies (which are represented by available green building regulations) is carried out by comparing the various options of passive strategies obtained in the literature with the passive strategies contained in the regulations. In this paper, the researcher limits comparative studies only on three green building regulations, namely the Governor DKI Regulation of Green Building No. 38 of 2012 concerning Green Building and Ministry of Public Works and Public Housing Regulation No.02 / PRT / M / 2015 concerning Green Building and Bandung's Major Regulation of Green Building No.1023 of 2016 concerning Green Building.

3. Results and discussion

According to Sharma [2], the use of passive systems could be done through understanding climate climates and zones, identifying comfort zones, identifying heat sources, optimizing microclimatic conditions, and defining characteristics needed for building configurations and building envelopes. Complementing this, King [3] mentions several things that need to be considered when designing buildings in humid tropical climates such as laying buildings, building forms, natural air conditioning, building interior layout, rain, openings and imagery, and thermal mass. Furthermore, Sharma [2] states that one of the important things in passive systems is passive cooling. The strategy that can be done to achieve the optimal passive cooling system is the right window placement and natural lighting design, the selection of the right glass material for windows and skylights, proper imaging of the glass when heat is not desired, the use of light-colored materials for building envelopes and roofing, proper building and orientation, and good landscape design. In addition, it can also be done by installing windows that can be operated (open-closed), using landscape elements to direct the wind, placing windows with the following window size for cross ventilation purposes, reducing internal obstacles (e.g. walls) so that wind can still flow, using wing walls when crossing is not possible, and using stack ventilation to obtain a chimney effect.

Sarte [4] added that the use of passive strategy can improve the performance of buildings in the energy field. Passive strategies can be done through: building form and orientation processing, landscape arrangement, choosing the right building envelope, designing natural ventilation, selecting building materials, using low-emittance glass, installing radiation barriers, using bright/reflective colors. According to Jones [5], buildings that considering their microclimate conditions through orientation processing, building shapes and sheaths can save energy use in the amount of five to six times conventional buildings during their life cycle periods.

Various passive strategy options for climate transfer according to the climate are outlined in a scheme created by The Low Energy Architecture Research Unit (LEARN) of the London Metropolitan University [6]. They divided the passive strategy options into three climatic conditions, namely hot humid, hot arid and cold. The difference is that in hot climates, both humid and dry, the strategy used is resist gain (heating avoidance) and promoting loss (buildng cooling). In cold climates the strategy used is to promote gain (hastening heat gain) and resist loss (resisting heat release). The four strategies can be done in four ways, namely conduction, convection, radiation and evaporation. Conduction method can be done for resist loss and resist gain by minimizing heat flow (minimize conductive heat flow), while to promote loss can be done by earth cooling. Convection can be done in hot and cold climates with minimizing infiltration for resist lost or resist gain.

Table 1 shows various passive strategy options that can be applied to buildings, both low, medium and high-rise buildings. The strategy includes a strategy for buildings to optimize building design (arrangement of orientation, shape, sheath, material and interior) before utilizing electrical and mechanical equipment. In addition, the building also needs to optimize all the potential that exists in the surrounding environment in order to create an environment that is comfortable for the occupants and economical in the utilization of energy and not to damage the environment.

Table 1. Summary of passive strategy options

No	Definition of Passive Strategy	Strategy	Keywords	References
1	Architectural Design (The absence of electrical or mechanical equipment)	Determine orientation and elaborate the shape of the building Designing the building envelope and selects the materials Layout design and selects the material Optimizing building design (orientation, shape, envelope, interior)	• Natural lighting • Natural ventilation • Shadow • Heating avoidance • Building Cooling	[1] [2] [3] [4] [5] [6] [7] [8]
2	Utilize the potential environmental conditions around the building	Utilize the surrounding landscape Utilize the surrounding buildings Utilize the local micro and macro climate		

The various options available in Table 1 are then compared with three green building regulations, namely Governor Regulation (Pergub) of DKI Jakarta No. 38 year of 2012 concerning Green Building and Ministry of Public Works and Public Housing Regulation No. 02 / PRT / M / 2015 concerning Green Building and Mayor Regulation (Perwal) of Bandung No.1023 year of 2016 concerning Green Building. The three regulations were chosen because they also represent three levels of government, namely the central government (Permen), the province (Pergub) and the city (Perwal) while representing various diversity of environmental conditions that need to be considered by the government in setting regulations.

Governor DKI Regulation No. 38 year of 2012 concerning Green Building divides the criteria of green buildings into two main categories, namely the new building category and the existing building category. This study will only limit the analysis of the criteria for green buildings for new buildings with the consideration that new buildings provide more opportunities for the adoption of passive strategies than existing buildings. The Governor DKI Regulation only applies to buildings with certain functions and certain areas (minimum 10,000 sqm).

The criteria listed on the Governor DKI Regulation for new buildings include energy efficiency criteria (building envelope systems, ventilation systems, air conditioning systems, lighting systems, in-building transportation systems, and electrical systems), water efficiency criteria (design of water-saving sanitary equipment and design of water use), indoor air quality criteria (content of fresh air, harmful gases and dangerous refrigerants), criteria for land management and waste (spatial requirements, supporting facilities, solid and liquid waste management), criteria for implementing construction activities (safety, health work and environment, water conservation during construction activities, management of waste from construction activities).

The contents of regulation related to the implementation of passive strategies for the thermal and visual environment are listed in the energy efficiency criteria for building envelope systems by specifying a 45 watt/sqm Overal Thermal Transfer Value (OTTV) limitation without giving details of what strategies can be done to obtain the value; energy efficiency criteria for natural gas systems with limits as long as it is possible to reduce the cooling load with reference to SNI 03-6572; energy criteria for air conditioning systems with a room temperature limit of 25° C and relative humidity of 60% ± 10%; energy criteria for natural lighting systems with restrictions being an integral part of the building lighting system according to SNI 03-2396; Indoor air quality criteria take into account the indoor rate of air change and fresh air input; criteria for land and waste management for reforestation by mentioning the limits of a certain minimum area of the ground floor area for buildings with a certain number of floors and possible to be carried out on the roofs and walls of buildings and inside and outside buildings (the regulation recommend to use tropical, local vegetation and consume less water); criteria for land management and waste for pavement using permeable material. The seven passive strategy criteria are summarized in Table 2 below:

Table 2. Passive strategy identification at Governor DKI regulation of green building (Pergub)

Criteria Category	Passive Strategy	Parameter
Energy	Building envelope	OTTV $\leq$ 45 watt/sqm
	Natural Ventilation	Used as long as it is possible to reduce the cooling load Comfortable standard: Temperature 25°C Humidity 60%$\pm$10%
	Daylighting	Become an integral part of the building lighting system
Indoor Air Quality		The indoor rate of air change and fresh air input
Land and Waste	Greenery	Inside and outside the building Green roof, green wall Tropical vegetation, local, minimal water consumption Minimum area limit according to the number of floors
	Pavement	Permeable material

In the Ministry of Public Works and Public Housing Regulation No.02 / PRT / M / 2015 concerning Green Building, the buildings affected by the requirements of this regulation include new buildings and existing buildings. Buildings that are subject to the requirements of building green buildings are divided into three categories, namely the mandatory category, recommended, and voluntary. This study will only limit the analysis of the mandatory green building criteria at the technical planning stage, namely site management criteria including building orientation, site processing including accessibility/circulation, management of B3 waste contaminated land, private green open space (RTH), pedestrian lane provision, management of building sites, provision of parking lots, outdoor lighting systems, construction of buildings on and/or under the ground, water and/or infrastructure/public facilities; the efficiency criteria for energy use include building envelopes, ventilation systems, air conditioning systems, lighting systems, in-building transportation systems, and electrical systems; the criteria for water use efficiency include water sources, water use, use of water fixtures; indoor air quality criteria include prohibiting smoking, controlling carbon dioxide and carbon monoxide, and controlling the use of refrigerant; the criteria for using environmentally friendly materials include controlling the use of hazardous materials, and the use of certified environmentally friendly materials (eco labeling); the waste management criteria include the application of the 3R principle (reduce, reuse, recycle), the application of a waste handling system, and the implementation of a waste generation record system; criteria for wastewater management include the provision of facilities for managing solid waste and liquid waste before being discharged into municipal sewers, recycling water from gray water, implementing the Occupational Health and Safety Management System, and implementing environmentally friendly behavior; green supply chain criteria include the use of construction materials, selection of suppliers and/or sub-contractors, and energy conservation.

The contents of regulation related to the implementation of passive strategies for thermal and visual environments are listed in: site criteria for the orientation and shape of buildings with limits to maximizing natural lighting and minimizing heat propagation and according to site context; site criteria for site greening with restrictions according to local regulations; energy criteria for enveloping buildings with RTTV and OTTV limits of at least 35 watts/sqm; energy criteria for ventilation systems with limits to minimizing cooling loads; Energy criteria for natural lighting systems with maximum use limits. The five passive strategy options are summarized in Table 3 below:

Table 3. Identification of passive criteria and strategies in Ministry of Public Works and Public Housing Regulation (Permen)

Stage	Criteria Category	Passive Strategy	Parameter
Planning	Site	Orientation and shape of the building	Maximize natural lighting and minimize building heating and considering the site context
		Greenery	Refer to local regulations
	Energy	Building envelope	RTTV and OTTV min 35 watt/sqm
		Ventilation system	Minimize cooling loads
		Daylighting	Make the most of it

In Bandung's Major Regulation No.1023 year of 2016 concerning Green Building, the building affected by the requirements of the regulations covers all new buildings and/or additions which are divided into 2 categories, namely buildings with an area of less than 5,000 sqm or a minimum of 5,000 sqm so that all buildings located in the administrative area of the city of Bandung have to comply with these regulations, including residential houses. This regulation also listed an additional criteria to accommodate the wishes of building owners who try to exceed the minimum criteria so that the building in question has the right to receive two and three star awards along with incentives in the form of adding floor layers or reducing land and building taxes. This study will only limit the analysis of the criteria for buildings with a minimum area of 5,000 sqm, namely: energy efficiency criteria including building envelope systems, air conditioning systems, lighting systems, in-building transportation systems, and electrical systems; water efficiency criteria include water source planning, water use planning, water saving sanitary equipment planning; and planning for handling wastewater; criteria for building management systems (SPB) include lighting systems, air conditioning systems, and in-building transportation systems; the criteria for managing air quality in mechanical ventilation spaces and systems include controlling carbon dioxide (CO_2) in certain rooms, controlling carbon monoxide (CO) in closed parking areas, and using air conditioning refrigerants; the criteria for land management include the provision of open green space, the provision of supporting facilities, and the management of solid waste and garbage

The contents of regulation related to the implementation of passive strategies for thermal and visual environments are contained in: energy criteria for building envelope systems with RTTV and OTTV limitations of at least 45 watts/sqm; energy criteria for air conditioning systems with a room temperature limit of 25 ° C ± 1 ° C; energy criteria for natural lighting systems with limits to optimizing their use; air quality criteria in the space for ventilation systems with limitations if the use of natural ventilation is not possible; land management criteria for green open space (RTH) and green building area (DHB) with the limitation of the provision of green open space refer to permits and DHB (green roof and vertical garden) not to exceed 25% of green open space. These five passive strategy option are summarized in Table 4 below:

Table 4. Identification of passive criteria and strategies in the BGH Head of Bandung city

Category Criteria	Passive Strategy	Parameter
Energy	Building Envelope System	RTTV and OTTV min 45 watt/sqm
	Air Conditioning System	Room temperature limit 25°C ± 1°C
	Lighting System	Optimize natural lighting
Indoor Air Quality	Ventilation	Mechanical ventilation is used if natural ventilation is not possible

Category Criteria	Passive Strategy	Parameter
Land Management	RTH and DHB	RTH according to rules DHB does not exceed 25% RTH

When we compared between the literature study and the three available green building regulations, not all passive strategy options have been explicitly stated in the green building regulations:

1. The strategy of orientation and shape of buildings is only mentioned in Ministry of Public Works and Public Housing Regulation with the aim of maximizing natural expediency and minimizing heat.
2. The building envelope strategy is mentioned explicitly in all three regulations including the expected achievement number parameters. However, the regulation does not specify how to do it to get the desired value.
3. The interior strategy is not mentioned in the Governor DKI Regulation of Green Building, but it is mentioned in the Ministry of Public Works and Public Housing Regulation, although it is not specifically mentioned what strategies can be done along with the parameters.
4. Greening strategies are mentioned in all three regulations and have included various variations not only on the building site but also on the roof (green roof) and on the wall (vertical garden) or balcony (terrace garden). The Governor DKI Regulation even encourages the use of pavement which allows the absorption of water into the soil.

   The five comparison points are summarized in Table 5 below:

Table 5. Passive strategy accommodation on green building regulations

Passive Strategy Options	Permen PUPR	Pergub DKI Jakarta	Perwal Kota Bandung
Building Orientation	Maximum lighting	-	-
Building Shape	Minimal heating	-	-
Building Envelope	RTTV and OTTV 35 watt/sqm gradually	OTTV 45 watt/sqm	RTTV and OTTV 45 watt/sqm
Building Interior	Listed but not specific	-	-
Greenery	According to the provisions	Greening and Pavement	RTH and DHB comply to the rules

4. Conclusion

Based on the results of passive strategy identification in Governor DKI Regulation of Green Building, Ministry of Public Works and Public Housing Regulation, and Bandung's Major Regulation, it can be concluded that:

- Passive strategies are still hidden behind OTTV parameters as part of the energy criteria for building envelope systems. The hidden position should be clarified because:
  a. The application of passive strategies is precisely the main thing that must be explicitly written to emphasize the importance of enforcement efforts. However, the achievement of the value of OTTV and or RTTV can still be included as a parameter of achieving requirements.
  b. If OTTV is used as a parameter, what happens then is an effort to meet the minimum requirements. Even though what is expected is an effort to optimize the application of passive strategies so that the results can be more than just meeting the minimum requirements.
- The passive strategy option has not been fully described as one of the main strategies in implementing the green building concept. Impressed that passive strategies are only attempts to get natural lighting and natural ventilation. This impression is obtained because explicitly written information and natural energy are part of the criteria for green buildings but other strategies are not conveyed, for example the strategy of building envelope material selection, interior spatial planning strategies, strategies to optimize building design through processing building orientation, shape and envelope. The lack of explicit passive strategies should be clarified because:

a. Giving the impression of ignoring various passive strategy options in addition to natural lighting and ventilation
b. Giving the impression of weak government efforts to encourage the optimization of passive strategies implementation even though it was less encouraging for designers (architects) and various parties involved to optimize its application.

- Passive strategies are more often placed on the criteria of energy and indoor air quality because of their relevance to natural lighting and ventilation even though the application of passive strategies is also part of the criteria for reducing environmental impacts, criteria for the construction phase and criteria for building utilization.

References

[1] Yeang, Ken (2005) Green Design in the Hot Humid Tropical Zone.
[2] Sharma, Anupama., Tiwari, K. K. Dhote R. (2002): Climatic Responsive Energy Efficient, Passive Techniques in Building, Eighteenth National Convention of Architectural Engineers, Jaipur.
[3] King, Steve (1991) Building for Conservation: Appropriate Design for Environmental Controls in the Tropics, Proceedings of Symposium Cultural Heritage in Asia dan the Pasific: Conservation dan Policy, Hawaii.
[4] Sarte, S Bry (2010) Sustainable Infrastructure: The Guide to Green Engineering Design, John Wiley & Sons Inc, Hoboken, New Jersey.
[5] Jones, D.L. (1998) Architecture and the Environment, Laurence King, London.
[6] LEARN, London Metropolotan University.
[7] Aun, Ar Chan Seong (2009) Applying MS1525:2007 Code of Practice on Energy Efficiency dan Use of Renewable Energy for Non-Residential Buildings, CPD Seminar 14th February 2009
[8] Hegger, Fuchs dan Zeimmer, Stark (2008) Energy Manual, Sustainable Architecture, Birkhauser, Basel.

Acknowledgement

We convey our gratitude and appreciation to the Directorate of Research and Community Service, Directorate General of Research Strengthening and Development, Ministry of Research, Technology and Higher Education of the Republic of Indonesia for the research grant given to us through the 2019 National Competitive Grant scheme with contract numbers: 2886/L4/PP/2019.

IOP Conf. Series: Materials Science and Engineering **669** (2019) 012048 doi:10.1088/1757-899X/669/1/012048

The dominant factor causing potential evapotranspiration in Malang raya using geographic information systems

T Rahayuningsih [1], G Idfi [1], A Anindyasari [1] and Sukriyadi [2]

[1] Lecturer in Civil Engineering Universitas Negeri Malang
[2] Civil Engineering Students Universitas Negeri Malang

titi.rahayuningsih.ft@um.ac.id

Abstract. The water resources management in Malang Raya is very important, because this area still has a lot of agricultural land. Agricultural land needs evapotranspiration maps. By now, there has been no research on the dominant determinant of the value of potential evapotranspiration in the Malang Raya area. Therefore research on the main factors that determine the value of potential evapotranspiration is needed. Purpose research: 1). Discovering the irrigation area that has the lowest and highest potential evapotranspiration value in Malang Raya; 2). Determine the dominant factors that affect the potential evapotranspiration value of irrigation areas in Malang Raya. The method to predict and calculate the amount of Potential Evapotranspiration (ETo) is Penman Method. The main determinants of potential evapotranspiration are mapped using Geographic Information Systems (GIS). Determination of the dominant factors that influence potential evapotranspiration was tested using SPSS. Conclusion: 1). The lowest potential evapotranspiration value were in Blimbing and Kedungkandang Districts from 3.46 to 3.61 mm / day. While, the biggest values were in Batu and Junrejo Districts with values of 6.32 to 6.33 mm / day; 2). The dominant factor affecting the potential evapotranspiration value is the air humidity (RH) factor of 46.4%.

1. Introduction

The rain water that falls to the surface of the earth does not entirely fall to the ground, some of it fall into the plants. Water that falls on plants, partially will be held up by plants where some of it will evaporate and some will fall or flow through branches to the ground

In managing evapotranspiration water resources, evapotranspiration is a very important process because it requires total water loss data caused by total evaporation. The value of evapotranspiration is affected by various factors found in nature. The factors that affect evaporation and evapotranspiration are water temperature, air temperature (atmosphere), humidity, wind speed, air pressure, solar radiation and others which are interconnected one another [1].

There are various potential evapotranspiration measurement methods, one of them is the Penman Modification method. The needed data to determine potential evapotranspiration is by measuring temperature, wind speed, average relative humidity and the relative duration of solar radiation.

Malang Raya consists of Malang Municipality, Batu Municipality, and Malang Regency. The Malang Raya region spreads from the coast to the mountains. Its height starts from 0 meters to 2000 meters above sea level. Its topographical condition causes the Malang Raya is a potential agricultural area to plant, rice, fruit, flowers, and vegetable. The water resources management in Malang Raya is very important, because this area still has a lot of agricultural land. Agricultural land needs evapotranspiration maps. Evapotranspiration becomes one of the parameters needed in calculating irrigation water requirements for plants. The various determinants of evapotranspiration, there is one dominant factor that affects the value of evapotranspiration [2].

By now, there has been no research on the dominant determinant of the value of potential evapotranspiration in the Malang Raya area. Therefore research on the main factors that determine the value of potential evapotranspiration is needed. The main determinants of potential evapotranspiration are mapped using Geographic Information Systems (GIS) [3-4].

2. Purpose of research

- Discovering the irrigation area that has the lowest and highest potential evapotranspiration value in Malang Raya
- Determine the dominant factors that affect the potential evapotranspiration value of irrigation areas in Malang Raya

3. Methods

The data which used were taken from meteorological, climatology and geophysics stations in Karangploso, Karangkates and AR. Saleh Military Airport. The method to predict and calculate the amount of Potential Evapotranspiration (ETo) is Penman Method. The Penman method can be used if climatology data is available. The potential evapotranspiration based on the Penman formula that has been modified for calculation in Indonesia are as follows [5]:

$$ETo = C\,(W.Rn + (1 - W)(ea - ed).f(U)) \quad (1)$$

As:

ETo = potential evaporation (mm/day)

C = An adjusted factor of day and night conditions (correction number).

W = The Factors that depend on average temperature (temperature) and altitude (elevation).

Rn = The total of net radiation (mm / day)

$= 0{,}75 \,.\, Rs - Rn1$ .... (2)

Rs = The amount of shortwave radiation reaching the earth's surface

$= (0.25 + 0.5\, n/N)\, Ra$ (mm/hari).. (3)

Ra = Shortwave radiation that fills the outer of the Earth's atmosphere (Angot value), in (mm/day).

n = The average of actual length of the sun (mm/day)

N = The maximum possible amount of sunlight (mm/day)

Rn1 = Net long wave radiation (mm/day)

$= f(t) \,.\, f(ed) \,.\, f(n/N)$.. (4)

f(t) = temperature function of the table of relation between temperature (t) with a value of f (t)..

f(ed) = the function of steam pressure.

$= 0.34 - 0.044 \times ed^{0.5}$.. (5)

f(n/N) = the function of sun brightness

= (0,1 + 0,9 x n/N)... (6)

f(U) = The function of average wind speed during the day time at an altitude of 2 meters (m/second)

= 0,27 . (1 + U/100)... (7)

U = The average of wind speed (m/second)

ea–ed = The saturated vapor pressure with actual vapor pressure at average of air temperature (mbar)

ed = ea × RH... (8)

ea = The actual vapor pressure.

RH = The relative humidity (%)

Rns = The shortwave radiation that goes reflected

= (1-α) x Rs... (9)

After obtaining data calculated using formulas (2) through (9), the data is entered into a table in the Geographical Information System (GIS). GIS is a system designed to capture, store, manipulate, analyze, organize and display all types of geographic data. (Irwansyah, E, 2013). The process of calculating the potential evapotranspiration value of formula (1) is conducted in the GIS table. The next is making a map of potential evapotranspiration patterns using GIS. The data interpolation method from the three climatology and geophysics stations using the Inverse Distance Weighted method (IDW).

The potential evapotranspiration of temperature (T), humidity (RH), solar radiation (n/N), and wind speed (u) Determination of the dominant factors that influence potential evapotranspiration was tested using SPSS [6].

4. Similar research

Similar studies are in Table 1

Table 1. Similar research

No.	Name	Year	Title
1	Herianto, et al [7]	2016	Evapotranspiration using two regions in West Java as references to analyze of irrigation water needs planning.
2	Supangat, A.B. [8]	2016	Analysis of the changes of estimated value of potential evapotranspiration affected by climate change in the Eucalyptus Pellita plantation forest area.
3	Singal, R.Z. [9]	2017	Study on the Development of Potential Evapotranspiration Maps with Geographic Information Systems for the East Java Region

5. Research result

In this research was used data from 3 Climatology and Geophysics Stations, namely the Karangploso Climatology Station, Karangkates Geophysical Station, and the AR Saleh Military Airport

Climatology Post. The coordinate data consists of latitude, longitude, and altitude above sea level are shown in Table 2.

Table 2. Position of climatology and geophysics stations

No	Name of Station	Coordinate		Altitude
		latitude	longitude	(m)
1	Karangploso Climatology Station	-7.90080	112.59790	590
2	Karangkates Geophysical Station	-8.15000	112.45000	285
3	AR. Saleh Military Airport Climatology Post	-7.92639	112.71444	526

The data used as a parameter in the calculation of potential evapotranspiration was data with 11 years span starts from 2007 to 2017. The data consists of air temperature, humidity, solar radiation, and wind speed of Karang ploso Station are shown in Table 3. The Table 4 and 5 contain about data from the Karangkates climatology station, and AR Saleh Military Airport climatology post.

Table 3. Climatology data and potential evapotranspiration at Karangploso Station

Month	Air Temperature	Humidity	Solar radiation	Wind speed	Potential Evapotranspiration
January	23.72	80.95	41.84	6.01	4.79
February	23.75	80.52	44.20	5.12	4.83
March	23.78	80.51	46.94	4.95	4.33
April	23.93	79.50	58.56	6.32	4.12
May	23.84	75.87	72.22	6.10	4.16
June	23.15	74.82	70.43	4.93	3.77
July	22.27	74.00	71.96	6.08	3.94
August	21.96	72.30	76.45	6.70	4.91
September	23.16	69.13	79.71	7.81	6.42
October	24.46	71.02	71.83	6.34	6.35
November	24.40	75.97	60.74	6.32	5.79
December	24.05	80.35	41.60	4.24	4.63

Table 4. Climatology data and potential evapotranspiration at Karangkates Station

Month	Air Temperature	Humidity	Solar Radiation	Wind Speed	Potential Evapotranspiration
January	25.67	83.68	50.19	5.14	5.05
February	25.39	80.30	50.26	4.51	5.17
March	25.75	83.80	55.26	3.51	4.55
April	25.61	81.90	58.97	3.23	4.01
May	25.75	81.70	66.39	2.91	3.80
June	24.89	79.70	74.00	3.24	3.72
July	24.31	79.30	75.63	4.29	3.86

Month	Air Temperature	Humidity	Solar Radiation	Wind Speed	Potential Evapotranspiration
August	24.37	77.90	80.43	4.23	4.86
September	24.81	75.29	74.13	4.20	5.76
October	26.37	76.30	73.56	3.99	6.20
November	25.95	80.40	54.11	3.60	5.25
December	25.90	84.53	49.89	3.65	4.90

Table 5. Climatology data and potential evapotranspiration at AR Saleh Climatology Post.

Month	Air Temperature	Humidity	Solar Radiation	Wind Speed	Potential Evapotranspiration
January	24.37	85.40	27.00	8.48	4.31
February	23.89	85.40	31.00	7.40	4.34
March	23.99	85.90	52.00	7.03	4.44
April	24.17	85.90	51.00	6.46	3.70
May	24.17	84.30	85.00	5.91	4.16
June	23.41	83.30	64.00	11.33	3.88
July	22.62	82.30	46.00	5.44	3.12
August	22.52	80.10	66.00	5.84	4.32
September	23.48	76.44	71.00	6.62	5.67
October	24.56	77.33	71.00	7.12	6.15
November	24.60	84.44	49.00	6.36	4.98
December	23.91	87.00	31.00	66.87	4.16

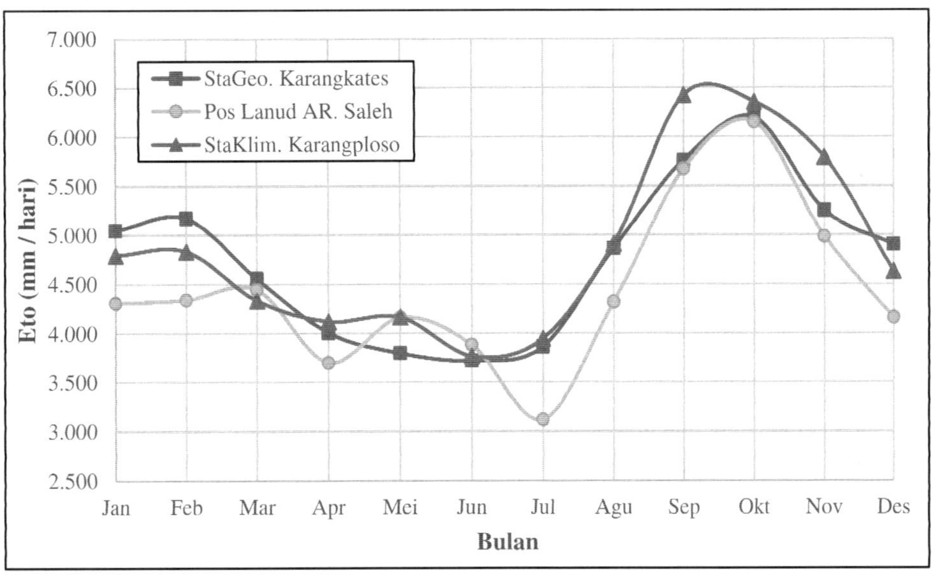

Figure 1. The potential evapotranspiration chart of Karang Ploso, Karangkates, and AR Saleh Climatology Station areas.

Based on the potential evapotranspiration data in Tables 3, 4, and 5, a potential evapotranspiration chart can be made from the three climatology stations, namely Karangploso Station, Karangkates Climatology Station, and AR Saleh Climatology Post. The potential evapotranspiration chart can be made as shown in Figure 1

The results of the calculation of the SE values to the Temperature (T), humidity (RH), solar radiation (n/N), and wind speed (u) using SPSS are shown in Table 6

Table 6. The results of calculation using SPSS

Variable	Regression Coefficient (Beta)	Correlation Coefficient	R Square	Se
Temperature	0.514	0.261	52.2%	13.4%
Humidity	-0.888	-0.522		46.4%
Solar radiation length	-0.270	0.238		-6.4%
Wind speed	0.125	-0.094		-1.2%

6. Discussion

From Table 3, the highest potential evapotranspiration value was in September with an Eto value of 6.42 mm / day and the lowest was in June with an Eto value of 3.77 mm / day. In September, the position of the sun was in the equator and not too far from the research location so that the research area had a relatively large solar irradiation time of 79.71%. The lowest potential evapotranspiration value was in June, when the position of the sun was around the northern turning line and that was the farthest position from the research area and was in the dry season. This condition affected the air temperatures turned relatively low, and humidity was also relatively small. The highest potential evapotranspiration value was in October with an Eto value of 6.15 mm / day and the lowest was in July with an Eto value of 3.12 mm / day.

Figure 1 shows that the three climatology stations have the same potential evapotranspiration pattern, which was the lowest in July and was in the dry season. The highest potential evapotranspiration was in September for Karangploso climatology station. Karangploso Station has an altitude of 529 m above sea level, so the area has entered the rainy season earlier compared to Karangkates and AR Saleh stations.

7. The value of potential evapotranspiration of irrigation areas

Malang Raya Region has several irrigation areas, such as Donomulyo District, Turen District, Gondanglegi District, Pagelaran District, and Kepanjen District. The irrigation in Malang City covers Sukun District, Blimbing District, Lowokwaru District, and Kedungkandang District. Batu City covers Batu District, Junrejo District and Bumiaji District. The following are the coordinates and area of irrigation land in several regions in Malang Raya. Data of Malang Raya irrigation area is shown in Table 7.

Table 7. Location and land area of Malang Raya irrigation

No	District	Coordiante		Irrigation
		Latitude	Longitude	*(Hectare)*
1	Donomulyo	-8.321968	112.405305	2017
2	Turen	-8.176870	112.697335	2248
3	Gondanglegi	-8.155192	112.631784	2969
4	Pagelaran	-8.197451	112.607946	2650

IOP Conf. Series: Materials Science and Engineering **669** (2019) 012048 doi:10.1088/1757-899X/669/1/012048

No	District	Coordiante		Irrigation
		Latitude	Longitude	(*Hectare*)
5	Kepanjen	-8.159373	112.560269	2152
6	Batu	-7.883065	112.533449	650.78
7	Junrejo	-7.910549	112.542389	1062
8	Bumiaji	-7.794130	112.536429	687.2
9	Sukun	-7.993945	112.613905	251
10	Blimbing	-7.953594	112.649578	87
11	Lowokwaru	-7.937013	112.613906	241
12	Kedungkandang	-8.011014	112.655622	591

Source : **Google Earth*

** Central Bureau of Statistics of Malang Regency 2017, CBS of Batu City 2017 dan KMDA (a set of statistic data named Kabupaten Malang Dalam Angka/ Malang Regency in Numbers) of Malang City 2016

The map which shows the lowest potential evapotranspiration conditions for irrigation areas is Figure 2. Figure 3 shows the highest potential evapotranspiration map (September) for each irrigation area.

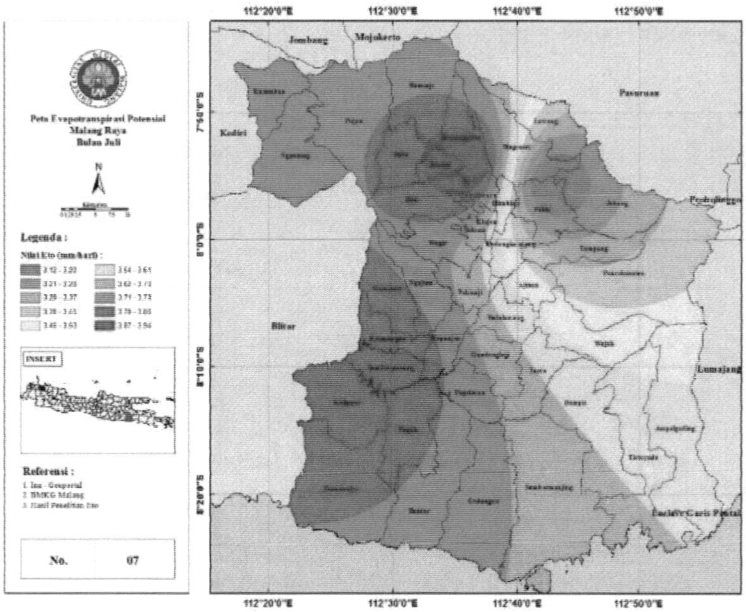

Figure 2. The Map of potential evapotranspiration in July

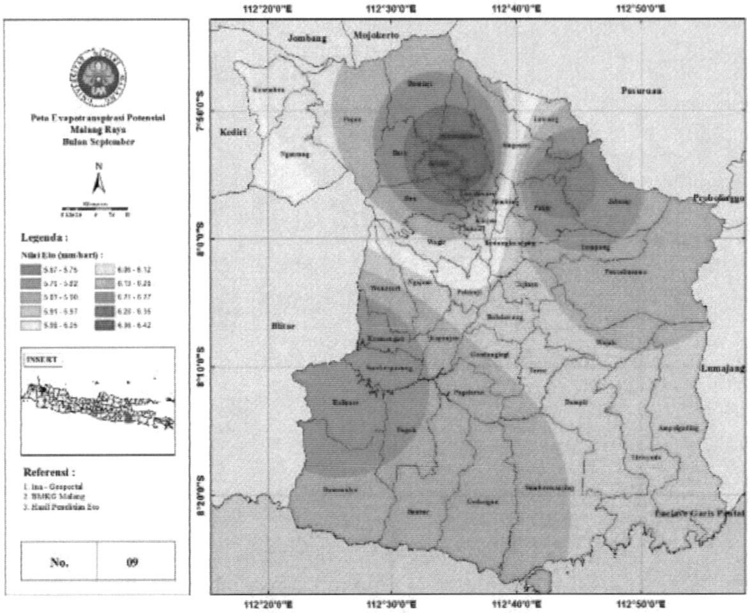

Figure 3. The map of potential evapotranspiration in September

The value of potential evapotranspiration of the irrigation area for January to June is shown in Table 8, while for July to December is shown in Table 9.

Table 8. The value of potential evapotranspiration of irrigation areas for January to June

No	District	The Value of Potential Evapotranspiration (Eto) mm / day					
		Jan	Feb	Mar	Apr	May	Jun
1	Donomulyo	4.91 - 4.97	5.01 - 5.09	4.52 - 4.53	3.96 - 3.99	3.88 - 3.91	3.74 - 3.75
2	Turen	4.69 - 4.75	4.76 - 4.84	4.45 - 4.47	3.92 - 3.95	4.03 - 4.05	3.79 - 3.80
3	Gondanglegi	4.76 - 4.82	4.85 - 4.92	4.48 - 4.49	3.92 - 3.99	3.95 - 4.02	3.77 - 3.78
4	Pagelaran	4.83 - 4.90	4.93 - 5.00	4.50 - 4.51	3.96 - 3.99	3.92 - 3.94	3.76 - 3.77
5	Kepanjen	4.91 - 4.97	5.01 - 5.09	4.50 - 4.53	3.96 - 3.99	3.88 - 3.91	3.74 - 3.75
6	Batu	4.69 - 4.82	4.76 - 4.84	4.33 - 4.38	4.05 - 4.12	4.14 - 4.16	3.78 - 3.79
7	Junrejo	4.76 - 4.82	4.76 - 4.84	4.33 - 4.38	4.05 - 4.12	4.14 - 4.16	3.78 - 3.79
8	Bumiaji	4.69 - 4.75	4.68 - 4.84	4.36 - 4.40	4.00 - 4.04	4.14 - 4.16	3.79 - 3.80
9	Sukun	4.61 - 4.68	4.68 - 4.75	4.36 - 4.40	3.92 - 3.99	4.10 - 4.16	3.79 - 3.82
10	Blimbing	4.47 - 4.60	4.51 - 4.67	4.39 - 4.42	3.84 - 3.95	4.14 - 4.16	3.81 - 3.83
11	Lowokwaru	4.69 - 4.82	4.68 - 4.84	4.33 - 4.38	4.00 - 4.12	4.14 - 4.16	3.78 - 3.80
12	Kedungkandang	4.61 - 4.68	4.51 - 4.67	4.41 - 4.42	3.84 - 3.91	4.10 - 4.16	3.83 - 3.84

Table 9. The value of potential evapotranspiration of irrigation areas for July to December

No	District	The Value of Potential Evapotranspiration (Eto) mm / day					
		Jul	Aug	Sep	Oct	Nov	Dec
1	Donomulyo	3.79 - 3.86	4.80 - 4.85	5.76 - 5.82	6.20 - 6.23	5.24 - 5.31	4.76 - 4.83
2	Turen	3.54 - 3.70	4.62 - 4.73	5.91 - 5.97	6.22 - 6.23	5.24 - 5.31	4.54 - 4.60

No	District	The Value of Potential Evapotranspiration (Eto) mm / day					
		Jul	Aug	Sep	Oct	Nov	Dec
3	Gondanglegi	3.62 - 3.70	4.68 - 4.79	5.83 - 5.97	6.22 - 6.23	5.32 - 5.39	4.61 - 4.68
4	Pagelaran	3.71 - 3.78	4.74 - 4.79	5.76 - 5.82	6.22 - 6.23	5.24 - 5.31	4.69 - 4.75
5	Kepanjen	3.79 - 3.86	4.80 - 4.85	5.76 - 5.82	6.20 - 6.23	5.24 - 5.31	4.76 - 4.83
6	Batu	3.79 - 3.94	4.80 - 4.91	6.28 - 6.35	6.32 - 6.33	5.64 - 5.71.	4.61 - 4.68
7	Junrejo	3.79 - 3.94	4.80 - 4.91	6.28 - 6.35	6.32 - 6.33	5.64 - 5.79	4.61 - 4.68
8	Bumiaji	3.71 - 3.78	4.74 - 4.79	6.13 - 6.27	6.30 - 6.31	5.56 - 5.63	4.61 - 4.68
9	Sukun	3.61 - 3.70	4.74 - 4.79	6.06 - 6.20	6.26 - 6.29	5.40 - 5.55	4.54 - 4.60
10	Blimbing	3.46 - 3.61	4.60 - 4.67	5.91 - 6.12	6.22 - 6.27	5.24 - 5.47	4.32 - 4.46
11	Lowokwaru	3.79 - 3.94	4.74 - 4.91	6.21 - 6.42	6.30 - 6.35	5.56 - 5.79	4.61 - 4.68
12	Kedungkandang	3.46 - 3.61	4.56 - 4.61	5.91 - 6.05	6.22 - 6.25	5.24 - 5.39	4.32 - 4.46

The highest potential evapotranspiration value for Batu District was in October of 6.32 - 6.33 mm / day and the lowest was in June was 3.78 - 3.79 mm / day. The highest potential evapotranspiration value for Kepanjen district was in October at 6.20 - 6.23 mm / day and the lowest was in June was 3.74 - 3.75 mm / day. The highest potential evapotranspiration value for Lowokwaru district was in October at 6.30 - 6.35 mm / day and the lowest was in June was 3.78 - 3.80 mm / day. The lowest potential evapotranspiration value were in Blimbing and Kedungkandang districts at 3.46 to 3.61 mm / day. While, the biggest value were in Batu and Junrejo districts at 6.32 to 6.33 mm / day.

The results of the dominant factor test on potential evapotranspiration from elements of Temperature (T), humidity (RH), solar radiation (n / N), and Wind Speed (u) obtained data

From Table 6, it is known that the largest effective contribution to the Eto Value is given by the Air Humidity Parameter (RH) which is 46.4% or 0.464. Beside that, there are two parameters that have a negative value, namely Solar Radiation (-6.4%) and Wind Speed (-1.2%), but for the used effective contribution is the absolute value.

8. Conclusion

- The lowest potential evapotranspiration value were in Blimbing and Kedungkandang Districts from 3.46 to 3.61 mm / day. While, the biggest values were in Batu and Junrejo Districts with values of 6.32 to 6.33 mm / day.
- The dominant factor affecting the potential evapotranspiration value is the air humidity (RH) factor of 46.4%

9. Suggestion

For research on potential evapotranspiration values in other areas, it should be conducted using a Geographic Information System so that the results will be more precise and easier to interpret because the research location has spatial data (coordinate)

10. References

[1] Sosrodarsono, S., & Takeda, K. 2003. *Hidrologi Untuk Pengairan*. Jakarta. Penerbit PT Pradnya Paramita.
[2] Kabupaten Malang Dalam Angka (KMDA). 2008-2018. "*Malang Regency In Figures*". Badan Pusat Statistik, Kabupaten Malang.
[3] Handareni, W., 2015. *Interpolasi Data Metode IDW dan Kriging*. Bogor. Departemen Geofisika dan Meteorologi IPB.
[4] Irwansyah, E. 2013. Sistem Informasi Geografis: Prinsip Dasar dan Pengembangan Aplikasi. Yogyakarta. Penerbit Digibooks.

IOP Conf. Series: Materials Science and Engineering **669** (2019) 012048 doi:10.1088/1757-899X/669/1/012048

[5] Ariyani, D. 2015. *Hidrologi*. Jakarta. Universitas Pancasila
[6] Nuryanto, D.E., & Rizal, J. 2013. Perbandingan Evapotranspirasi Potensial Antara Hasil Keluaran Model ReGCM 4.0 dengan Perhitungan Data Pengamatan. Jakarta Pusat. Puslitbang BMKG
[7] Herianto., H.A.K., & Romdani, A. 2016. *Evapotranspirasi Referensi Dua Daerah Di Jawa Barat Untuk Analisis Perencanaan Kebutuhan Air Irigasi*. Tasikmalaya. Jurusan Teknik Sipil, Fakultas Teknik, Universitas Siliwingi
[8] Supangat, A.B. 2016. Analisis Perubahan Nilai Pendugaan Evapotranspirasi Potensial Akibat Perubahan Iklim di Kawasan Hutan Tanaman EUCALYPTUS PELLITA. Balai Penelitian dan Pengembangan teknologi Pengelolaan DAS. Universitas Muhammadiyah Surakarta.
[9] Singal, R. Z. 2017. Studi Pengembangan Peta Evapotranspirasi Potensial Dengan Sistem Informasi Geografis (SIG) untuk Wilayah Jawa Timur. Surabaya. Institut Teknologi Sepuluh Nopember.

IOP Conf. Series: Materials Science and Engineering **669** (2019) 012049 doi:10.1088/1757-899X/669/1/012049

Tensile test analysis of cold-formed column in channel profile

A A Aziz[1], N Nindyawati[1] and K Karyadi[1]

[1]Department of Civil Engineering, State University of Malang, Indonesia

nindyawati.ft@um.ac.id

Abstract. This study analyzes the joint capacity of cold-rolled steel frame structure for multi-level buildings to withstand axial loads. The column frame joint structures will be modeled, analyzed and tested. In this paper, the study is about cold-formed channel profile that will be tensile tested with 2 variations of the joint. The research objectives are (1) to determine the stress and strain, (2) the damage pattern. The test object is modeled using FEM with a scale of 1: 1 and the loading used is linear static loading. Simulation with FEM use a trial and error method by entering the parameters obtained. The draft and design of channel profiles using cold rolled steel material refers to several guidelines including SNI-7971-2013 on Cold Formed Steel Structures.

1. Introduction

Cold Formed steel is a structural component made of steel sheets that is processed with certain profile shapes using press-braking or roll forming processes. Cold-formed steel or also called as cold-formed because it does not require high temperatures in the manufacturing process as it does in making hot rolled steel. In general, cold-formed steel is a component that is lightweight, thin, easy to work on compared to hot-rolled steel [1].

The advantages of cold-formed steel are found in the production process which is easier and environmentally friendly because it does not produce waste that is harmful to the environment. In the production process, steel is formed in such a way at room temperature by using bending brakes, press brakes, and roll-forming machines. With cold-formed method, the desired profile cross section can be produced economically so that an increase in the ratio of strength to weight can be obtained [2]. According to cold-rolled steel has excellency in strength and stiffness that is quite high in value, does not shrink and expand easily when affected by weather, easy in fabrication and production, uniform in quality, non-flammable, more accurate in detailing, anti-termite and environmental friendly material, could be recycled, and have a high structural lifetime [3].

Tensile testing is used to determine the thermal and mechanical properties of an object that is deforming as material. Tensile testing is testing an object so that its strength can be known by pulling on both sides of the object tip slowly until it breaks. The material strength is when the material receives a stress/tension but does not experience a friction, so that the tensile strength and the compressive strength of the tested specimen can be determined. The test process is that the object is given a load so that it experiences stress strain and the extension of the material increases [4].

In tensile test, the results of the steel stress are given a symbol (σ) and strain with (ε) with the following formula: Stress/tension is also called the ratio of tensile loads acting against the cross-sectional area.

IOP Conf. Series: Materials Science and Engineering **669** (2019) 012049 doi:10.1088/1757-899X/669/1/012049

Tensile stress formula:

$$\sigma = \frac{F}{A}$$

σ = Tensile stress (MPa)
F = Pull Load (kN)
A = Initial Cross-sectional Area (mm)

Strain is a change in the shape of a material after it is broken so that the material rod occurs an extension. To get the strain value, the final length minus the initial length of the test object then it is divided by the initial length.

$$\varepsilon = \frac{lf - lo}{lo} \text{ x } 100$$

ε = Strain (%)
Lo = Initial length (mm)
Lf = Final Length (mm).

2. Methodology

This study analyzes the tensile strength of the joints due to the effect of the joint shape on the column frame. The type of joint is varied into 2 parts, face-to-face and opposite joint. The study was conducted by modeling and analyzing joint using Finite Element Method (FEM)-based software. The purpose of this study is to obtain the tensile strength that occurs in each type of joint, determining the pattern damage of joint, determining the appropriate type of joint for the column frame. The test object is modeled with a scale of 1: 1 and the loading used is linear static loading. Simulation with FEM software uses a trial and error method by entering the parameters obtained. The draft and design of channel profiles using cold-formed steel material refers to several guidelines including SNI-7971-2013 on Cold-Formed Steel Structures.

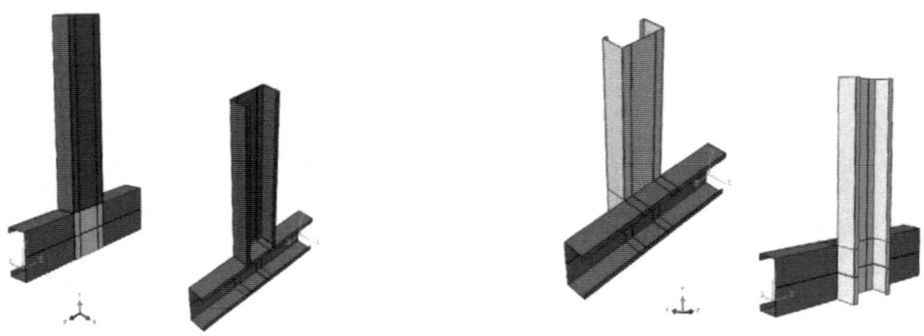

(a) (b)

Figure 1. (a) Face-to-face (b) Opposite

The modeling of test object in FEM software is adjusted to the profiles of cold-rolled steel used in experimental tests. Joint variations are varied into face-to-face (Type A) and opposite (Type B). The mechanical properties of cold-formed in modulus of elasticity are 200000 MPa the, the cold-formed steel melt stress value (Fy) extends from 165 MPa to 552 MPa [2], in this modeling the melting tension/stress is set at 550 MPa.

The restrain used are fixed support on the lower section of the joint and is loaded evenly on the top end of the connection with a total load of 100 kg as shown below.

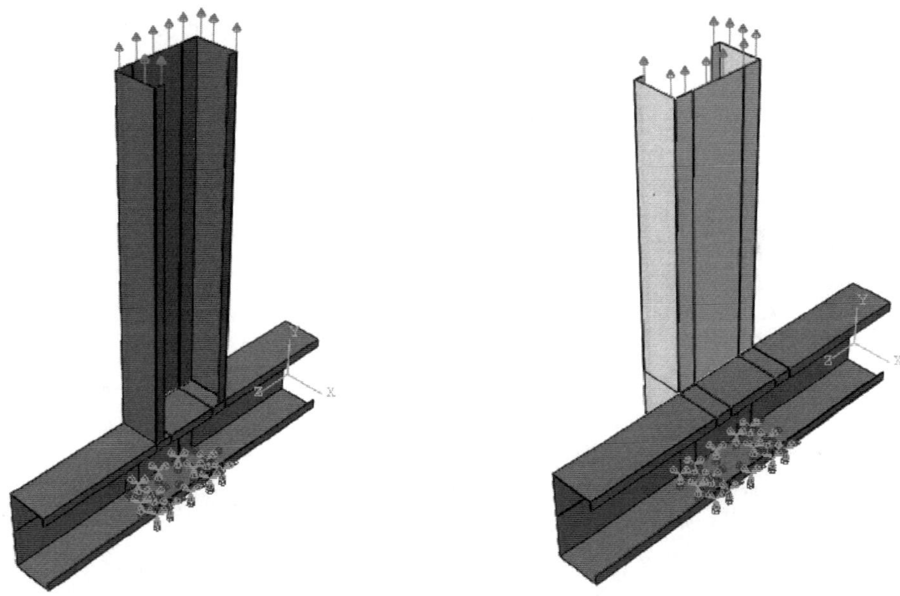

Figure 2. Restrain and loading

The following is a specification of the cold-formed C channel profile that will be used as a test object, as follows [5]:

Name : Truss C75.75
Thickness : 0,75 mm
Joint tool : Self driving screw type 12-14 x 20
Steel Grade : G 550 (fy = 0,9 x 550 MPa).

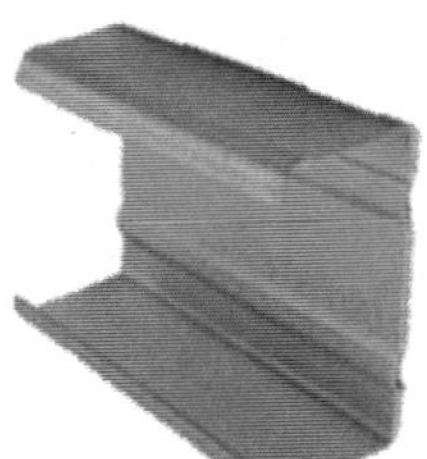

Figure3. Cold-formed

Figure 4. Self-driving screw

The joint tool used are self driving screw bolt with a type 12-14 x 20. Here are the specifications:

Table 1. Specification of self-driving screw

Nominal Size or Basic Screw Diameter	Threads Per Inch	T Major Diameter		t Minor Diameter		P Protrusion Allowance	
		Max.	Min.	Max.	Min.	#2 Pt.	#3 Pt.
12 0.2160	14	0.215	0.209	0.164	0.157	-	0.353

Data Processing

Data that is obtained includes:

- Strain &Stress, to find out the capacity of the C channel profile according to the modeling.
- Damage pattern on channel profile C.

Structural Modeling

After obtaining the data, the next step is making structural model. Structure modeling are done by using FEM. being modeled according to building structure data, material specifications, and structural element data used. Structure modeling are made into 2 type of model that is joint type A and type B. Structure modeling figure are as follow:

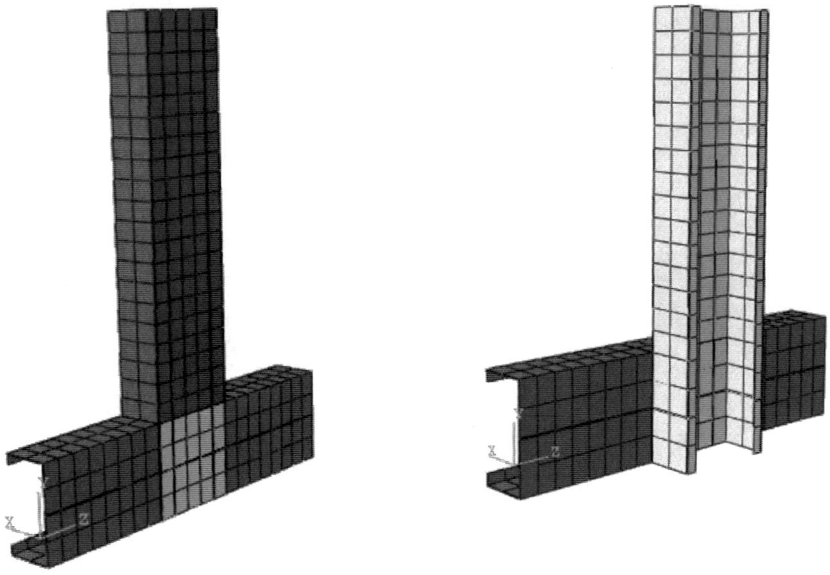

Figure 5. Structure modeling using FEM

3. Results and discussions

Modeling are done by simplifying the model using a vertical and transverse channel profiles then analyzed by using FEM

3.1. *The results of strain-stress that occur in the screw joint area can be seen in the following tables and graphs*:

Table 2. Stress and strain result type-A joint

Stress (MPa)	Strain
0	0
316,86	15,66
316,86	16,61
313,53	31,31
162,24	32,33
208,67	33,74
135,14	34,68
77,10	48,43
77,24	49,33
50,34	50,09
50,37	50,96

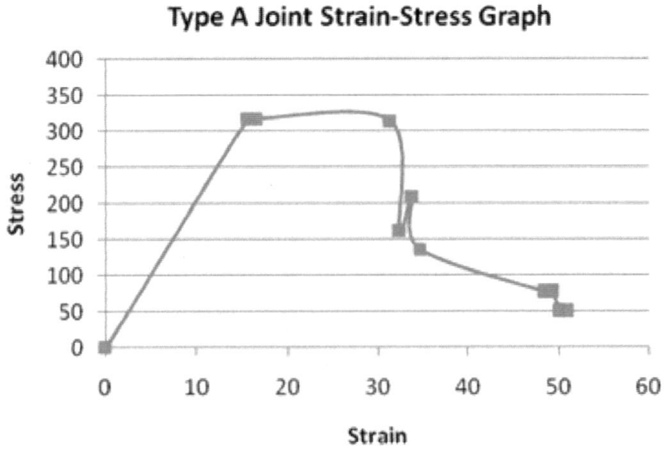

Figure 6. Type-A joint strain-stress graph

From the graph above it can be explained that the melting stress at the joint profile of type A channel is 316.86 MPa. The stress that occurs in type A joint move linearly toward its melting limit then slowly stretches to 313.53 MPa which then decreases where the joint has begun to experience damage and losing its strength. The strain on the test object are obtained from the result of the increase in the final length of 50.96 mm from the initial length of 300 mm, so that the strain obtained is 16.98%.

Table 3. Stress and strain result type-B joint

Stress (MPa)	Strain (mm)
-1,02	0,00
0,01	25,39
3,24	48,99
25,77	74,33
-0,40	102,88
0,31	128,01
1,22	129,08
-3,30	142,85
17,84	144,16
84,31	145,70
84,31	169,44

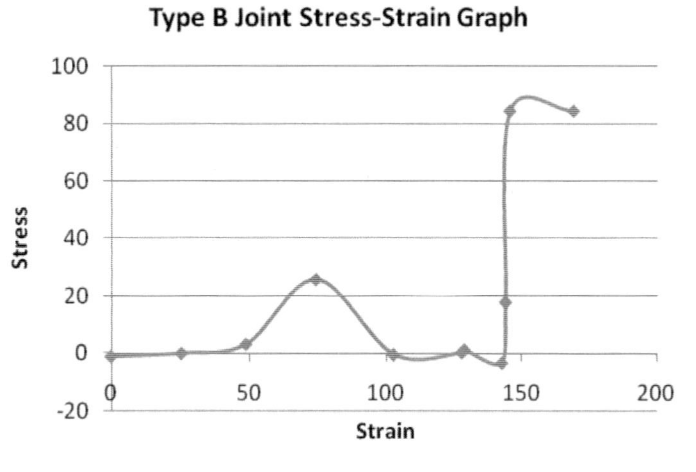

Figure 7. Type-B joint strain-stress graph

The type B joint graph shows the up and down movement in stress value with the melting stress reaching 25.77 MPa which then slowly decreases and returns to reach the peak at 84.31 MPa and finally the joint loses its strength. Strain on the test object are obtained from the comparison of the final length increase of 169.44 mm with an initial length of 300 mm so that the strain obtained is 56.48%.

3.2. Visual analysis of the damage patterns at the joint results from modeling on FEM

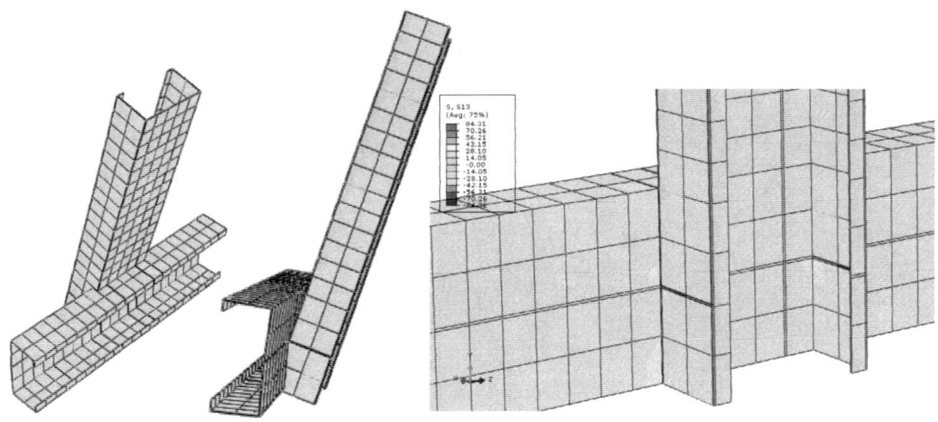

Figure8. Type-A joint damage pattern

The damage pattern of the test object that occurred after being analyzed by FEM resulted in damage form of local buckling. In type A joints, the screw section experiences a high stress accompanied by damage to the surrounding cold-formed support. Damage also occurs in the lower base of the upright profile which is stretched and bent so that the profile has a fracture potential.

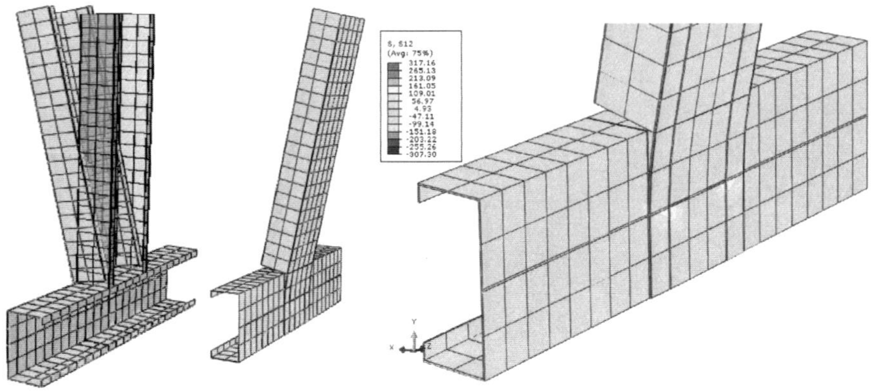

Figure9. Type-B joint damage pattern

Whereas in type B joints the damage pattern that occurs is a local buckling in the cross section profile due to tensile loads. The pattern of buckling opens out with a pressure that occurs at the base of the upright profile. In the joint area there is a large buckling in the wing section of the channel section c. This section is a critical part that is prone to break. The weakness of cold-rolled steel structures is local buckling due to its thin cross-section elements. This causes structural failure to occur before reaching the highest load capacity [6].

4. Conclusion

From the results of this study it can be concluded that each joint has their own advantages. Type A joint has a melting stress of 316.86 MPa with a strain of 16.98% and a type B joint produces a stress of 25.77 MPa with a strain of 56.48%. Meanwhile, when analyzed from the damage pattern, both type local buckling occurs in the screw junction area, so that the tensile rod capacity is determined by bending local elements where the type B joints are more resistant to overcome the tensile load on the cold-formed column frame structure.

References

[1] Mutawalli, M., 2007. Stability of cold formed steel frame connection type t tocyclicloading on the knockdown of a simple house. Postgraduate Thesis, Universitas GadjahMada, Yogyakarta, Indonesia

[2] Yu Wei-Wen, Ph.D., P.E. 2000.Cold Formed Steel Design Third Edition. Canada

[3] Meiyalagan, M. 2010.Investigation on cold – formed csection long column with intermediate stiffner & corner lips-under axial compression. Internatonal journal of apllied engineering research, Dindigul. Vol.1 No.1.

[4] Murtiono, A. 2012. Effects of quenching and tempering on hardness and tensile strength and micro structure of medium steels for palm harvesting blades. *Jurnal e-Dinamis,* September, Volume II, No.2.

[5] Standar Nasional Indonesia. 7971:2013. *Struktur Baja Canai Dingin.* Jakarta : Badan Standarisasi Nasional

[6] Sanjaya.2018.Experimental study of cold rolled cold pressed steel rods partially. *Jurnal Teoretis dan Terapan Bidang Rekayasa Sipil.* Vol. 25 No. 1.

Acknowledgement

This work was supported by PNBP research grant of Universitas Negeri Malang.

IOP Conf. Series: Materials Science and Engineering **669** (2019) 012050 doi:10.1088/1757-899X/669/1/012050

Achieving open defecation free in Surabaya city by 2019

E Soedjono[1], N Fitriani[2], F R E Santoso[3], R Destio[1], I Fahmi[1], A Gemardi[1] and D A Ningsih[1]

[1]Environmental Engineering Department, Institut Teknologi Sepuluh Nopember, Surabaya, Indonesia

[2]Study Program of Environmental Engineering, Department of Biology, Universitas Airlangga, Indonesia

[3]Civil Engineering, Narotama University

nurina.fitriani@fst.unair.ac.id

Abstract. Achieving open defecation free (ODF) in Indonesia by 2019 is closely monitored by the Government of Indonesia (GoI) as it is targeted in the RPJMN (national mid-year term). This achievement is also targeted at Surabaya City as the second largest city in Indonesia. There are still 15.527 houses in Surabaya which do not have proper sanitation system or still practicing open defecation. This study is aimed to assess how ODF would be achieved in Surabaya by 2019; including what type of toilets are to be used. A household survey was conducted to locate and assess the condition of those houses. Survey results show that 12.772 houses still practice defecation in the open area. These houses comprise of legal and illegal houses which are 5.586 and 7.186, respectively. Priority is given to legal houses as they are manageable to be given toilet under Surabaya's regulation procedure. Based on the available technologies to construct sanitation system then there will be 2.833 individual septic tank, 419 communal septic tanks, 15 communal toilet with bathing and washing (MCK), 12 communal wastewater treatment, and 1 centralized system in Keputih Sub-district of Surabaya to be constructed to set free these legal houses from open defecation practices.

1. Introduction

Open defecation free (ODF) environment is explicitly stated in the Indonesian Ministry of Health Regulation number 3/2014 regarding Community Based Sanitation (STBM). According to Indonesia Mid-Year Plan 2015, ODF target in Indonesia must be achieved in 2019 as though the ODF target of SDGs is by 2030. The second one hundred of Indonesia Universal Access of 100.0.100 is to show ODF by 2019 while the first is for total coverage of water supply and zero is to show no slump areas in cities. ODF by 2019 seems ambitious as compared to SDG's target of by 2030. It is, however, obvious by Indonesia Long-Year Plan as ODF is already targeted in 2009 within the first Indonesia Mid-Year Plan of 2005. It is then also obvious that Mayor of Surabaya wants to have ODF in Surabaya by 2015 as mentioned in the Mayor's Decree of 443/310/436.6.3/2015 which was still based on the Millennium Development Goals (MDGs). Furthermore, according to Surabaya Sanitation Strategy [14], Santoso [11], Destio [1], Fahmi [2], and Gemardi [4], the total sanitation coverage (ODF) in 2021 is shown in Figure 1. Sanitation services are in the form of off-site and on-site sanitation system. On-site system (blue color) is the old part of Surabaya Communities while the rest is off-site system either runs by cities

IOP Conf. Series: Materials Science and Engineering **669** (2019) 012050 doi:10.1088/1757-899X/669/1/012050

or communities. Out of 154 sub-districts (kelurahan) in Surabaya: 32 are covered with communal system and only 3 kelurahans with the on-site system. The rest is in the form of off-site system runs by Surabaya Government.

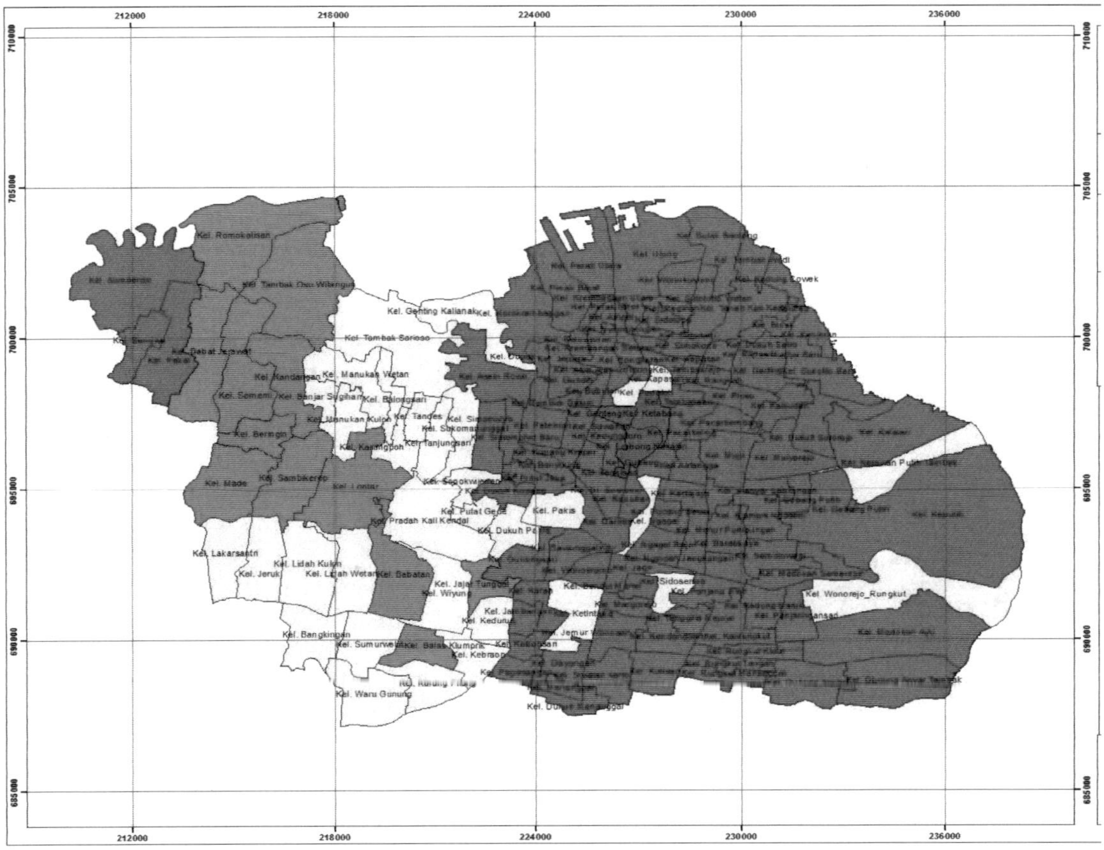

Figure 1. Sanitation coverage in Surabaya by 2021 (SSK 2017-2021)

At the moment, Surabaya City as the second largest city in Indonesia which is also the capital of East Java Province has an area of 350.5 km2 with more than 21% as open spaces. According to Surabaya Environmental Health Risk Assessment (EHRA 2015): 91% of the population has already individual septic tanks (92% in the form of squatting toilets and the rest is sitting toilets) and 7.6% uses public toilets. There is still of 2.9% of Surabaya people which still practices open defecation (OD) in the open fields and ditches, rivers, and even shorelines (EHRA 2015).

According to Surabaya Health Agency 2016, there were 15.527 houses (out of about seven hundred thousand houses) without access to toilet, as though Surabaya has its owned septic tank sludge treatment plant (IPLT) with capacity of 400 m3/day. Distribution of the houses are shown in Table 1. Surabaya, however, does not have scheduled desludging services (LLTT) yet. By 2017, the OD houses reduced to 12.772 houses, 5.586 are in the form of legal houses.

Table 1: House distribution without septic tank in Surabaya

No	Region	The Number of houses (Unit)		
		Legal	**Ilegal**	**Total**
1	West Surabaya	505	1.885	2.390
2	Central Surabaya	1.350	1.226	2.576
3	South Surabaya	1.441	1.345	2.786
4	East Surabaya	1.628	1.536	3.164
5	North Surabaya	2.808	1.803	4.611
	Total	**7.732**	**7.795**	**15.527**

Source: Surabaya Health Agency 2016

Based on Table 1: half of the OD houses are in illegal property which the land is owned by Indonesia Train Company (PT KAI), Irrigation Office, Surabaya Local Government, and Housing Developers. Furthermore, ODF covers all houses in legal and illegal properties. Especially in East of Surabaya, the highest number of OD is in Gubeng Sub district and the lowest is in Tambaksari Sub district (Purnama et al., 2019)

Based on Minister Regulation of Public Works number 4/2017 regarding Management of Domestic Wastewater there are four technologies can be used to achieve ODF, i.e. individual septic tanks, communal septic tanks, communal wastewater treatment, and conventional wastewater treatment. A hundred percent sanitation coverage within the Mid-Year Plan 2014 consists of 95% on-site sanitation system (85% of individual system plus 10% of communal system) and 5% off-site sanitation system. Besides, pit latrines are still allowed as sanitation ladder in the ODF simply notifies not to defecate in the open.

Using the financial support from the central and local government like APBN and APBD and other financial support like corporate social responsibility or CSR (Gunawan et al, 2014) then the leading sectors in providing toilet in Surabaya are Public Works Agency (Dinas PUCK) for communal system, Health Agency (Dinkes) of individual system, and Social Agency of Rehabilitasi Sosial Daerah Kumuh (RSDK) Program (Dinsos). The objective of this study is to access how ODF can be achieved in Surabaya based on the best suited technologies for the specific sites.

2. Materials and method

Surabaya City has still some problems with sanitation in the term of unhealthy community behavior [12]. Communities are still defecating in the open area such as in river, drainage system, and open field. Field survey was conducted to find out the distribution of any houses with no septic tanks for both houses in legal or illegal properties. A massive support was given by Dinas PUCK to conduct the survey as toilet construction, especially the ones with communal and conventional wastewater treatment plants, had to be provided by the Dinas. A massive support was also given by Dinsos. Sanitarians in each Puskesmas (community health center) was the one responsible for data and locations of houses with no septic tanks. There were 63 puskesmas in Surabaya. Based on Indonesia minimum standard services (SPM), each district (kecamatan) must have at least one puskesmas. As there were already 32 ODF sub-districts (kelurahans), toilet survey excluded these sub-districts. Survey results and Surabaya SSK 2016 were referred to decide which type of wastewater systems to be provided. As the number of OD houses were still lots in number, then toilet provision was to be prioritized for legal housings. Further studies

IOP Conf. Series: Materials Science and Engineering **669** (2019) 012050 doi:10.1088/1757-899X/669/1/012050

on people's knowledge and understanding on healthy sanitation were also needed as having a proper toilet was not caused by poverty but was more towards awareness and changing of attitudes (Soedjono, et.al, 2018).

3. Results and discussion

The survey results show there were still 12.772 houses without septic tanks: 5.586 were in legal properties while 7.186 houses in illegal properties as shown in Table 2. Some photos from the sites are shown in Figure 2 to 4.

Tabel 2 The Number of OD that have been surveyed

No	Region	The Number of houses (Unit)		
		Legal	Ilegal	Total
1	West Surabaya	354	2.000	2.354
2	Central Surabaya	1070	870	1.940
3	South Surabaya	1464	860	2.324
4	East Surabaya	1.167	1.909	3.076
5	North Surabaya	1531	1.547	3.078
	Total	**5.586**	**7.186**	**12.772**

Source: Survey result, 2017

Figure 2. Condition of toilets in the Kedungdoro Sub-district

Figure 3. Land condition in Menur Pumpungan Sub-district

Figure 4. Condition of MCK in Wonokusumo Sub-district

Based on Ministry of Public Works Regulation (2017), the selected technologies to achieve ODF in Surabaya would be individual septic tanks, communal septic tanks, communal wastewater treatments, and public toilets with bathing and washing (MCK). MCK was selected as some public places like traditional markets and bus terminals were used for domestic purposes. The selected technologies for Surabaya is shown in Table 3.

Tabel 3. Distribution of disposal system

Region	Individual septic tanks (Unit)	Communal septic tanks (Unit)	Conventional wastewater treatments (Unit)	MCK (Unit)	Number of OD houses (legal)
West Surabaya	267	20	0	2	354
Central Surabaya	455	75	3	9	1.070
South Surabaya	911	116	1	0	1.464
East Surabaya	674	92	0	1	1.167
North Surabaya	526	116	11	0	1.531
Total	**2.833**	**419**	**15**	**12**	**5.586**

Since the field survey has already been conducted to determine the appropriate wastewater technology, individual septic tanks is the most suitable technology to reduce the number of OD. This choice was chosen as the communities did not have area for both communal septic tanks and conventional wastewater treatment. Usually the community gets support funding from the local government, academic institution, and association of sanitation management and empowerment (APPSANI) with the cost per septic tank is about IDR 1,650,000. The payment mechanism was by giving the down payment about 40% and the rest of the cost was paid by installment for four month [12].

There were several houses that did not have space even for septic tanks but they had quite large foot-path in front of their house. These areas was suitable for communal septic tank system. The minimum width of the foot path is 2-3 meters. This values were also used for the minimum depth of the water table. Surabaya city is in coastal area and water table is quite high [3].

Surabaya has several existing communal wastewater treatment, but in these cases this technology was not recommended. It is because there are several previous communal system have not been optimized yet especially in South of Surabaya. Community empowerment and technical aspect are the most problem caused. From technical perspective, then the problem is usually for the communal system unit that was built not in accordance with the design criteria [8].

4. Conclusion

There were 5.586 legal houses in Surabaya with proper disposal of blackwater (illegal houses were excluded yet) which had to be provided with sanitation system before 2019. The facilities to be provided were 2.833 individual septic tanks, 419 communal septic tanks, 15 communal wastewater treatment plants, and 12 MCK. This visualizes that infrastructures construction has to be accelerated with massive financial support as compared with current condition which development was slow with poor financial support.

References

[1] Destio, Rizki. 2018. *Perencanaan Sistem Penyaluran Air Limbah Domestik dan Instalasi Pengolahan Air Limbah di Kelurahan Putat Jaya Kota Surabaya.*

[2] Fahmi, Iqbal. 2018. *Perencanaan Sistem Penyaluran Air Limbah Domestik dan Instalasi Pengolahan Air Limbah di Kelurahan Bulak Banteng Kota Surabaya.*

[3] Fitriani N, Isnadina, D R W, Yufansyah, M, Ningsih, D A, Soedjono, E S 2019 *Communal Septictanks as Appropriate Technology to Achieve Surabaya Open Defecation Free (ODF).* International Journal of Geomate **17** 62 pp: 260-265. doi: 10.21660/2019.62.90711.

[4] Gemar, Iqbal. 2018 *Perencanaan Sistem Penyaluran Air Limbah Domestik dan Instalasi Pengolahan Air Limbah Industri Tahu di Kota Probolinggo.*

[5] Gunawan J, Rumiati AT, Trisunarno L, Soedjono E. 2014. *Corporate social responsibility practices in Indonesia: Intermediary matters-case studies from Indonesia (Book Chapter) .Corporate Social Responsibility and Local Community in Asia.*

[6] Mayor's Decree Surabaya of 443/310/436.6.3/2015 about *Millennium Development Goals (MDGs).*

[7] Purnama M Y, Soedarti T, Oktavitri N I, Fitriani N 2019 *Distribution Mapping of Open Defecation (OD) houses in East Surabaya based on Geographic Information System (GIS).* IOP Conf. Series: Earth and Environmental Science, 259, 012023. doi:10.1088/1755-1315/259/1/012023

[8] Purnawanti, E., Oktavitri, N.I., Isnadina, D, R, M., Fitriani, N. 2019. *Evaluation of communal wastewater treatment plan based on technical, institutional, an community participation aspects in South Surabaya.* IOP Conf. Series: Earth and Environmental Science, 259, 012016. doi:10.1088/1755-1315/259/1/012016.

[9] Regulation of Ministry of Health Regulation No. 3 about *Regarding Community Based Sanitation (STBM). 2014.* (Jakarta: Ministry of Health)

[10] Regulation of Ministry of Public Works No. 4 about *Operation of Domestic Wastewater System. 2017. Jakarta: Ministry of Public works.*

[11] Santoso, R.E. 2017. *Pengembangan model risiko kesehatan lingkungan untuk penetapan system sanitasi perkotaan. Disertasi – ITS.*

[12] Soedjono E S, Fitriani N, Yuniarto A, and Wijaya I M W 2017 *Provision of healthy latrine for low income communities based on community empowerment in Kelurahan Kebonsari, Surabaya City, towards Indonesia ODF in 2019.* AIP Proceeding 1903, 040012.

https://doi.org/10.1063/1.5011531.
[13] Soedjono E S, Budiani J R, Marsono B D, Ulfin I, and Tukunan KP 2018 *Providing Healthy latrine based on community empowerment in Gebang Putih Village and Kejawan Putih Tambak Village.*
[14] Surabaya Sanitation strategies 2016 *Sanitation Strategies in Surabaya 2017-2021.*
[15] Surabaya Environmental Health Risk Assessment 2015 *Environmental Health Risk Assessment*
[16] United Nations. *Sustainable Development Goals 2030.* New York: United Nations

Compression test analysis of cold-formed column in channel profile connection

D F Rachmahwati[1], N Nindyawati[1] and K Karyadi[1]

[1] Departement of Civil Engineering, State University of Malang, Indonesia

nindyawati.ft@um.ac.id

Abstract. This study analyzes the load capacity of the cold-formed steel column frame structure for multilevel buildings to withstand axial loads. The column frame structure will be modeled, analyzed and tested. In this paper, the study is about a cold-formed channel profile that is compressed. The research objectives are to determine (1) stress and strain, (2) damage patterns, and (3) research development. The test object is modeled by using finite element method with a scale of 1:1 and the loading used is linear static loading. The simulation with finite element method uses atrial and error method by entering the obtained parameters. The draft and design of channel profiles used in cold-formed steel material refers to a several guidelines including SNI-7971-2013 on Cold Formed Steel Structures.

1. Introduction

Cold-formed steel havean advantages in production process which is easier to produce and also environmentally friendly because it does not produce waste that is harmful to environment. In the production process, steel are formed in such a way at a room temperature by a process of bending brakes, press brakes, and roll-forming machines. With cold formed method, the desired cross-section profile could be produced economically so that an increase in the ratio of strength to weight could be obtained [9]. According to [5], cold-formed steel has excellency in strength and stiffness that is quite high in value, does not shrink and expand easily when affected by weather, easy in fabrication and production, uniform in quality, non-flammable, more accurate in detailing, anti-termite and environmental friendly material, could be recycled, and have a high structural lifetime.

Research on light steel has improved greatly indetermining the right analysis and design. However the focus of attention when designing light steel are the ability to carry a compressive load that is buckling factor. Buckling is a phenomenon in construction materials, especially in structures that have a large slenderness ratio and a thin cross-section thickness. Buckling is a structural failure where the cross-section changes shape when or before the structure reaches its melting capacity. There are two types of buckling that often occurs in light steel and it is called local buckling and global buckling whereas one of the factors that made it happen is the geometry or cross-section of profile. According to [7] local buckling is a failure that often occurs in light steel compressed elements even when the light steel are given stiffener.

Some form and shape of cross-section of light steel that is available in the market are channel sections and box sections with varying sizes and needs for construction use. Channel shape is often used in the structural frame. According to [6] the same channel profile has different pressure capacities and bending behaviors. This is due to the modification of some of these channel profiles. Furthermore, according to [4] the pressure capacity calculated by the direct approach of a double channel cross section has overestimate differences but it is still conservative when used. Of course, based on the theory that has been explained before, the cross sections provide different pressure capacities and behaviors. Analytically, the compressive capacity of a light steel section can be calculated based on SNI 7971: 2013 article 3.4. However, in essence the calculation of the critical buckling stress is not fully in accordance with the design requirements. Many parameters are in need to be found first for a more conservative calculation approach.

Based on the introduction above, it is necessary to carry out a finite element analysis. The analysis of light steel are pressure test to obtain the pressure behavior of several cross-sections of light steel by calculating the pressure capacity of cold-formed steel based on SNI 7971: 2013. The light steel profile that are tested is the C channel profile available on the market. The profile will be connected to a double channel back to back and a box using screws. This test uses a universal testing machine. Thus, the results of this study can be used as parameters for designing a lightweight steel structures, especially pressure elements.

2. Methodology

This study analyze the load capacity of the cold-formed steel column frame structure for multi level buildings to withstand axial loads. The column frame structure will be modeled, analyzed and tested experimentally using one type of test object. The test specimens consist of cold-formed canal profiles are tested in pressure test. This research objectives are to determine the stress and strain, to determine the damage pattern, and to find out the development in research. The test object is modeled using finite element method with a scale of 1: 1 and the loading used is linear static loading. Simulation with finite element method uses a trial and error method by inputting the parameters that are obtained. The draft and design of cannel profiles using cold-formed steel material refers to several guidelines including SNI-7971-2013 on Cold-formed Steel Structures, [9] on Cold-Formed Steel Design.

2.1. Modeling of test object

The Cold-formed steel used isconsisting of a single channel profile joint (T1). The model will be given a load according to the results of the finite element method preliminary calculation.

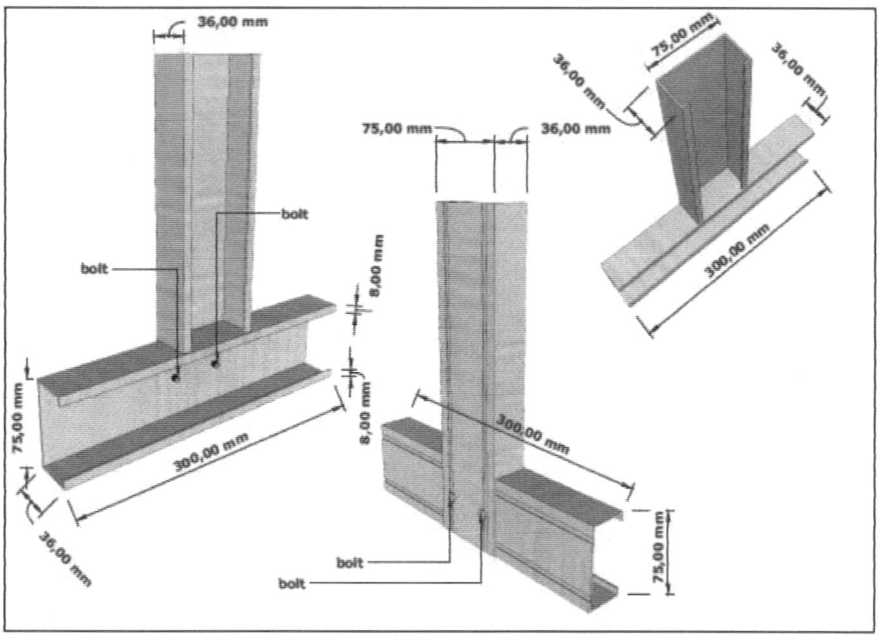

Figure 1. Test Objects

The following is a specification of the cold-formed C cannel profile that will be used as a test object, as follows

Figure 2. Cold-formed

- Name : Truss C75.80 SNI
- Thickness : 0,8 mm
- Length : 6 m
- Material : Cold-Formed Steel

2.2. Joint tool

The joint tool used are self driving screw bolt with a type of 12-14 x 20. Here are the specifications:

Figure 3. Joint tool

Table 1. Specifications of self driving screw bolt

Nominal Size or Basic Screw Diameter	Threads Per Inch	T Major Diameter		t Minor Diameter		P Protrusion Allowance	
		Max.	Min.	Max.	Min.	#2 Pt.	#3 Pt.
6 0.1380	20	0.139	0.135	0.104	0.099	0.190	-
8 0.1640	18	0.166	0.161	0.122	0.116	0.211	-
10 0.1900	16	0.189	0.183	0.141	0.135	0.235	0.300
12 0.2160	14	0.215	0.209	0.164	0.157	-	0.353
1/4 0.2500	14	0.246	0.240	0.192	0.185	-	0.393

2.3. Data processing

Data that is obtained includes:

- Strain &Tension/Stress, to find out the capacity of the C channel profile according to the modeling. It could be applied to pressure and pull tests.
- Damage pattern on channel profile C, to find out how much damage has occurred

2.4. Structural modeling

After obtaining the data, the next step is making structural model. Structure modeling are done by using finite element method. Structures being modeled according to building structure data, material specifications, and structural element data used. Structural modeling is made into 2 models namely face-to-face press and opposite press. Both models are made with channel profiles. Structure modeling figure as follows:

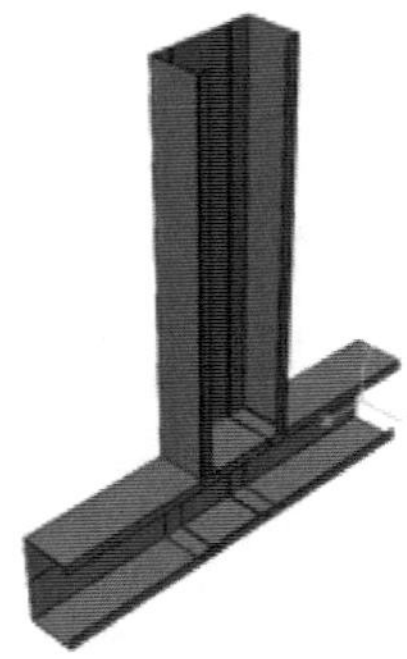

Figure 4. Structure modeling using finite element method

3. Results and Discussions

3.1. Stress/Tension and Strain Analysis in finite element method

Structural modeling drawings were made into 2 models namely face-to-face press and opposite press afterward they are analyzed to find out the yield stress and ultimate stress and how strong it is to be used as a channel profile connection

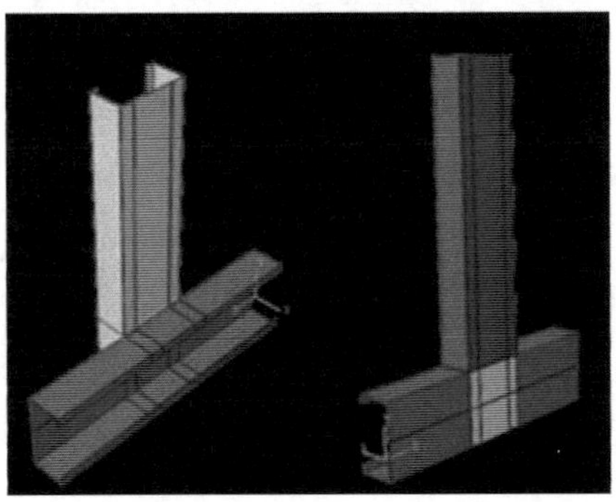

(a) (b)

Figure 5. (a) Modeling of opposite press and (b) face-to-face press

The following are the graphic results of channel profiles modeling with opposite press and face-to-face press

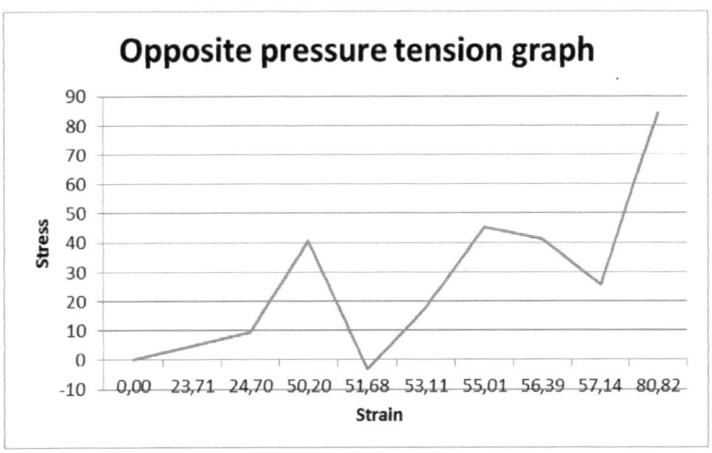

Figure 6. Opposite pressure tension graph

From Figure 6. above it can be explained that the yield stress at this channel profile connection is 45.06 N/mm^2 and for the ultimate range is 84.31 N/mm^2. Test object experienced an increase and decrease in stress averaging of 26.57 N/mm^2, with a compressive load of 100 N. The strain on the object was obtained from the result comparison of the final length increase of 80.82 mm with an initial length of 300 mm so the strain value result are 26 %

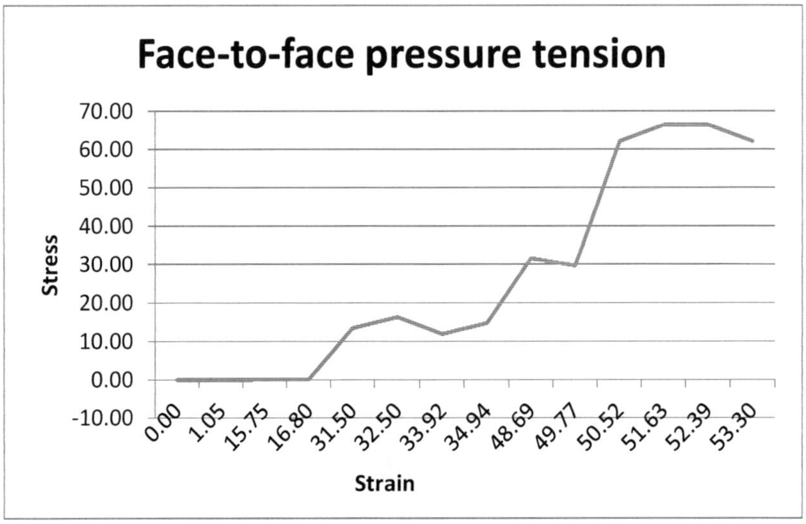

Figure 7. Face-to-face pressure tension

From the figure 7. above it can be explained that the yield stress at this channel profile connection is 31.64 N/mm^2 and for the ultimate range is 66.36 N/mm^2. Test object experienced an increase and decrease in stress averaging of 26.73 N/ mm^2, with a compressive load of 100 N. The strain on the test object was obtained from the result comparison of the final length increase of 53.30 mm with an initial length of 300 mm so the strain value result are 18 %

3.2. Damage Pattern

The damage pattern of the test object analyzed by finite element method occur a buckling factor. The weakness of cold-formed steel structures is local buckling due to it's thin cross-section elements. This causes structural failure to occur before reaching the highest load capacity. To reduce the likelihood of local buckling of cold-formed steel press structure components, a structured cross-section is used [7].

The following is a picture of damage pattern in the compressed face-to-face modeling which occurs a local buckling. Local buckling happen because of the large slenderness ratio of the cross-section element, so that the compressive stem capacity is determined from the bending of the element (local buckling) first and not from the global bending of the compressive stem.

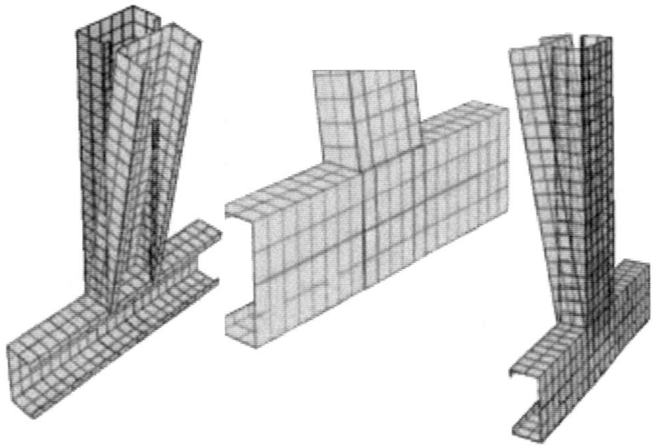

Figure 8. Damage pattern on face-to-face press

For the joint in face-to-face pressure in Figure 8. When the ultimate loads are reached, the local buckling begins to occur in the wing side on the right and left. The wing side also began to deform towards the front, because the front wing was cut off. When the loads are increasing, bending begins to occur at the bottom of the sample and is followed by bending toward the inside of the wing. Increase in strength occurs due to increased stiffness in the part that has experienced bending. By the time the shortening reaches the displacement limit of 20 mm, the wing has been torn.

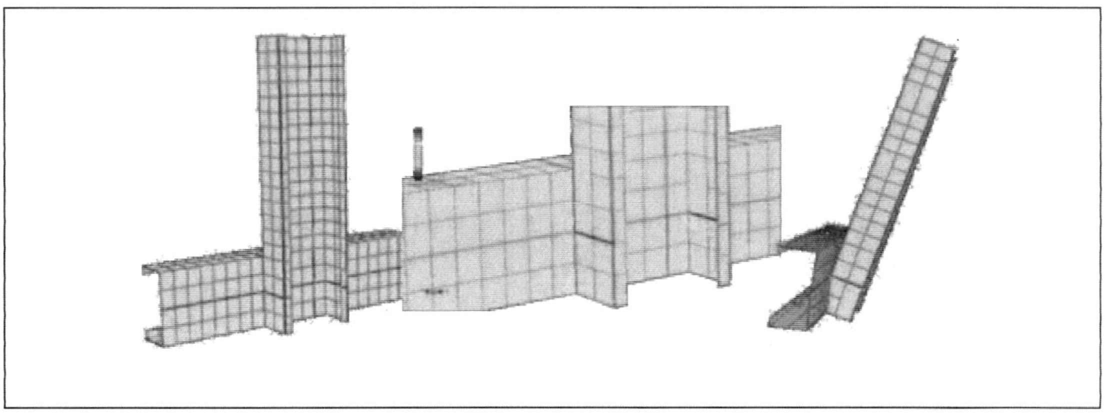

Figure 9. Damage pattern on opposite press

For the opposite join press in figure 9. When the ultimate loads are reached, local buckling begins to occur in the middle outer wing on the right side. When the loads are increasing, the wing on the lower side begins to bend and the outer wing begins to look wider. Increase in strength occurs due to the resistance of the screw. By the time the shortening reaches the displacement limit of 20 mm, the underside of the lightweight steel wing has experienced a huge buckling widening and the screw part has failed.

Every test object failure (face-to-face press and opposite press models) happen because a local buckling in the channel profile. This is caused by the changes in stress due to changes in channel cross-sectional area. finite element method analysis used are buckling analysis, with the assumption of using elastic material so that the critical load obtained is far greater than the failure load obtained from experimental studies. Even so, the finite element method results show the same tendency.

4. Conclusion

It can be concluded that the modeling of opposite pressure channel profiles has an ultimate stress/tension of 84.31 N/mm2 which is greater than that of the compressed channel profile face-to-face models which has an ultimate stress/tension of 66.36 N/mm2. So a suitable connection for channel profile modeling is the opposite pressure which is strong enough to hold the load and has a high ultimate stress.

In the damage pattern of the opposite pressure channel profile and face-to-face pressure, there occurs a buckling factor. In both models there has been a local buckling due to the large slenderness ratio of the cross-section element, so that the compressive stem capacity is determined from the bending of the element (local buckling) first and not from the global bending of the compressive rod.

References

[1] American Iron and Steel Institute 2007, *North American Specification for Design of Cold-Formed Steel Structure Members* (Mexico)
[2] AISI STANDARD 2007 *Supplement No. 2 to the North.* American Specification *for the. Design of Cold-Formed Steel. Structural Members*
[3] British Standards and codes of practice BS 5950 Part 5 *Code of practice for design of cold formed thin gauge sections*
[4] Kadir A M 2013 *Investigasi Eksperimental Perilaku Profil Baja Ringan Berpenampang Persegi Akibat Interaksi Beban Tekan Memusat dan Momen Lentur* Jurnal Metalurgi Universitas Indonesia Vol 28 No 1 Hal 33-48 (Depok)
[5] Meiyalagan M 2010 *Investigation on Cold–formed C section Long Column with intermediate Stiffner & Corner Lips-under Axial Compression* Internatonal Journal of Apllied Engineering Research, Dindigul. Vol.1 No.1.
[6] Ruus K 2017 *Pengaruh Bentuk Profil Baja Ringan Terhadap Kuat Tekan.* Jurnal Sipil Statik Vol 5 No. 5 Hal 249-262 (Manado)
[7] Sandjaya, Arif; dan Suryoatmono B 2018 *Studi Eksperimental Batang Tekan Baja Canai Dingin Diperkaku Sebagian* . Jurnal Teknik Sipil Institut Teknologi Bandung Vol 25 No. 1.Hal 19-24 (Bandung)
[8] Standar Nasional Indonesia 7971 2013. *Struktur Baja Canai Dingin (*Jakarta: Badan Standarisasi Nasional)
[9] Yu Wei-Wen, Ph.D., P.E. 2000.*Cold Formed Steel Design Third Edition.*(Canada)

Acknowledgement

This work was supported by PNBP research grant of Universitas Negeri Malang.

The addition of lamella in anaerobic baffled reactor used for decentralized municipal wastewater treatment

A Yulistyorini[1], M C Yuaniar[1], M Mujiyono[1], N Suryoputro[1] and S Sukarni[2]

[1] Department of Civil Engineering, Faculty of Engineering, Universitas Negeri Malang, Indonesia
[2] Department of Mechanical Engineering, Faculty of Engineering, Universitas Negeri Malang, Indonesia

anie.yulistyorini.ft@um.ac.id

Abstract. An anaerobic baffled reactor (ABR) represents a low-cost system for decentralized wastewater treatment in a developing country. Despite the advantages of ABR units such as low operating and maintenance costs, this system requires sufficient land to construct it. In particular of an urban area, land availability is limited and therefore acquire modification in ABR design. This study aimed to test the addition of lamella for further reduction of solids particle in domestic wastewater, and therefore it could reduce the size of ABR settler. A lab-scale of an ABR lamella with a dimension of 75 cm length, 20 cm width, and 40 cm height used in the experiment. A 20 L/d of raw municipal wastewater taken from the existing ABR in Kampong Bareng, Malang (Indonesia) used to feed the lab-scale of the ABR lamella. The experiment was conducted in two months and the sample of raw and treated wastewater collected from the ABR lamella for ten times from September to October 2018. The samples were taken from the inlet, lamella compartment, and the outlet. The results showed that the addition of lamella reduced the concentration of TSS and BOD for 95% and 90% respectively. While these results could help further reduction of organic and solids particle of wastewater that meet the treated wastewater regulation. The concentration of TSS and BOD as treated wastewater were 19 mg/L and 18 mg/L respectively. Although the ABR could reduce the nutrient concentration, the level still higher than the global standard for N and P.

1. Introduction

The smart environment is waste processing management in the city by providing of sanitation system, namely septic tank, toilet, and decentralized wastewater treatment plant (DWWTP). Most of the DWWTP in Indonesia use ABR. This indicator is realized through a program, carried out by the government, called KOTAKU which primary aim is to achieve No Slums Cities [1], in which public DWWTP procurement becoming one of the attempts.

However, several technical issues found during the application of ABR in Indonesia; for instance, the effluent of domestic wastewater does not meet the wastewater quality standard. Hence, achieving the stable processing of wastewater to fulfil the standard requires further process [2]. Some countries have implemented wastewater management technology with simple maintenance, i.e., lamella. Lamella consists of two types: a plastic tube and a stainless-steel plate. The benefits of using plate lamella encompass high efficiency, low maintenance, easy application, and higher durability compared with PVC lamella, which is prone to damage when applied in construction [3]. Therefore, plate lamella uses in the development of further processing or design modification in ABR.

The flow concept of plate lamella uses gravitation to pull the solids down so that they can be settled. The angle of the lamella is designed to prevent solids from being washed together with water. Additionally, lamella sedimentation tub can reduce the average water turbidity 5–12 times better than the conventional sedimentation unit [4]. Mathematical modeling and numerical simulation were used to observe if using lamella can boost efficiency. As a result, lamella indeed rises efficiency and makes effluent quality parameters meet the water quality standard [5]. Based on that descriptions, the author aims to develop ABR for DWWTP to improve wastewater treatment process by the addition of lamella. We also applied U-shape flow pattern to ease the maintenance. The pattern, however, is rarely applied in ABR; instead, in general, DWWTPs apply zig-zag or S-shape pattern, which is quite hard in both construction and maintenance process.

2. Materials and Methods

2.1. Anaerobic Baffled Reactor with lamella

Design calculation of ABR-lamella referred to BOD concentration, retention time, and flow rates. Raw wastewater was taken from DWWTP of KPP Purworupo in Malang that containing BOD concentration of 558.4 mg/L. The retention time of three days and up-flow velocity of 2 m/h were applied in the design calculation [6], and 0.231 ml/s of wastewater flowrate. The ABR design was calculated based on the analytical calculation of wastewater processing technology [7], literature review related to SSWM [6], and basic theory of decentralized wastewater treatment in developing countries (DEWATS) [8]. The U-turn flow pattern used to ease maintenance [9]. The choice of lamella material refers to several studies which focused on the stainless-steel plate shape as it is more efficient than PVC tube [3], the distance between lamella was 2–4 cm [10], and the angle of 60° was applied to enhance higher sedimentation rate [11].

The ABR made from plastic fiber. The dimension of the first compartment was 30 cm x 10 cm x 40 cm. While 15 layers of the lamella, having area of 20 x 10 cm, arranged in the compartment with a dimension of 45 x cm x 10 cm x40 cm, 60° of angle, and 2 cm of distance between the layers. The other three compartments have dimension of 20 cm x 10 cm x 40 cm, and the last one (outlet) was 15 cm x 10 cm x 40 cm. The overall reactor size was 75 cm x 20 cm x 40 cm. The pipe diameter of 0.5 used to connect each compartment in ABR.

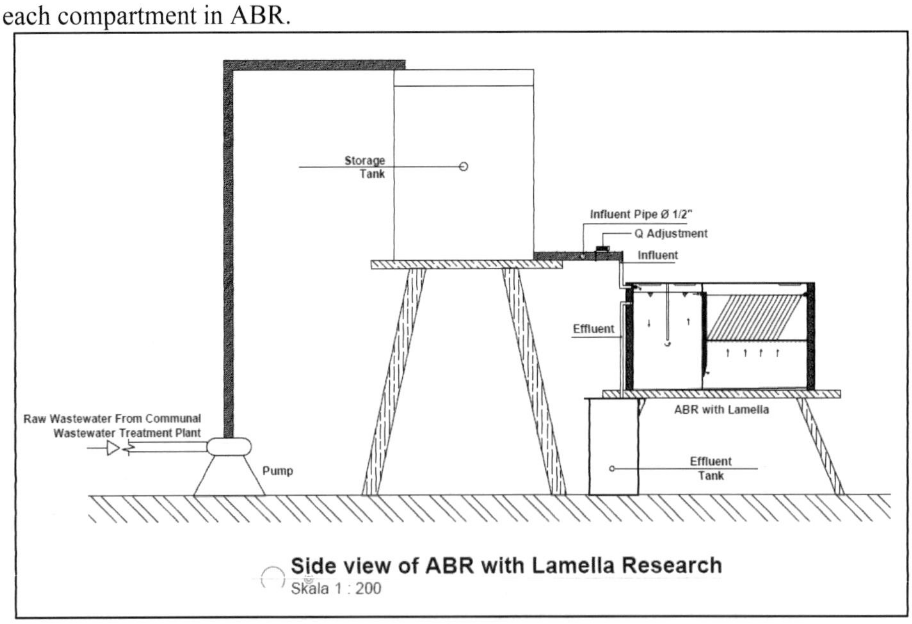

Figure 1. Scheme of ABR with lamella in laboratory-scale

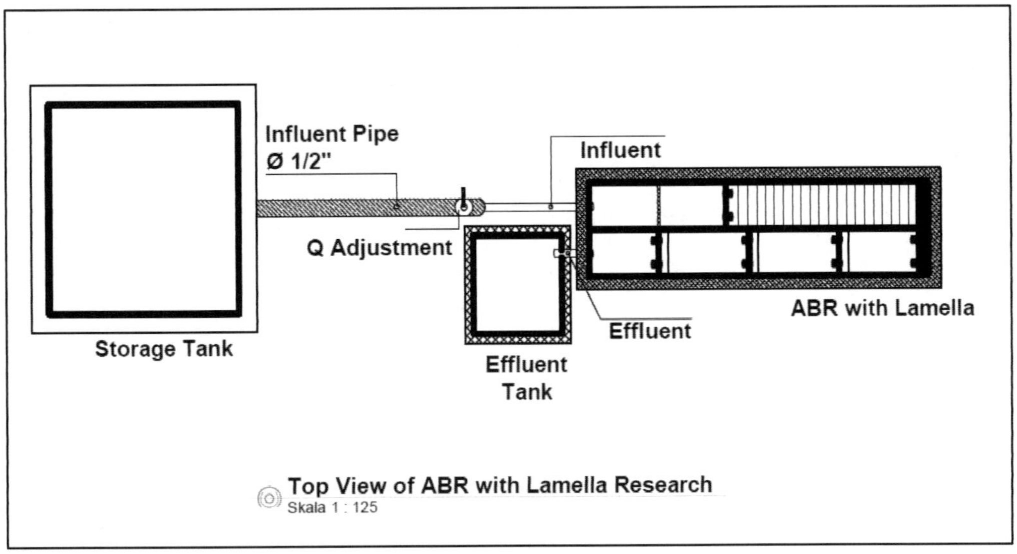

Figure 2. Top view of ABR-lamella

The inlet or sedimentation compartment has a function as a divider between solid and liquid by gravitation, and this can eliminate solids in the water [12]. The water then flew down and passed through the lamella compartment in which they can control the up-flow water to the next compartment, hence reduce more solids in the water [13].

The concept of flowing against the stream works when the up-flowing water goes through the angle of 60; when the water flows up, solids carried by water will collide with the lamellas hence fall and create more sedimentations [14]. The distance between lamella plates should be between 2–4 cm [10] however, to obtain the most optimum TSS removal efficiency, it was decided that the best distance was 2 cm, and we made the height of lamella to be half of the reactor's height (20 cm).

After passing through the lamella, the water cascades towards baffled compartment. The baffled able to reduce the water velocity and lead longer retention time which triggers in organic matter reduction [15] as the baffled also create up-flow water movement.

2.2. Data Analysis

Biological and physicochemical parameters like BOD, TSS, TKN, TP, and TC were analyzed after sample collection in Jasa Tirta laboratory following the standard method for the examination of wastewater samples. T, pH, and DO were measured directly on the site using portable equipment. The removal values of BOD, TSS, TKN, TP, and TC represented the reduction efficiency of the influent (raw wastewater) and expressed it in percentage.

3. Results and discussion

3.1. pH and temperature

The observation revealed that the pH in the influent was around 6 to 7 while in the compartment after lamella and effluent the pH was about 7.8 to 8.2. These number indicated that the treatment of wastewater using ABR inserted with lamella showed very good performance. Under laboratory condition, pH parameter met the Indonesian wastewater quality standard of 6–9. The increase of pH occurred because oxygen used in degraded the organic matter; thus, CO_2 concentration fell. Increasing pH value to neutral is directly proportional to the decreasing concentration of BOD [16]. This statement is in line with research by Budiwati [17] reported that while the concentration of CO_2 declines, the pH will incline. Supporting this, Muson [18] also stated, when pH value decreases, CO_2 concentration increases, as a further consequence, concentration of BOD increases as well. High pH value of 7.5–10 in wastewater revealed the pollution level decreases; the higher the value of pH in wastewater, the lower concentration of BOD contains in wastewater [19].

Temperature and pH are essential parameters which influence the growth and performance of microorganisms inside the reactor [20]. The temperatures at three sampling points (inlet, after lamella, outlet) were stable, averaging at 24°C; meanwhile, the temperature of water in the outlet from 22°C to 26°C. Water analysis results obtained from the experiment are presented in the following graphs.

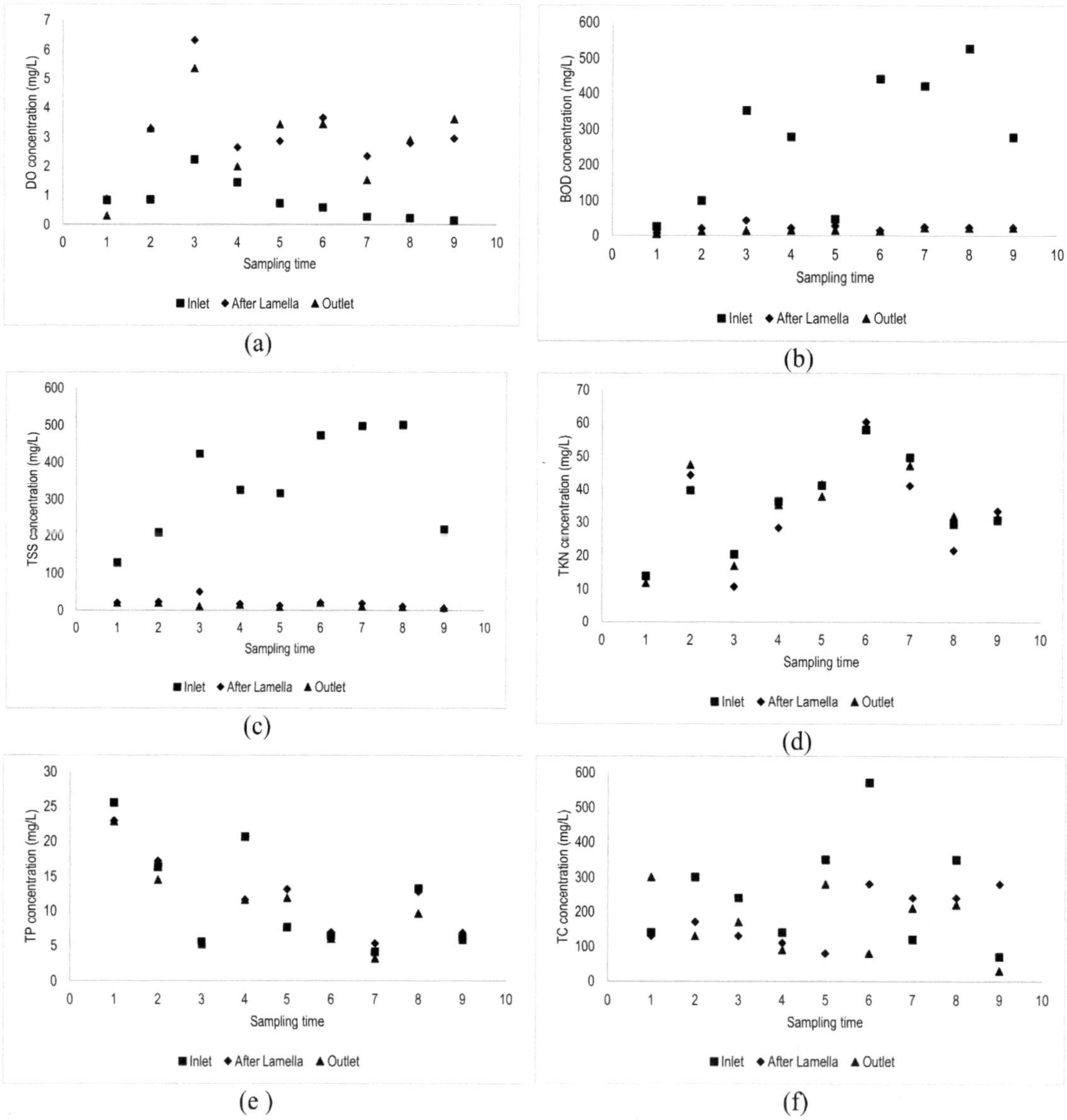

Figure 3. Water analysis results of domestic wastewater treated with ABR which inserted by lamella

3.2. DO

DO is an important compound which supports aquatic organisms to survive, such as for breathing, metabolism process, and substance exchanges which then generates energy to grow and breed [21]. The oxygen concentration of the raw wastewater ranging from 0.1 mg/l to 2.1 mg/l and then increased in the effluent from 2 mg/l to 5.4 mg/l (Fig. 3a). The concentration of DO has already met the minimum standard for fish breeding. The DO concentration after-lamella, and outlet point was higher compared with the inlet in which it was followed by the decreased of BOD concentration. Similar results of decreasing BOD proved by Aldo [23].

3.3. BOD

BOD concentration shows the amount of dissolved oxygen concentration required by microorganisms during the degradation. Hence, the BOD concentration is equivalent to organic pollutant concentration in wastewater [23]. Regarding the lamella performance, this study showed that BOD concentration removal efficiency from the inlet to lamella (A) was 89%, it was higher compared to point B (inlet to outlet) at 82%. On average BOD of the influent was 275 mg/l, and it decreased to 23 mg/l after passed through the lamella. The addition of lamella helped the system to reduce at 84% of the organic matter as the lamella reduced the velocity of water and increased the settling of the solids particle, and therefore reduce BOD concentration (Fig. 3b).

High removal of BOD level at point A illustrated the ability of lamella inside ABR on enhancing sedimentation, which optimally reduced the BOD concentration in the domestic wastewater. It was in line with lamella's purpose to reduce dissolved solids within the water, where 80% of BOD consists of dissolved and suspended solids [24]. This lead to the final treatment, which BOD concentration removed about 90% and met the standard quality of effluent.

3.4. TSS

The result of TSS removal efficiency was similar to the previous result of BOD, in which the average removal at point A was 93% and in the effluent was 95%. TSS concentration decreased at 343 mg/l, 20 mg/l and 14 mg/l respectively for inlet, after lamella and outlet (Fig. 3c). The concentration of TSS in the effluent has met the standard of discharge consent with a maximum value of 30 mg/l. The decrease of TSS proved that lamella triggers the solids particle to be settled in the first part of the ABR. Lamella (tilted plate) supported in increasing organic solids secondary sedimentation performance and worked effectively in removing suspended solids and BOD content [25].

3.5. TKN

The concept of ABR lamella performance is sedimentation oriented, which leads to removing BOD and TSS concentration. On the other hand, ineffective in reducing nutrient compound, e.g., nitrogen [6]. Figure 3d shows that the concentration of TKN fluctuated over 30 days. TKN concentration in raw wastewater ranging from 14 mg/l to 58 mg/l, and the average was 35.4 mg/l. After passed the lamella TKN level decreased with the average value of 32 g/l and slightly increased in the effluent at 34 mg/l.

It was assumed that the decline of TKN was because of organic nitrogen sedimentation along with solids particle. Besides, the decrease of nitrogen in the effluent was also be triggered by nitrification and denitrification process in ABR (Yulistyorini et al., 2019). Also, lamella can help the removal of nitrogen in ABR for 9.7%. Generally, nitrogen is essential for organisms as nutrition and mineral trace, but in large amount, it pollutes the environment, e.g., anaerobic nitrogen and ammonium nitrate (NH_4NO_3) which consumes oxygen hence interferes ecosystem balance and inhibits aquatic animals from living and metabolizing [26]. A high concentration of ammonium inhibits further processing, and therefore pretreatment is required [27].

Although the reactor was only sedimentation-based, the additional lamella in ABR reduced better TKN than research by McKean [28] which total removal efficiency of nitrogen using lamella clarifier method in domestic wastewater was only 1%.

3.6. TP

Phosphor concentration reduction showed a similar pattern in which the ABR was ineffective in reducing nutrient (phosphor). However, in comparison to TKN concentration (Fig 3d), the value of TP from the inlet to the outlet was gradually declined over time (see Fig 3e). TP concentration in raw wastewater was very high with a minimum and maximum value at 4.1 mg/l and 26 mg/l respectively, with the average TP concentration of raw wastewater was 12 mg/l. The addition of lamella did not seem to reduce phosphor concentration in wastewater as the average TP concentration after passed lamella compartment was 11.4 mg/l. There was only 11% removal of total phosphorus from the lamella compartment to the outlet which reaches 10 mg/l.

Although the performance of ABR lamella was not good in reducing phosphorus, the 11% reduction was still higher than another study conducted by McKean [28] which obtained 6% reduction of total phosphor removal using lamella clarifier in domestic wastewater. This reduction was possible due to the phosphorus bond with organic matter and settled in the sedimentation process. High phosphorus concentration in raw and treated wastewater was probably due to excessive household laundry waste. Phosphorus comes from sodium tripolyphosphate (STPP), which is one of the detergent ingredients. Excessive phosphorus leads to eutrophication (nutrient enrichment) and to prevent it requires further wastewater polishing [29] by adding further processes such as phytoremediation or constructed wetlands.

3.7. Total Coliform

TC concentration of raw wastewater in this study was below the discharge standard of domestic wastewater where the concentration still lower than 3000 /100 ml. The maximum TC concentration in raw wastewater was 571/100 ml. Moreover, on average TC concentration reduced from 255/100 ml in influent to 168/100 ml in effluent with total removal of 34%. Since the lamella assists the ABR to enhance the settling of the solids particle, it helped to reduce TC about 27% from the raw wastewater. During sedimentation, bacteria bind with solids particle and trigger the sedimentation process [30,31].

4. Conclusion

Domestic wastewater treated in ABR-lamella to meet the discharge consent. On average, the removal efficiency of BOD and TSS were 90% and 95% respectively. While nutrient removal (N and P) were 9.7% and 11% respectively. Overall, temperature, pH, BOD, TSS, and TC met the quality standard; however, the TKN and TP concentration above the standard limit. Additional lamella in ABR reactor proved that it was lead to enhance sedimentation that removed TSS of raw wastewater at 94%.

References

[1] PUPR. 2016. Program Kota Tanpa Kumuh (Kotaku) Kementerian PUPR tahun 2016. Kementerian PUPR (online), (http://kotaku.pu.go.id), diakses 16 Agustus 2018.
[2] Hastuti, Elis., Reni, Nuraeni., Sri Darwati. 2017. Pengembangan Proses Pada Sistem Anaerobic Baffled Reactor Untuk Memenuhi Baku Mutu Air Limbah Domestik. Jurnal Permukiman, Vol. 12 No.2. Dari http://jurnal permukiman.pu.go.id.
[3] Disingger, Kerry. 2017. Plate Settler Vs. Tube Settler: Best Choice For A Sedimentation Basin, (online), (www.jmsequipment.com/tube-settlers-vs-plate-settlers), diakses 9 Oktober 2018.
[4] Vasconcelos, Jose G., Carmen D. Chosie, Holly M., & Andrew C. Patrick. 2014. Lamella Settlers: Sediment Removal From Roadway Construction Runoff. (online), (https://pdfs.semanticscholar.org), diakses 15 Agustus 2018.
[5] Moga, I. C., Bogdan D. N. G., Aurelia C. 2011. Modeling and simulation of the settling process. Proceedings of the 12th WSEAS international conference on Mathematics and computers in biology, business and acoustics ISBN: 978-960-474-293-6. Dari https://www.researchgate.net/publication.
[6] Eawag & Spuhler, Dorothee. 2015. Anaerobic Baffled Reactor (ABR), (online), (http://www.sswm.info/), diakses 16 Agustus 2018.
[7] Said, Nusa I. 2017. Teknologi Pengolahan Air Limbah. Jakarta: Erlangga.
[8] Sasse, Ludwig. 1998. DEWATS: Decentralized Wastewater Treatment in Developing Countries. (online) (https://sswm.info/sites/default/files/reference_attachments/), diakses 15 Agustus 2018

[9] Kristianto, Lukas.2014. Perancangan tata letak lantai produksi baru CV. Yudha Havana. (online), (http://e-journal.uajy.ac.id), diakses 16 Agustus 2018.

[10] Pratiwi, Kasih D.S., Joni Hermana. 2014. Efisiensi Pengolahan Limbah Cair Mengandung Minyak Pelumas pada Oil Separator dengan Menggunakan Plate Settler. Jurnal Teknik ITS Vol 3, No 1. DOI: 10.12962/j23373539. v3i1.5379

[11] Robescu, Diana., Corina Mandis, & Robescu.2010. Design lamellar secondary settling tank using numerical modeling. UPB Scientific Bulletin, Series D: Mechanical Engineering 72(4). Dari www.researchgate.net/publication.

[12] Sirampun, Maksi., Saparudin & Tunas, I Gede. 2014. Perencanaan bak pengendapan dan penampungan air yang berasal Dari mata air di kecamatan lamala. Jurnal Teknik Sipil Vol 3, No 2 ISSN: 2088-6004. http://jurnal.untad.ac.id/jurnal/index.php/JTSI/article/view/2411.

[13] Pontius, Frederick W., 1990. Water Quality and Treatment. US of Amerika: Mc Graw-Hill, Inc.

[14] Droste, Ronald L.1997. Theory and Practice of Water and Wastewater Treatment. Canada: John Wiley & Sons. Inc.

[15] Kaswinarni, Fibria. 2007. Kajian Teknis Pengolahan Limbah Padat Dan Cair Industri Tahu. (online), (http://eprints.undip.ac.id/), diakses 16 Agustus 2018.

[16] Nugrahini, Panca., Rizki, Habibi, T.M., & Anita, D. S. 2008. Penentuan Parameter Kinetika Proses Anaerobik Campuran Limbah Cair Industri Menggunakan Reaktor Upflow Anaerobic Sludge Blanket (UASB). Jurusan Teknik Kimia Universitas Lampung ISBN: 978-979-1165-74-7. Dari https://dokumen.tips/documents/pengolahan-limbah-kelapa-sawit.html.

[17] Budiwati, Tuti. 2009. Analisis hujan asam dan CO2 atmosfer. (online), (https://eprints.uny.ac.id/12393/1/34.%20Tuti%20Budiwati%201.pdf), diakses 04 Desember 2018.

[18] Muson, B. Hermanus., Bobby, Polii., Lucia, C. Mandey. 2015. Pengaruh perlakuan aerob dan anaerob terhadap variabel bod, cod, ph, dan bakteri dominan limbah industri desiccated coconut pt. Global coconut radey, minahasa selatan. Jurnal ilmu dan teknologi pangan Vol 3 No 2. Dari https://ejournal.unsrat.ac.id/index.php/itp/article/view/13370.

[19] Isyuniarto, Widdi Usada dan Agus Purwadi. 2007. Degradasi Limbah Cair Industri Kertas / Menggunakan Oksidan Ozon Dan Kapur. Prosiding PPI-PDIPTN, ISSN 0216- 3128. Dari https://inis.iaea.org/collection/NCLCollectionStore/_Public/45/095/45095780.pdf.

[20] Indrayani, L., & Rahmah, N. 2018. Nilai Parameter Kadar Pencemar Sebagai Penentu Tingkat Efektivitas Tahapan Pengolahan Limbah Cair Industri Batik. Jurnal Rekayasa Proses Vol 12, No 1. DOI: doi.org/10.22146/ jrekpros.35754.

[21] Salmin. 2005. Oksigen Terlarut (DO) Dan Kebutuhan Oksigen Biologi (BOD) Sebagai Salah Satu Indikator Untuk Menentukan Kualitas Perairan. Oseana Vol xxx No 3: 21-26, ISSN 0216-1877. Dari http://oseanografi. lipi.go.id/dokumen/oseana_xxx(3)21-26.pdf.

[22] Aldo, Suryaldi., Saberina, Hasibuan., & Syafriadiman. 2015. Changes In The Value Of Cod And Bod In The The Media Life Of Fish Snakeskin Gourami (Trichogaster trichopterus Blkr) By Given Palm Oil Waste With Different Doses. Jurnal Online Mahasiswa Vol 2, No 2. Dari https://jom.unri.ac.id/ index.php/JOMFAPERIKA/article/view/6949/6637.

[23] Saputri, Endah S.H. 2014. Performance Evaluation Of Wastewater Treatment Plant In Rusunawa Tanah Merah II Surabaya. (online), (http://repository.its.ac.id/48901/1/3310100100-Undergraduate-Theses.pdf), diakses 15 November 2018.

[24] Crites, R, & Tchobanoglous. 1998. Small and Decentralized Wastewater Management System. (online), (https://sswm.info/node/285) diakses 04 Desember 2018.

[25] Kamalpoor, Abdulsalam Mohamed Saleh. 1996. Application of Tube Settlers in Secondary Wastewater Treatment. (online), (https://scholarworks.uaeu. ac.ae), diakses 16 September 2018.

[26] Hasanuddin. 2011. Kondisi Ph Terhadap Denitrifikasi Air Limbah Nitrogen Menggunakan Reaktor Berbahan Isian Batu Belerang Dan Batu Kapur. Jurnal Inovasi, Vol. 08, No. 02. Dari http://ejurnal.ung.ac.id/index. php/JIN/article/view/747.

[27] Zhang, Tao., Ding, Lili., Ren, Hongqiang. 2009. Pretreatment of ammonium removal from landfill leachate by chemical precipitation. Journal of hazardous materials, Vol 166, hal 911-915. DOI: https://doi.org/10. 1016/j.jhazmat.2008.11.101.

[28] McKean, Tony., Brian, Bourke., Warren, Mitchell., Lara, Caplygin., Rory, McKeown. 2010. Novel Application Of A Lamella Clarifier For Improved Primary Treatment Of Domestic Wastewater. Annual Water Industry Engineers and Operators' Conference. (Vol:123). Dari: http://www. wioa.org.au/conference_papers.

[29] Sabli, Edy T & Zahrah, Siti. 2015. Reduksi kandungan fosfat dalam air limbah deterjen menggunakan sistem rawa bambu. Jurnal Dinamika Pertanian Vol 30 No 2 Dari http://journal.uir.ac.id/index.php/dinamika pertanian/article/view/802/480.

[30] Suhardjo, Dradjat. 2008. Penurunan cod, tss dan total fosfat pada septic tank Limbah mataram citra sembada catering dengan menggunakan wastewater garden. Jurnal Manusia & Lingkungan Vol 15 No 2. DOI: https://doi.org/ 10.22146/jml.18681.

[31] Yulistyorini, Anie., Miller Alonso Camargo-Valero, Sukarni-Sukarni, Nugroho Suryoputro, Mujiyono-Mujiyono, Hadi Saputro, and Endang Tri Rahayu. 2019. Performance of Anaerobic Baffled Reactor Decentralised Wastewater Treatment in Urban Malang, Indonesia. Processes, Vol. 7, No. 184. https://doi.org/10.3390/pr7040184

Acknowledgement

The authors are very grateful to the Ministry of Research Technology and Higher Education of the Republic of Indonesia through grant number no. 1.3.55/UN 32.14/LT/2018 for the financial support given to this study.

Structural performance of multi-storey building using flat slab and conventional slab to seismic loads (Case study: Faculty building of Sport Science in Universitas Negeri Malang, Indonesia)

R A Pradhana[1], M M A Pratama[1], E Santoso[1] and A Karjanto[1]

[1] Department of Civil Engineering, Faculty of Engineering, Universitas Negeri Malang, Indonesia

mirza.abdillah.ft@um.ac.id

Abstract. Innovations to increase the effective height of storeys in buildings are using a flat slab system. Flat slab system is a slab that is only supported by columns without any supporting beam. By the abstain of beams, designers are able to reduce the storey height and to reduce the structural weight. Other than that, flat slabs offer a more flexible installation of building utilities, simpler reinforcement, and simpler installation of scaffolding and formwork. This study was conducted to analyse the performance of flat slabs when applied to buildings compared with the conventional slabs. The flat slab system is expected to be able to withstand gravity and earthquake loads. The Building of the Faculty of Sport Science, Universitas Negeri Malang, with a total of 7 floors and 1 semi-basement was used as research object. This research was conducted using programme simulation, ETABS and SAP2000. The research was intended to analyse the resulting seismic base shear, the fundamental period, the structural stiffness, the ultimate performance evaluation, and the storey drift if the building designed using flat slab system. Results showed that (1) Building with flat slab system has a lighter weight than the one with conventional slab system. Building weight affects the seismic base shear value of the building. The heavier a building is, the greater the seismic base shear value of a building; (2) The fundamental period analysis resulted the structural vibration time (Tc) in flat slab structure is greater than in the conventional slabs. The heavier the building weight, the structural vibration time (Tc) that occurs will be smaller; (3) The structural stiffness of building using flat slab system is less than in the conventional slabs system; (4) Both of building using flat slab system and conventional slabs meet the requirement of service performance and ultimate performance; and (5) The storey drift in building using flat slab system is larger than in the conventional slabs. This is due to the conventional slab structure has a greater stiffness than flat slab structure.

1. Introduction

The slab is a structural component in a building that functions to withstand the load transversely through bending action to each support. There are several types of slab construction systems, some of which are flat slabs and conventional slabs. The conventional slab is a slab supported with beams and columns and widely used in construction. Columns on conventional slabs function as elements that carry the load to the foundation, while the beam functions as a horizontal reinforcement framework of the loads.

However, the presence of beams in conventional slab systems will reduce the effective height of storeys in buildings [1-3].

Innovations to increase the effective height of storeys in buildings are using a flat slab system. Flat slab system is a slab that is only supported by columns without any supporting beam. The load from the slabs will be distributed directly to the supporting columns [4]. Flat slabs are often used in offices, residences, or other industrial facilities with medium height (medium-rise office) because of the efficiency of the span or thickness ratio; and economic aspects because it reduces floor height [5]. By the abstain of beams, designers are able to reduce the storey height and to reduce the structural weight. Other than that, flat slabs offer a more flexible installation of building utilities, simpler reinforcement, and simpler installation of scaffolding and formwork. However, due to the absence of beams, it reduces the ability of column-slab in holding the shear force that can cause a horizontal spread of damage and cause the slab to collapse (punching shear) [6-8]. It agrees with previous research findings stating that the behaviour of the flat slab structure is proven to be able to withstand the gravity load, but it is weak in accepting lateral loads [9-10].

Research shows that flat slab is still rarely used in buildings. This is because the plates in the flat slab system are thicker than the plates in conventional slab systems [2-3]. Therefore, this study was conducted to analyse the performance of flat slabs when applied to buildings. The flat slab system is expected to be able to withstand gravity and earthquake loads, like a diaphragm structure, thereby reducing structural failure.

Therefore, the Building of the Faculty of Sport Science, Universitas Negeri Malang is a tall building with a total of 7 floors and 1 semi-basement. In the current design, it uses conventional reinforced concrete slabs. With all these considerations, this building will be used as research object. The researchers intended to analyse the resulting seismic base shear, fundamental period, structural stiffness, ultimate performance threshold, storey drift if the building designed using flat slab system. The building is modelled using ETABS and SAP2000 with accommodating Indonesian code as provision (SNI 1727:2013; Peraturan Pembebanan Indonesia; SNI 1726:2012).

2. Research methodology

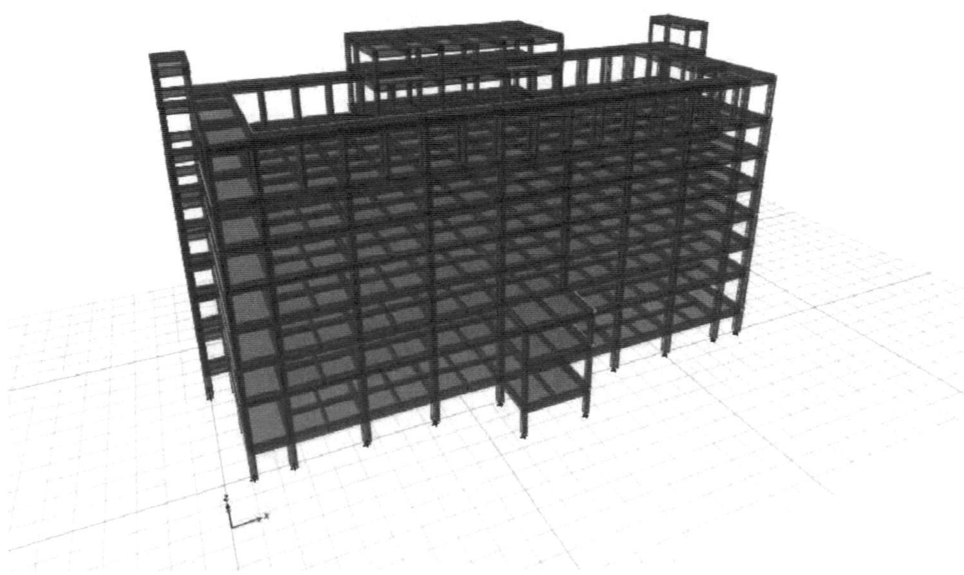

Figure 1. Building structure using conventional slabs

This research was conducted using programme simulation, ETABS. The research method was initiated with the literature review, defining design criteria, structural modelling, data analysis, and drawing conclusion.The steps of building modelling in ETABS was begun with defining provision, material input, element properties, creating structures, defining working loads, and analysing the model.

Faculty building of Sport Science Universitas Negeri Malang consists of 7 storeys with a typical height of 4 metres. The building is designed for teaching purpose. All of the building members is designed with 30 MPa concrete strength and 400 MPa of steel strength. The building structure designed with conventional slabs are shown in Figure 1; meanwhile the building structure designed with flat slabs are shown in Figure 2.

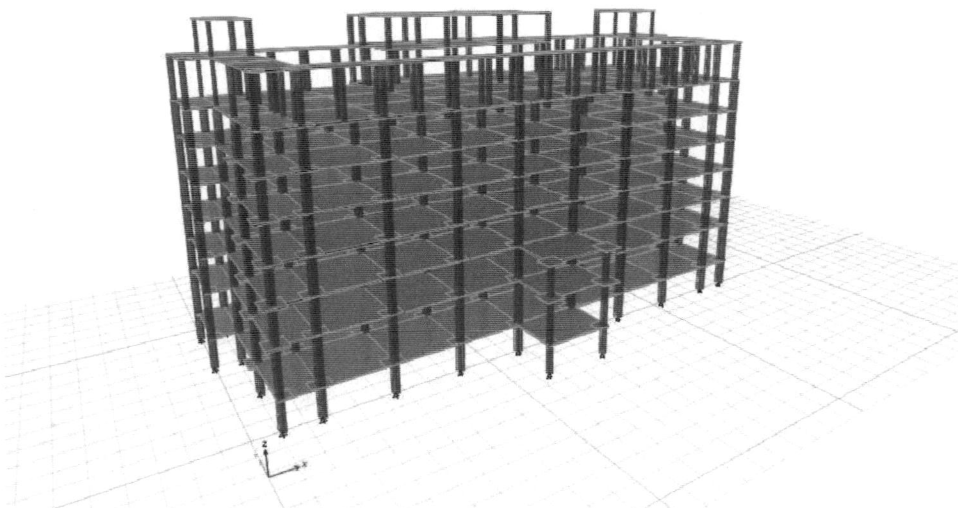

Figure 2. Building structure using flat slabs

The earthquake load on this building structure is designed based on SNI 1726: 2012. The building is located in Malang, Indonesia. The earthquake analysis is approached with response spectrum analysis. The soil mechanics data and earthquake acceleration data are approached using data in the Malang area. The slabs are considered as a rigid diaphragm.

3. Results and discussion

3.1. Seismic base shear

The result of seismic base shear of building using conventional slab and flat slab system is shown in Table 1. Building with flat slab system has a lighter weight than the one with the conventional slab system. The conventional slab system is 12% heavier than the flat slab. This is because the flat slab system does not use beams. Building weight affects the seismic base shear value of the building. The heavier a building is the greater the seismic base shear value of a building.

Table 1. Seismic base shear of building using conventional slab and flat slab system

Slab system	Building weight (kg)	$85\%V_{static}$ (kg)	$V_{dynamic}$ X (kg)	$V_{dynamic}$ Y (kg)
Conventional slab	10 084 090	903.219	3 705. 061	2 111. 518
Flat slab	8 914 885	795.654	2 352. 817	1 375. 845

The dynamic response of the building structure must not be less than 85% of the static shear force of the structure. The x-direction of dynamic shear force on conventional slabs is 35% larger than the flat slab, while the y-direction dynamic shear force on conventional slabs is 33% greater than the flat slab. These findings agree with More's research stating the results of the dynamic base shear values in the x and y directions of conventional slab systems larger than flat slab structures. More [11] also stated that the increase in the basic shear force with changes in building weight is crucial.

3.2. Fundamental period

The fundamental period of the building using conventional slab and flat slab system is shown in Table 2. The maximum period of a structure is influenced by the height of the building, the type of structure, and the earthquake spectral acceleration, while the vibrational time of the structure is influenced by the elevation and weight of the building. Basically, the vibrational period of the structure is strongly influenced by mass and stiffness.

Table 2. The fundamental period of building using conventional slab and flat slab system

Slab system	T_{max} (seconds)	Tc (seconds)
Conventional slab	13.511	1.669
Flat slab	13.511	5.168

The fundamental period analysis resulted in the structural vibration time (Tc) on the conventional slabs equal to 1.669 seconds, 67% smaller than on the flat slab system that is equal to 5.168 seconds. This is caused by the weight of the building on the conventional slabs system is greater than the weight of the flat slab. The heavier the building weight, the structural vibration time (Tc) that occurs will be smaller. The structure vibration time (Tc) must not exceed the maximum structure vibration time (Tmax). If structural vibration time exceeds the maximum vibration time of the structure (Tmax), there will be a resonance in the building, and if there is a resonance the building will collapse. A building using a flat slab system has a greater vibration time than a building using a conventional plate structure.

3.3. Structural stiffness

The structural stiffness of building using conventional slab and flat slab system is shown in Table 3.

Table 3. The structural stiffness of building using conventional slab and flat slab system

Storey level	X-direction			Y-direction		
	Conventional slab N/mm	Flat slab N/mm	Ratio	Conventional slab N/mm	Flat slab N/mm	Ratio
(1)	(2)	(3)	(4) = (2)/(3)	(5)	(6)	(7) = (5)/(6)
7	803.827	69.404	12x	755.933	74.019	10x
6	957.923	81.308	12x	923.550	105.150	9x
5	995.167	84.344	12x	958.469	109.300	9x
4	1.014.538	90.227	11x	980.176	113.393	9x
3	996.631	93.051	11x	984.357	136.915	7x
2	622.190	86.546	7x	677.081	140.400	5x
1	2.107.832	260.842	8x	2.255.716	426.199	5x

The average floor stiffness from 1st to 7th level in the x-direction of conventional slab system is 10 times greater compared to the flat slab system, whereas in the y-direction the conventional slab system is 8 times larger compared to the flat slab. This is because the conventional slab has a greater weight than the flat slab. Beams in conventional slabs have a big influence on structural stiffness. In the calculation of structural stiffness, it requires data of column stiffness and beam stiffness. In a flat slab system, the plate replaces the role of the beam. The stiffness of the x-direction and y-direction in both conventional and flat slab has significant differences. This is because the configuration of different dimensions between columns. The buildings using conventional slab system have a greater stiffness than buildings using flat slab. He also stated that the stiffness is influenced by the weight of the building and the configuration of the structure.

3.4. Ultimate performance limit

The performance evaluation to service performance limit and ultimate performance limit of building using conventional slab and flat slab system is shown in Table 4 and Table 5.

Table 4. Performance evaluation to service performance limit

Slab system	Top displacement (mm)	Service performance limit (mm)
Conventional slab	3.138	15
Flat slab	4.263	15

The top displacement of the conventional slab is 28% smaller than the flat slab. This is because the conventional slab has a greater weight than the flat slab. The great building weight causes large structural stiffness and small structural vibrational time (Tc). If the structural vibration time (Tc) is small, the displacement that occurs is small. Based on the results of the analysis, the performance in service stage is still safe both in conventional slabs and in the flat slabs.

Table 5. Performance evaluation to the ultimate performance limit

Slab system	0,02h (mm)	Ultimate performance limit (mm)
Conventional slab	80.00	16.453
Flat slab	80.00	25.553

Based on Table 5, the performance of building in the ultimate limit on the conventional slabs is 36% smaller than in the flat slab. This is because the top displacement on conventional building is smaller than the displacement value of the flat slab structure. The function of the ultimate limit performance is to find out the maximum drift of structures.

3.5. Storey drift

The storey drift of building using conventional slab and flat slab system is shown in Table 6. The storey drift in the conventional building is 56% smaller than in the building using flat slab. This is due to the conventional slab structure has a greater stiffness than flat slab structure. The stiffness is inversely proportional to the value of the storey drift. The occurring storey drift does not exceed the limit, indicating that the structural performance per storey is still safe.

The graph representing the relationship of drift on each storey level is shown in Figure 3. Figure 3 shows that building with the conventional slab and the flat slab exhibits a significant increase in the 4th to 7th storey. It is due to the structural stiffness of the upper floor decreased significantly. This finding agreed with Mohana [12] which stated that buildings using conventional slab structures have smaller floor deviations than buildings using flat slab.

Table 6. Storey drift of building using conventional slab and flat slab system

Storey level	h (mm)	Δ flat slab (mm)	Δ conventional slab (mm)	(0.03/R)*h (mm)
7	4000	14.631	11.506	15
6	4000	12.511	9.658	15
5	4000	10.010	7.025	15
4	4000	6.879	4.961	15
3	4000	5.269	2.658	15
2	5500	3.197	1.599	15
1	3500	1.478	0.961	15

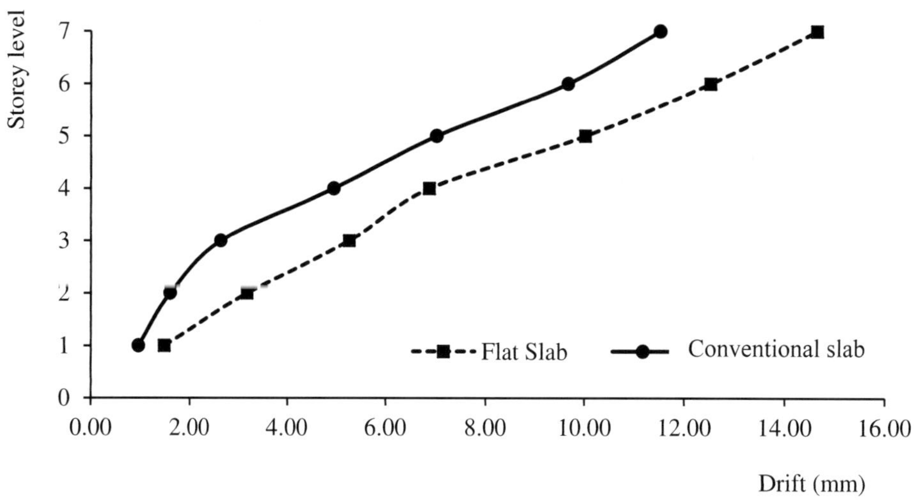

Figure 3. Storey drift of building using conventional slab and flat slab system

4. Conclusion

- Building with flat slab system has a lighter weight than the one with the conventional slab system. Building weight affects the seismic base shear value of the building. The heavier a building is the greater the seismic base shear value of a building.
- The fundamental period analysis resulted in the structural vibration time (Tc) in flat slab structure is greater than in the conventional slabs. The heavier the building weight, the structural vibration time (Tc) that occurs will be smaller.
- The structural stiffness of building using flat slab system is less than in the conventional slabs system.
- Both of building using flat slab system and conventional slabs meet the requirement of service performance and ultimate performance.
- The storey drifts in the building using a flat slab system are larger than in the conventional slabs. This is due to the conventional slab structure has a greater stiffness than flat slab structure.

References

[1] Bhaduria SS, Chhugani N. Comparative Analysis and Design of Flat and Grid Slab System With

Conventional Slab System. Int Res J Eng Technol. 2017;
[2] B.Anjaneyulu KJP. Analysis And Design Of Flat Slab By Using Etabs Software. Int J Sci Eng Adv Technol. 2016;
[3] Khan KM, Jeelani M. Analysis And Design Of Flat Slab By Using Etabs Software. Int Res J Eng Technol. 2018;
[4] Dilger W. Flat slab-column connections. Prog Struct Eng Mater. 2000;
[5] Manvi A, Gouripur S, Sambrekar P. Cost Comparison Between Conventional and Flat Slab Structures. Int Res J Eng Technol. 2015;
[6] Mirzaei Y, Sasani M. Punching shear failure in progressive collapse analysis. In: Structures Congress 2011 - Proceedings of the 2011 Structures Congress. 2011;
[7] Nguyen-Minh L, RovňÁk M, Tran-Quoc T, Nguyen-Kim K. Punching shear resistance of steel fiber reinforced concrete flat slabs. In: Procedia Engineering. 2011;
[8] Meenathethil Alex S, Lekshmi S. Study of Progressive Collapse Analysis of Flat Slab Building. International Journal of Science and Research. 2013;
[9] Hemawan L. Studi Perumusan Lebar Efektif Pelat pada Struktur Flat Plate Akibat Beban Gempa. Undergrad Theses, Civ Eng RSS 624177 65 Hem s, 2007. 2008;
[10] Rao R, Gowda R. Comparative Study of Equivalent Static Analysis and Response spectrum analysis on Flat Slab Using Etabs. Int Res J Eng Technol. 2016;
[11] More RS, Sawant VS, Suryawanshi YR. Analytical Study of Different Types of Flat Slab Subjected to Dynamic Loading. International Journal of Science and Research. 2013;
[12] Mohana HS, Kavan MR. Comparative Study of Flat Slab and Conventional Slab Structure Using ETABS for Different Earthquake Zones of India. Int Res J Eng Technol. 2015.

IOP Conf. Series: Materials Science and Engineering **669** (2019) 012054 doi:10.1088/1757-899X/669/1/012054

Flexural behaviour of the functionally graded concrete beams using two-layers and three-layers configuration

M M A Pratama[1], T Vertian[1], B S Umniati[1] and W H Yoh[1]

[1] Department of Civil Engineering, Faculty of Engineering, Universitas Negeri Malang, Indonesia

mirza.abdillah.ft@um.ac.id

Abstract. Studies related to the bending behaviour of functionally graded concrete (FGC) beams are still investigated to explore the possibility of the more cost-efficient casting techniques and to produce high-performance FGC beam elements simultaneously. Previous researches show that the manufacture of FGC beams by combining two concrete mixes in two layers can increase the stiffness of the elements but it reduces the beam ductility. Research concentrating on the FGC beams with additional numbers of concrete layers was analysed in this study. Four pieces of FGC beams measuring 120 x 240 x 2200 mm were prepared with concrete strength of 25 MPa (NA); 30 MPa(NB); 30-20 MPa (GBA); and 30-20-30 MPa (GBAB). The specimens were tested in the laboratory using a four-point bending method. The experimental data consisting of the load increments and the mid-span deflections are further analysed to compare the resulting (1) maximum load-carrying capacity; (2) maximum deflection; (3) beam stiffness; (4) moment-curvature; and (5) ductility. The result shows that (1) the GBAB exhibited 0.83% greater maximum loads than the GBA; (2) the GBAB deflected 29.52% lesser than the GBA; (3) the GBAB posses stiffness 12.03% greater than the GBA; (4) the GBAB exhibited higher resulting yield points and ultimate state than the GBA at the moment-curvature relationship; and (5) the GBAB was 20.06% less ductile than the GBA.

1. Introduction

Research related to functionally graded concrete (FGC) is still continuing as environmental issues require human-being to reduce the use of cement in construction and to reduce the contribution of carbon dioxide to the air [1-3]. FGC is one of the concrete innovations that is proven to reduce the use of cement while still maintaining the performance of the structures. Applying FGC in building structures could reduce 6.48 – 9.72 % of cement use. The FGC is created by combining 2 (two) or more concrete mixes that possess a distinct concrete strength into structural members [4]. This concrete technology optimises the use of concrete materials according to its characteristics. The FGC configures the higher concrete strength to concentrate on the extreme compression fibre and the lower concrete strength on the extreme tensile fibres. In structural analysis. the tensile strength of concrete is neglected as the presence of longitudinal rebar that will replace the concrete in withstanding the tensile stress [5-10].

Previous studies showed that FGC beam created from two layers of concrete mixes with different strength resulted in higher structural stiffness than the beam cast from lower concrete strength. The increase of beam stiffness is accompanied by the increase of resulting maximum loads and the decrease of midspan deflection [11-12]. Unfortunately, the increase of structural performance is somewhat not significant and requires further investigation of this research programme.

In this research, authors would like to investigate the possibility to increase the structural performance of the FGC beam by adding the number of concrete layers from two to three layers but still using two concrete mixes. The higher concrete strength is planned to be cast at both of the compression and the tensile fibre of beams to improve the bonding strength between the longitudinal rebar and the concrete; meanwhile the lower concrete strength is situated at the middle depth area of the beam as the lesser stress occurs around neutral axis so it does not require high reinforcement from concrete and rebars. In the end, the authors would like to compare the resulting data from 3 layers FGC beam with the 2 layers FGC beam in terms of (1) maximum load-carrying capacity; (2) maximum deflection; (3) beam stiffness; (4) moment-curvature; and (5) ductility

2. Research methodology

2.1. Specimens preparation
This research was conducted experimentally at the Laboratory of Material Testings and Structure. Universitas Negeri Malang, Indonesia. Two groups of specimens were prepared, which were controlling specimens and testing specimens. The controlling specimens consisted of conventional reinforced concrete (RC) beams for the bending test; meanwhile, the testing specimens only consisted of FGC beams created from two layers of concrete mixes and FGC beams from the layers of concrete mixes. Details of specimens used in this research are shown in Table 1.

Table 1. Specimens detail

Specimen Group	Notation	Description	Amount (pieces)
Controlling specimens	NA	Conventional RC beams 25 MPa	2
	NB	Conventional RC beams 30 MPa	2
Testing specimens	GBA	Reinforced FGC beams with 2 layer configuration (30 – 25 MPa)	2
	GBAB	Reinforced FGC beams with 3 layer configuration (30 – 25 – 30 MPa)	2

2.2. Reinforced concrete beam design
The beams for controlling specimens and the testing specimens were expected to undergo failure in flexure due to the design of a/d = 3.04. All of the RC beams ware prepared to have 2200 mm in length and 120 mm x 240 mm for the cross-sectional dimension. The beams were doubly reinforced with 2D10 and 2D10 for the longitudinal rebars and were installed with d6-100 for stirrups in supports and d6-175 in midspan.

2.3. FGC casting configuration
Both GBA and GBAB specimens are made of two concrete mixes. those were 30 MPa and 25 MPa. The gradation pattern was created vertically. Further, for the GBA. the 30 MPa concrete mix was placed at the compression fibre while the 25 MPa mix was at the tensile fibre. The two concrete mix shares the depth of beams by half. Intensive compaction procedures were conducted using shaft vibrator to eliminate distinct transition of two different concrete mixes. For the GBAB, the concrete casting was initiated by placing the 30 MPa mix by one-third part of beam depth, continued by placing the 25 MPa mix on the previous layer by another one-third part of beam depth, and was finished by pouring the 30 MPa mix until full. The vibration was also applied to each stage of concrete casting to create a smooth transition of concrete strength. The illustration of concrete casting configuration for GBA and GBAB are displayed in Figure 1.

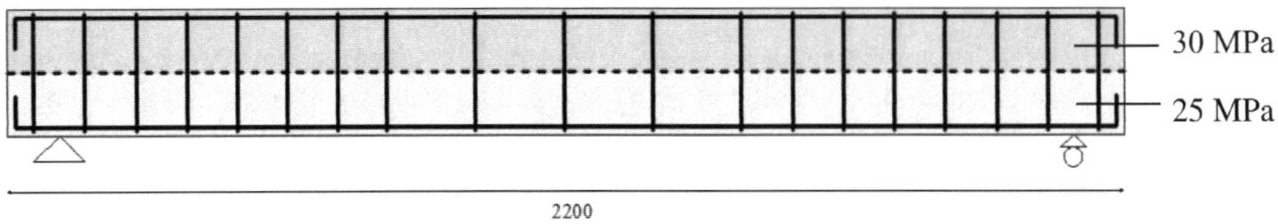

(a) GBA

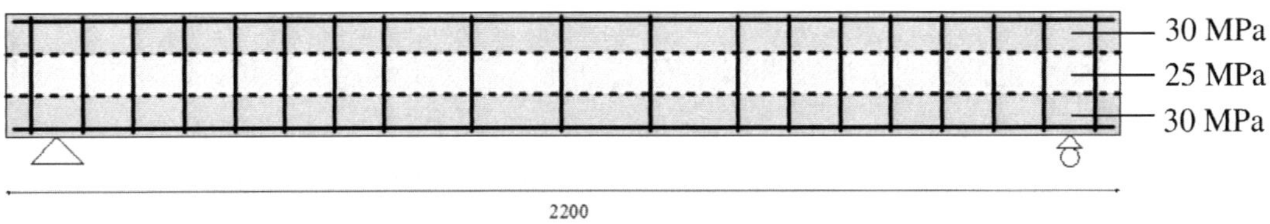

(b) GBAB

Figure 1. Casting configuration of (a) GBA and (b) GBAB

2.4. Four-point bending test

The 28 days old RC beams were set on the loading frame for conducting four-point bending test. To ease the observation of crack propagation during the test, the beams were all dyed white and grid using a marker. The author utilised hydraulic jack, load cell, loading head, data logger, and dial gauges during the test. The data consisting of midspan deflection and incremental loads were analysed from the reading of the dial gauges and the data logger. The test setup is illustrated in Figure 2.

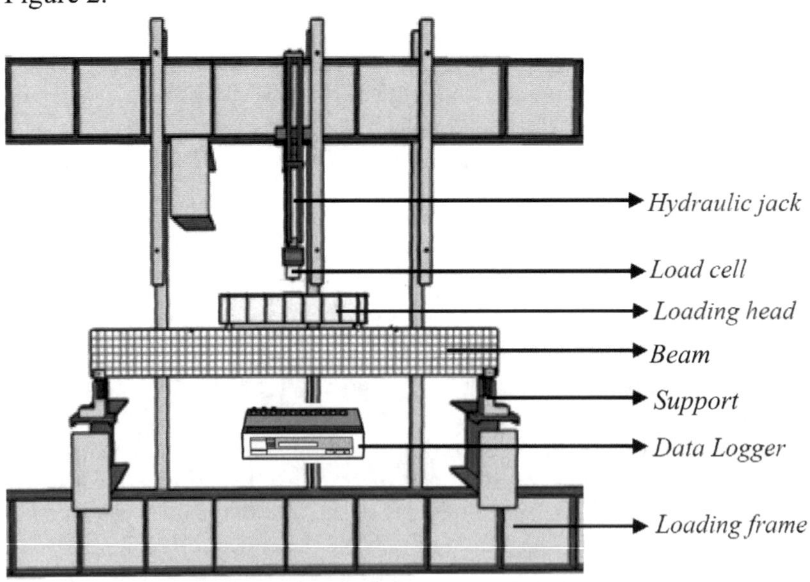

Figure 2. Test setup for bending test

3. Results and discussion

3.1. Load – midspan deflection relationship

The load during the testing is obtained by the reading of the load cell that was connected to the data logger. The rupture condition was obtained by observing the beam surface during the testing. The first rupture occurred at the midspan of all of the RC beam which confirmed that the beams underwent the failure in flexure. The load-deflection interaction of beams is displayed in Figure 3.

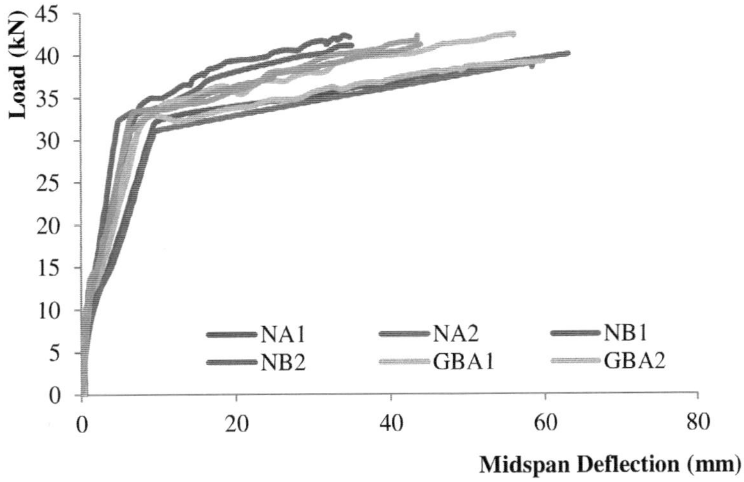

Figure 3. Load – deflection of FGC beams and conventional beams

The resulting loads at rupture, yield point, and ultimate condition are presented in Table 2 and Table 3.

Table 2. Load at rupture, yield, and ultimate condition

Specimens	Load (kN)		
	Rupture	Yield	Ultimate
NA1	10.81	31.22	39.39
NA2	8.56	30.50	39.45
Average	9.68	30.86	39.42
NB1	10.58	32.35	42.28
NB2	12.30	32.42	41.11
Average	11.44	32.29	41.70
GBA1	9.60	32.19	39.19
GBA2	11.29	31.51	42.41
Average	10.45	31.85	40.80
GBAB1	11.12	32.59	40.59
GBAB2	10.81	31.28	41.68
Average	10.97	31.94	41.14

Table 3. Maximum load ratio

Specimens	NA		NB		GBA	
	Load difference (kN)	%	Load difference (kN)	%	Load difference (kN)	%
GBA	1,38	3,50	-0,90	-2,16	-	-
GBAB	1,72	4,36	-0,56	-1,35	0,34	0,83%

IOP Conf. Series: Materials Science and Engineering **669** (2019) 012054 doi:10.1088/1757-899X/669/1/012054

Table 3 shows that the addition of a concrete layer in FGC beams (GBAB) increases the load-carrying capacity by 0.83% as compared with the GBA and by 4.36% as compared with the NA. The present of the higher concrete strength in tensile fibre increases the bond strength and increase the beam resistant to cracking. The resulting midspan deflection at rupture, yield point, and ultimate condition are presented in Table 4 and Table 5.

Table 4. Midspan deflection at rupture, yield, and ultimate condition

Specimens	Midspan deflection (mm)		
	Rupture	Yield	Ultimate
NA1	1.31	9.22	87.65
NA2	1.20	9.33	58.41
Average	1.26	9.28	73.03
NB1	0.74	4.95	34.07
NB2	0.94	6.92	35.14
Average	0.84	5.94	34.61
GBA1	0.95	12.76	59.90
GBA2	1.11	7.50	56.09
Average	1.03	10.13	58.00
GBAB1	1.13	6.41	41.87
GBAB2	0.56	6.71	38.89
Average	0.85	6.56	40.88

Table 5. Maximum deflection ratio

Specimens	NA		NB		GBA	
	Deflection difference (mm)	%	Deflection difference (mm)	%	Deflection difference (mm)	%
GBA	-15,03	-20,58	23,39	67,58	-	-
GBAB	-32,15	-44,02	6,27	18,12	-17,12	-29,52

Table 5 shows that the GBAB reduces the midspan defection by 29.52% as compared with the GBA. The higher concrete strength on tensile fibre contributes to the increase of structural stiffness and to strengthen the resistant of deflection.

3.2. Beam stiffness

The beam stiffness was analysed from the load-deflection relationship of each beam specimens. The beam stiffness is calculated by dividing load to deflection at each stage of failure, which was at rupture, yield, and ultimate condition. Table 6 shows the resulting beam stiffness on each failure stage. Table 6 shows that the more percentage of higher concrete strength in structural members, it increases the resulting stiffness during rupture, yield, and ultimate condition. The highest stiffness was found at the NB regarding it was cast using a full 30 MPa concrete mix. For the FGC specimens, the highest stiffness was on the GBAB due to the use of 2 layers of 30 MPa mix on outer fibre that improve the structural rigidity. The beam stiffness is incrementally decreasing as the crack propagate and reducing the moment inertia of beam cross-section.

Table 6. Beam stiffness at rupture, yield, and ultimate condition

Specimens	Beam stiffness (kN/mm)		
	Rupture	**Yield**	**Ultimate**
NA1	8.25	3.39	0.48
NA2	7.13	3.27	0.68
Average	7.69	3.33	0.58
NB1	14.3	6.54	1.24
NB2	13.09	4.68	1.17
Average	13.69	5.61	1.21
GBA1	10.11	2.52	0.65
GBA2	10.17	4.2	0.76
Average	10.14	3.36	0.71
GBAB1	9.84	5.08	0.97
GBAB2	12.87	4.66	1.04
Average	11.35	4.87	1.01

3.3. Moment – curvature relationship

Moment – curvature represents the incremental curvature of structures when a bending moment is applied. For this research, the graph of moment-curvature is generated by analysing the load-deflection data obtained during experimental testing. The moment-curvature interaction of beams is presented in Figure 4.

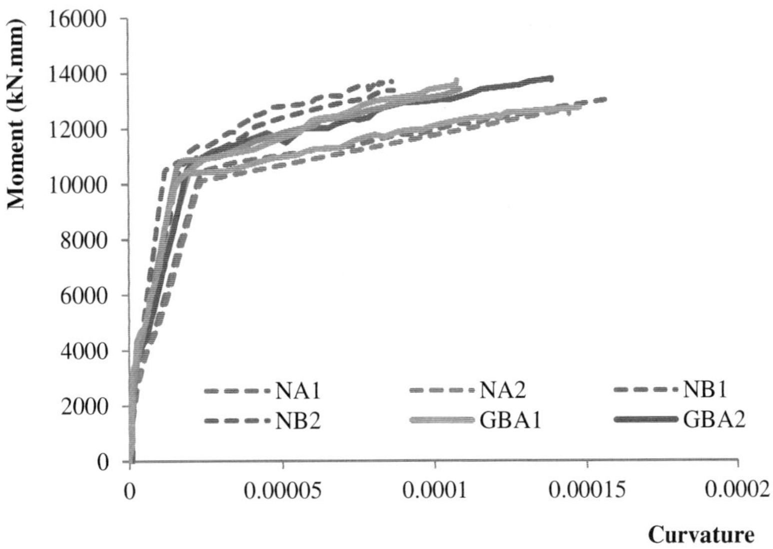

Figure 4. Moment – curvature relationship

Figure 4 shows that the moment-curvature interaction resulted from the bending test follow the type of load-deflection data. The data show that the GBAB exhibited lesser curvature from the beginning of the loading phase until the beams failed. The GBA exhibited moment-curvature interaction that is similar to the NA.

3.4. Structural ductility

Ductility is the structural ability to withstand an inelastic response and is calculated by comparing the curvature at the ultimate point to the yield point. The curvature data is obtained from the moment-curvature relationship in the preceding part. The structural ductility of each beam is shown in Table 7.

Table 7. Ductility of beams

Specimens	Ductility
NA	7.87
NB	5.83
GBA	7.48
GBAB	6.23

Table 7 shows that the addition of high strength concrete layer in the GBAB decreases the ductility of the structures by 20.06% than the GBA and alter the characteristic to more brittle in nature. The data also shows that the ductility of GBAB is comparing the ductility of NB.

4. Conclusion

The research resulted that (1) the GBAB exhibited 0.83% greater maximum loads than the GBA; (2) the GBAB deflected 29.52% lesser than the GBA; (3) the GBAB posses stiffness 12.03% greater than the GBA; (4) the GBAB exhibited higher resulting yield points and ultimate state than the GBA at the moment-curvature relationship; and (5) the GBAB was 20.06% less ductile than the GBA.

References

[1] Berardi U. Sustainable Construction: Green Building Design and Delivery. Intell Build Int. 2013;
[2] Bignozzi MC. Sustainable cements for green buildings construction. In: Procedia Engineering. 2011.
[3] Pratama MMA, Pratiwi YD, 'Ilmi MA, Mahardika A. GECOBRIDGE: A concept of a smart eco-friendly integrated bridge utilizing geopolymer and graded concrete technology. MATEC Web Conf [Internet]. 2018;204. Available from: https://doi.org/10.1051/matecconf/201820406005
[4] Pratama MMA. An experimental study and finite element approach to the behaviour of graded concrete. Master's thesis. Diponegoro University. 2015.
[5] Gan BS, Aylie H, Pratama MMA. The Behavior of Graded Concrete, an Experimental Study. Procedia Eng [Internet]. 2015;125:885–91. Available from: http://linkinghub.elsevier.com/retrieve/pii/S1877705815033937
[6] Han A, Gan BS, Pratama MMA. Effects of Graded Concrete on Compressive Strengths. Int J Technol [Internet]. 2016 Jul 28;7(5):732. Available from: http://www.ijtech.eng.ui.ac.id/index.php/journal/article/view/3449
[7] Pratama MMA, Aylie H, Gan BS, Umniati BS, Risdanareni P, Fauziyah S. Effect of concrete strength gradation to the compressive strength of graded concrete, a numerical approach. AIP Conf Proc [Internet]. 2017 Sep 29;1887(1):20029. Available from: https://aip.scitation.org/doi/abs/10.1063/1.5003512
[8] Pratama MMA, Umniati BS, Arumsari Mutiara Wulandari B, Han AL, Sthenly Gan B, Zhabrinna Z. Modulus elasticity of the graded concrete, a preliminary research. MATEC Web Conf [Internet]. 2018;195. Available from: https://doi.org/10.1051/matecconf/201819501005
[9] Pratama MMA, Sthenly Gan B, Han Ay Lie HAL, Nur Rahma Putra AB. A Numerical Analysis of The Modulus of Elasticity of The Graded Concrete. In: 2nd International Conference on Vocational Education and Training (ICOVET 2018) [Internet]. Atlantis Press; 2019. Available from: https://doi.org/10.2991/icovet-18.2019.29
[10] Pratama MMA, Umniati BS, Prasiwi G, Han AL, Sthenly Gan B, Risdanareni P, et al. Effect of monotonic lateral load on the performance of reinforced graded concrete column, an

experimental study. MATEC Web Conf [Internet]. 2018;195. Available from: https://doi.org/10.1051/matecconf/201819502022

[11] Aylie H, Gan BS, As'ad S, Pratama MMA. Parametric Study of the Load Carrying Capacity of Functionally Graded Concrete of Flexural Members. Int J Eng Technol Innov [Internet]. 2015;5(4):233–41. Available from: http://www.scopus.com/inward/record.url?eid=2-s2.0-84957600233&partnerID=MN8TOARS

[12] Pratama MMA, Suhud RK, Puspitasari P, Kusuma FI, Rahma Putra ABN. Finite element analysis of the bending moment-curvature of the double-layered graded concrete beam. IOP Conf Ser Mater Sci Eng [Internet]. 2019 Mar 29;494:012064. Available from: http://stacks.iop.org/1757-899X/494/i=1/a=012064?key=crossref.e0f2a62cfeeaf5c05c43077bf777a94b

Acknowledgement

This work was supported by PNBP research grant of Universitas Negeri Malang, under the contract number: 29.11.231/UN32.14.1/LL/2019. Authors would also deliver thank you in advanced for PT. Sika Indonesia for the material support to complete this research.

IOP Conf. Series: Materials Science and Engineering **669** (2019) 012055 doi:10.1088/1757-899X/669/1/012055

Numerical and experimental investigation on the shear resistance of functionally graded concrete (FGC) beams

M M A Pratama[1], W Arifanda[1], K Karyadi[1], N Nindyawati[1], R Sulaksitaningrum[1] and H Prayuda[2]

[1] Department of Civil Engineering, Faculty of Engineering, Universitas Negeri Malang, Indonesia
[2] Department of Civil Engineering, Faculty of Engineering, Universitas Muhammadiyah Yogyakarta, Indonesia

mirza.abdillah.ft@um.ac.id

Abstract. Functionally graded concrete (FGC) is one of the most innovative materials combining two or more different concrete strength in building elements to optimise the construction cost. Studies focussing on the behaviour of graded concrete in flexural strength have been conducted by several researchers, whilst the research on the shear behaviour of FGC beam has not been initiated yet. Investigation on the shear behaviour of RC graded concrete beam is necessary due to the fabrication process could create a cold-joint that lower the load-carrying capacity of the element. Delamination possibly presents on the casted concrete layer during the loading and turns the beam behaviour more brittle in action. To further investigate these hypotheses, a laboratory testing and a numerical analysis are conducted in this research. Three pieces of RC beams with an identical dimension of 120 x 240 x 2200 mm are prepared with the concrete strength of 20 MPa, 30 MPa, and 20-30 MPa. The specimens are then tested using a four-point bending method in the age of 28 days. A set of strain gauge is mounted in stirrups surface to record the strain response on each incremental load. To validate the result, the specimens are also modelled in three-dimension using Abaqus. The results show that there is a significant difference in the stress-strain curve resulted from the experimental works and the model simulation. It is due to the limitation of programme to model the real interaction between the rebars and the concrete. In reality, the presence of aggregate interlocking, concrete contribution in withstanding compression stress, dowel effect improves the shear resistance of beams. Additionally, the interfacial failure did not present during the testing owing to the support of stirrups that confine the transition zone of concrete layers.

1. Introduction

Research related to functionally graded concrete (FGC) is continuing both in the material scope and its application to building elements. In previous research, investigations regarding the behaviour of FGC have been carried out and resulted in the finding that FGC can increase the rigidity of the material so as to increase the serviceability of structural elements. With the increased serviceability of the structure, the structure is able to undergo a minimum deflection as compared to structures composed of low strength concrete. The advantage of FGC is to reduce construction costs while still maintaining the level of serviceability. The FGC beams only require half the volume of high strength concrete and are combined with low strength concrete to produce serviceability that is identical with the beams composed of fully high strength concrete. This scheme absolutely could save the use of cement as an expensive construction material that further optimizes the use of concrete in structural elements [1].

Preliminary research to investigate the compressive strength and modulus of elasticity of FGC has been carried out both experimentally and numerically [2-7]. Further, investigations regarding flexural behaviour that encompass the load-deflection, moment-curvature, stiffness, and ductility of FGC structures have been carried out through several research schemes. The results of the research show that the application of FGC to the structural beams can improve the performance of the structure, especially under flexural conditions [8-11].

The implementation of FGC casting is a crucial stage because the researchers should ensure that the transition of concrete layers bonded one another in creating a monolith structure. The provision of intensive mechanical compaction is carried out to accommodate these technical obstacles. Although the concrete layers transition has been eliminated through a compaction scheme, investigations regarding the shear resistance of the beam structure need to be studied to determine the shear behaviour of the beams in withstanding the working loads.

In this study, the authors intend to investigate further about the shear behaviour of FGC beams in withstanding the loads experimentally and validated using a finite element simulation program. FGC beams that have been designed are equipped with testing instruments to read shear strain during the testing. Beams with identical properties are also modelled in programme simulation and the resulting shear strain is compared with the experimental works. This research is expected to be able to complete studies related to FGC beams especially in terms of shear resistance, compared to conventional reinforced concrete beams.

2. Experimental programme

2.1. Specimens preparation

This research is split into two programmes, there were experimental study and finite element modelling. The laboratory testing was conducted at the Laboratory of Material Testing and Structure, Universitas Negeri Malang, Indonesia. The beams were modelled using Abaqus by inputting properties material obtained from the experimental test. Two groups of specimens were prepared, which were controlling specimens and testing specimens. The controlling specimens consisted of concrete cylindrical specimens for a compression test, rebars for tensile test, and conventional reinforced concrete (RC) beams for flexural test; meanwhile the testing specimens only consisted of FGC beams created from two layers of concrete mixes. The detail of specimens used in this research is shown in Table 1.

Table 1. Specimens detail

Specimen Group	Notation	Description	Amount (pieces)
Controlling specimens	CA	Cylindrical concrete specimen 25 MPa	3
	CB	Cylindrical concrete specimen 30 MPa	3
	SM	Steel bars sample of longitudinal rebars	3
	SV	Steel bars sample of stirrups	3
	NA	Conventional RC beams 25 MPa	2
	NB	Conventional RC beams 30 MPa	2
Testing specimens	GBA	Reinforced FGC beams with 2 layers configuration (30 – 25 MPa)	2

2.2. Beam design

The beams were expected to undergo failure in flexure as the design of $a/d = 3.04$ which fulfil the range of $2.5 < a/d < 6.0$. All of the RC beams ware prepared to have 2200 mm in length and 120 mm x 240 mm for the cross-sectional dimension. The beams were doubly reinforced with 2D10 at the compression fibre and 2D10 at the tensile fibre as the longitudinal rebars. To record the reading strain of stirrups, a strain gauge was installed on the stirrups locating at 650 mm from the supports precisely at the mid-depth of the beams. The stirrups were d6-100 for supports and d6-175 in midspan. The reinforcement ratio used was 0.6% and concrete cover = 15 mm. The detail of beam design is shown in Figure 1.

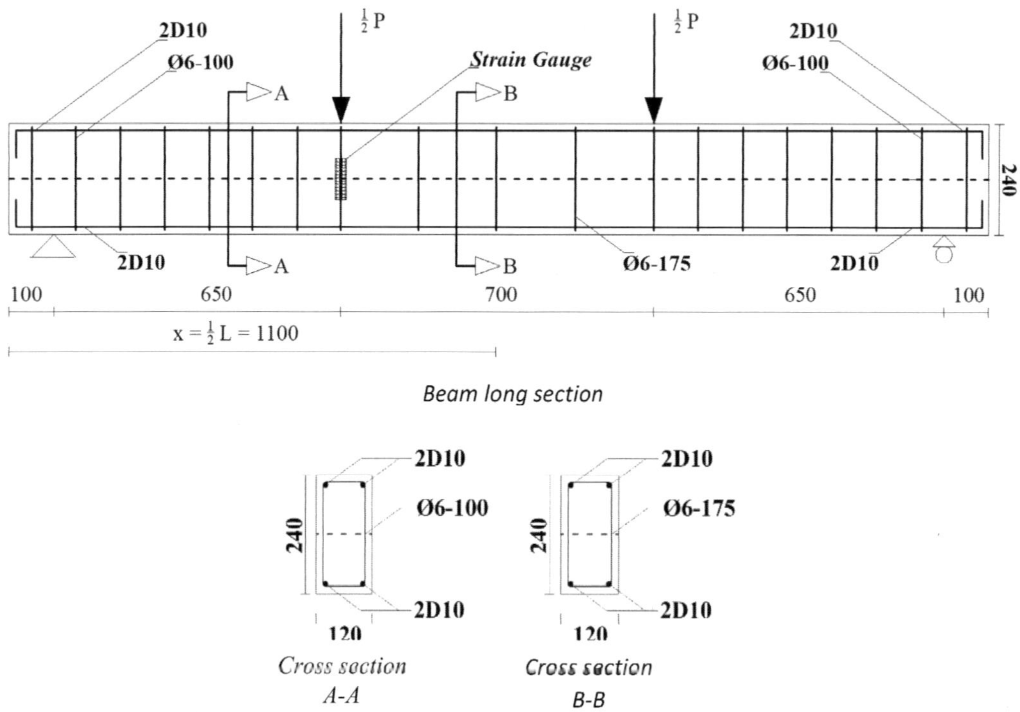

Figure 1. Beam design for the experimental programme

3. Finite element modelling

The finite element modelling is intended to verify the findings on the experimental testing. The modelling was divided into 3 (three) phase, which was preprocessing, simulation, and post-processing. In the preprocessing stage, the data of concrete compression damage, concrete tension damage, and stress-strain relationship of concrete were input to the programme as material properties. The stress-strain relationship of steel is generated from experimental data, while the stress-strain of concrete is generated from CEB FIB 2010 based on the actual strength obtained from experimental testing in laboratory. In beam modelling, the beams were modelled in 3 dimensions using Abaqus. To model the graded structure, the beam is idealized into two layers of concrete with two different strength, which was 25 MPa and 30 MPa. The rebars were modelled as was applied in experimental testing. The interaction of intra-layers of concrete is set using tie to assure that the layers stay intact and were not slip during the bending; while the rebars – concrete were set to have a perfect bond. In simulation phase, the model is run using the parameter determined. The loading step used during the iteration is static, Riks. Maximum number of increments = 100; Arc-length increment = 10-15 to 1036. In post-processing, the resulting data from program simulation is analysed to generate the load-deflection relationship and the record of stirrups strain. Illustration of beam modelling in Abaqus is shown in Figure 2.

Figure 2. Beam modelling in Abaqus

4. Results and discussion

4.1. Steel tensile strength

Two types of rebars were used in this research programme. which were 6 mm in diameter (plain) and 10 mm (deformed). For this test, three pieces of rebar samples were prepared for each rebar type. The universal tensile test was conducted using Universal Testing Machine (UTM). The resulting yield stress and ultimate stress of rebars are shown in Table 2. The result shows that the average yield stress of 6 mm rebars is 312.99 MPa. while the 10 mm is 345.55 MPa. Next. the average ultimate stress of 6 mm rebars is 418.31 MPa. while the 10 mm is 489.09 MPa

Table 2. Yield stress and ultimate stress of rebars

Specimens Notation	Yield stress (MPa)	Average yield stress (MPa)	Ultimate stress (MPa)	Average ultimate stress (MPa)
SV1	311.42	312.99	425.98	418.31
SV2	311.42		418.04	
SV3	316.12		410.90	
SM1	334.04	345.55	486.95	489.09
SM2	358.86		496.11	
SM3	343.77		484.22	

4.2. Concrete compressive strength

The cylindrical concrete specimen size of 150 x 300 mm were tested to verify the actual concrete strength cast in RC beams. Three pieces of concrete specimens were tested for each of concrete strength at the age of 28 days using UTM. The result of the concrete compression test is shown in Table 3.

Table 3. Compressive strength of concrete

Specimens Notation	Compressive strength (MPa)	Average compressive strength (MPa)
CA1	20.94	24.55
CA2	26.47	
CA3	26.25	
CB1	37.17	31.56
CB2	29.50	
CB3	28.01	

The test showed that the actual strength of 25 MPa concrete specimens was 24.55 MPa. while the 30 MPa had actual strength of 31.56 MPa.

4.3. Load – midspan deflection

The flexural test on controlling beams and testing beams resulted in the load – midspan deflection graphs that are depicted in Figure 3 - 5. Figure 3 - 5 explains that all of the specimens of NA, NB, and GBA exhibited a similar achieving load – deflection in pre-cracked condition. It corresponded to the use of concrete strength that having small difference in strength; therefore, it resulted a similar trend. A noticeable contrast is found on the yield and ultimate condition which the NB withstanding a higher load capacity with less deflection on the elements. The contribution of higher concrete strength on the GBA was proven to reduce the resulting deflection by 14% and increase the load-carrying capacity by 5.07%.

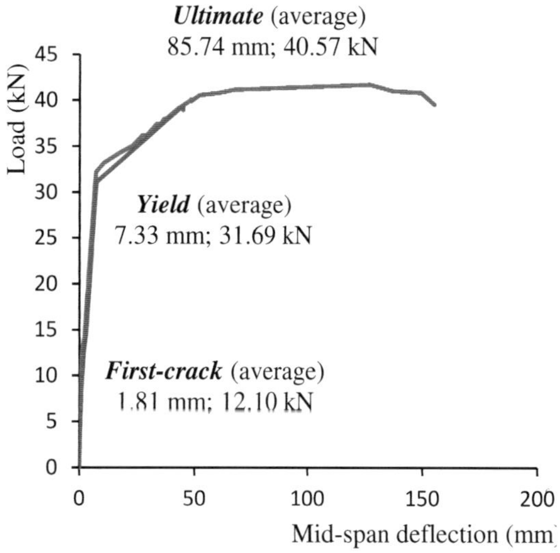

Figure 3. Load – mid-span deflection of NA

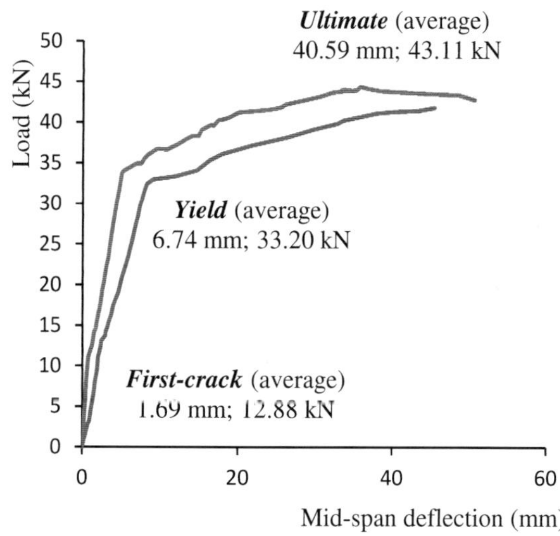

Figure 4. Load – mid-span deflection of NB

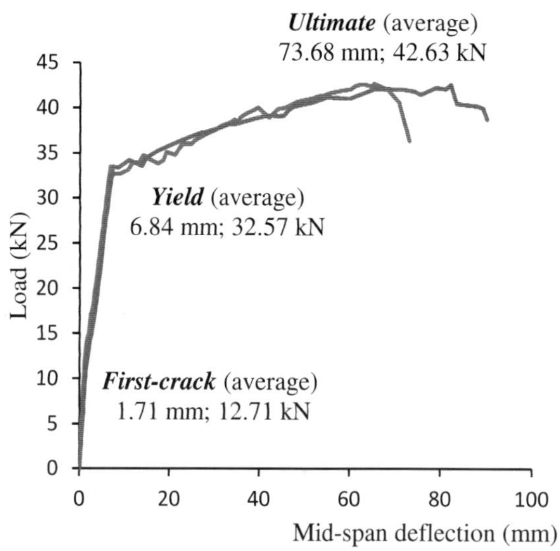

Figure 5. Load – mid-span deflection of GBA

4.4. Stress-strain response of stirrups

4.4.1. Experimental works. In experimental works, the strain was retrieved from the reading of strain gauge installed on the stirrups; while in programme simulation, the strain was generated from an element constructing the stirrups member. Figure 6 recaps the stress-strain relationship of the stirrups from experimental works. The resulting stress-strain graphs of each specimen's variable exhibited a different characteristic from one to another. The NA resulted in a less steep curve as compared with the GBA and the NB. In higher stress points, the curves changed its pattern from linear to more fluctuated due to materials behaviour and numbers of technical issues. The strain gauge was suspected not installed correctly on the stirrups surface so it generated a bias data reading. To improve the strain reading during the testing, the preparation of stirrups surface should provide a smooth area for strain gauge to adhere perfectly until the testing ended. The adhesive compound should also maintain the flexibility of strain gauge while it provided a good stickiness between the strain gauge filaments and the contacting area. The use of small strain gauge, i.e. FLA 5-11, as testing device requires good preparation and maintenance due to its characteristics that are more sensitive to environmental condition and handling.

The fabrication and installation of rebars contributed to this result. During the fabrication process, the bending angle and parts length of steel bars were not uniform. It was expected that the assembling procedure could create a perfect spacing and a perfect bond between the longitudinal bars and the stirrups. In fact, it was indicated that some part of stirrups did not installed perfectly; resulting in a noticeable gap to the longitudinal bars. These findings, of course, interfered the obtaining strain data during the bending test. This gap would allow the stirrups to produce a high reading of strain in low loading attempt. In higher level of loading, the strain to stress pattern changed and indicated a strengthening behaviour regarding when the longitudinal bars reach and adhere the stirrups, it provided higher resisting respond of the stirrups during the loading. These findings are supported by the stress-strain data that changed from a leaning pattern to a steeper trend (Figure 6).

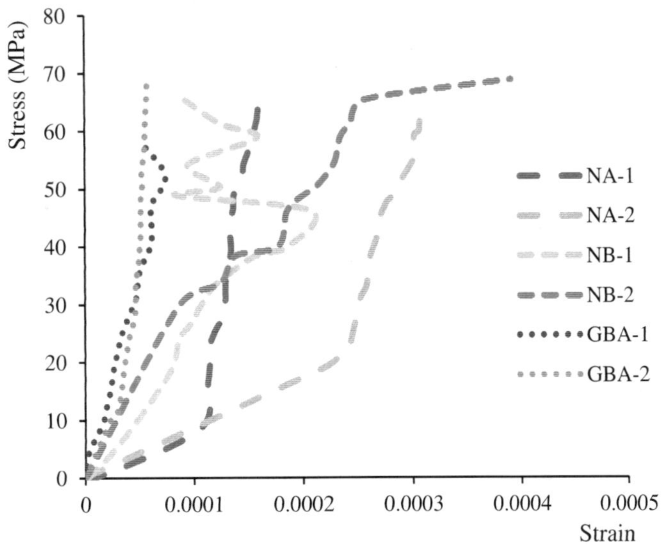

Figure 6. Stress-strain relationship of NA, NB, and GBA from experimental test

The respond of GBA in Figure 8 deviated from the controlling specimens. It was hypothetically due to the placement of the strain gauge device at the transition area of the higher concrete strength mix and the lower concrete strength mix; and the loading scheme that was not symmetrical. The asymmetric loading procedure leads the beam to deflect more at one side of beams resulting the strain gauge positioned at the other side underwent less response of strain at high increment of loading. This consideration is supported by the condition of beams after conducting a bending test (Figure 7). Figure 7 shows that the beams deflected more on the left side and created large crack opening than the other side. In this case, the strain gauge was installed in the right side which was affected by small deflection so the less strain reading at that part. From Figure 7, it is observed that the interfacial failure that is predicted does not present. It was due to the support of stirrups to maintain the bond and the compacting procedure during the concrete making is robust in creating a good adhesiveness between the concrete layers.

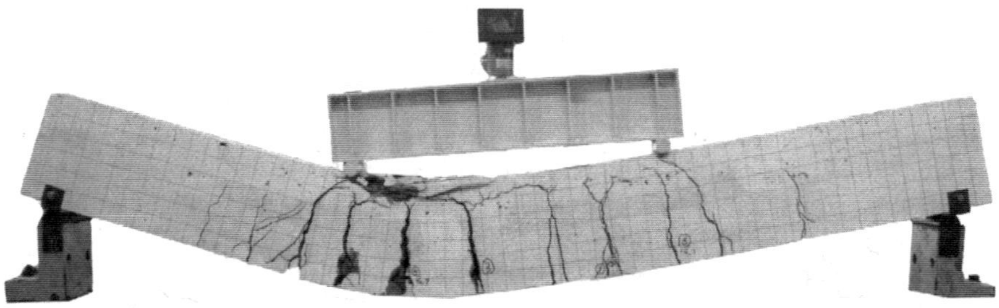

Figure 7. Asymmetrical beam deflection after bending test

4.4.2. Programme simulation. Figure 8 shows a stress-strain relationship of stirrups of NA, NB, and GBA from programme simulation. A different pattern of the graph is observed from this data to the experimental works. All of stirrups were in elastic condition indicated from the resulting stress that below the yield stress. All of the beams followed a linear pattern in the stress-strain relationship. By using identical configuration rebars, the higher concrete strength helps the stirrups to withstand less strain the beams composed of the lower concrete strength. Interestingly, the stress-strain relationship of the GBA was in between the NA and the NB; and followed the inclination of the NA the most than the NB. It shows that the presence of higher concrete strength mix could improve the structural performance of beam and replicate the behaviour of normal beam composed of fully high strength concrete mix. This scheme will reduce the construction cost while still maintaining the structural performance.

The comparison of strain – strain relationship obtained from experimental testing and programme simulation are displayed in Figure 9 – 11. Figure 9 – 11 show that there are several differences in the pattern observed from the graph. The resulting graph from experimental data rather fluctuates than the programme simulation results. It is due to the limitation of programme to model the real interaction between the rebars and the concrete. In reality, the presence of aggregate interlocking, concrete contribution in withstanding compression stress, dowel effect improves the shear resistance of beams. These factors could be mimicked by programme modelling.

5. Conclusion

- In experimental works, the stress-strain curves changed its pattern from linear to more fluctuated due to materials behaviour and numbers of technical issues. The strain gauge was suspected not installed correctly on the stirrups surface so it generated a bias data reading.
- The gap between stirrups and longitudinal bars allow the data to produce a high reading of strain in low loading attempt. In higher level of loading, the strain to stress pattern changed and indicated a strengthening behaviour regarding when the longitudinal bars reach and adhere the stirrups, it provided higher resisting respond of the stirrups during the loading.
- The asymmetric loading procedure leads the beam to deflect more at one side of beams resulting the strain gauge positioned at the other side underwent less response of strain at high increment of loading.

- It is observed that the interfacial failure that is predicted does not present. It was due to the support of stirrups to maintain the bond and the compacting procedure during the concrete making is robust in creating a good adhesiveness between the concrete layers.
- The resulting graph from experimental data rather fluctuates than the programme simulation results. It is due to the limitation of programme to model the real interaction between the rebars and the concrete. In reality, the presence of aggregate interlocking, concrete contribution in withstanding compression stress, dowel effect improves the shear resistance of beams.

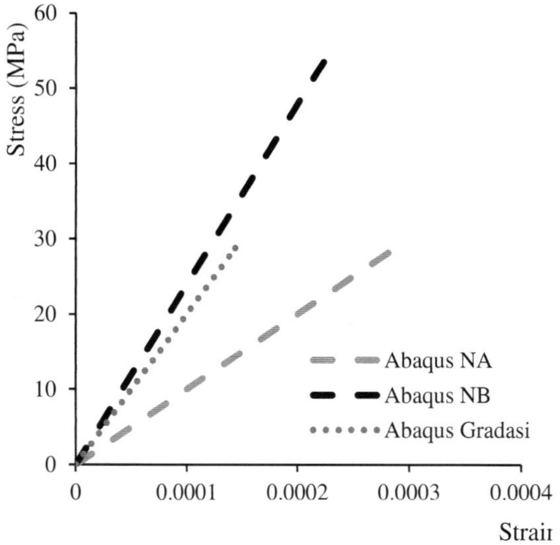

Figure 8. Stress-strain relationship of NA, NB, and GBA from programme simulation

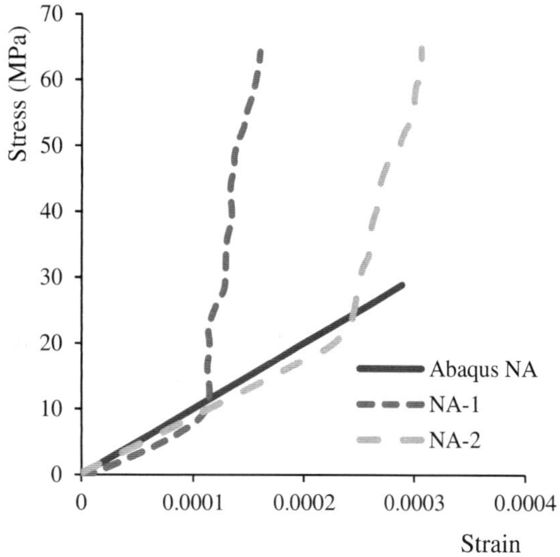

Figure 9. Stress-strain relationship of NA

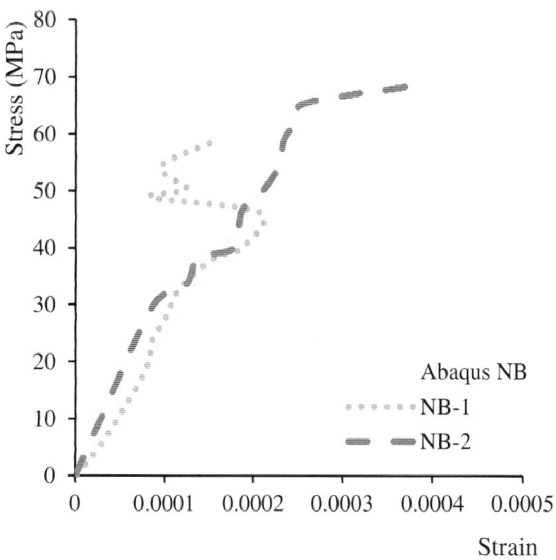

Figure 10. Stress-strain relationship of NB

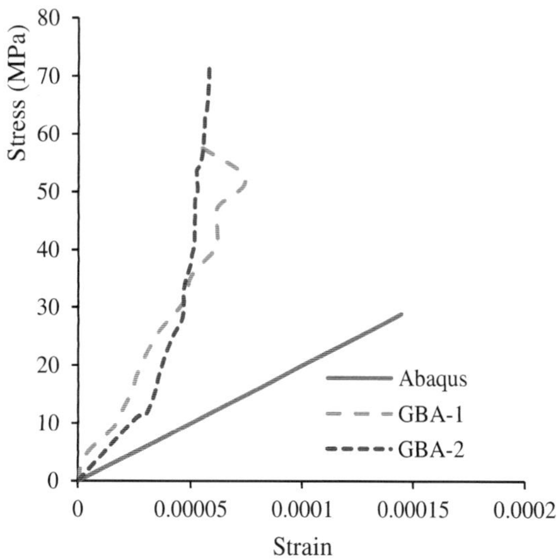

Figure 11. Stress-strain relationship of GBA

References

[1] Han A, Gan BS, Kristiawan SA, Trtnh TH. The behavior of functionally graded concrete and the prospect of material optimization. In: CESB 2016 - Central Europe Towards Sustainable Building 2016: Innovations for Sustainable Future. 2016.

[2] Pratama MMA. An experimental study and finite element approach to the behaviour of graded concrete. Master's thesis. Diponegoro University. 2015.

[3] Gan BS, Aylie H, Pratama MMA. The Behavior of Graded Concrete, an Experimental Study. Procedia Eng [Internet]. 2015;125:885–91. Available from: http://linkinghub.elsevier.com/retrieve/pii/S1877705815033937

[4] Han A, Gan BS, Pratama MMA. Effects of Graded Concrete on Compressive Strengths. Int J Technol [Internet]. 2016 Jul 28;7(5):732. Available from: http://www.ijtech.eng.ui.ac.id/index.php/journal/article/view/3449

[5] Pratama MMA, Aylie H, Gan BS, Umniati BS, Risdanareni P, Fauziyah S. Effect of concrete strength gradation to the compressive strength of graded concrete, a numerical approach. AIP Conf Proc [Internet]. 2017 Sep 29;1887(1):20029. Available from: https://aip.scitation.org/doi/abs/10.1063/1.5003512

[6] Pratama MMA, Umniati BS, Arumsari Mutiara Wulandari B, Han AL, Sthenly Gan B, Zhabrinna Z. Modulus elasticity of the graded concrete, a preliminary research. MATEC Web Conf [Internet]. 2018;195. Available from: https://doi.org/10.1051/matecconf/201819501005

[7] Pratama MMA, Sthenly Gan B, Han Ay Lie HAL, Nur Rahma Putra AB. A Numerical Analysis of The Modulus of Elasticity of The Graded Concrete. In: 2nd International Conference on Vocational Education and Training (ICOVET 2018) [Internet]. Atlantis Press; 2019. Available from: https://doi.org/10.2991/icovet-18.2019.29

[8] Pratama MMA, Umniati BS, Prasiwi G, Han AL, Sthenly Gan B, Risdanareni P, et al. Effect of monotonic lateral load on the performance of reinforced graded concrete column, an experimental study. MATEC Web Conf [Internet]. 2018;195. Available from: https://doi.org/10.1051/matecconf/201819502022

[9] Aylie H, Gan BS, As'ad S, Pratama MMA. Parametric Study of the Load Carrying Capacity of Functionally Graded Concrete of Flexural Members. Int J Eng Technol Innov [Internet]. 2015;5(4):233–41. Available from: http://www.scopus.com/inward/record.url?eid=2-s2.0-84957600233&partnerID=MN8TOARS

[10] Pratama MMA, Suhud RK, Puspitasari P, Kusuma FI, Rahma Putra ABN. Finite element analysis of the bending moment-curvature of the double-layered graded concrete beam. IOP Conf Ser Mater Sci Eng [Internet]. 2019 Mar 29;494:012064. Available from: http://stacks.iop.org/1757-899X/494/i=1/a=012064?key=crossref.e0f2a62cfeeaf5c05c43077bf777a94b

[11] Pratama MMA, Pratiwi YD, 'Ilmi MA, Mahardika A. GECOBRIDGE: A concept of a smart eco-friendly integrated bridge utilizing geopolymer and graded concrete technology. MATEC Web Conf [Internet]. 2018;204. Available from: https://doi.org/10.1051/matecconf/201820406005

Acknowledgement

This work was supported by PNBP research grant of Universitas Negeri Malang, under the contract number: 29.11.231/UN32.14.1/LL/2019. Authors would also deliver thank you in advanced for PT. Sika Indonesia for the material support to complete this research.

IOP Conf. Series: Materials Science and Engineering **669** (2019) 012056 doi:10.1088/1757-899X/669/1/012056

The optimal damper placement configuration for three-dimensional RC building

R Sulaksitaningrum[1], B S Umniati[1], I P Ellsa Sarassantika[2], C B Casita[3], M M A Pratama[1], E Santoso[1], and M Sulton[1]

[1] Department of Civil Engineering, Faculty of Engineering, Universitas Negeri Malang, Indonesia

[2] Department of Civil Engineering, Faculty of Engineering, Universitas Warmadewa, Indonesia

[3] Department of Civil Engineering, Faculty of Engineering, Universitas Pembangunan Nasional "Veteran" Jawa Timur, Indonesia

roro.ningrum.ft@um.ac.id

Abstract. During an earthquake, high-rise buildings attempt to dissipate energy by generating swaying motion, which causes inevitable building damage. The dampers or passive energy dissipation devices can significantly diminish the energy emitted by earthquakes and reduce building damage accordingly. Especially for high-rise building structures constructed in an earthquake-prone area, it is necessary to install a sufficient number of dampers to reduce the building response. The passive energy dissipation devices present such an optimal performance depends on their locations in the building structures. Therefore, the objective of this paper is to observe an optimal placement configuration of passive energy devices lest excessive installment of the devices. The proposed approach is performance-based optimizations, which seismic performance of buildings considered is story drift of the building. This study focuses on optimizing fluid viscous dampers (FVD) placement incorporated into three-dimensional (3D) regular RC building structure. The result shows that fluid viscous dampers (FVD) when incorporated into three-dimensional of regular RC building structures can modify the performance of the building to fulfill the code standard of allowable drift ratio. It will be then suggested to be prevention acts to improve the performance of high-rise buildings under earthquakes excitation. Hence, the protection of the occupants therein can be accomplished.

1. Introduction

There are three main periods of structural design. The first is the classical era, where only vertical loads are considered in structural design. The second is the modern era, where dynamic loads have been used in design. The third is the postmodern era when energy is absorbed by active, passive or hybrid systems using technological tools and methods. Passive energy dissipation systems include a set of materials and tools to increase damping, stiffness and strength and achieve the best performance when exposed to wind, earthquakes and explosions or other types of vibrational effects. [1]

The traditional earthquake resistant design approach relies on inelastic deformation behavior in several major structural zones to take in earthquake energy that subjected to, typically, the beam ends and column ends. Conversely, in structures integrated with passive dampers, earthquake energy is directly transferred to its dissipative mechanism, thereby being able to reduce the burden received by the main structure and reduce the possibility of structural damage [2-3]. While absorbing energy, the damper slowly reduces the amplitude caused by the earthquake [4].

Installation can be easily integrated in diagonal bracing, chevron brace or with other configurations. Can be an integration of new structural design (planning phase), it can also be a method of strengthening the structure in the service period (strengthening), as well as a strategy for post-earthquake building retrofitting. Passive dampers are very widely applied in structures because they are effective with affordable price [5-7].

Passive earthquake dampers as earthquake mitigation have been developed in recent years, accompanied by rapid application in construction, this provides an evolution in the structure design guidelines to supplement additional damping in buildings [8-11]. However, the efficiency of passive energy dissipation placement configuration still can be thoroughly investigated and improved. Furthermore, current codes, especially Indonesian codes (SNI), do not provide a method or guidelines for optimal passive energy dissipation device configuration.

This paper is intended to investigate the optimal placement configuration of passive energy dissipation device, or commonly called as damper, based on the story drift of structure. This study focuses on optimizing fluid viscous dampers (FVD) placement configuration when incorporated into three-dimensional of regular RC building structures.

2. Research methodology

The model of the building used in this study is a reinforced concrete (RC) open frame system. Due to symmetrical design, it is regular building with 3 bays in both X and Y directions. The height of each story is 4.5 m and the span of each bay is 5 m. The cross section of the columns is 400 mm by 400 mm and the dimension of frame beams is 250 mm by 350 mm. To evaluate the seismic responses of the building, with and without damper, the structure of the building is modelled and analyzed using ETABS software, as shown in figure 1.

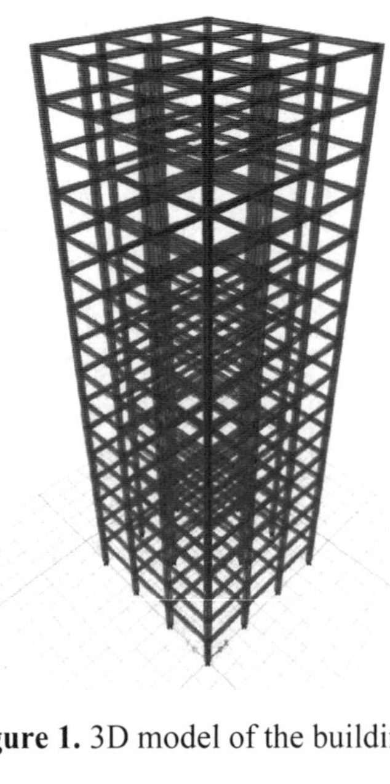

Figure 1. 3D model of the building

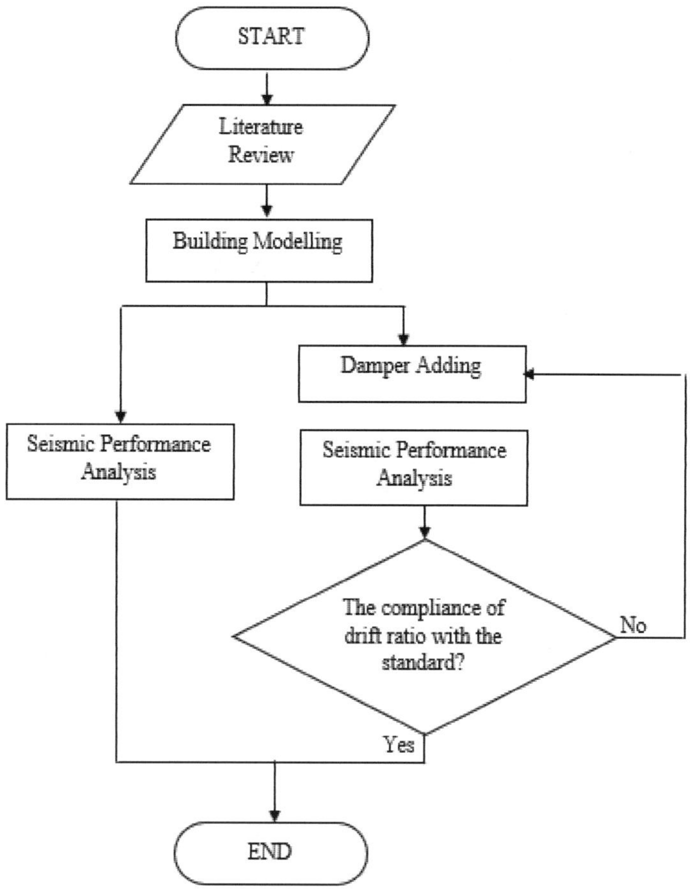

Figure 2. Flowchart of the proposed method

Figure 2 shows the process of this proposed methodology to obtain optimal damper placement configuration. In order to create the proposed method, literature review is the first step taken in this study. Then, time history analysis has been conducted to validate the design regarding the seismic response. Three ground motions, comprising Altadena, Sylmarr, and Newhall earthquake, are used in this time history analysis. The graphs of acceleration versus time for component 0° and 90° are shown in figure 3 and figure 4, respectively.

The compliance of story drift with the code standard, ASCE 7-10 or SNI 1726-2012, is examined. When the drift ratio of the structure is found more than the allowable one, the iteration of adding damper need to be carried out. The damper will be placed at the floor whose drift ratio is the highest.

3. Results and Discussion

The building structure used in this study is assumed as Risk Category III. The S_{DS} and S_{D1} used in this study is 0.62 and 0.38, respectively. Hence, seismic Design of this building is categorized as D. In accordance to ASCE 7-10 or SNI 1726-2012, the allowable story drift of this structure is 51,92 mm.

The result of seismic performance analysis of the building without damper is shown in figure 5 and figure 6 for displacement and story drift, respectively. In figure 5, the displacement of the building due to Newhall earthquake is the greatest, it is approximately five times of the response caused by Altadena and Sylmarr earthquake. Figure 6 shows also the story drift of the building due to the excitation of three ground motions, Altadena, Newhall, and Sylmarr. It can be seen Newhall earthquake causes the story drift of the building exceed the maximum limitation.

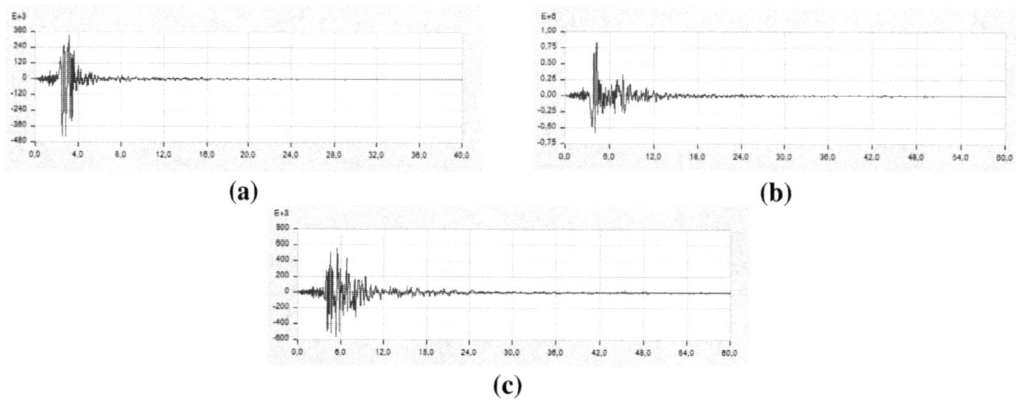

Figure 3. Acceleration versus time data, component 0° (a) Altadena Earthquake, (b) Sylmar Earthquake, (c) Newhall Earthquake records

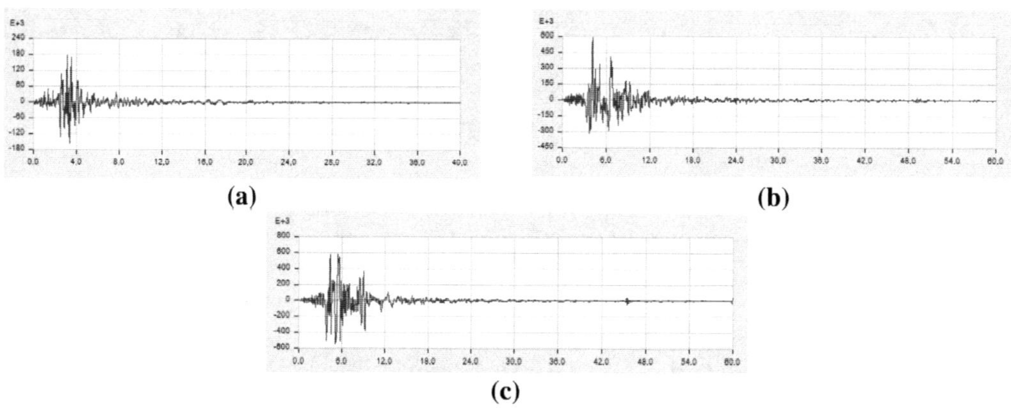

Figure 4. Acceleration versus time data, component 90° (a) Altadena Earthquake, (b) Sylmar Earthquake, (c) Newhall Earthquake records

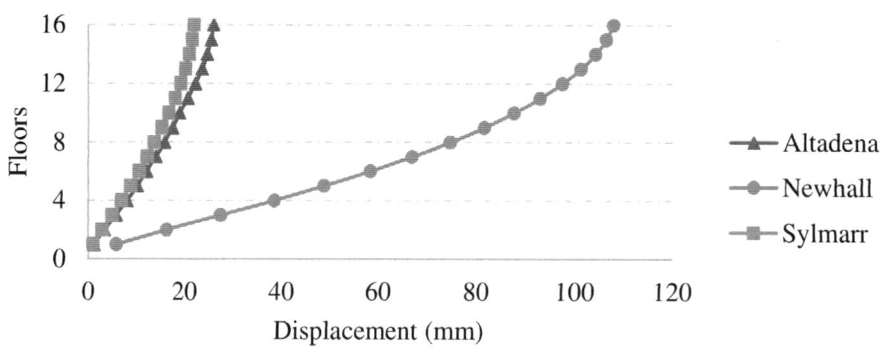

Figure 5. Displacements of the building without damper

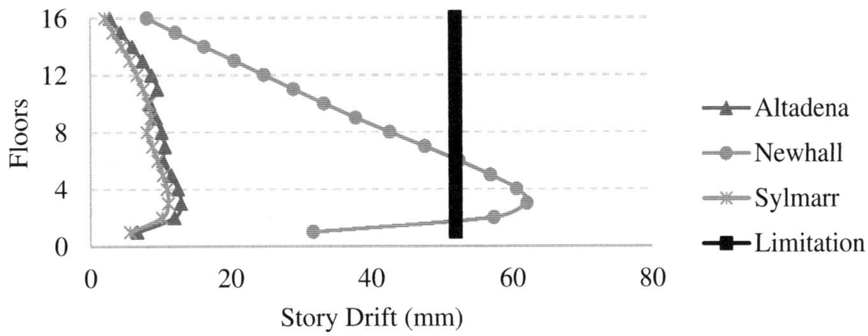

Figure 6. Story drift of the building without damper

In the figure 6, the highest point of story drift is placed in 3rd floor. Therefore, the first iteration of adding damper is needed. Considering torsional responses [12], the configuration of the dampers is placed in the corner bay of each floor. The illustration of the first iteration is shown in figure 7.

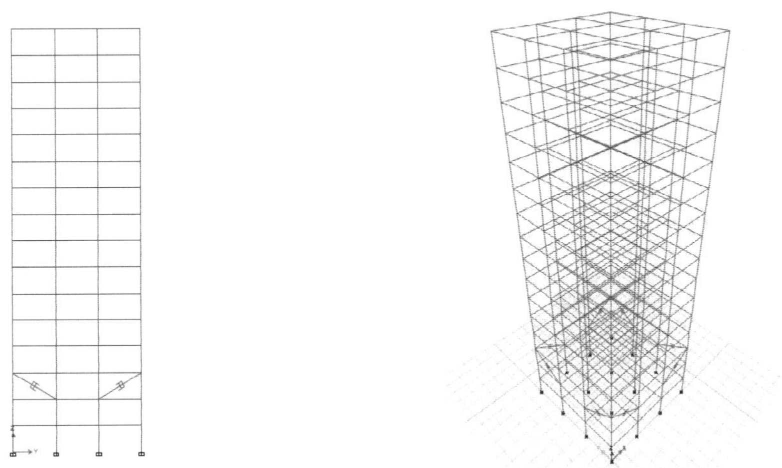

(a) **(b)**

Figure 7. First iteration of added damper configuration (a) 2D, (b) 3D

The result of seismic performance analysis of the first iteration is shown in figure 8. After first iteration, the behavior of building is changing. On the 3rd floor where the damper added, the drift ratio of the structure decrease. However, the seismic response of the building caused by Newhall earthquake still beyond the allowable story drift.

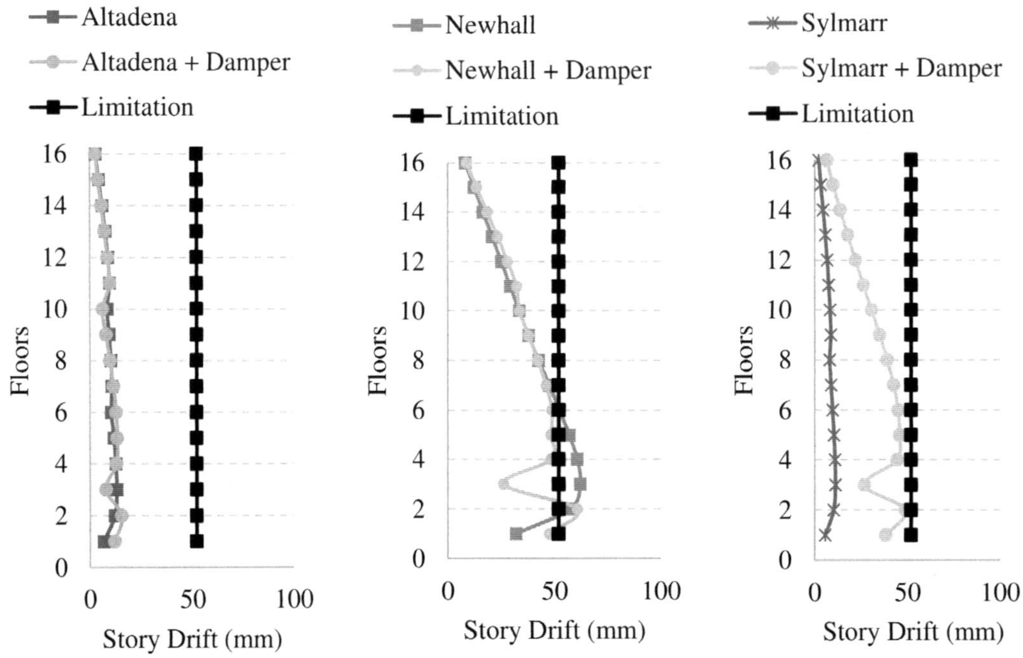

(a) **(b)** **(c)**

Figure 8. Story Drift of the building with and without damper due to (a) Altadena Earthquake, (b) Newhall Earthquake, (c) Sylmarr Earthquake

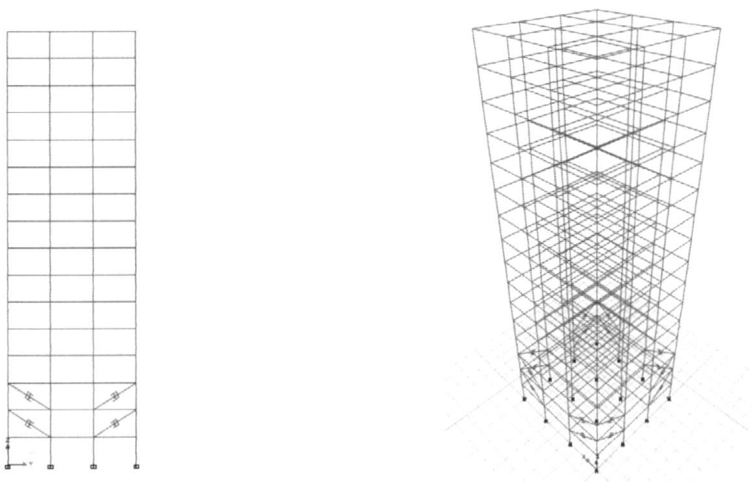

(a) **(b)**

Figure 9. Second iteration of added damper configuration (a) 2D, (b) 3D

In order to decrease the remaining exceeding story drift, the second iteration of adding dampers is needed. As seen in figure 8(b), the second added dampers should be located at 2nd floor. The 2D and 3D illustration of this second added damper is shown in figure 9.

IOP Conf. Series: Materials Science and Engineering **669** (2019) 012056 doi:10.1088/1757-899X/669/1/012056

Figure 10 shows the result of seismic performance analysis of the building after second iteration of adding dampers. In figure 10 (a), it can be seen the story drift caused by Newhall earthquake is under the limitation. The displacement, as shown in figure 10(b), is also reduced by 10%.

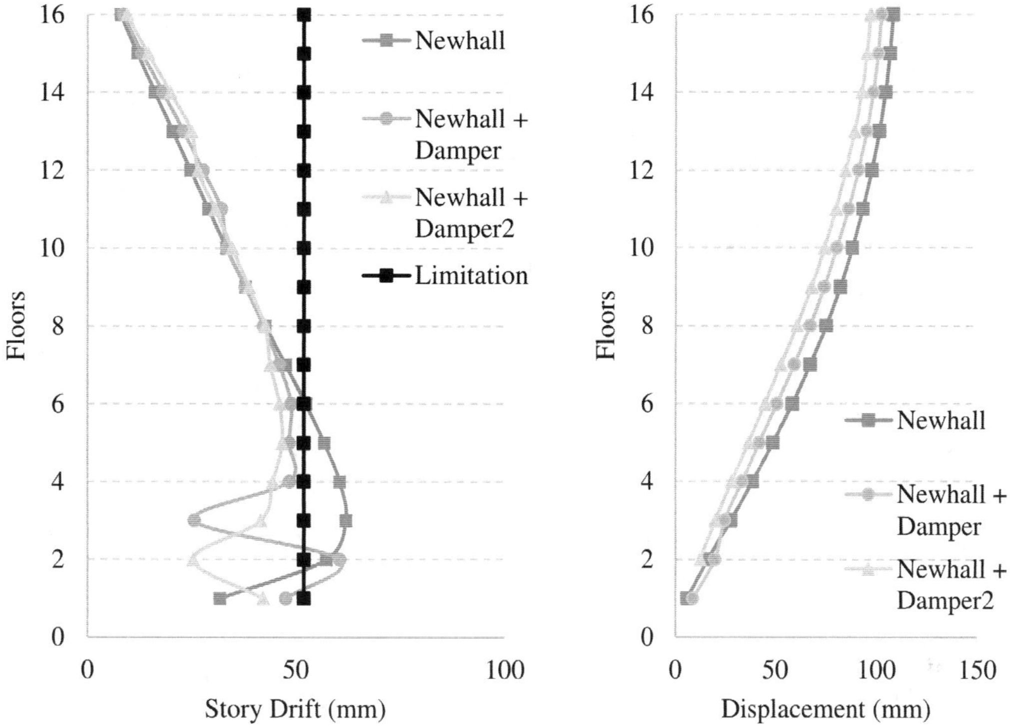

Figure 10. The seismic response of the building after second iteration of added damper

4. Conclusion

The proposed methodology for optimizing damper placement configuration for three-dimensional RC building based on the story drift of structure is investigated. The fluid viscous dampers (FVD) when incorporated into three-dimensional of regular RC building structures can modify the performance of the building to fulfill the code standard of allowable drift ratio.

References

[1] Aydin E, Öztürk B and Dutkiewicz M, "Analysis of Efficiency of Passive Dampers in Multistorey Buildings," Journal of Sound and Vibration, 2018.

[2] Tchamo JM and Ying Z, "An alternative practical design method for structures with viscoelastic dampers," Earthquake Engineering and Engineering Vibration, vol. 17, no. 3, pp. 459-473, 2018.

[3] Symans M, Charney FA, Whittaker A, Constantinou MC and Kircher CA, "Energy Dissipation Systems for Seismic Applications: Current Practice and Recent Developments," Structural Engineering, vol. 134, no. 1, pp. 3-21, 2008.

[4] Shakibabarough A, Valinejadshoubi M and Bagchi A, "Effects of Damper Locations and Base Isolators on Seismic Response of a Building Frame," International Journal of Civil and Environmental Engineering, vol. 10, no. 6, pp. 739-744, 2016.

[5] Benavent-Climent A, "A brace-type seismic damper based on yielding the walls of hollow structural sections," Engineering Structures, vol. 32, no. 4, pp. 1113-1122, 2010.

[6] Qamaruddin S, "Seismic Response Study of Multi-storied Reinforced Concrete Building with Fluid Viscous Dampers," Chaitanya Bharathi Institute of Technology, India, 2016.
[7] Dehghan MJ and Soleymannejad M, "Improving Seismic Performance of Concrete Buildings with Special Moment Frames Using Viscous Damper," International Journal of Engineering Research, vol. 4, no. 7, pp. 339-345, 2015.
[8] Ezz El-Arab I, "Earthquake analysis of a high-rise building in Dubai Strengthening by FVD dampers," in Insights and Innovations in Structural Engineering, Mechanics and Computation - Proceedings of the 6th International Conference on Structural Engineering, Mechanics and Computation, SEMC 2016, London, 2016.
[9] Omika Y, Koshika N, Yamamoto Y, Kawano K and Shimizu K, "High-rise Reinforced-concrete Building Incorporating an Oil Damper in an Outrigger Frame and Its Vibration Analysis," International Journal of High-Rise Building, vol. 5, no. 1, pp. 43-50, 2016.
[10] Bose AK and Prasad Varma Thampan CK, "Optimal Positioning of X Plate Damper in Concrete Frame Building," International Research Journal of Engineering and Technology (IRJET), vol. 5, no. 5, pp. 1873-1878, 2018.
[11] Quintana HC and Petkovski M, "Optimum performance of structural control with friction dampers," Engineering Structures, vol. 172, pp. 154-162, 2018.
[12] Wu B and Ou JP, "Optimal placement of energy dissipation devices for three-dimensional structures," Engineering Structures, vol. 19, no. 2, pp. 113-125, 1997.

Acknowledgments
This research work was supported by PNBP research grant of Engineering Faculty, Universitas Negeri Malang.

IOP Conf. Series: Materials Science and Engineering **669** (2019) 012057 doi:10.1088/1757-899X/669/1/012057

The evaluation of green performance of Miftahul Huda Islamic boarding school, Malang

A A Sari[1], A E Winahyo[1], D Ariestadi[1] and Imam Alfianto[1]

[1]Department of Civil Engineering, State University of Malang, Indonesia

astri.anindya.ft@um.ac.id

Abstract. Green building is believed to be one of the solutions to resolve the problem of the world's climate change and energy crisis. The commitment of the government of Indonesia in increasing the number of green building has resulted in some regulation related to green certification that should be achieved by new buildings and also existing building in Indonesia. This research intends to evaluate the green performance of Miftahul Huda Islamic boarding school based on Greenship Existing Building version 1.1 developed by Green Building Council of Indonesia. It was found that the green performance of Miftahul Huda Islamic boarding school still needs to be optimized further. The main factor to be the cause is the lack of commitment of the board managerial of Miftahul Huda Islamic boarding school in achieving optimum performance of the green building criteria.

1. Introduction

Since the last decades, climate change and energy crisis have been the main issues of the worlds. International Energy Agency [1] stated that the building sector is the biggest consumer of energy in the worlds of more than one-third of the world's energy consumption. Those amounts of energy consumption will indirectly affect the emission of green gas houses produce by this sector thus made this sector seen as one of the main causes of the world's problem of climate change and energy crisis. Yet, on the other hand, that fact also reveals the opportunity of a significant energy saving potential in the building sector.

World Resources Institute [2] stated that Indonesia ranks fifth as the largest emitter of greenhouse gases in the world, accounted for 4.63%. Even though greenhouse gas emissions are not entirely derived from the impact of the development of the construction sector, efforts to reduce emissions must still be optimized. The development of infrastructure that has become the main priority of the Indonesian Government since the last fifth year has affected the rapid infrastructure development and the increasing number of construction project throughout Indonesia. The impact of the development process that is not environmentally friendly of course will cause a bad impact on the environment and might lead to the problem of climate change in the long term. That is why the environmental awareness and knowledge of the actors in the building and construction is very much needed so that the development process could result in an environmentally friendly infrastructure that brings a benefit for the society and the environment.

The government of Indonesia has already set a commitment in reducing the carbon emission and greenhouse gases. It is proven by the reduction of emission by 26% in 2007, and this achievement leads to an improvement to set the target of reduction of 29% by 2020 [3]. Promoting the 'green building could be one of the efforts of Indonesian Government to reduce the carbon emission and improve the environmental condition of Indonesia. Green building is a concept that is believed as a way in resolving the environmental problems in the building sector. A 'green' building means that the entire process of a building lifecycle from design, construction, operation and the demolition is conducted in an environmentally friendly way thus reduces or eliminates negative impacts to the environment and surrounding, and can create positive impacts, on our climate and natural environment. Green buildings preserve precious natural resources and improve our quality of life. There is some regulation related to Green Building that is already appointed by the government to support this commitment.

The government of Indonesia works together with the Green Building Council Indonesia (GBCI), a non-profit organization that concerns in the field green building and construction to conduct a building certification for both new and existing building in Indonesia to make sure that buildings are designed, built, and operate in an environmentally friendly way. In order to conduct the evaluation process, Green Building Indonesia develop a tool called Greenship Rating Tools [4]. This tool was developed based on another green rating tools that used internationally but with some adaptation to the Indonesian context.

This research intends to evaluate the green building performance of the facility of Miftahul Huda Islamic Boarding School in Malang based on Greenship Existing Building 1.1 version by Green Building Council Indonesia. Miftahul Huda Islamic Boarding School is among the oldest Islamic Boarding School in Malang, thus this research is hoped to provide some inputs about the strategy to improve the green performance of Miftahul Huda Islamic Boarding School.

2. Research Methods

This research is designed to be conducted in a qualitative method. The data collection is conducted by survey, documentation, literature review, and interview with the board members and students of Miftahul Huda Islamic Boarding School. The research focus was the sixth variable of Greenship rating tools including appropriate site development, energy efficiency and conservation, water conservation, material resources, and cycle, indoor health comfort, and building environment management. The existing data of the sixth greenship variable of Miftahul Huda Islamic Boarding School then being compared with the standard mentioned in Greenship Existing Building version 1.1 by Green Building Council of Indonesia to know the green performance of this building.

3. Research Findings

Miftahul Huda boarding school is one of the oldest Salafiah Islamic boarding School that was founded in I700 M. Located in Jl Gading Pesantren Malang, this Islamic boarding school now has 915 students. All of the students live in one of the twelve boarding houses that are owned by this Islamic boarding school with 6484.34 m^2 total area of the boarding house.

As stated before, Greenship Existing building version 1.1 will be used to assess the green performance of this boarding school complex. Greenship Existing building is the rating tools which made to assess the performance of building which already been built, while Greenship New building is used when the building to be assessed is a new or still in the design phase. The greenship performance of Miftahul Huda Islamic Boarding school can be described as follow.

3.1. Appropriate site development

The variable of appropriate site development consists of nine sub variable including site management policy, motor vehicle reduction policy, community accessibility, motor vehicle reduction, site landscaping, heat island effect, stormwater management, and site management.

From the interview that has been conducted to explore the policy of site management and motor vehicles, it can be stated that there are no written rules that especially made in that case, yet there are general rules to maintain and clean the environment surrounding periodically in the form of "environmental caring" activities. As for to reduce the use of motor vehicle, the board of Miftahul

Huda Islamic Boarding School has been stated a rule that students do not allow to bring their own vehicles unless they are already in the second year of their study or more.

After conducting a survey for the sub variable of community accessibility, we can mention that there is only one public transportation that can access the site within the distance of 500 m. There is no bus stop and special pedestrian that specifically built for that purpose. In the sub variable of landscaping, we can conclude that the area in this Islamic boarding school that is dedicated to the green area is only less than 20%. There is some local and productive vegetation including mango, star fruit, bamboo, and some local flower. Most of the vegetation is planted in the pot instead of directly grown in the ground. As for the heat island effect sub variable, this complex building of Islamic boarding school does not get any score since the albedo of the roof material is lower than standard. In the storm water management, it is noticeable that some effort in reducing the storm water flow to the environment drainage had been conducted by conducting a rain harvesting by collecting it and used it to wash the vehicles and watering the plant, yet the amount is still below 75% of all the storm water.

The sub variable of site management focuses on the animal habitation on site and the management in pest and weed disease. It is noticeable from the observation that there is some chicken found on the site but the area for their activities is less than 5% of the total area. As for the prevention of the pest and weed disease, there is no special treatment, from the interview we can get the information that they never use poison form of treatment to prevent pest and weed disease.

The last sub variable in the appropriate site development is building the neighborhood. From the interviewed conducted, we know that some effort has been conducted by this Islamic boarding school to improve the quality of life of the society periodically by providing the toilet improvement program and sharing a place for praying, also to serve some training for the society as a community service.

3.2. Energy efficiency and conservation

This sub variable mainly focuses on the effort of reducing the energy consumption of the building. It consists of nine sub variable including policy and energy management plan, minimum building energy performance, optimized efficiency building energy performance, testing, recommissioning or retro-commissioning, system energy performance, energy monitoring, and control, operation and maintenance, on-site renewable energy, and less energy emission. From the interview conducted to get some information on the policy and energy management plan, we know that the commitment of reducing energy consumption is only conducted spontaneously without any structured plan. Some campaign related to energy saving is conducted by putting some sticker in some strategic place reminding to turn off the lamp when not in use.

In the sub variable of minimum building energy performance, optimized efficiency building energy performance, and testing, recommissioning or retro-commissioning, Miftahul Huda Islamic boarding school does not get any score. It is because the information on the detail of energy consumption per month until a year is not provided. Hence, by looking only at the last three-month bills of energy consumption provided, we can still assume that the energy consumption of the complex building is still meet the required standard of green building. The amount of energy consumed is still fluctuated month by month and does not show a gradual reduction. Testing, recommissioning or retro-commissioning is also never been conducted in this building since they do not use any air conditioning.

In the variable of system energy performance, it was found that nearly all lamp for lighting in the Miftahul Huda Islamic boarding school is already replaced with an LED lamp to save the energy. Only a few lamps in the canteen, parking area counted for 10% of all the lamps uses. While for the mechanical air conditioning and ventilation systems since as stated before that this building does not use any air conditioning but the fan, it can be stated that the efficiency target is already reached at this stage.

In the sub variable of operation and maintenance, on-site renewable energy, and less energy emission, the complex building of Miftahul Huda Islamic boarding school again gets no score. There are never been conducted an energy audit in this complex building, there is no use of air conditioning so there is no need to maintain the AC system. There is no on-site renewable energy gain in this site, and they never conduct any audit of CO2 emission gained.

The last sub variable in the appropriate site development is building the neighborhood. From the interviewed conducted, we know that some effort has been conducted by this Islamic boarding school to improve the quality of life of the society periodically by providing the toilet improvement program and sharing a place for praying, also to serve some training for the society as a community service.

3.3. Water conservation

The variable of water conservation consists of eight sub variable which is water management policy, water sub-metering, water monitoring control, freshwater efficiency, water quality, recycled water, potable water, deep well reduction, and water tap efficiency.

In terms of water management policy, again there is still no real commitment shown by the Miftahul Huda Islamic boarding school management as it is shown that there is no written regulation about water management system neither structured plan or auditing plan in the water-saving management system. Yet, some campaign had been conducted spontaneously by placing some stickers reminding to use water wisely.

There is no water sub-metering that was found on the site. As for the water monitoring control, the checking maintenance of the water and plumbing system does not conduct periodically but only when there is a problem found. The amount of water consumed every month was also never being recorded thus they never know if there is a reduction in water consumption or not. The quality of the water in the Miftahul Huda Islamic boarding school is also never been tested in the laboratory to get a water quality certification.

In terms of recycled water sub variable, the effort that being conducted in the Miftahul Huda Islamic boarding school is using the rainwater collecting for the purpose of plant watering and vehicle washing. They do not use any recycled water for WC flushing. They also do not have any standardize water recycle system, thus they do not use recycled water for consumption.

There is no deep well reduction effort in the site since 100% of the consumption is obtained from the deep well. Hence, this Islamic boarding school uses a water tap efficiency strategy since they use water tap with an auto stop in more than 50% of the water tap in the site.

3.4. Material resource and cycle

The variable of material resource and cycle consist of seven sub variables of fundamental refrigerant, material purchasing policy, waste management policy, non-ODS usage, material purchasing practice, waste management practice, hazardous waste management, and management of used good.

Since as stated before that the complex of Miftahul Huda Islamic boarding school does not use any air conditioning system, so they do not have any concern with anything related to the sub variable of fundamental refrigerant and non-ODS usage.

In terms of the material purchasing policy and waste management policy, from the interview, we get the information that there is no statement of commitment from the management related to that matters. Yet, campaign and some effort of waste separation and recycle have been conducted regularly. They also encourage the use of food-grade packaging and prohibit the use of Styrofoam for food packaging. There is no information about material purchasing practice that we can obtain from the interview, but we notice that in some part of the building they use reuse material and pre-fabricated material.

In the hazardous waste material variable, Miftahul Huda Islamic boarding school is again cannot get any point since they never conduct any training related to hazardous waste management. Yet, in terms of management of used good, even though there is no written standard procedure related to that, there are unwritten rules to donate the uses goods that can still be used by other people.

3.5. Indoor health and comfort

The variable of indoor health and comfort is including nine sub variable of no-smoking campaign, outdoor air introduction, environmental tobacco smoke control, CO^2 and CO monitoring, physical, chemical, and biological pollutants, visual comfort, thermal comfort, acoustic level, and building user survey.

In the no-smoking campaign sub variable, it was found that there is no written statement of commitment related to that matter. There is also no campaign in terms of poster or sticker that stated "no smoking". Hence there is some oral regulation to not smoking in a certain area. This condition is also applicable to the sub variable of environmental tobacco smoke control since there is no specific regulation for a place that can be used as a smoking area.

In terms of outdoor air introduction, it is noticeable from the survey that not all of the room in the complex building of Miftahul Huda Islamic Boarding School get an outdoor air introduction. There are some rooms including the rooms in the dormitory that are not really well ventilated. Hence, the thermal measurement shows that the thermal condition of the room is already meet the standard of 24°C – 27°C with the relative humidity accounts for 60% + 5%.

In terms of visual comfort, the measurement from the lux meter showed that in general the illumination in the working room is already meet the standard. Even though in some cases the illumination obtained from the natural lighting is not enough, but after the lamp is turned on, there is a significant increase of illumination obtained to meet the standard.

As for the acoustic level, there had been conducted measurement of the noise level in a random location in the site, and the result stated that the noise is still considered to be acceptable according to SNI 03-6386-2000.

Sub variable of physical, chemical, and biological pollutants focus on the measurement of air pollution level and kinds of air pollutant on site. The management of Miftahul Huda Islamic boarding school had never conducted any measurement on the air pollution level, so there is no information regarding this subject that we can get from the interview and observation.

The last sub variable in the variable of indoor health and comfort is the sub variable of building user survey. In this sub variable, Miftahul Huda Islamic boarding school management cannot get any point since they had never conducted this kind of survey yet.

3.6. Building environment management

The variable of building environment management consists five sub variable including operation maintenance and policy, innovations, design intent and owner project requirement, green operational and maintenance team, green occupancy or lease, and operation and maintenance training.

Based on the survey and interview that already conduct, we can state that there are no single criteria in this variable that is already being conducted in the Miftahul Huda Islamic boarding school.

4. Conclusion

From the findings stated above, it can be seen that no single variable of the greenship rating tools that can meet the optimum requirement the greenship criteria. It might be due to the lack of commitment of the management related to the target to achieve a green performance of the building. It can be noticed that the lack of commitment of the management resulted in no written statement nor clear regulation related to green building. A written statement of commitment and regulation might function as the guideline to make the plan to achieve the real target. From the plan, then we might evaluate and analyze the action to do in order to achieve the target. The plan then functions as a guideline of the activities at the operational level.

It can be stated that without the clear commitment from the managerial board, it might be impossible that Miftahul Huda Islamic boarding school can optimize the green performance of their building. Therefore, the commitment of the management team should be stated clearly first, before creating the plan of action in optimizing the green building performance.

It is suggested that the board management of Miftahul Huda Islamic boarding school conduct such building user survey to know the opinion of the building user related to the preference and facility comfort. The result of the survey than can be used as a reference in determining the priority variable that needs to be improved.

References

[1] International Energy Agency. 2017. World Energy Outlook. http://dx.doi.org/10.1787/weo-2005-en.
[2] Worlds Research Institute, cait.wri.org.
[3] Nur.M, Emma. R, Yulia. S, Hanny.S, Farid. S, and Nur. S. 2017 Strategi Implementasi NDC (Nationally Determined Contribution). http://ditjenppi.menlhk.go.id
[4] Green Building Council Indonesia. 2019 Rating Tools. gbcindonesia.org/greenship/rating-tools

Acknowledgement

This research is funded by PNBP research grant of Engineering Faculty of The State University of Malang in 2019.

The role of the fire safety management in providing a guarantee of a fire protection: The case of Graha Rektorat building of State University of Malang

A A Sari[1], F K Rafrita[1], T Rahayuningsih[1] and I Alfianto[1]

[1] Department of Civil Engineering, State University of Malang, Indonesia

astri.anindya.ft@um.ac.id

Abstract. The safety of building's occupants is one of the most important things that should be considered in the design and construction process of a building. The provision of a reliable fire protection system could be the best way to provide a guarantee for fire safety. However, in a high-rise building, the evacuation process could be a problem when the occupants do not really understand about the evacuation procedure. Thus, a good fire safety management system is also needed to ensure that both the fire protection system and the evacuation procedure could run well when the accident happen. Through observation, documents review, and interview with the member of fire protection management as well as the building's occupants, this paper intends to assess the fire safety management organisation and its role to support the fire safety protection system of the 9th-floor building of Graha Rektorat of the State University of Malang. The assessment conducted will be based on regulation of the Ministry of Public Works no. 29/PRT/2006 on the fire safety protection and evacuation. It was found that the organisation of the fire safety management of the Graha Rektorat building is need to be optimized to meet the requirement of the regulation of the Ministry of Public Works no. 29/PRT/2006 as to serve a better support for the fire safety protection system in that building.

1. Introduction

The fire protection system is one of the important factors in the building to serve a guarantee for fire safety. According to the regulation of the Ministry of Public Works no. 29/PRT/2006, the fire protection system is included in the set of safety requirement that should be provided in the building. In the time being, when the accident happened, people still rely the most on the firefighters while the readiness of the fire safety protection system in the building is sometimes still being neglected [1]. This statement is proven by the case of a fire accident in the Karya Building, the office of the Ministry of transportation on the 8th of July 2018. What happened in that accident is actually a small scale of a fire accident, but the smoke is already spread everywhere that caused the death of many people that mostly being founded in the emergency stairs [2]. This case tells us that good fire safety management is needed to support the performance of the fire safety protection system. The fire safety management will ensure that the fire protection system could run well when the accident happens, besides ensuring that the building users have adequate knowledge on what to do during the fire accident.

A research on the fire safety management had been conducted by Kristyanto (2012) [3] with the case of the 1st -4th floor of the Rectorate building of Brawijaya University. It was found that there are some weaknesses in the fire safety management system on the case study building including; the team that put as the person in charge in the fire accident is the member of security, there are no evacuation maps that put on every floor of the building; not all of the building users are given a training in the management of fire safety [3]. Therefore, the finding of that research stated again the importance of the fire protection management system to support the reliability of the fire protection system.

The regulation of the Ministry of Public Works no 20/M/2009 [4] stated that there are three most important things in the management of fire safety protection which are; (1) the organization that include the fire safety manager, the personnel of communication, fire extinguisher, paramedic, engineer, floor warden, and security; (2) the operational governance which include the activities of developing the team planning, the development of building's risk analysis in fire hazard, the development of a fire safety plan and the fire emergency plan; (3) human resources, this factor is very important to determine the effective and efficient result. Hence, the fire safety management should be supported with people who have background knowledge in the fire protection field including an expert in fire safety, paramedic, and management.

This research intends to assess the fire safety management in Graha Rektorat building of the State University of Malang as a high rise building that should be occupied with a good fire protection system that supports by a good fire safety management. The assessment will be conducted based on the criteria of fire safety management system mentioned in the regulation of the Ministry of Public Works no 20/PRT/M/2009 [4].

2. Research methods

This research is designed as qualitative research. The data collection is conducted by observation, interview, and document review. Observation serves to identify the fire safety protection system which is provided in the Graha Rektorat building of the State University of Malang. Interview with the member of fire safety management and the building's occupants serve to get the detail information on the role of fire safety management in the case study building. Documents review serves to gain secondary information and standard operating procedure related to the fire safety management of the Graha Rektorat building of the State University of Malang.

3. Research findings

Graha Rektorat building of the State University of Malang is a high rise building consisting of nine-floor and a basement with a total area of 16.37 m^2 built in the 18.000 m^2 of site area. Since the regulation of the Ministry of Public Works, no 20/PRT/M/2009 stated that a building with the minimum area of 5000 m^2 or consist of a minimum eight-floor height is required to provide a fire safety protection management, therefore Graha Rektorat building has to implement this policy. In order to implement that regulation mentioned, firstly Graha Rektorat should be equipped with a standardized fire safety protection system based on the National Standard of Indonesia with the support of a good the fire safety protection management as required by the regulation of the Ministry of Public Works, no 20/PRT/M/2009.

3.1. Fire safety protection system of Graha Rektorat building

Based on the requirement of the regulation of the Ministry of Public Works, no 20/PRT/M/2009 the fire safety management system provided in the Graha Rektorat building of State University of Malang could be categorized as the fire safety protection infrastructure and the fire safety protection facilities.

3.1.1. Fire safety protection infrastructure. The infrastructure of fire safety protection provided in the Graha Rektorat building of the State University of Malang are including water resource, access for the fire engine and the evacuation route

Water is taken from the PDAM and been collected in the ground water tank. The information getting from the interviewed is already confirmed that amount of water collected is considered to be enough for the purpose of fire safety protection when the accident happens.

The access for the fire engine is provided around the building of Graha Rektorat. Besides, this access, actually the National Standard of Indonesia number 03-1735-2000 [5] requires some other access including access for the firefighters to go inside the building, including the fire lift, fire safety stair, and shaft for the fire engine. Yet, Graha Rektorat building does not provide those access.

Figure 1. Access for the fire engine

There are two emergency stairs provided, one emergency stair is located in the north zone, and the other is located in the south zone. The emergency stairs connect the ninth-floor until basement with the exit door located on the ground floor.

Figure 2. Emergency stairs and exit

As the guideline and information for the building user, this building is also been occupied with the maps of evacuation route which located on every floor of the building. Moreover, a sign of an exit route has also been placed to complete the information.

3.1.2. Fire safety protection facilities. The fire safety protection facilities provided in Graha Rektorat building of the State University of Malang are including fire alarm, detection system, emergency communication system, fire extinguisher system. Actually, there is a facility of fire safety protection which are required by the regulation of the National Standard of Indonesia number 03-1735-2000 [5] but not provided in the building. This facility is smoke detector.

The fire extinguisher system available in the Graha Rektorat building of the State University of Malang is including the lightweight fire extinguisher, fire hydrant, and sprinkler system.

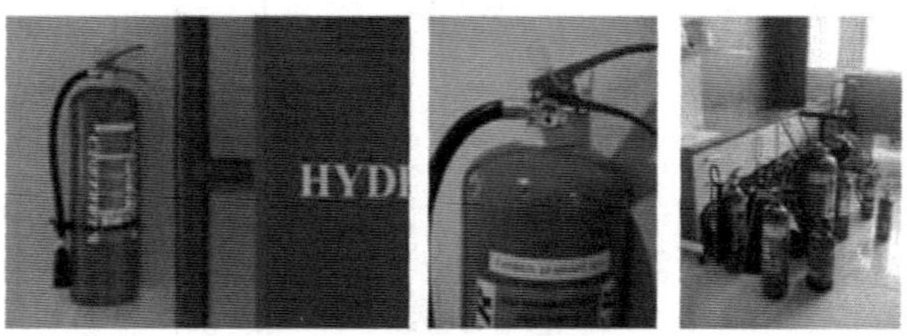

Figure 3. Fire extinguisher

Fire hydrant system provided outside and inside the building. There are six outside fire hydrants which are placed around the building, four fire hydrants located in the semi-basement, and two fire hydrants in each floor.

Figure 4. Fire hydrant

Glass bulb sprinkler which will explode when the temperature goes up until 57°C provided in each floor of the Graha Rektorat building of State University of Malang.

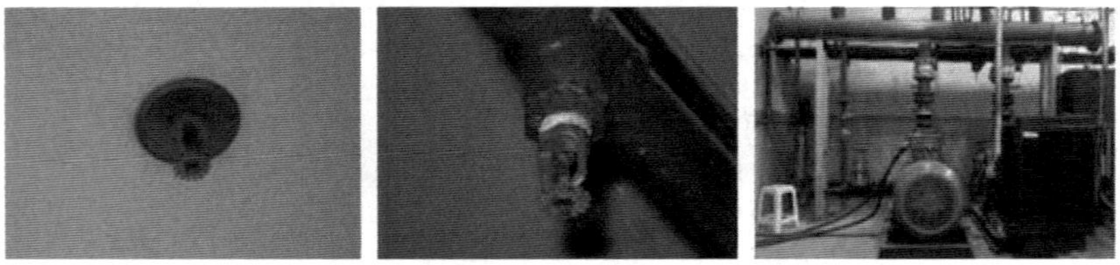

Figure 5. Sprinkler

Overall infrastructure and facilities of fire safety protection are already provided in a good condition as the required by the regulation of the Ministry of Public Works no 20/PRT/M/2009. However, those infrastructures and facilities cannot be functioned well without the support of a qualified fire safety management.

3.2. Fire safety management of Graha Rektorat building of the State university of Malang

This following review will discuss the condition and the role of the fire safety management of Graha Rektorat of the State University of Malang based on the regulation of Ministry of Public Works no 20/PRT/M/2009 [4].

3.2.1. Organization. The fire safety management team of Graha Rektorat Building of the State University of Malang is a part of the team of Occupational Health and Safety (OHS) which is formed based on SK Rector no 28.1.232.UN32/KP/2013 [6]. The organization and job descriptions of that team can be described as follow; (a) health and Accident Protection (HAP) is responsible in handling the work accident in terms of the campaign, prevention, curative, and rehabilitative, that occur in the scope of responsibility of the Occupational Health and Safety team of State University of Malang; (b) safety tool and maintenance (STM) is responsible in handling and maintain the tools related to the work scope of OHS so that the tools be ready whenever needed; (c) safety patrol and Education (SPE) is responsible for both direct and indirect monitoring of OHS activities and training; (d) data and Information Department (DID) has the responsibility in handling all of the data and information related to OHS of The State University of Malang

That organization of OHS team mentioned above is not only working for the Graha Rektorat building, but also being responsible as the standard planner of OHS in the entire complex building in The State University of Malang. As for the fire safety protection management in the Graha Rektorat, consists of; (a) evacuation coordinator; (b) fire safety protection coordinator; (c) occupational health safety coordinator; (d) emergency response officer, the person in charge are security in charge in each floor; (e) emergency response officer, the person in charge are the operator in the first-floor; (f) health care officer.

From the interview with one of the members of fire safety management team of State University of Malang, we got the information that the development of the fire safety management team is not based on the regulation of Ministry of Public Works no 20/PRT/M/2009 but based on the act year of 1970 on occupational safety. That is why the team that was being developed is the OHS team instead of the fire safety protection management. In the future, they expect to build the team of fire protection team for a building which consists of a fire engine, evacuation team, and safety team.

3.2.2. Operational governance. Some activities that already being conducted by the OHS team of the State University of Malang are including plan, standard, and analysis development, workshop and training on fire safety and evacuation, and simulation of evacuation in the Graha Rektorat building.

Plan, standard, and analysis that already developed by the OHS team of the State University of Malang are including analysis of the facility of fire safety protection system, the document of fire safety plan, the document of fire emergency plan. Besides, the team was also had already developed some operational procedure standard related to fire safety including the operational procedure standard for emergency evacuation, heat detector and sprinkler check list, hydrant check list, lift cleanliness and tidiness check list, cleanliness and tidiness of operator and panel room checklist, the fire extinguisher maintenance checklist.

As stated before, the OHS team of State University of Malang has already conducted some workshop and seminar on OHS, fire safety system, and evacuation procedure. Unfortunately, not all of the building user of the Graha Rektorat of State University of Malang are invited as the participant of that workshop. Hence, since all the security team are belonging to the team of emergency that being responsible to the evacuation process in the building, all of them are already participated on the workshop.

Another activity that had already being conducted by the OHS team is a simulation of the evacuation process in the Graha Rektorat building of the State University of Malang. This activity was held on 26th April of 2019 but unfortunately, this activity only involving the 1-4th floor since at the same time there is another important event that was held on the eighth-floor.

3.2.3. Human resource. From the interview, we got the information that the member of the OHS team of the State University of Malang is already involving some people with the background and experience on the fire safety protection and evacuation. Some expertise that involves in the OHS team is; (1) expertise in fire safety protection, expertise in emergency rescue, and expertise in the occupational health and safety management.

In the future, the OHS team of State University of Malang plan to (1) complete the facilities of fire safety protection system in the Graha Rektorat building; (2) develop the operational team of Occupational Health and Safety ; (3) encouraging the certificate of building proper function of the Graha Rektorat building ; (4) develop the team of building fire safety protection as required by the regulation of Ministry of Public Works no 20/PRT/M/2009.

4. Conclusion

This research found that the fire safety protection system that provided in Graha Rektorat building of the State University of Malang is quite sufficient as required by the regulation of the Ministry of Public Works no 20/PRT/M/2009. On the other hand, the fire safety system management is still needed to be optimized as it still does not meet the requirement of the Ministry of Public Works no 20/PRT/M/2009.

In order to meet the requirement of the regulation of the Ministry of Public Works no 20/PRT/M/2009, Graha Rektorat building of the State University of Malang should be equipped with the fire safety protection management that is dedicated for this building only, instead of just rely on the Occupational Health and Safety unit that works for the whole complex building of the state university of Malang.

Without the support good of fire safety protection management, it might be impossible that the fire safety protection system and the evacuation process can run well when the accident takes place.

Therefore, it is essential to optimize the function, organization and the operational governance of the fire safety management of Graha Rektorat building of the State University of Malang.

It is suggested that the training of fire safety protection and evacuation procedure also involves all of the building users as the participant so that when the accident happens, the building users know exactly what to do. Relying on all the evacuation process only to the security team that have no background knowledge on the fire safety protection and evacuation might be considered as too risky. It is also suggested that the simulation be conducted regularly involving not only the security from the first until the fourth floor but also involving all of the building users of the building. It is important to make sure that all of the building users be familiar with the evacuation process and do not feel panic when the accidents take place.

References

[1] Lestari F and Amaral Y 2008 Audit Sarana Prasarana Pencegahan Penanggulangan Dan Tanggap Darurat Kebakaran Di Gedung Fakultas X Universitas Indonesia Tahun 2006 Makara Teknol.

[2] Kustin. A. 2018 Korban Kebakaran Gedung Kemenhub Diduga Tewas Karena Asap. https://www.cnnindonesia.com/nasional/20180708105039-20-312444/korban-kebakaran-gedung-kemenhub-diduga-tewas-karena-asap pada tanggal 5 September 2019

[3] Kristiyanto A 2012 Evaluasi Sistem Manajemen Kebakaran Gedung Rektorat Universitas Brawijaya (Lt. 1 S.d 4) Erud. J. Educ. Innov.)

[4] Ministry of Public Works, Regulation no 20/PRT/M/2009. 2009 . Pedoman Teknis Pedoman Teknis Manajemen Proteksi Kebakaran di Perkotaan

[5] National Standard of Indonesia number 03-1735-2000

[6] Admin. 2009 UM Tingkatkan Kompetensi K3 k3.um.ac.id.

IOP Conf. Series: Materials Science and Engineering **669** (2019) 012059 doi:10.1088/1757-899X/669/1/012059

Confinements of concrete cylinders using bamboo spiral stirrups

N Nindyawati[1], M Sulton[1], M M A Pratama[1] and T Rahayuningsih[1]

[1]Civil Engineering, Universitas Negeri Malang, Malang, Indonesia

nindyawati.ft@um.ac.id

Abstract. Confinement affects the increase in beam flexural capacity. The strain stress relationship of concrete cylinder compressive strength is used to analyze the beam and column capacity. The area of the curve of the concrete cylinder core with confinements is wider than the curve without confinements. Confinements can also increase ductility and shear strength of reinforced concrete structural elements. The paper aims to determine the effect of bamboo spiral stirrups distance on normal strength concrete. Concrete cylinder size 150 mm x 300 mm. Variation in crossing distance of 30 mm, 50 mm, 70 mm and without confinement. Confinements on cylinders are bamboo and steel spiral stirrups. The results show that concrete cylinders with steel confinements increase the compressive strength by 45% of the concrete cylinder without confinements. Concrete cylinders with bamboo confinements increase the compressive strength by 28 %.

1. Introduction

Stress-strain relations for confined and unconfined concrete are aim to overcome some shortcomings of existing commonly used models. Existing models are neither easy to integrate to obtain equivalent rectangular stressblock parameter for analysis and design. The stress-strain relations proposed are validated for a whole range of concrete strenght and confining stresses. Then confinement are derived for the equivalent rectangular stress-block parameters. The wider the curve, the greater the concrete compressive strength for beams and columns [1][2].

Kent and Park [3][5] developed a stress-strain relation of confined concrete from the stress-strain relation of unconfined concrete. Mander at al. [4] prposed a strss-strain relation of confined concrete with according the confinement effects to the various configurations of lateral ties. Park at al. [7] modified the stress-strain relation proposed by Kent and Park [5].

Several classical studies [6][7]8][10] about flexural behavior of reinforced concrete beams section is permit the simulation but, in these classical studies the concrete confinement is often neglected. The spiral reinforcement or the rectilinear ties in reinforced concrete play an important role in enhancing the strength and ductility. Under axial loads, concrete pressure in lateral direction of the sections acts on the lateral ties and resistance of the ties may restrain the core of concrete to a degree. The mechanical behaviour of the confined concrete is characterized by the increase in strength and ductility. Beams with bamboo stirrups affect the flexural strength that occurs. Flexural strength increases with increasing beam stirrup distance. The stirring distance of 5 cm shows optimal flexural strength [1].

IOP Conf. Series: Materials Science and Engineering **669** (2019) 012059 doi:10.1088/1757-899X/669/1/012059

The objective of this study is to present an empirical stress strain relation according the confinement effect to various parameter variables (stirrup type, stirrup material and stirrup distance). This paper presents an empirical and experimental relationship to describe the behavior of confined concrete. Confinement material is iron and bamboo.

2. Confinement

Concrete cylinder compressive test have proposed stress-strain relationships for concrete. Figure 1 pesents typical stress-strain curves obtained from concrete cylinders loaded in compression test [5]. The curves are almost linier up to about one –half the compressive strenght. The peak of curve for high-strength concrete relatively sharp, but for low-strength concrete the curve has a flat top. The strain at the maximum stress is 0,002.

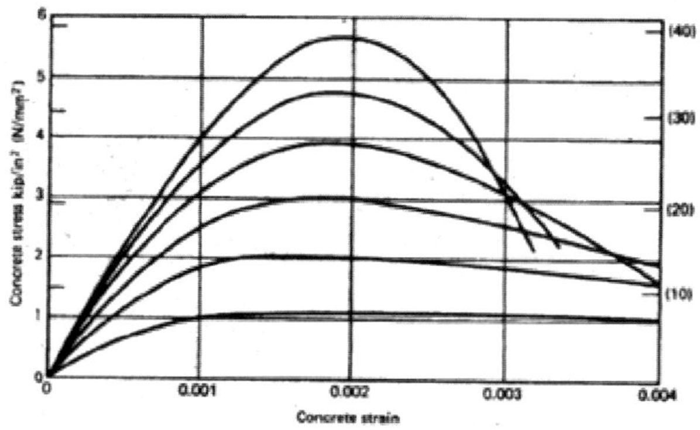

Figure 1 Stress-strain curves obtained from concrete cylinders loaded in compression test [5]

The circular section in figure 2 to the same of effectivelly confined concrete, steel/bamboo spiral and unconfined concrete.

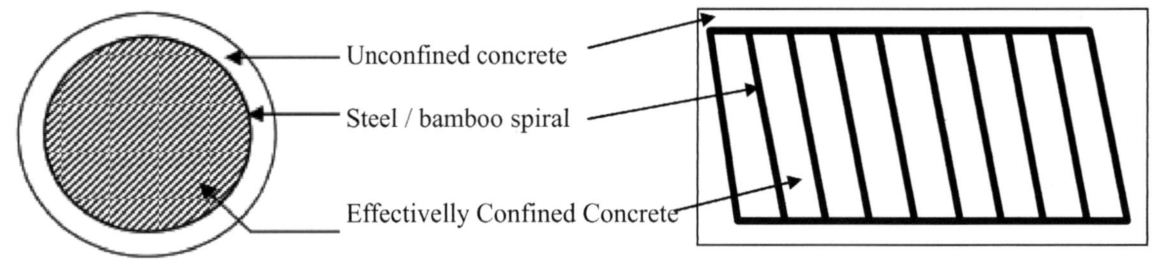

Figure 2. Flowchart of Idriss and Boulanger's Method

The compressive strength is developed by Mander at al. [4] and it can be defined by:

$$f_{cc} = f_{c0}\left(-1{,}254 + 2{,}254\sqrt{1+\frac{7{,}94 f'_l}{f_{c0}}} - \frac{2f'_l}{f_{c0}}\right) \tag{1}$$

The confinement coefficient is defined by Mander at al. [6] as

$$k_e = \frac{\left(1-\dfrac{s'}{2d_s}\right)}{1-\rho_{cc}} \tag{2}$$

The loads equilibrium in the section can be written as

IOP Conf. Series: Materials Science and Engineering **669** (2019) 012059 doi:10.1088/1757-899X/669/1/012059

$$2f_{yh} \times A_{sp} = f_l \times s \times d_s \quad (3)$$

The confinement stress is given by

$$f'_l = \frac{1}{2}\rho_s \times k_e \times f_{yh} \quad (4)$$

f_{c0} : Compressive strength of unconfined concrete.
ε_c : Confined concrete strain
ε_{65} : The strain corresponding to the stress equal $0.65f_{cc}$.
E_{bc} : Initial confined concrete Young modulus
σ_{cc} : Confined concrete stress
f_l : Lateral confinement stress.
f_{yh} : Yield strength of lateral ties.
ke : Effective confinement coefficient.
ρ_{cc} : Longitudinal reinforcement ratio.
A_{sp} : Transverse reinforcement area
s : Distance between longitudinal steels bounded by perimeter ties
ρ_s : Volumetric ration of circular ties.

3. The Results and Discussion

The test specimen is a concrete cylinder with a diameter of 15 cm with a height of 30 cm to test the compressive strength of concrete. Figure 3 pesents variation of the test specimens consisted of 3 concrete cylinders without stirrups, each 3 concrete cylinders with iron and bamboo stirrups distance away 3 cm, 5 cm and 7 cm. The total number of test specimens is 21 concrete cylinders. Steel and bamboo for reinforcement measuring Ø 8 mm, steel and bamboo spirals used for stirrups measuring Ø 6 mm as confinement on reinforced concrete and testing was carried out on the 28th day.

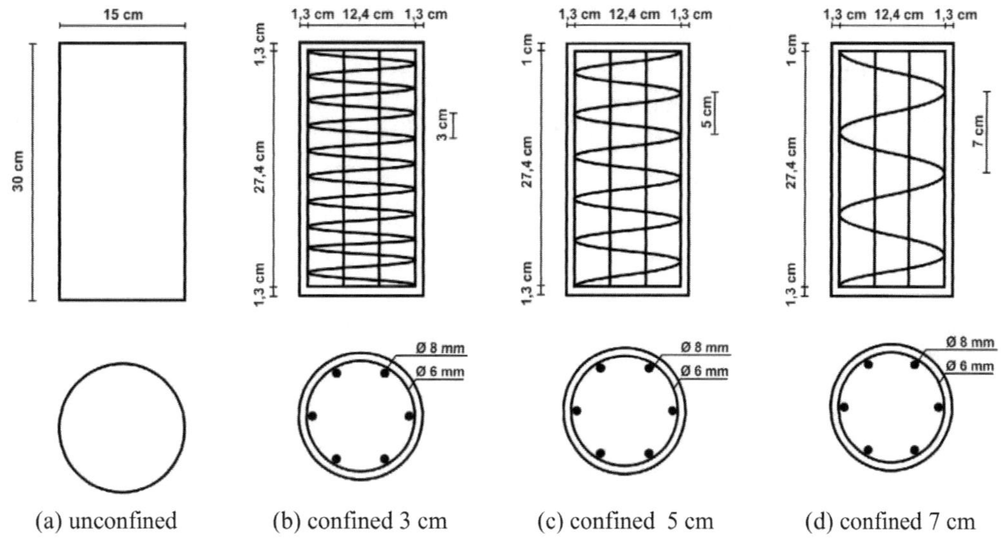

Figure 3.Variations of sample

The concrete cylinder compressive test was carried out using the universal testing machine as shown in Figure 4. The Compressive testis performed only on the core area with a diameter of 12 cm. The results of the compressive strength test will be divided by the area of concrete contact with an instrument that is equal to 11304 mm^2

(a) Sample

(b) Universal Testing Machine

Figure 4. Compression Test

Table 1. Compresion strength of concrete cylinder with steel stirrup

Stirrup Distance	CODE	Diameter (mm)	H (mm)	Load (N)	Compression Strength (N/mm2)	Average
Unconfinement	k1	120	293	226883	20,071	19,78
	k2	120	290	225865	19,981	
	k3	120	292	218179	19,301	
Confinement bamboo 30 mm	C3.1	120	295	327646	28,985	28,77
	C3.2	120	300	323690	28,635	
	C3.3	120	300	324187	28,679	
Confinement bamboo 50 mm	C5.1	120	290	268255	23,731	23,89
	C5.2	120	295	280701	24,832	
	C5.3	120	287	261122	23,1	
Confinement bamboo 70 mm	C7.1	120	295	246427	21,8	21,50
	C7.2	120	300	239645	21,2	
	C7.3	120	293	243036	21,5	

Table 2. Compresion strength of concrete cylinder with bamboo stirrup

Stirrup distance	CODE	Diameter (mm)	H (mm)	Load (N)	Compression strength (N/mm) of research results	Average
Unconfinement	k1	120	293	226883	20,071	19,78
	k2	120	290	225865	19,981	
	k3	120	292	218179	19,301	
Confinement bamboo 30 mm	C3.1	120	295	301817	26,7	25,37
	C3.2	120	300	280339	24,8	
	C3.3	120	300	278078	24,6	
Confinement bamboo 50 mm	C5.1	120	290	233993	20,7	21,80
	C5.2	120	295	246427	21,8	
	C5.3	120	287	258862	22,9	
Confinement bamboo 70 mm	C7.1	120	295	232862	20,6	20,33
	C7.2	120	300	221558	19,6	
	C7.3	120	293	235123	20,8	

There is a difference between confined and unconfined on a concrete cylinder. For steel material as confinement with a distance of 3 cm increase 45 % compared to concrete cylinders unconfinement in Table 1 (unconfinement = 19,78 N/mm^2 and confinement = 28,77 N/mm^2). The longer the stirrups, the effect of the confinement does not affect.

Spiral stirrup with bamboo material increased by only 28% in Table 2. The bamboo stirrup in confine is not very good. This is proven when applied to bamboo stirrup for beam, it does not greatly increase the flexural capacity than steel stirrup [1][2]. While the beam with steel stirrup increased flexural strength.

Table 3. Comparison of Experimental Results With Mander Calculations (Steel Stirrup)

distance stirrup (mm)	Steel Stirrup		
	Compression Strength (N/mm) of research results	Compression Strength (N/mm) by Mander	Different
no stirrup	19,78		5%
30	28,77	27,23	4%
50	23,89	24,82	6%
70	21,50	22,83	5%

Table 4. Comparison of experimental results with Mander calculations (bamboo stirrup)

Distance stirrup (mm)	Bamboo stirrup		
	Compression strength (N/mm) of research results	Compression strength (N/mm) by Mander	Different
No stirrup	19,78		4%
30	25,37	24,27	6%
50	21,80	20,59	1%
70	20,33	20,04	4%

Confinement increases the compressive strength of concrete cylinders. Confinement affects the capacity of beams and columns. Calculations using the formula from Mander produce different values from the results of the experimen (table 3 and table 4). There is a difference of 4% -5% between the Manders formula [4][5] and the results of the compressive test. So the Manders formula can be used to predict the contribution of stirrups, with a difference of approximately 4%-5%.

The smoothing surface of concrete cylinder can improve stngth of concrete by 1%-21% [9]. Porosity of concrete was also predicted to be another factor that led decreasing strength of concrete. Previous research stated that the higher porosity precentage of concrete, the lower compressive strength owned [11]

4. Conclusions

- Steel material as confinement with a distance of 3 cm increase 45 % compared to concrete cylinders unconfinement. The longer the stirrups, the effect of the confinement does not affect.
- Spiral stirrup with bamboo material increased by only 28%

References

[1] N Nindyawati et al. 2017 Effects of Variations in Overlap Length and Stirrup Spacing on Flexural Capacity of Bamboo Reinforcement Concrete Beams. *AIP Conference Proceedings* 1887, 020044 (2017); doi: 10.1063/1.5003527.

[2] N Nindyawati, Risdanareni P and Umniati 2017 Flexural Test of Fly Ash Based Geopolimer Concrete Beam. *MATEC Web of Conferences* · January 2017. I: 10.1051/matecconf/20179701030.

[3] Kent DC and Park R 1971 Flexural members with confined concrete. *J. Struct.* Div., ASCE 1971 ; 97(7): 1969-90.

[4] Mander JB, Priestley MJN and Park R 1988 Theoretical stress-strain model for confined concrete. J. Struct. Engin., ASCE 1988 ; 114(8): 1804-26

[5] Park R, Priestley MJN and Gill WD 1982 Ductility of square-confined concrete columns. J. Struct. Div., ASCE 1982 ; 108(4): 929-50

[6] Saatcioglu and Razvi SR 1982 Strength and ductility of confined concrete. *J. of Structural Engineering*, vol. 118, n° 6. 1590-1607.

[7] Sargin M 1968 Stress-strain relationship for concrete and analysis of structural concrete sections. *PHD Thesis*, University of Waterloo, Ontario, Canada. March 1968.

[8] Sheikh SA and Uzumeri SM 1982 Analytical model for concrete confined in tied *J. Struct.* Div.,ASCE 1982; 108(12): 2703-22

[9] SushilKD, Sajal KA, Abu ZM,MosaddekHM and GiashD 2012 Effect Capping on compressive Strength of Concrete (ICETCESD 2012)

[10] Yong YK, Nour MG and Nawy EG 1988 Behaviour of laterally confined concrete highstrength concrete under axial load. *J. Struct. Engin.*, ASCE 1988 ; 114(2): 333-51.

[11] Chen X, Huang W and Zhou J 2012*Indian J. Eng. Mater.Sci*, 19(6) 427 (2012)

Acknowledgement

This work was supported by PNBP research grant of Universitas Negeri Malang

IOP Conf. Series: Materials Science and Engineering **669** (2019) 012060 doi:10.1088/1757-899X/669/1/012060

The application of fuzzy logic for profit optimization to contractor project cash flow

B Djatmiko[1], A Yulistyorini[1], R M Sugandi[1] and E Setyawan[1]

[1] Civil Engineering Department, Faculty of Engineering, Universitas Negeri Malang, Indonesia

bambang.djatmiko.ft@um.ac.id

Abstract. Profit has become the main goal for a contractor without neglecting other aspects in the content of contract. According to previous empirical studies, profit has been optimized. However, it was optimized not as an output variable. Therefore, in this study we were going to try to position profit as an output variable using Fuzzy logic. The research aimed to: 1) describe cash in and cash out to prepare empirical Project Cash Flow (PCF) statement, 2) develop Fuzzy Mamdani set model on PCF 3) formulate manual method and Fuzzy Mamdani computational programming on PCF, 4) validate the results of manual simulation using computation, and 5) make profit which can be simulated. The results of the study were: 1). Fuzzy Mamdani set model with 4 input variables (%), those were: Direct Cost (7-75), Indirect Cost (2-25), Informal cost (1-15) and Progress (0-100) and single output which was profit (0 - 10); 2). The manual method consisted of 4 stages: (1) creating a fuzzy set of 4 input variables and 1 output variable, (2) making the Application Function Implications based on 2 rules, (3) Composing all output variables with the MAX operation, and (4) Defuzzification using the method of centroid of area, 3). computational way with indicators: Trapezoid membership function, two-type linguistic set (minimum - maximum) and (slow - fast), MISO (Multiple Input Single Output) implication functions, "MAX - DOT" operation system, fuzzy rules (if ... and ... then) = 16 rules, 4). The output of the manual method and computational method resulted the same profit of 4.46, so it was declared valid, and 5). The simulation revealed that minimum profit can be made if the cash out allocation and progress on input variables are not balanced and in contrast, we will obtain maximum profit if the allocation of cash out and progress are balanced with input variable.

1. Introduction

The condition of PCF in every construction contractor varies especially buildings owned by the government which is pretty costly. In contrast, in private-owned buildings informal cost is relatively small. Despite, the offering nomenclature is strict, the quality demands is high, and moreover, tenders are done on every type of work [1]. As a result, contractors experience difficulties in making optimal profit [2]. Other factors of PCF vary, such as material raising prices, unexpected cost overrun, weather factors, and construction failures, so that those lead to progress delay which causes profit to fall or even the contractor fails in making any [3].

Profit is the primary goal of entrepreneurs in general and construction contractors in particular. This is because profits are used for company survival, development, trust and working capital to get the next project [4]. Profit is also a reward for contractors during and after completing their works [3].

Various ways to get optimal profit include: (1) using fuzzy sets, it must be detailed in identifying the duration of the project main cost, calculating cash flow periodically and cumulatively, and drawing a progress curve precisely which, if those are applied, can increase profits by 2-3% [5], (2) Using fuzzy PCF: material and builder salary overrun increases by 7% and so as not to decelerate the profit rate, the allocation of non-technical funds must be minimized [6], (3) In complex construction company, applying fuzzy logic approach helps make decision easier for work priority, thus can save expenses up to 2% [7]., (4) application of fuzzy nets in project scheduling optimizes construction progress and minimizes the risk of cost overrun [8], (5) by working with fuzzy project cash flow approach, project costs are easily evaluated starting from the feasibility study, planning, and execution [9], (6) by using fuzzy Mamdani cash flow cycle, a contractor company can minimize human error to 2% [10], (7) by adopting Fuzzy Mamdani, optimization of partial productivity of furniture companies can increase profits up to 2.5% [11], (8) With fuzzy mamdani ready-mix concrete delivery setting can be optimized up to 2.7% of the plan capacity [12], (9) With fuzzy Mamdani, procurement of goods is based on what is sold, so that buildup and damage can be avoided, risk factors can be minimized to 4% and productivity/ progress increases [13], (10) With Fuzzy Mamdani, PCF process management can be controlled and save cost to 2% [14], (11) Fuzzy Mamdani simulation on large-scale project work utilities saves 4% of the planned time [15], and (12) Mamdani optimization simulation on raw, semi-finished, and finished materials at the project site saves 3.4% of the total material expenses [16].

According to the empirical studies above, profit was optimized not as an output variable. Therefore, in this study profit would be tested as a single output variable using MISO system of Fuzzy Mamdani model [17]. While the input variables were: Direct Cost, Indirect Cost, Progress and Informal cost. Continued to compile a manual and computational model of the Fuzzy Mamdani MATLAB R2017b program including: (1) input and output variables, (2) membership functions, (3) rules making, (4) the relationship among inputs and outputs and 3 dimensional 3 surfaces [10] . The input and output variables were in the form of empirical data/ PCF of construction contractors in East Java (Surabaya, Malang and Nganjuk). For this reason, the study was considered important to be done so that the profit which status was fuzzy in the project cost plan for the wholesale project is not allocated explicitly [1], so that profit can be optimized without neglecting other aspects.

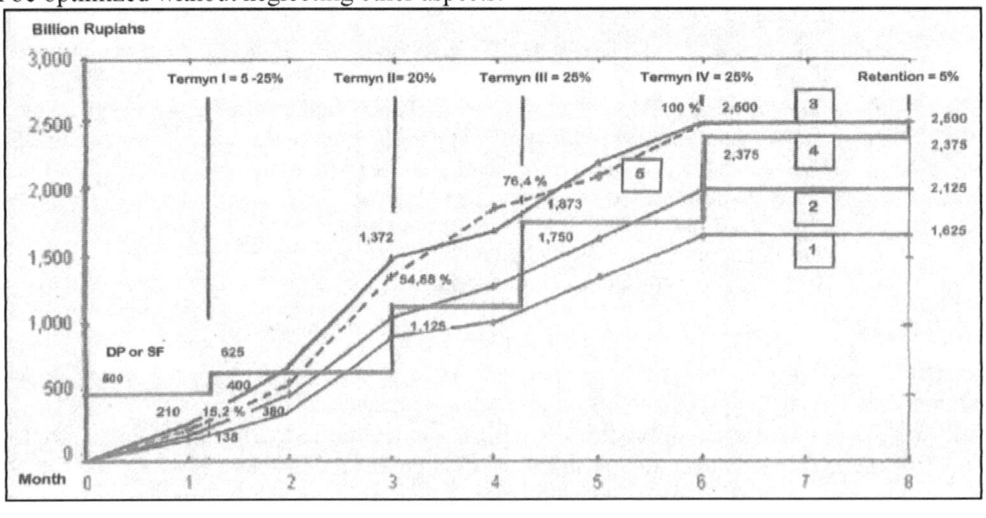

Figure 1. Model PCF [1]

Notes:
1) Direct cost
2) Indirect cost
3) Mark Up = contractor's profit + informal cost
4) Term = Periodic payment from the owner based on the progress
5) Progress = contractor's performance

PCF is the flow of cash in and cash out per period of time in a project work [18]. Furthermore, cash in consists of: 1. Own capital, 2. Loan funds, and 3. Initial and terms funds [3]. Meanwhile, cash out consists of: (1) Indirect cost, (2) Direct costs, and (3) Mark Up (contractor's profits and informal costs) [19,1]. The preparation of PCF can be done with 2 (two) methods, those are: (1) Table Method, and (2) Graphical Method. One of the graphical methods is the PCF model [1] as shown in Figure 1.

Profit is the main goal for entrepreneurs, particularly, construction contractors. This is because profits are used for company survival, development, trust and working capital to get the next project [4]. Profit is also a reward for contractors during and after completing their work [3]. The explicitly fuzzy profit is not allocated in the project budget plan (PPP) [20].

S curve is a type of time schedule or PCF in the form of linear graphical correlation between percentage of work and time allocation which creates letter-S form [21]. Time schedule and S curve become one unit which is easier to do and function on the construction project to make sure that the contractor is doing right according to what is planned by comparing with the progress curve which can be seen in Figure 2 [20].

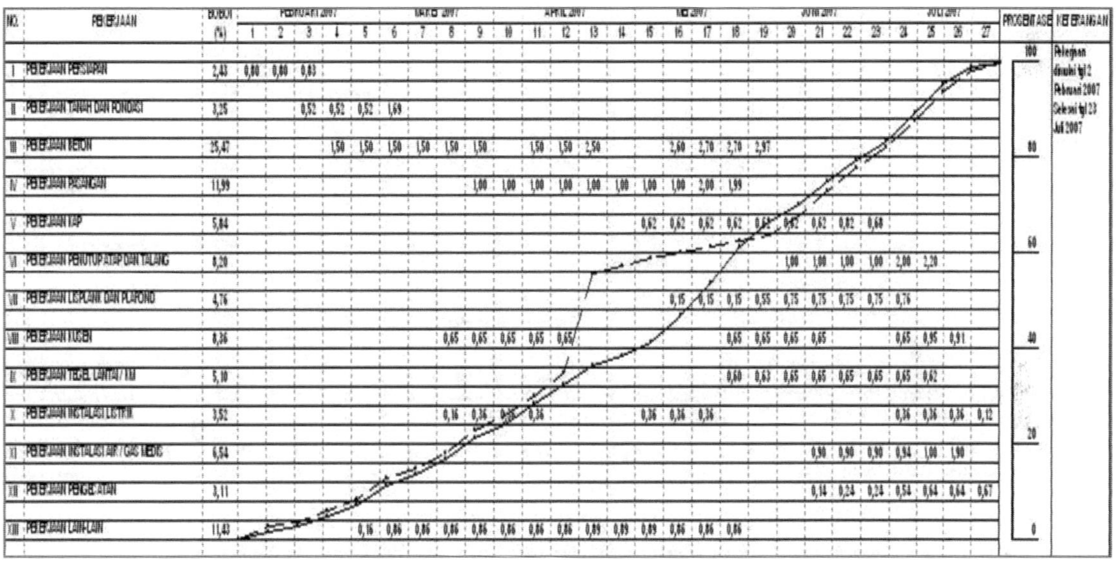

Figure 2. Time Schedule, S-Curve and Progress Curve [20]

Notes:
1. ⟶ = The S curve is made by a planning consultant
2. - - - - - - - - = The progress curve is made by a supervisory consultant

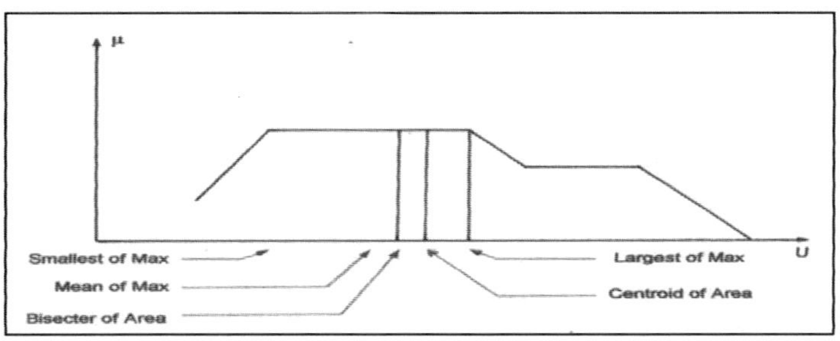

Figure 3. Defuzzification Process [17]

Fuzzy Mamdani Logic is an approach to map an input space into the output space using IF-THEN rules [22]. According to Kusumadewi and Purnomo [10], it takes 5 steps to get an output: (1) Fuzzy sets making which consists 2 attributes: linguistic and numerical attributes, (2) application of Implication Functions that is IF x is A THEN y is B, (3) Composition of Rules which consists of 3 methods (max, additive and probabilistic (OR)), (4) If the implication function used is MIN, then this composition (MAX-MIN) and (5) Affirmation (Defuzzification) results in numbers in the fuzzy set domain which can be achieved in 5 ways as shown in Figure 3.

According to [10], fuzzy systems consist of: 1) Fuzzy variables, variables to be discussed, for instances: input variables (direct cost, indirect cost, informal cost, progress) and output variables (profit), 2) Fuzzy sets are groups which represent a condition of fuzzy variables (input variables, fuzzy variables, linguistics and symbols), 3) Value talks universe (min-max), 4) Domain which means all values allowed in the universal set and may be operated [0 0 7 75].

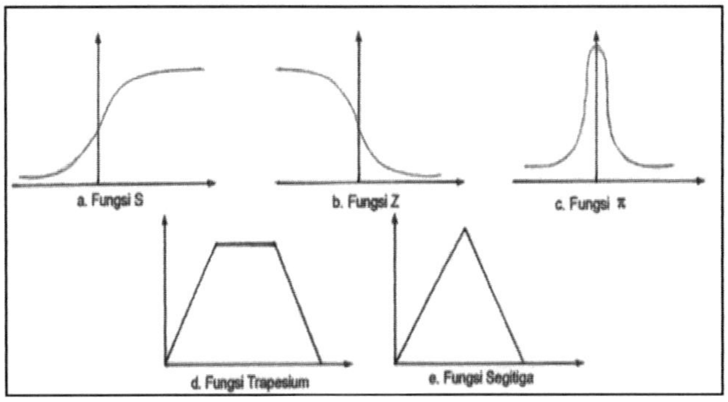

Figure 4. Membership Function Forms [17]

Software used in this research is MATLAB R2017b. According to [23], MATLAB (Matrix Laboratory) is a program for numerical analysis and computation, an advanced mathematical programming language created with the principle of properties and matrix form use. MATLAB language supports vector operations which are basics of data technique scientific problems. Fuzzy Logic toolbox provides Graphical User Interface (GUI) facility to make building fuzzy system easier. GUI consists of 5 tools which can be seen in Figure 5.

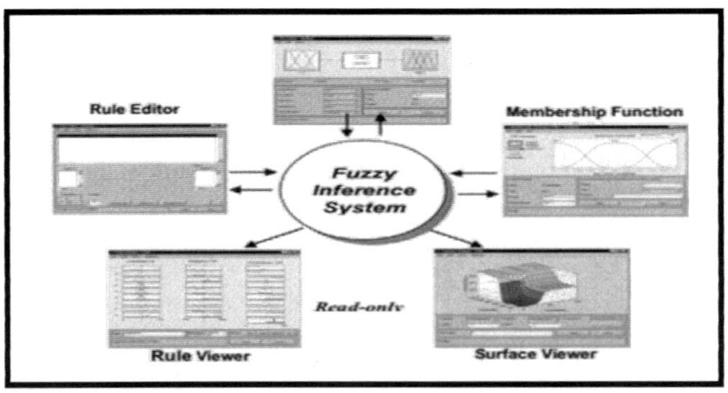

Figure 5. Fuzzy Inference System [24].

2. Method

TResearch samples were determined purposively [25], those are buildings in three cities in East Java: Surabaya, Malang, and Nganjuk, with 30 samples in total. Data collection methods used are: 1) Interview which aimed to get informal cost and profit information and 2) Documental study which aimed to get information about PPP, time schedule, progress, S curve, cash in and cash out

Other paragraphs are indented (BodytextIndented style).

Table 1. Profit optimization variables breakdown

No	Concept	Variable	Sub variable	Indicator	Instrument
I	Arranging empirical PCF	1.1. Cash in		1. Own capital	1. Interview guideline
				2. Loan	
				3. Initial capital from *owner*	
				4. Term	2. questionnaire
		1.2. Cash Out	1.2.1. Indirect cost	5. Tax and formal cost	
				6. Operational cost	3. a.PPP b.Financial Recapi-tulation
			1.2.2. Direct cost	7. Material cost	
				8. Salary	
				9. Equipment + rental/ quality test cost	
				10. Paying sub-contractor	4. S-curve
				11.Unforeseen cost	
				12. Contractor's profit	5. Progress curve
			1.2.3. Pro In	13. Informal cost	
II	Arranging Fuzzy app MAMDANI For Profit Optimization	2.1. Input	2.1.1. Direct cost	1. Value [min–max] empirical data	Fuzzy MAMDANI MATLAB R2017b Software
			2.1.2. Indirect cost	2.Value [min–max] survey result	
			2.1.3. Informal cost	3. Value [min–max] empirical data	
			2.1.4. Progress	4. Value [min–max] empirical data	
		2.2. Output	Profit	5. Value [max-dot]	

Research analysis was done by validating output [10, 14] which is comparing the manual method output with computational method output. If those results are in the same number, it is considered valid.

3. Results and discussion

Data obtained through this study is shown in Table 2 below.

Table 2. Cash in and cash out in construction contractor

No	PCF	Min – Max (%)
I	**1. Cash in :**	**10 – 30**
	1.1. Own capital	0
	1.2. Loan	0
	1.3. Initial capital	10 – 30
	1.4. 1st Term	15 – 25
	1.5. 2nd Term	20 – 25
	1.6. 3rd Term	10 -15
	1.7. 4th Term (retention)	5
II	**2. Cash out :**	**0 - 100**
	2.1. Indirect cost	**2 – 25**
	2.1.1. Tax and formal cost	11,5 – 13
	2.1.2. Operational cost	5 – 10
	2.2. Direct cost	**7 – 75**
	2.2.1. Material expense	50 – 60
	2.2.2. Salary	10 – 20
	2.2.3. Cost of equipment + rent / quality test	0,5 – 2
	2.2.4. Paying sub-contractor	10 – 30
	2.2.5. Unforeseen expenses	2 – 5

No	PCF	Min – Max (%)
	2.3. Profit and Informal cost	**1 – 25**
	2.3.1. Informal cost	1 -15
	2.3.2. Contractor's profit	0 – 10
III	**Progress**	**0 - 100**

Based on the data presented in Table 2, fuzzy variables were arranged in Table 3

Table 3. Fuzzy membership function trapezoid variable

No	Function	Variable	Grouping	Symbol	Universal set (%)	Domain (%)
1	Input	Direct cost	minimum	min	7 - 75	0 0 7 75
			maximum	max		7 75 75 75
		Indirect cost	minimum	min	2 - 25	0 0 2 25
			maximum	max		2 25 25 25
		Informal cost	minimum	min	1 - 15	0 0 1 15
			maximum	max		1 15 15 15
		Progress	slow	slow	0 - 100	0 0 0 100
			fast	fast		0 100 100 100
2	Output	Profit	minimum	min	0 - 10	0 0 0 10
			maximum	max		0 10 10 10

Output validity was tested by [10]: 1) The Manual method using Fuzzy Mamdani set and deffuzzyfication using the method of centroid of area. Figure 6 illustrates results of the test

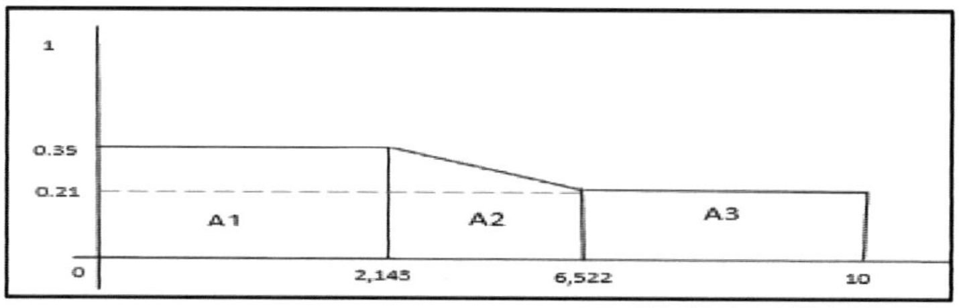

Figure 6. Centroid of Profit Area

$\sum$ Moment = M1 + M2 + M3 = 0.798 + 5.332 + 6.157 = 12.28
$\sum$ Area = A1 + A2 + A3 = 0.745+ 1.230 + 0.745 = 2.72
Z = Profit = ($\sum$ Moment of area)/($\sum$ Area) = 12.28/2.72 = 4,56.

Calculation using the computational method with Fuzzy Mamdani MATLAB 2017b software obtains profit output as shown in Figure 7.

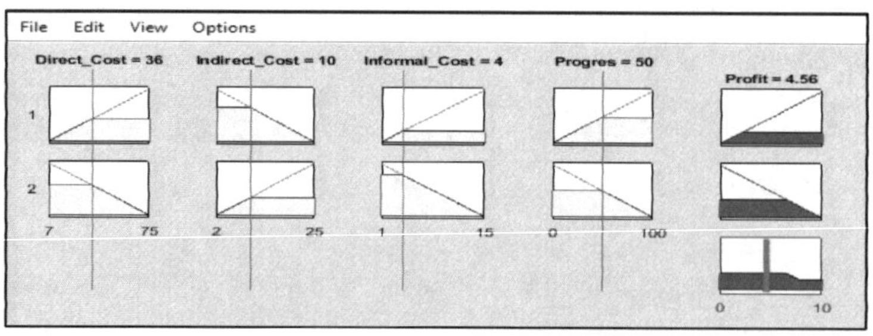

Figure 7. Result of Computation

Manual calculation = computational calculation = 4.56 (valid).
Based on data presented in table 3, we obtained simulation result as shown below.

Example 1: Progress position simulation results = 25% and profit = 6.16%

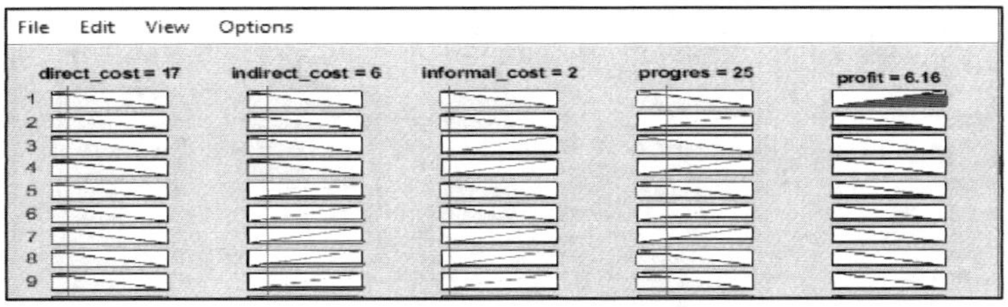

Figure 8. 6.16% Profit rate simulation result

Example 2: Progress position simulation results = 50% and profit = 5 %

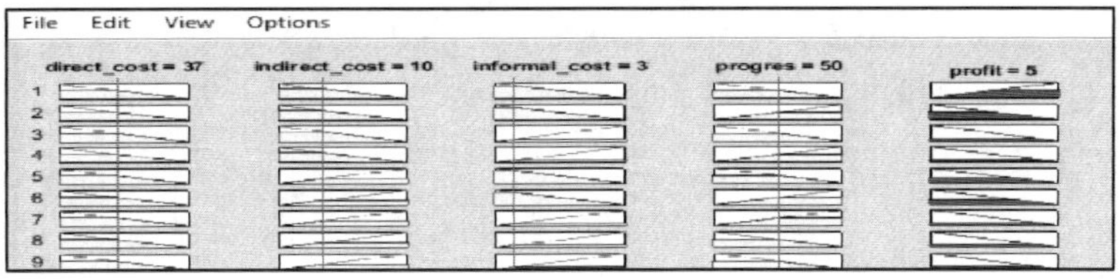

Figure 9. 5% Result of Profit rate simulation

Example 3: Progress position simulation results = 75 % and profit = 5,75 %

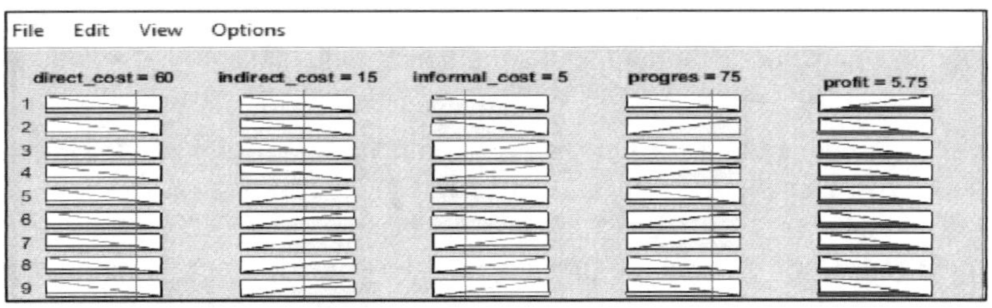

Figure 10. result of profit Value simulation= 5 ,75 %

Example 4: Progress position simulation results = 100 % and profit = 4,67 %

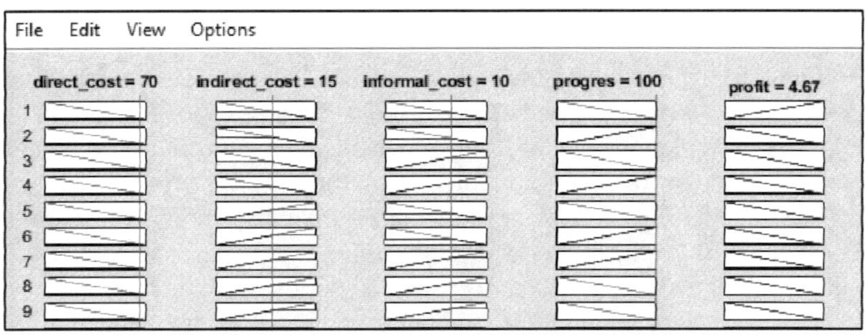

Figure 11. Result of simulation of profit = 4,67 %

Example 5: Progress position = 100 % and profit = 5 %

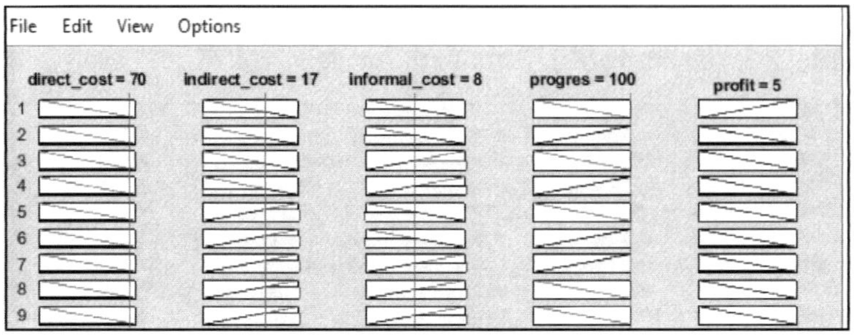

Figure 12. Result of simulation of profit = 5 %

The example simulation results 1, 2, 3, 4 and 5 above are summarized in Table 4

Table 4. Results of profit optimization simulation

No	Multiple Input (%)				Single Output (%)	Notes
	Direct cost	Indirect cost	Informal cost	Progress	Profit (max)	
1	17	6	2	25	6,16	Realistic
2	37	10	3	50	5	Realistic
3	60	15	5	75	5,75	Unrealistic
4	70	15	10	100	4,67	Unrealistic
5	70	17	8	100	5	Realistic

On the Table 4 row 1, on the input variable, some certain values are added orderly such as direct cost value = 17%, indirect cost value = 6%, informal cost value = 2% and progress value = 25%, then on the output variable was obtained profit value = 6.6%. Which means that with the progress value = 25% the contractor in the qualified position to get the first term. At the same time obtaining profit / profit = 6.16% is relatively high and unrealistic due to the long work journey and the informal cash cost is relatively small, so that the contractor can enjoy it or be prepared for further work costs. Profit is relatively easy to obtain, because construction work is not yet complex, informal costs are small and costs are borne by the DP or if there is a delay in DP payments, the contractor can use a letter of work order (LWO) to guarantee a short-term loan with low interest [1]

On the Table 4 row 2, on the input variable, some certain values are added orderly such as direct cost = 37%, indirect cost = 10%, informal value = 3% and progress value = 50%, then the output variable is obtained the profit = 5%. Which means with the progress value = 50% of the contractor are in the qualified position to get the term II (second), while the contractor gains profit = 5%. This simulation is more realistic that the contractor gets profit = 5% which is the maximum profit that must be achieved as their target progress = 50%. If the contractor able to reach progress as target (50%, 75% and 100%) and get a profit of 5%, and the maintenance period plus retention payment is 5%, the contractor will get a maximum profit of 10% according to empirical data, namely maximum profit = 10%. Such realistic and ideal conditions according to [3] are called better progress.

On the Table 4 row 3, on the input variable, some certain values are added orderly such as direct cost value = 60%, indirect cost = 15%, informal value cost = 5% and progress value = 75%, then the output variable obtained profit value = 5.75 %. This simulation, according to the comments of some respondents, is unrealistic, because in the progress position 75% of the informal cost expenditure = 5% is too small, and not according to the facts in the field. The results of the study, in the progress position 75% informal cost of minimum expenditure 7% [1].

On the Table 4 row 4, on the input variable, some certain values are added orderly such as direct cost value = 70%, indirect cost = 15%, informal value = 10% and progress value = 100%, then the output variable obtained profit value = 4.67 %. This simulation contrasts with table 4 line 3 above, that is the informal cost expenditure is increased by 10%, so that the profit gain decreases by = 4, 67% <5%, namely the maximum profit that must be sought and maintained by the contractor, especially on the final target of 100% progress, if they want get maximum profit. This less ideal allocation of PCF according to [26] is called indicates a negative cash flow.

On the Table 4 row 5, on the input variable, some certain values are added orderly such as direct cost = 70%, indirect cost = 17%, informal value = 8% and progress value = 100%, so the output obtained profit value = 5%. This simulation is in the same principle as table 4 row 2: the implementation of the work according to the target of progress = 100%, the contractor earns a profit of = 5% and the maintenance period is added plus retention payments = 5%. Total profit contractor becomes = 10% is the maximum profit [3].

The cash out value in table 4 line 2 and table 4 row 5 are just a simulation example so that the contractor's progress position is in accordance with the ideal progress, so that the contractor can get term payment and the amount of profit in ideal conditions. There is still the possibility of allocating variations in cash out without changing the purpose of the simulation, so that it will be controlled all time, and avoiding the contractor for experiencing a loss for no irregularities [10, 14].

4. Conclusion

The conclusion of this study can be stated that: 1) Fuzzy Mamdani set model for profit optimization has 2 functions, those are: 1) input function and 2) output function. Input function consists of 4 variables: (1) Direct cost with linguistic set (min-max) and universe of discourse (7-75%), (2) Indirect cost with linguistic set (min-max) and universe of discourse (2-25%), (3) Informal cost with linguistic set (min-max) and universe of discourse (1-15%) and (4) Progress with linguistic set (slow-fast) and universe of discourse (0-100%). Output function is only one variable, that is Profit with linguistic set (min-max) and universe of discourse (0-10%), 2) The manual way of Fuzzy Mamdani logic application for profit optimization consists of 4 stages: (1) Making fuzzy set 4 input variables (Direct cost, Indirect cost, Informal cost and Informal cost) and 1 output variable (Profit), (2) Creating Application Function Implications based on 2 rules (rules), (3) Making Composition of All Output variables with MAX and (4) operations Defuzzifcation uses the centroid of profit area method 3) Computation method for profit optimization with Mamdani MATLAB R2017b fuzzy software with indicators, (1) Trapezoid membership function, (2) Linguistic set (min-max) and (slow-fast), (3) MISO implication function (4 input variables; direct cost, indirect cost, informal cost, progress and 1 output variable, namely profit), (4) "MAX - DOT" and (5) operating systems. Fuzzy rule (if ... and ... then) consists of 16 rules, 4) The level of validation by manual method by computation method resulted the same that is profit = 4, 46 (valid) so that it can be used for simulation, and 5) Simulation results will be obtained by minimum profit if in allocating cash out and the progress of the input variable are not balanced and vice versa, it will be obtained maximum profit if the allocation of cash out and progress in the input variable are balanced.

References

[1] Djatmiko, Bambang 2017 *Modelling of Project Cash Flow on Construction Projects in Malang City* AIP Conference Proceedings 1887 020014 (2017) https: // doi.org/10.1063/1.5003497

[2] Gurcanli Emre G, Harun, Turkoglu., Senem, Bilir 2017 *Heavy Equipment Scheduling for Horizontal Construction Projects* Procedia Engineering 265, 265-266

[3] Lu W and Jian Liu. 2014 *Research in to the moderating effects of progress and quality performance in project dispute negotiation.* International Journal of Project Management .

Letters **32** 654-662

[4] Djatmiko, Bambang 2016 *Analisa Project Cost Pada Pengembangan Rumah Type 21 dan 36.* Jurnal Teknologi dan Kejuruan **2** FT Universitas Negeri Malang.

[5] Chen, MT. 2012 *ABC of Cash Flow Projections AACE,International Transaction*. Cochran WG. 2015. Teknik Penarikan Sampel Penerjemah: Rudiansyah (Harvard University, Harvard, 2015), pp 21 – 22

[6] Sukma, KE., Diputro, SH., dan Wiguna, IPA. 2014 *Pemodelan Schedule dan Cost Pada Pekerjaan di Proyek Pembangunan Jalan Tol Surabaya-Mojokerto Seksi IV*

[7] Fu Liang, T. 2009 *Fuzzy Multy-Obyective Project Management Decisions Using Two-hase Fuzzy Goal Programming Aproach*. Computers & Industrial Engineering **57** 1407–1416

[8] Ucal, I and Kuchta, D. 2016 *Project Scheduling to Maximize Fuzzy Net Present Value*. Proceedings of the World Congress on Engineering **II** WCE, July 6 - 8, London, U.K.

[9] Maravas, A and Pantaouvakis, JP *2014 A Study of Cash Flows in Projects with Fuzzy Activity Duration.* International Journal of Project Management. Letters 30, 374-384

[10] Kusumadewi, S & Purnomo, H. 2013 *Aplikasi Logika Fuzzy Untuk Pendukung Keputusan, Edisi Kedua.* Yogyakarta: Graha Ilmu

[11] Bidiawati dan Nasutioan. 2013 *Aplikasi Logika Fuzzy Dalam Pengukuran Produktivitas Parsial Perusahaan*, JurusanTeknik Industri, FTI, Universitas Bung Hatta, Padang.

[12] Dichu Bao, Kam Chan C, Weining, Zhang 2013 *Asymmetric cash flow sensitivity of cash holdings.* Journal of Corporate Finance . Letter 18, 290-202.

[13] Saputro, A. 2013 *Sistem Pendukung Keputusan Penentuan Jumlah Pengadaan Barang Pada CV Roda Jaya Mandiri Semarang Dengan Metode Mamdani.*

[14] Mohagheghi, Vahid, Seyed, Meysam, Mousavi., Behnam, Vahdani. 2015 *A New Uncertain Modeling of Production Project Time and Cost Based on Mamdani Fuzzy Sets. Journal of Quality Engineering and Production Optimization* **1** no. 2 pp 57-70

[15] Purnus A and Bodea CN, 2016 *Multi-Criteria Cash Flow Analysis in Construction Projects by Mamdani Fuzzy Logic. Procedia Engineering* Letters 164 pp 98-105

[16] Sharifi MM and Baghepor M. 2016 *Optimizing Cash-Flow-at-Risk in Construction Projects: A Cost Reduction Approach*. Periodica Polytechnica Civil Engineering. Letters 60, 337-338

[17] Widodo, TS 2005. *SistemNeuro Fuzzy* (Yogyakarta : Graha Ilmu)

[18] Santoso, B 2015 *Manajemen Proyek* (Surabaya : Guna Widya)

[19] Halpen, DW and Woodhead, Ronald W 2014 *Construction Management* (New York. USA. : John Wiley & Sons. Inc.)

[20] Djatmiko, Bambang. 2018. *Optimasi Progres Pada Kontraktor Jasa Konstruksi Dengan Sofware Fuzzy MATLAB R2014a* Jurnal Bangunan **23** no 1 43-54

[21] Noldi, 2016 *Cash Flow Proyek.* (Konsultan Perencana ONLINE, Jakarta) pp 28-3

[22] Naba, A. 2009 *Belajar Cepat dan Mudah Fuzzy Logic Menggunakan Matlab.* (Malang: ANDI)

[23] Kusumadewi, Sri. 2008. *Analisis dan Desain Sistem Fuzzy Mengunakan Toolbox*

[24] *Matlab* (Yogyakarta : Graha Ilmu)

[25] Pusadan, M., Y. 2014. *Pemrograman Matlab Pada Sistem Pakar Fuzzy Edisi Pertama.* (Yogyakarta: Deepublish)

[26] Cochran WG. 2015. *Teknik Penarikan Sampel* Penerjemah: Rudiansyah (Harvard University, Harvard, 2015), pp 21-22

[27] Reyers, Michelle., Cornelis, Hendrik van Schalkwyk., Daniël, Gerhardus, Gouws. 2015. *Rational and behavioural predictors of pre-retirement cash-outs* Journal of Economic Psichology. Letter 47, pp 23-3

IOP Conf. Series: Materials Science and Engineering **669** (2019) 012061 doi:10.1088/1757-899X/669/1/012061

Effects of segregation on reinforced concrete column, a numerical approach

W Damayanthi[1], Y Novitasari[1] and M M A Pratama[1]

[1] Department of Civil Engineering, Faculty of Engineering, Universitas Negeri Malang, Indonesia

mirza.abdillah.ft@um.ac.id

Abstract. Columns are parts of building members that are exposure by the risk of segregation due to the elements' dimension ratio, characteristic of concrete mix, and the concrete casting method. The segregation will create a higher density of coarse aggregate in lower part of the columns and result non-uniform distribution of concrete strength along the columns' height. In previous research, the segregation on RC column were proven to decrease the performance of the member in withstanding the working load. As a result, the resulting ultimate load decreased up to 4% compared to the normal column. The segregation also decreases the level of ductility of the column, which was supported by the decrease of top drift on the segregated column. The researchers also found that segregation on RC column could change the failure mechanism from tension-controlled failure to compression-controlled failure. The failure on segregated column was initially begun by the crushing of concrete on the compression fibre and followed by the yield of tensile rebar. On the normal column, on the other hand, the failure on was initially begun by the yield of tensile rebar and followed by crushing of concrete on the compression fibre. In this paper, researchers would like to investigate the behaviour of segregated RC column using numerical approach. The results showed that (1) the segregated column creates a linear drift pattern, while the normal column shows a graph with double curvatures at the ultimate condition; (2) the resulting stress contour showed a similar pattern in the segregated column and in the normal column, while the normal column could withstand working stress up to 25% higher than the segregated model; and (3) column drift from numerical analysis is more reliable than the experimental works as the experimental data only provided the reading of drift at two level of column height.

1. Introduction

In structural analysis, building elements, such as reinforced concrete beams, columns, and plates, are assumed to have a homogeneous strength on the entire part of the elements. It ignores the reality that concrete undergoes the segregation and/or bleeding that change the material distribution on elements' height. The use of coarser aggregate and the higher water-to-cement ratio will decrease the viscosity of concrete mix and increase the possibility of concrete to occur segregation. The segregation will create a higher density of coarse aggregate in lower part of cast element and result less concentration in the upper part. This effect will be more obvious in the depth elements such as columns. In the same time, bleeding will separate the cement-water from the concrete mix and leave the water pockets under the coarse aggregates that reduce the concrete strength.

Hidayat [1] stated that segregation is influenced by the concrete mix proportion, casting methods, and concrete compacting methods. The present of segregation causes the concrete to create a different concrete strength pattern as a function of elements' height. It was found that the segregation created a higher concrete strength laid at the bottom while gradually decrease to the upper part. This finding was supported by the reading of hammer test procedure conducted on the deep beam in the existing building.

Corinaldesi [2] also conducted research regarding segregation phenomenon on the concrete specimens sized of 80 x 80 x 1000 mm. He utilised Ultrasonic Pulse Velocity (UPV) apparatus to inspect the wave velocity on each specimens' height. The research resulted that the higher the reading position is, the lower velocity is recorded from the testing. The highest velocity is found the extreme lower fibre of the specimens. The velocity of UPV is highly related to material density. The higher material density increases the resulting velocity reading. The higher material density in concrete is indicated by the increase of coarse aggregate percentage at the inspected part, meaning that the higher velocity in the lower part of concrete member is caused by the high concentration of coarse aggregate. The lower velocity in upper part of concrete element is caused by the less percentage of coarse aggregate. It is concluded that deep elements are prone to segregation and change the material homogeneity that effects the performance of the designed elements.

Columns are parts of building members that function to withstand axial loads besides shear and moment that arise due to lateral loads. The columns that are insufficiently designed to lateral loads will undergo shear failure and create a brittle failure mechanism that endangers the building host [3-4]. The experimental works focusing on the behaviour of RC column undergone segregation had been conducted by Pratama [5]. The result showed that the segregation on RC column decreases the performance of the element. The segregation leads the coarse aggregate to sink at the bottom of column and creates a gradual strength pattern from the upper to the lower part of the column. As a result, the achieving ultimate load decreased up to 4% compared to the normal column. The segregation also decreases the level of ductility of the column, which is supported by the decrease of top drift on the segregated column. The researchers also found that segregation on RC column could change the failure mechanism from tension-controlled failure to compression-controlled failure which was evidenced by the reading of strain gauge installed on tensile rebars and on the concrete compression fibre. The failure on segregated column was initially begun by the crushing of concrete on the compression fibre and followed by the yield of tensile rebar. On the normal column, on the other hand, the failure on was initially begun by the yield of tensile rebar and followed by crushing of concrete on the compression fibre.

In this paper, researchers would like to investigate the behaviour of segregated RC column using numerical approach. The resulting data, particularly consisting load – top drift data and stress concentration will be compared with the experimental data on the previous research. This paper aims to complement the research findings regarding the effect of segregation on RC column.

2. Finite element modelling

The detail of the testing columns followed the specification of Pratama's research [5]. The RC columns size of 150 x 150 x 1000 mm with reinforcement configuration 4Ø8 mm for the longitudinal bars (reinforcement ratio = 1.17%) and Ø6-100 mm for the stirrups were modelled in Strand7. The reinforcement detailing of columns is displayed in Figure 1.

The idealization of column specimen underwent segregation also adapted from the research of Pratama [5]. In that research, the observation was focused on the upper section of the column that possessing lower strength and was exposed to the higher risk of shear force concentration when the lateral load occurs. The illustration of column modelling is depicted in Figure 2.

In finite element programme, the segregated column was modelled by creating multiple layers of concrete with strengths that gradually change with the function of column height. In this case, the researchers applied a linear pattern to relate the strength on the extreme top fibre to the extreme bottom fibre of the column. This modelling approach is so-called with 'layerwise homogenisation' method [6-13]. The layerwise homogenization method is the best solution to reduce the degree of freedom on the model constructed from heterogeneous materials and will decrease the running time during the analysis.

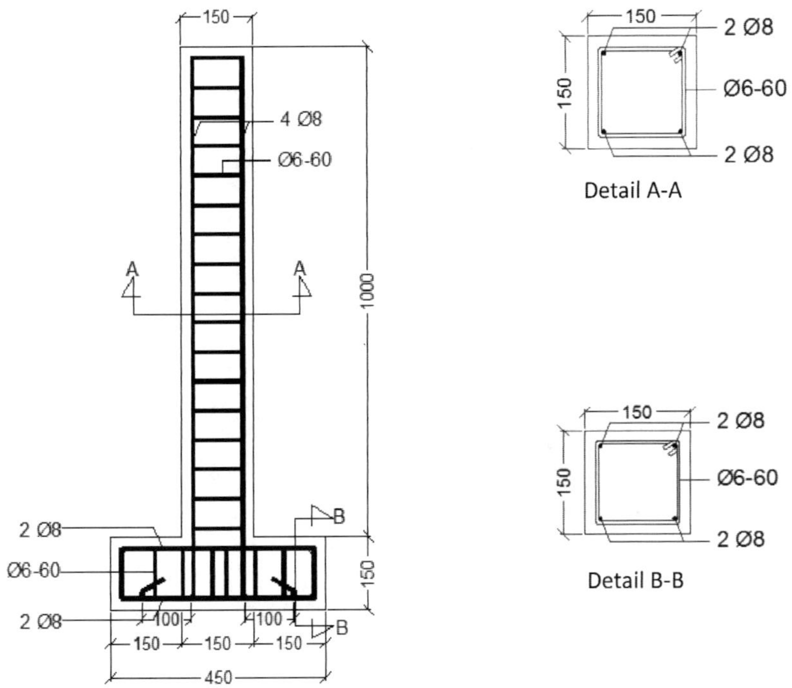

Figure 1. Detail of RC columns [5]

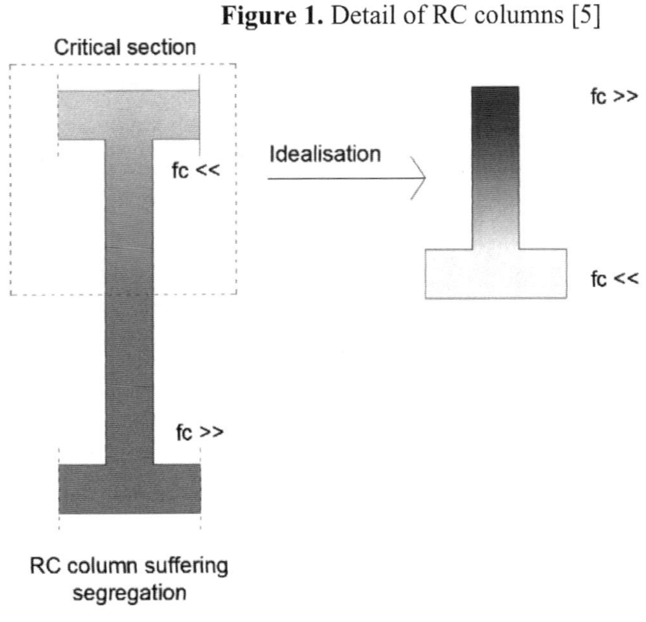

Figure 2. The idealisation of segregated column [5]

During the modelling, material parameters consisting of, at least, Poisson ratio, density, stress-strain table, and failure criteria are input to the programme. The failure criteria used in this research was Max-Stress so the failure of nodes is evaluated by the maximum stress of materials. The materials were set as isotropic. To model the interaction of rebars to concrete, there are three alternatives, that are the discrete model, the embedded model, dan the smeared model. In this research, researchers used the discrete model, which idealized the rebars as beam elements connected to concrete nodes. The concrete elements were modelled using plate. The illustration of discrete model and the model of segregated column is shown in Figure 3.

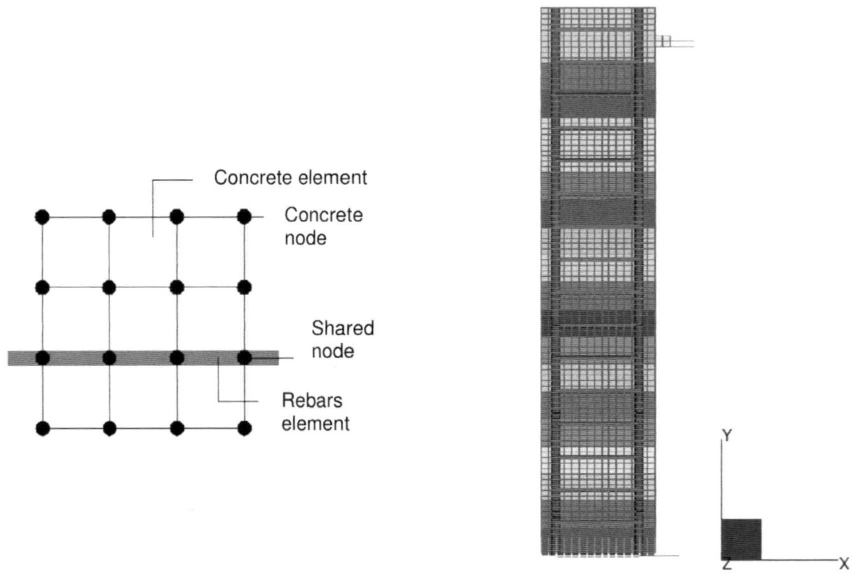

Figure 3. (a) Discrete modelling [6], (b) Column modelling in Strand 7

3. Results and discussion

3.1. Drift ratio – column height relationship

Figure 4 shows the drift ratio of the segregated column and the normal column under the lateral loads at the ultimate condition. The drift ratio was obtained by comparing drift on each level with the ultimate drift of each model. Figure 4 presents that the segregated column creates a linear drift pattern, while the normal column shows a graph with double curvatures.

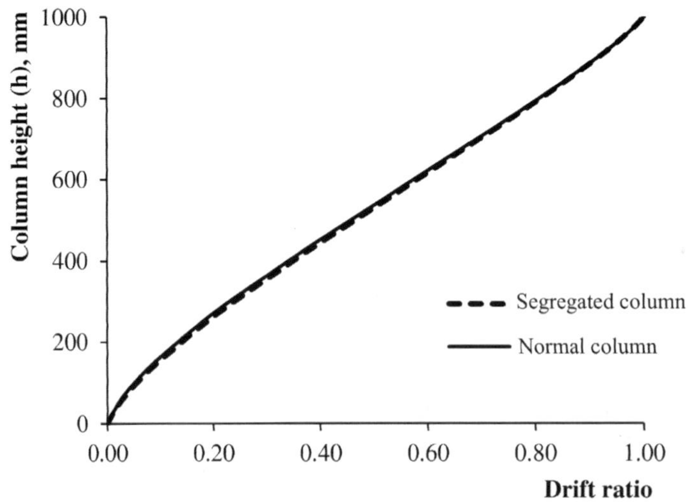

Figure 4. Drift ratio of the segregated column and the normal column

IOP Conf. Series: Materials Science and Engineering **669** (2019) 012061 doi:10.1088/1757-899X/669/1/012061

3.2. Stress contour

The appearance of stress contour will help the researchers to investigate the stress distribution and the stress concentration on the testing models. The resulting stress contour in Figure 5 shows a similarity in the segregated column and in the normal column. The lateral loads created tensile stresses and compressive stresses area on the models. Based on the data of plate stress, the highest stress is found in the normal concrete model with a difference up to 25% than the segregated model. A small difference is found in the area of tensile stress that is lesser in the segregated column. This phenomenon will lead the segregated column to undergo compression failure rather than the tensile controlled failure. It supports the findings on the previous research [5] that the segregation alters the failure mode and also decrease the structural ductility of the column.

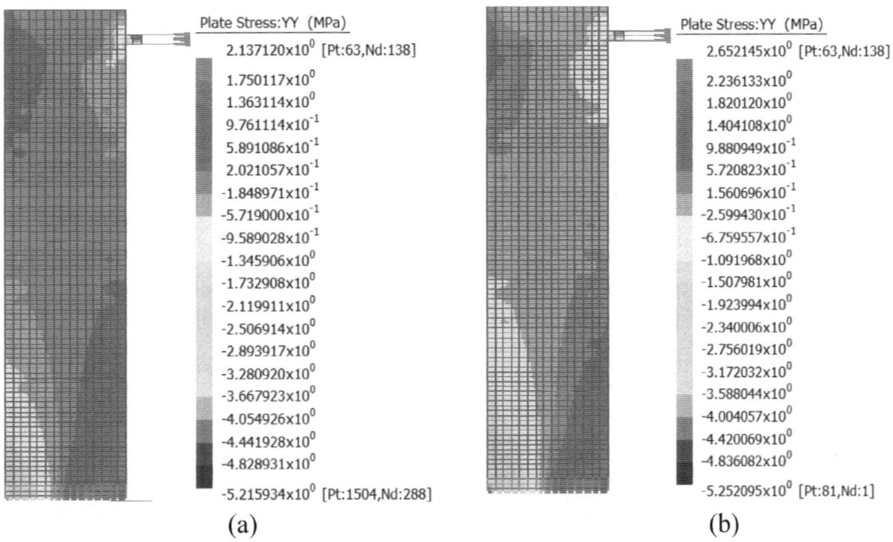

Figure 5. Stress contour on (a) segregated column, and (b) normal column

3.3. Drift comparison of the segregated column from experimental works and finite element analysis

Figure 6 shows the data of the drift pattern of the segregated column from the numerical analysis and the previous research [5]. In terms of drift pattern, the experimental data shows a nearly linear pattern which the column drifts with the same inclination from lower part to the mid-top part. Meanwhile, the numerical analysis shows a less drift at support area and begin to sway in linear pattern to the upper section.

In experimental works, three pieces of dial gauge were installed at the bottom-end, mid-height, and top part of the column. The dial gauge set at the bottom part was aimed to control the stability of testing specimens during the loading. This testing setup only provided the reading of column drift at two levels, which were at the mid-height and the top part. Therefore, the drift data from experimental works was not able to represent the actual drift pattern during the loading. In numerical simulation, the column drift is generated from the displacement of nodes along the centroid of column height. In this case, drift data of column from numerical analysis is more reliable than the experimental works. It is supported with the drift data on the lower part of column that is lesser due to the presence of restraints positioned at the bottom-end part of column.

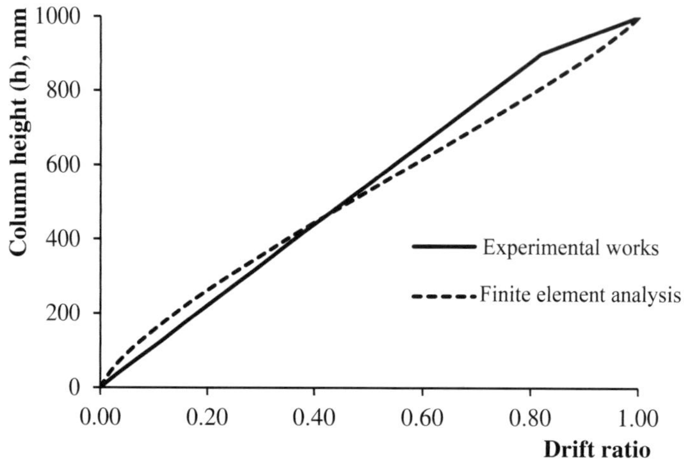

Figure 6. Drift comparison on the segregated column from experimental works and finite element analysis

References

[1] Hidayat A, Puspowardojo J, Abdul F. The Influence of Graded Concrete Strength on Concrete Element. Procedia Eng [Internet]. 2015 [cited 2017 Jun 6];125:1023–9. Available from: http://dx.doi.org/10.1016/j.proeng.2015.11.157

[2] Corinaldesi V, Moriconi G. Self-Compacting Concrete: a Great Opportunity for Recycling Materials. 2004. 1-10.

[3] Seible F, Priestley MJN, Hegemier GA, Innamorato D. Seismic retrofit of RC columns with continuous carbon fiber jackets. J Compos Constr. 1997;

[4] Sezen H. Seismic Behavior and Modeling of Reinforced Concrete Building Columns. Vasa. 2000;

[5] Pratama MMA, Umniati BS, Prasiwi G, Han AL, Sthenly Gan B, Risdanareni P, et al. Effect of monotonic lateral load on the performance of reinforced graded concrete column, an experimental study. MATEC Web Conf [Internet]. 2018;195. Available from: https://doi.org/10.1051/matecconf/201819502022

[6] Aylie H, Gan BS, As'ad S, Pratama MMA. Parametric Study of the Load Carrying Capacity of Functionally Graded Concrete of Flexural Members. Int J Eng Technol Innov [Internet]. 2015;5(4):233–41. Available from: http://www.scopus.com/inward/record.url?eid=2-s2.0-84957600233&partnerID=MN8TOARS

[7] Gan BS, Aylie H, Pratama MMA. The Behavior of Graded Concrete, an Experimental Study. Procedia Eng [Internet]. 2015;125:885–91. Available from: http://linkinghub.elsevier.com/retrieve/pii/S1877705815033937

[8] Han A, Gan BS, Pratama MMA. Effects of Graded Concrete on Compressive Strengths. Int J Technol [Internet]. 2016 Jul 28;7(5):732. Available from: http://www.ijtech.eng.ui.ac.id/index.php/journal/article/view/3449

[9] Pratama MMA, Aylie H, Gan BS, Umniati BS, Risdanareni P, Fauziyah S. Effect of concrete strength gradation to the compressive strength of graded concrete, a numerical approach. AIP Conf Proc [Internet]. 2017 Sep 29;1887(1):20029. Available from: https://aip.scitation.org/doi/abs/10.1063/1.5003512

[10] Pratama MMA, Pratiwi YD, 'Ilmi MA, Mahardika A. GECOBRIDGE: A concept of a smart eco-friendly integrated bridge utilizing geopolymer and graded concrete technology. MATEC Web Conf [Internet]. 2018;204. Available from: https://doi.org/10.1051/matecconf/201820406005

[11] Pratama MMA, Umniati BS, Arumsari Mutiara Wulandari B, Han AL, Sthenly Gan B, Zhabrinna Z. Modulus elasticity of the graded concrete, a preliminary research. MATEC Web Conf

[Internet]. 2018;195. Available from: https://doi.org/10.1051/matecconf/201819501005

[12] Pratama MMA, Sthenly Gan B, Han Ay Lie HAL, Nur Rahma Putra AB. A Numerical Analysis of The Modulus of Elasticity of The Graded Concrete. In: 2nd International Conference on Vocational Education and Training (ICOVET 2018) [Internet]. Atlantis Press; 2019. Available from: https://doi.org/10.2991/icovet-18.2019.29

[13] Pratama MMA, Suhud RK, Puspitasari P, Kusuma FI, Rahma Putra ABN. Finite element analysis of the bending moment-curvature of the double-layered graded concrete beam. IOP Conf Ser Mater Sci Eng [Internet]. 2019 Mar 29;494:012064. Available from: http://stacks.iop.org/1757-899X/494/i=1/a=012064?key=crossref.e0f2a62cfeeaf5c05c43077bf777a94b

IOP Conf. Series: Materials Science and Engineering **669** (2019) 012062 doi:10.1088/1757-899X/669/1/012062

Concepts of constructability for project construction in Indonesia

M M A Ansyorie[1]

[1] Department of Civil Engineering, Faculty of Engineering, Universitas Negeri Malang, Indonesia

musthofansyorie.ft@um.ac.id

Abstract. A study of constructability has recently become one of the research themes that has received much attention. Constructability can be defined as the application and integration of construction knowledge and experience to the process of planning, designing, procuring and implementing construction. The concept of constructability itself began and developed in the United States in the 1970s. In Indonesia, development in the field of construction services is a priority, especially for infrastructure development. The process of construction work can't be separated from the process of planning and implementing construction which is part of the concept of constructablity. Through the application of constructability in construction work can be one analysis for the implementation of construction work to get results a good work in accordance with the desired quality. This study is a review of the concept of constructability and its application in building construction industry. The results of this review are expected to be a source of knowledge and reference regarding the concept of constructability that can be applied by stakeholders in construction work.

1. Introduction

The construction industry in Indonesia is currently growing rapidly, the growth of the construction industry from year to year is increasing. The growth of this industry needs to be balanced with knowledge in the implementation of construction work. One thing that is common in the implementation of construction work and is an important part of the process is construcability. Definition of constructability by the Construction Industry Institute (CII) as the optimum use of construction knowledge and experience in the planning, design, procurement and field operations to echo overall project objectives [1]. Constructability is also defined as the extend to the design facilitate ease of construction [2].

Constructability itself in Indonesia has not been fully implemented by the construction industry players to the full. Some problems in the application of constructability in Indonesia due to lack of understanding of the concept of constructability correctly. This is not only in the construction process in the field, but from the beginning the project design is also not balanced with the implementation of a good constructability concept. Therefore, knowledge about the concept of constructability needs to be understood by the construction industry players from the beginning of the construction project procurement. In many developed countries, much research effort has been directed at improving constructability through the integrated efforts of owners, designers and constructors [3].

An understanding of low constructability is a major factor in the less optimal application of constructability in construction. As the results of a study from Zolfagharian (2012) The major problems in this phase are; lack of knowledge about constructability concepts, unwillingness to devote extra money for this in the initial stages of projects, and shorten of team building and partnering [4]. In addition to the low level of understanding, other factors in procurement of construction projects are related to the procurement process of construction projects that is still simple. In this case, the project owner can determine to use the services of consultants or contractors in accordance with his wishes, without considering the quality of these service providers, both individuals and companies. However, this is quite different when the project owner is part of the government which requires it to be registered as a construction service company. However, not all construction service companies have the capability and a good understanding of constructability. In addition, the main constraint to the implementation of constructability is the problem of coordination and the lack of involvement of the contractor at an early stage, as well as the consultants on the bid build design method not involved in the procurement stage so that planning consultants cannot be involved in the selection of contractors who are really expected to complete the project in accordance with the wishes of the owner and consultant [5].

Some of the problems regarding constructability implementation are also influenced by the absence of good constructability concepts that are suitable for application in Indonesia. Research or studies related to constructability itself in Indonesia have not yet become something important to be carried out. So, to get the concept of constructability in accordance with the conditions of the procurement process of construction service projects in Indonesia. In addition, barriers in implementing improvement in constructability, particularly lack of awareness or appreciation of constructability issues and lack of qualified personnel [6]. Therefore, it needs some studies on the concept of constructability that can be applied to Indonesia optimally.

2. Discussion

2.1. Definitions of constructability

Many definitions of constructability (or buildability) The ability to construct a building efficiently, economically and to agreed quality levels from its constituent materials, components and sub-assemblies. Constructability is often portrayed as integrating construction knowledge, resources, technology and experience into the engineering and design of a project (Anderson et al., 2000). Constructability is a project management technique for reviewing construction processes from start to finish during the pre-construction phase. It will identify obstacles before a project is actually built to reduce or prevent error, delays and cost, The term "constructability" is referred to as [7]:

- the extent to which the design of the building facilitates ease of construction, subject to the overall requirements for the completed building
- a system for achieving optimum integration of construction knowledge and experience in planning, engineering, procurement and field operations in the building process, and balancing the various project and environmental constraints to achieve overall objectives
- a system for achieving optimum integration of construction knowledge in the building process and balancing the various project and environmental constraints to maximise achievement of project goals and building performance overruns.

Constructability or buildability is a project quality improvement technique that if implemented throughout the project delivery process, mitigates these challenges according to Bakti et al. (2003). Improving constructability of construction projects is the responsibility of all project stakeholders: owners, designers and contractors. However, as the owners have the most authority in enforcing the implementation of constructability, the owners' awareness of the benefit of improved constructability is the most important [6].

2.2. Concept of constructability

Constructability ini United State of Amerika (USA) start form a research in 1970's, with a themes The Construction industri cost effectiveness project. This studi carry out by team of the Business Rountable in 1983, and complate study a four year study into how to promote quality, efficiency, productifity and cost effectiveness in the construction industry [1]. Many focus of the study on constructability factor, between a process implementation, method of constructability, and etc. Based on study of a number of engineering projects, they conclude that benefit of 10 to 20 times the cost of implementing constructability could be acgieved, and it is essential that project owners participate [4]. Although Business Roundtable initiated the constructability study, it was the Construction Industry Institute (CII) which had been the principal proponent and sponsor of most of the constructability research in the US.

According to Griffith and Sidewell (1995), the American Society of Civil Engineers (ASCE) suggested six principles of constructability to include evaluating the various design configurations to optimize owner requirements, knowing the various project systems and their interface requirement with other project components, understanding trade skills and practices, construction methods, materials, labor and sub-contract resources and plant and equipment, appreciation of local climatic conditions, evaluating site conditions both above and below ground and realizing their possible implications upon construction and finally determining availability of space and access routes on site [8]. The CII identified that, following the Pareto Principle, the ability to influence cost reduces over the life of the project and that constructability saving should be investigated at various stages of the project life-cycle [1] namely,

- Conceptual planning
- Design and procurement
- Field operations

O'Connor and Davis (1988) in further research identified the constructability improvement
that can be made during field operation. The first two – project team responsible for constructability are identified early on and the second, advanced information technologies are applied throughout the project for improvement at the conceptual planning phase while the third principle - Design and construction sequencing should facilitate system turn-over and start-up [9].

In other countries, The concept of buildability originated in the United Kingdom and is defined as the extent to which the design of a building facilitates ease of construction, subject to overall requirements for the completed building [10]. In Australia, the term of constructability and buildability are used interchangeably and the concept has been re-defined as the extent to which decisions made during the whole building procurement process, in response to factors influencing the project and other project goals, ultimately facilitates the ease of construction and the quality of the completed project or alternatively the integration of construction knowledge in the project delivery process and balancing the various project and environmental constraints to achieve project goals and building performance at an optimal level [11]. Based on the constructability concepts in many countries, Nima (2001) make a formula to utilities a study for constructability concept in 23 concepts for pashes for constructions [12], as follows:

2.2.1. Project constructability enhancement during conceptual planning phase comprises of concepts C1 to C7

- Concept C1: The project constructability programme should be discussed and documented within the project execution plan, through the participation of all project team members.
- Concept C2: A project team that includes representatives of the owner, engineer and contractor should be formulated and maintained to take the constructability issue into consideration from the outset of the project and through all of its phases.
- Concept C3: Individuals with current construction knowledge and experience should achieve the early project planning so that interference between design and construction can be avoided.
- Concept C4: The construction methods should be taken into consideration when choosing the type and the number of contracts required for executing the project.

- Concept C5: The master project schedule and the construction completion date should be construction-sensitive and should be assigned as early as possible.
- Concept C6: In order to accomplish the field operations easily and efficiently, major construction methods should be discussed and analysed in-depth as early as possible to direct the design according to these methods.
- Concept C7: Site layout should be studied carefully so that construction, operation and maintenance can be performed efficiently, and to avoid interference between the activities performed during these phases.

2.2.2. Project constructability enhancement during design and procurement phase comprises of concepts C8 to C15

- Concept C8: Design and procurement schedules should be dictated by construction sequence. Thus, the construction schedule must be discussed and developed prior to the design development and procurement schedule.
- Concept C9: Advanced information technologies are important to any field including the construction industry. Therefore, the use of those technologies will overcome the problem of fragmentation into specialized roles in this field, and enhance constructability.
- Concept C10: Designs, through design simplication by designers and design review by qualified construction personnel, must be configured to enable efficient construction.
- Concept C11: Project elements should be standardized to an extent that will never affect the project cost negatively.
- Concept C12: The project technical specifications should be simplified and configured to achieve efficient construction without sacrificing the level or the efficiency of the project performance.
- Concept C13: The implementation of modularization and preassembly for project elements should be taken into consideration and studied carefully. Modularization and preassembly design should be prepared to facilitate fabrication, transportation and installation.
- Concept C14: Project design should take into consideration the accessibility of construction personnel, materials and equipment to the required position inside the site.
- Concept C15: Design should facilitate construction during adverse weather conditions. Efforts should be made to plan for the construction of the project under suitable weather conditions; otherwise, the designer must increase the project elements that could be prefabricated in workshops.

2.2.3. Project constructability enhancement during field operations phases on concepts C16-C23

- Concept C16: Field tasks sequencing should be configured in order to minimize damages or rework of some project elements, minimize scaffolding needs, formwork used, or congestion of construction personnel, material and equipment.
- Concept C17: Innovation in temporary construction materials/systems, or implementing innovative ways of using available temporary construction materials/systems that have not been defined or limited by the design drawings and technical specifications will contribute positively to the enhancement of constructability.
- Concept C18: Incorporating innovation of new methods in using off-the-shelf hand tools, or modification of the available tools, or introduction of a new hand tools that reduce labour intensity, increase mobility, safety or accessibility will enhance constructability at the construction phase.
- Concept C19: Introduction of innovative methods for using the available equipment or modification of the available equipment to increase their productivity will lead to a better constructability.

- Concept C20: In order to increase the productivity, reduce the need for scaffolding, or improve the project constructability under adverse weather conditions, constructors should be encouraged to use any optional preassembly.
- Concept C21: Constructability will be enhanced by encouraging the constructor to carry out innovation of temporary facilities.
- Concept C22: Good contractors, based on quality and time, should be documented, so that contracts for future construction works would not be awarded based on low bids only, but by considering other project attributes, i.e. quality and time.
- Concept C23: Evaluation, documentation and feedback of the issues of the constructability concepts should be maintained throughout the project to be used in later projects as lessons learned.

Many processes of constructability implementations for construction [13], followings:
- Organixing the design team
- Collecting data
- Identify the constrains
- Developing program
- Framing options evaluation
- Developing preliminary design
- Checking of options
- Final design process
- Developing bid package
- Procedure of bidding
- Procedure of fabrications
- Installation process

2.3. Construction Process in Indonesia

The implementation of the construction process in Indonesia cannot be separated from several stages that must be passed, both from the beginning of the design to the implementation and the end of the construction work. Of all these processes, of course, have a goal of good quality. In order to achieve the planned quality, construction projects should be carried out through systematic steps by combining several aspects. Providers of construction services especially for state buildings according to PERMEN PUPR No 22 Tahun 2018 [14], follow of:
- Construction planning service provider
- Provider of construction implementation services
- Provider of construction supervision services
- Provider of construction management services

Based on PERLEM LKPP No 7 Tahun 2018 regarding procurement planning activities, including goods services procurement activities [15]:
- Procurement Planning starts from the identification of the needs of goods / services based on the Work Plan of the Ministry / Institution / Regional Apparatus.
- Procurement Planning becomes input in the formulation of Work Plans and Budgets of Ministries / Institutions (RKA-K / L) and Work Plans and Budgets of Regional Apparatus (RKA of Regional Apparatus).
- Procurement Planning whose funds are sourced from the State Budget (APBN) is carried out in conjunction with the RKAK / L preparation process after the establishment of the Indicative Ceiling.
- Procurement Planning whose funds are sourced from the Regional Revenue and Expenditure Budget (APBD) carried out in conjunction with the preparation of the Regional Budget Work

Plan after the memorandum of agreement on the APBD General Policy and the Priority and Provisional Budget Ceiling (KUA-PPAS).

The procurement of product/services should be carried out through an identification mechanism to the needs. The identification of goods and services needs is carried out by taking into account the following aspects:

- the principle of efficient and effective procurement of goods / services
- aspects of sustainable procurement
- needs priority assessment
- product / services in the electronic catalog
- consolidation of procurement of product / services; and / or
- product / services that are available / owned / controlled.

Some of the aspects above can be implemented in the process of procurement of products / construction services. Related to construction services, procurement is carried out through several activities. Planning activities include:

- preparation of technical specifications / TOR
- preparation of estimated costs / Budget Plan (RAB)
- the procurement of products / services
- consolidation of procurement of products / services; and
- supporting costs.

Procurement of products and services can be done through a mechanism in the following flow:

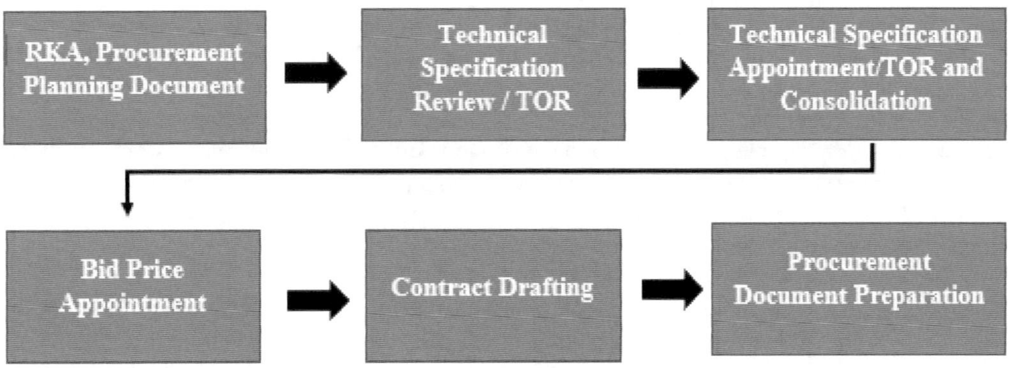

Figure 1. Procurement product and service

Related to construction project work, the flow of goods / services procurement is marked by a construction work contract [16]. Types of extracts for construction work consist of:

- Lump sum
- Work Prices Unit
- Combined lump sum and Unit Price for the job
- Turnkey
- Umbrella contract

To be able to contract in construction work, each construction work service provider must meet the technical qualification requirements which include:

- Experience of at least 1 (one) job in the last 4 (four) years, including experience of subcontracting except for Providers who have only been established for less than 3 (three) years.
- Have the Basic Ability (BA) for work that is intended for Medium Enterprises and Large Enterprises with a BA value equal to 3 NPt (the highest experience value) on subclassifications,

types of work, and / or key activities (major items) in accordance with the requirements in a period of time Past 10 years

- Have at least:
    - 1 (one) Skilled Certified Permanent Personnel (SKT) in accordance with the required SBU Classification (For Small Businesses)
    - 1 (one) Permanent Expert Certified (SKA) in accordance with the required SBU Sub-Classification (For Medium and Large Enterprises) (selected according to the required Qualifications)
- experience in similar work in the past 10 years with a weight of 25-40%
- appropriateness of the value of similar work in the last 10 years that has been completed with the value of the work to be competed with a weight of 35-45%
- work experience in the past 10 years in the same location at the Provincial or District / City level (selected) with a weight of 20-30%

2.4. Constructability Concepts for Project Construction In Indonesia

Based on Fischer and Tatum, 1997, constructability is the significant objective in construction projects [17]. Therefore we need in-depth knowledge of the concept of constructability as a basis in construction. There is a classification for knowledge of constructability theory to improve project performance: application heuristics, layout knowledge, dimensioning knowledge, detailing knowledge, and exogenous knowledge [18].

Constructability concepts for project construction in Indonesia can be done by considering related matters. The existence of owners from the government and the private sector can certainly show the difference in the application. But in general the concept of constructability implementation can be equally applied by maximizing the role of the owner's role. According to Trigunarsyah (2006) examined the role of clients. Clients have the authority in enforcing implementation therefore their role in constructability improvement is most important. The paper describes how construction project owners integrate construction knowledge and experience in planning and design in existing practice. The research was carried out by means of case studies of some notable projects in Indonesia. The results shows that project owners in Indonesia do have some understanding of the importance of constructability. The selection of project delivery method determined the type of constructability input for the respective projects.

Considering this, the implementation of constructability concepts for project construction in Indonesia can be implemented by combining the concepts of constructability and procurement construction services. Implementation of constructability concepts in Indonesia can follow the purpose of constructability. that purpose of constructability [19], following items:

- Fault of design either dimension or select materials indistinct features
- The specifications of the Project will be difficult or high cost to construct based on design
- Exceeding the project features in the capability of industry to properly build
- Project properties that are difficult to understand and also are hard to correctly bid

3. Conclusion

This papaer has discussed the concepts of Constructability which can be applied in Indonesia. The results of the study of several articles above can be concluded that:

- The concept of constructability that can be applied in Indonesia can refer to the concept of constructability that has been agreed upon in general.
- The concept of constructability can be combined with several methods of procurement of product / services construction available in accordance with applicable regulations in Indonesia.
- Knowledge of the concept of constructability needs to be conveyed in depth to all construction service performer, including the owner, contractor, consultant and other parties related to construction project work.

References

[1] The Construction Industry Intitute (CII). Constructability Concepts File. University of Texas, Austin-Texas. 1987.
[2] Construction Industry Research and Information Association (CIRIA). Buildability: An Assesment. London. 1993.
[3] Bakti ES, Trigunarsyah B. Constructability Implementation at Corporate Level The 19th East Asia-Pacific Conference on Structural engineering and construction (EASEC9) 19-18 Dec 2003, Bali Indonesia, pp. 88-93.
[4] Zolfagharian S, Nourbakhsh M, MydinSH, Zin RM, Irizarry J. A Conceptual Method of Constructability Improvement. International Journal of Engineering and Technology. 4(4):456-458. 2012.
[5] Dewi AADP, Diputra GA. Constructability Pada Proyek Dengan Metode Kontrak Design Bid Build. Prosiding Seminar Sains dan Teknologi IV. Bali. 2017.
[6] Trigunarsyah B. Case Studies on Implementation of Constructability Improvement by Construction Project Owners in Indonesia. Cooperative Research Centre for Construction Innovation. 2006.
[7] Practice Note 13. Constructability. IPENZ, Engineering New Zealand. ISSN 1176-0907. 2018
[8] American Society of Civil Engineers (ASCE). "Constructability and Constructability Programs: White Paper" Journal of Construction Engineering and Management. Vol. 117, No. i. 1991.
[9] O'Connor JT, and Davis VS. Constructability improvement during field operations. Journal of Construction Engineering and Management., 114(4):548–564. 1988.
[10] Construction Industry Research and Information Association (CIRIA). Buildability: An Assessment. CIRIA, London. P. 26. 1983.
[11] Construction Industry Institute Australia (CIIA). Constructability Principles File. Adelaide, Construction Industry Institute, Australia. 2nd Ed. 1992.
[12] Nima MA, Abdul-kadir M.R, Jaafar MS, Algbulami RG. Constructability implementation: A survey in the Malaysian construction industry. Construction management and economics. 19(8):819-829. 2001.
[13] Ruby DI. Award winning constructability (Lansing community college). Structure, pp. 20-25, 2006.
[14] Peraturan Menteri Pekerjaan Umum dan Perumahan Rayat (PERMEN PUPR) No 2 Tahun 2018 tentang Pedoman Pembangunan Bangunan Gedung Negara. Jakarta. 2018.
[15] LKPP. Perlem No 7 Tahun 2018 Tentang Pedoman Perencanaan Pengadaan Barang/Jasa Pemerintah. Jakarta. 2018
[16] LKPP. Perlem No 9 tahun 2018 tentang Pedoman Pelaksanaan Pengadaan Barang.Jasa Melalui Penyedia. Jakarta. 2018
[17] Fisher M, Tatum CB. Characteristics of Design- Relevant Constructability Knowledge. Journal of Construction Engineering & Management, Vol. 123. No 3, 253-260. 1997
[18] Tatum CB. Improving constructability during conceptual planning, Journal of Construction Engineering and Management ASCE, 113(2), 191–207, 1987
[19] Douglas EE. and Gransberg, D.D. (2009) Implementing Project Constructability. AACE International Recommended Practice No. 30R-03. TCM Framework: 11.5–Value Management and Value Improving Practices (VIPs). [Online]. Available from: www.aacei.org/technical/rps/30R-03.pdf (Accessed september 2019).

AUTHOR INDEX

Adi, R Y 238
Adu, M A J 173
Alfianto, Imam 445
Alfianto, I 451
Alias, N E B 229
Amara, S 165
Andaryati 55
Anggraheni, E 274, 285, 300
Anindyasari, A 375
Ansyorie, M M A 480
Antoni, A 181
Ariestadi, D 445
Ariesyady, H D 345
Arifanda, W 428
Arifi, E 338, 351
Arifin, Y F 197
Azis, S 125
Aziz, A A 385
Bachri, S 95
Baghdadli, T 165
Bolla, M E 110
Budihardjo, M A 159
Bukit, I N M 64
Casita, C B 437
Chen, Y W 246
Chin, Y H 220
Christiyanto, A 78
Damayanthi, W 473
Darmawan, V E B 246
Daryanto, D 238
Destio, R 391
Dewi, S M 71
Djatmiko, B 463
Driejana 345
Fahmi, I 391
Farichah, H 322
Fauziah, M 357
Fauziyah, S 253
Fitriani, N 391
Gan, B S 1
Gemardi, A 391
Hajji, A M 15, 246
Halim, H 10
Han, A 1
Hardjito, D 181
Haryadi 268
Hasan, N Y 345
Herwindo, W 95
Hidayat, Y M 95
Hsu, H L 10
Idfi, G 60, 133, 214, 229, 293, 375
Ignatius, S 300
Ishak, M F 21, 32
Jonbi 173
Karjanto, A 413
Karnisah, I 311
Karyadi, K 385, 398, 428
Khomsiati, N L 60, 133
Kiryu, S 1
Kistiani, F 268
Krisnayanti, D S 110
Laksono, Y A 60
Larasati, A 246
Larasati, D 149
Mahardika, A F P 246
Marzuki, P F 64
Mawarni, O S T 357
Meifrinaldi 64
Mintura, R 181
Misnawati 197
Muhammad, F I 159
Muhrozi 238
Mujiyono, M 405
Mujiyono 214, 293
Munandar, W A 78
Nasjono, J K 110
Nauri, I M 101
Niken, C 87
Nindyawati, N 385, 398, 428, 457
Nine, V 187
Ningsih, D A 391
Noorhidana, V A 261
Nordell, B 165
Normelani, E 197
Novitasari, Y 473
Nur, M T 119
Nuralinah, D 338, 351
Nurdiana, A 330
Paing, J 55
Partono, W 268
Patty, A H 140
Patulak, D C 15
Permana, A S 311
Pradhana, R A 413
Pradipta, A H 119
Prasetyo, B E 214

Prasetyo, F D 338
Pratama, M M A 1, 413, 420, 428, 437, 457, 473
Prawira, D 274
Prayuda, H 428
Priastiwi, Y A 238
Purba, R H 78
Purisari, R 368
Putro, E 268
Rachmahwati, D F 398
Rachmansyah, A 42
Rafrita, F K 451
Rahayuningsih, T 375, 451, 457
Ramadhan, W R 268
Risdianto, Y 119
Rizaldi, B 293
Rizaldianto, A R 159
Ruswandono 55
Sadukh, Y M I 173
Safarina, A B 311
Sahid 368
Salamsyah, Z 238
Santoso, E 413, 437
Santoso, F R E 391
Saputra, A 71
Sarassantika, I P Ellsa 437
Sari, A A 445, 451
Sari, U C 268
Setyawan, E 463
Setyowulan, D 338, 351
Sharaswati, D 285
Shoffan, I A N 101
Sholeh, M N 330
Simatupang, P H 173, 187
Sinaga, A E 187
Soedjono, E 391
Soehardjono, A 42
Soeryantono, H 274, 285, 300
Sofianto, M F 119
Sufianto, D 311
Sugandi, R M 463
Sukarni, S 405
Sukriyadi 375
Sulaeman, A 345
Sulaksitaningrum, R 428, 437
Sulton, M 437, 457
Sumarli, S 101
Sumiyati, Y 368
Surja, R T 181
Suryoputro, N 60, 229, 405
Susanti, L 71
Suseno, H 42
Sutjiningsih, D 274, 285, 300
Suwarno, E 214, 293
Suwarto, F 253
Tamin, R Z 64
Triyuliani, N I 71
Umniati, B S 420, 437
Vertian, T 420
Wahyono, I D 133
Wahyuni, Y S 149
Wangge, M J M 110
Wardhono, A 119
Wattimena, O K 181
Widarda, D R 205
Winahyo, A E 445
Yoh, W H 420
Yong, H N A 220
Yuaniar, M C 405
Yue, Huang 15
Yue, L J 21
Yulistyorini, A 15, 133, 229, 405, 463
Zaini, M S I 21, 32
Zolkepli, M F 21, 32
Zulkifli, E 205
Zuraida, S 149